# Student Solutions Manual

## Mark S. Erickson
*Hartwick College*

---

# INTRODUCTION TO ORGANIC CHEMISTRY

## Third Edition

## William Brown
*Beloit College*

## Thomas Poon
*Claremont McKenna, Pitzer, and Scripps Colleges*

WILEY

John Wiley & Sons, Inc.

Cover Photo:  Ray Coleman/Photo Researchers, Inc.

To order books or for customer service call 1-800-CALL-WILEY (225-5945).

ISBN 0-471-68263-2

Printed in the United States of America

10  9  8  7  6  5  4  3

# CONTENTS

# CHAPTER 1
## *Solutions to Problems*

**1.1**    *Write and compare the ground-state electron configurations for the elements in each set.*

(a) *Carbon:* $1s^2 2s^2 2p^2$          *Silicon:* $1s^2 2s^2 2p^6 3s^2 3p^2$

**Both carbon and silicon have four valence electrons.**

(b) *Oxygen:* $1s^2 2s^2 2p^4$          *Sulfur:* $1s^2 2s^2 2p^6 3s^2 3p^4$

**Both oxygen and sulfur have six valence electrons.**

(c) *Nitrogen:* $1s^2 2s^2 2p^3$          *Phosphorus:* $1s^2 2s^2 2p^6 3s^2 3p^3$

**Both nitrogen and phosphorus have five valence electrons.**

**1.2**    *Show how the gain of two electrons by a sulfur atom to form a sulfide ion leads to a stable octet:*

$$S\ +\ 2e^- \longrightarrow S^{2-}$$

**Elemental sulfur has six valence electrons. When it gains two additional electrons in its valence shell, a filled outer shell is achieved, thus satisfying the octet rule.**

S (16 electrons): $1s^2 2s^2 2p^6 3s^2 3p^4$ $+ 2e^- \longrightarrow$ $S^{2-}$ (18 electrons): $1s^2 2s^2 2p^6 3s^2 3p^6$

**1.3**    *Judging from their relative positions in the Periodic Table, which element in each pair has the larger electronegativity?*

**General Rule: Electronegativity increases from left to right across a row and from bottom to top within a column in the Periodic Table.**

(a) **_Lithium_** *or Potassium*
   **Lithium is higher up on the Periodic Table, thus more electronegative than potassium.**

(b) **_Nitrogen_** *or Phosphorus*
   Nitrogen is higher up on the Periodic Table, thus more electronegative than phosphorus.

(c) **_Carbon_** *or Silicon*
   Carbon is higher up on the Periodic Table, thus more electronegative than silicon.

**1.4** *Classify each bond as nonpolar covalent, polar covalent, or ionic.*
   *(a) S-H*   *(b) P-H*   *(c) C-F*   *(d) C-Cl*

| Bond | Difference in Electronegativity | Type of Bond |
|------|--------------------------------|--------------|
| *(a) S-H* | 2.5 – 2.1 = 0.4 | nonpolar covalent |
| *(b) P-H* | 2.1 – 2.1 = 0.0 | nonpolar covalent |
| *(c) C-F* | 4.0 – 2.5 = 1.5 | polar covalent |
| *(d) C-Cl* | 3.0 – 2.5 = 0.5 | polar covalent |

**1.5** *Using the symbols $\delta$- and $\delta$+, indicate the direction of polarity in these polar covalent bonds.*

   *(a) C-N*  $^{\delta+}$C-N$^{\delta-}$  **Nitrogen is more electronegative than carbon.**

   *(b) N-O*  $^{\delta+}$N-O$^{\delta-}$  **Oxygen is more electronegative than nitrogen.**

   *(c) C-Cl*  $^{\delta+}$C-Cl$^{\delta-}$  **Chlorine is more electronegative than carbon.**

**1.6** *Draw Lewis structures, showing all valence electrons, for these molecules.*
   *(a) $C_2H_6$*       *(b) $CS_2$*       *(c) HCN*

**Ethane**
6 H $6 \times 1\,e^- = 6\,e^-$
2 C $2 \times 4\,e^- = 8\,e^-$
Total valence $e^- = 14\,e^-$

**Carbon disulfide**
1 C $1 \times 4\,e^- = 4\,e^-$
2 S $2 \times 6\,e^- = 12\,e^-$
Total valence $e^- = 16\,e^-$

**Hydrogen cyanide**
1 H $1 \times 1\,e^- = 1\,e^-$
1 C $1 \times 4\,e^- = 4\,e^-$
1 N $1 \times 5\,e^- = 5\,e^-$
Total valence $e^- = 10\,e^-$

```
    H  H
    |  |
 H- C- C-H          ··      ··                 H-C≡N:
    |  |            S = C = S
    H  H            ··      ··
```

**For uncharged molecules, the total number of electrons described by a Lewis structure (the sum of two electrons for each bond and one electron for each dot) must be equal to the total number of valence electrons contributed by each atom.**

**1.7** *Draw Lewis structures for these ions, and show which atom in each bears the formal charge(s).*

*(a) $CH_3NH_3^+$ Methylammonium cation*
6 H $6 \times 1\,e^- = 6\,e^-$
1 C $1 \times 4\,e^- = 4\,e^-$
1 N $1 \times 5\,e^- = 5\,e^-$
(+) charge $= -1\,e^-$
Total valence $e^- = 14\,e^-$

*(b) $CH_3^+$ Methyl cation*
1 C $1 \times 4\,e^- = 4\,e^-$
3 H $3 \times 1\,e^- = 3\,e^-$
(+) charge $= -1\,e^-$
Total valence $e^- = 6\,e^-$

**Formal Charge (F.C.) = # of atom's valence e⁻ – (# unshared e⁻ + 1/2 # shared e⁻)**

H H
| |+
H–C–N–H      Formal Charge
| |          C:  4 - [0 + 1/2(8)] =  0
H H          N:  5 - [0 +1/2(8)]  = +1

H
|
H–C +        Formal Charge
|            C:  4 - [0 + 1/2(6)] =  +1
H

**For charged molecules, the total number of electrons described by a Lewis structure (the sum of two electrons for each bond and one electron for each dot) must be equal to the total number of valence electrons contributed by each atom (minus one electron for each positive charge and plus one electron for each negative charge).**

**1.8**    *Predict all bond angles for these molecules.*
*(a) CH₃OH*                    *(b) CH₂Cl₂*                    *(c) H₂CO₃ (Carbonic acid)*

**1.9**    *Both carbon dioxide, CO₂, and sulfur dioxide, SO₂, are triatomic molecules.  Account for the fact that carbon dioxide is a nonpolar molecule whereas sulfur dioxide is a polar molecule.*

**Although both C=O and S=O bonds are roughly the same in their polar covalent nature (both have a difference in electronegativity of 1.0 and significant bond dipoles), the molecular dipole is also influenced by a combination of bond dipole moments and molecular geometry.  CO₂ and SO₂ molecular shapes reveal a linear geometry for CO₂ and a bent geometry for SO₂.  The opposing C=O bond dipoles of CO₂ cancel each other out, thus resulting in no molecular dipole moment.  For SO₂, the O-S-O bond angle is approximately 120° and the bond dipole moments do not cancel, resulting in a non-zero dipole moment.**

**1.10**   *Which sets are pairs of resonance contributing structures?*

(a) $CH_3-C$ (with $\overset{\cdot\cdot}{O}$ double bond and $\overset{\cdot\cdot}{O}^-$) $\longleftrightarrow$ $CH_3-\overset{+}{C}$ (with $\overset{\cdot\cdot}{O}^-$ and $\overset{\cdot\cdot}{O}^-$)   (b) $CH_3-C$ (with $\overset{\cdot\cdot}{O}$ double bond and $\overset{\cdot\cdot}{O}^-$) $\longleftrightarrow$ $CH_3-\overset{-}{C}$ (with $\overset{\cdot\cdot}{O}$ and $\overset{\cdot\cdot}{O}$ both double)

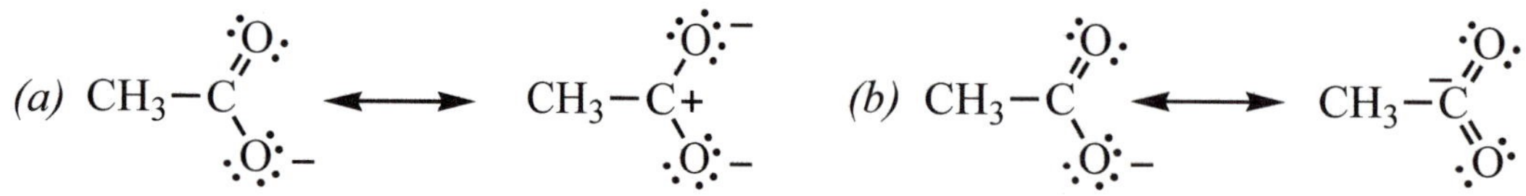

   *(a)* **Contributing structures. They differ only in the distribution of valence electrons.**

   *(b)* **Not a set of contributing structures. The octet rule is violated for the carboxylate carbon atom in the structure on the right (it has ten valence electrons).**

**1.11**   *Use curved arrows to show the redistribution of valence electrons in converting resonance contributing structure (a) to (b) and then (b) to (c).*

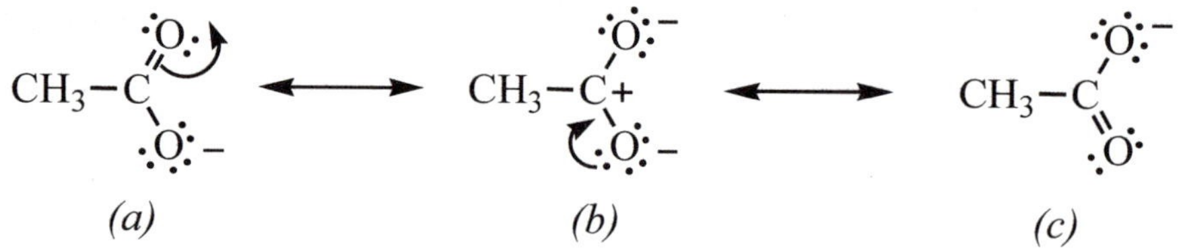

$CH_3-C$ $\longleftrightarrow$ $CH_3-\overset{+}{C}$ $\longleftrightarrow$ $CH_3-C$

   *(a)*                      *(b)*                      *(c)*

**1.12**   *Describe the bonding in these molecules in terms of the atomic orbitals involved, and predict all bond angles.*

   **Molecules involving second row elements involve *s* and *p* atomic orbitals, which can form *sp*, *sp$^2$*, and *sp$^3$* hybrid orbitals.**

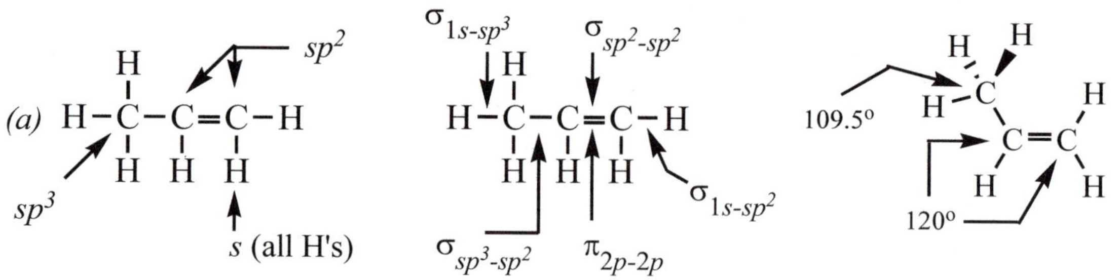

(a) $H-\overset{H}{\underset{H}{C}}-C=C-H$ with $sp^2$, $sp^3$, $s$ (all H's)

$\sigma_{1s\text{-}sp^3}$  $\sigma_{sp^2\text{-}sp^2}$

$H-\overset{H}{\underset{H}{C}}-C=\overset{H}{\underset{H}{C}}-H$ with $\sigma_{1s\text{-}sp^2}$, $\sigma_{sp^3\text{-}sp^2}$, $\pi_{2p\text{-}2p}$

$109.5°$, $120°$   (structure with $H-C$ and $C=C$)

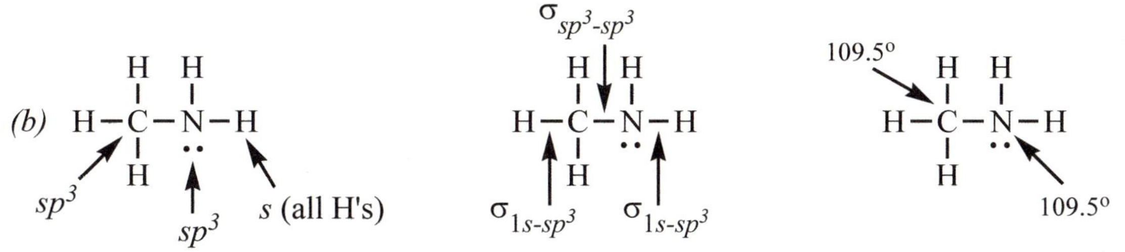

(b)

$$\sigma_{sp^3-sp^3}$$
$$\sigma_{1s-sp^3} \qquad \sigma_{1s-sp^3}$$
$$109.5°$$
$$sp^3 \qquad sp^3 \qquad s \text{ (all H's)}$$
$$109.5°$$

**1.13**  *Write condensed structural formulas for the four alcohols with molecular formula*
*$C_4H_{10}O$. Classify each as primary, secondary, or tertiary.*

$$CH_3 \cdot CH_2 \cdot CH_2 - CH_2 - OH$$
**Primary (1°) alcohol**

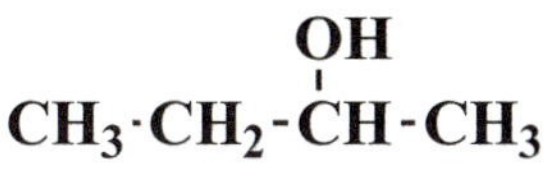

OH
$$CH_3 \cdot CH_2 - CH - CH_3$$
**Secondary (2°) alcohol**

$$CH_3$$
$$CH_3 \cdot CH - CH_2 - OH$$
**Primary (1°) alcohol**

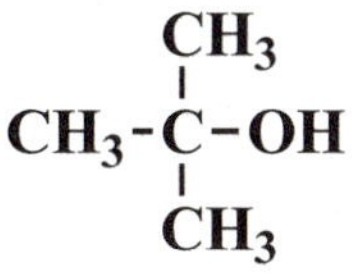

$$CH_3$$
$$CH_3 - C - OH$$
$$CH_3$$
**Tertiary (3°) alcohol**

**1.14**  *Write condensed structural formulas for the three secondary amines with molecular*
*formula $C_4H_{11}N$.*

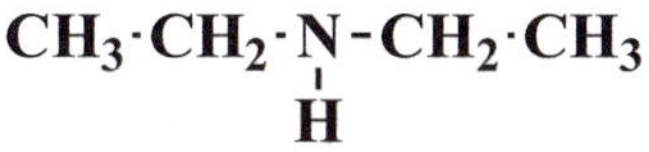

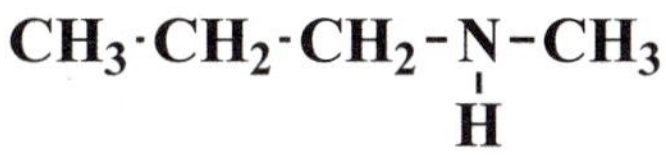

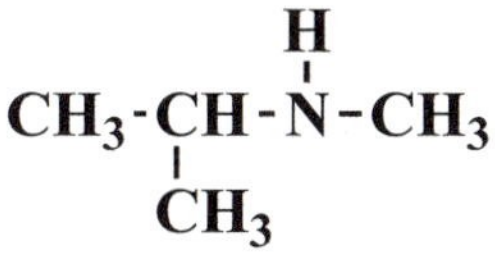

$$CH_3 \cdot CH_2 \cdot N - CH_2 \cdot CH_3$$
$$H$$

$$CH_3 \cdot CH_2 \cdot CH_2 - N - CH_3$$
$$H$$

H
$$CH_3 \cdot CH - N - CH_3$$
$$CH_3$$

**1.15**  *Write condensed structural formulas for the three ketones with molecular formula*
*$C_5H_{10}O$.*

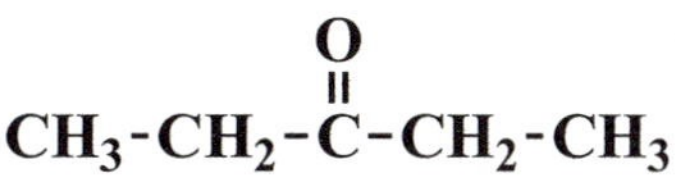

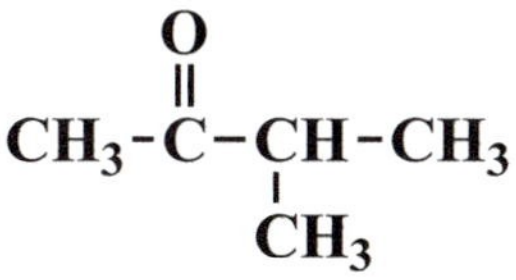

O
$$CH_3 - C - CH_2 - CH_2 - CH_3$$

O
$$CH_3 - CH_2 - C - CH_2 - CH_3$$

O
$$CH_3 - C - CH - CH_3$$
$$CH_3$$

**1.16**   *Write condensed structural formulas for the two carboxylic acids with molecular formula*
*$C_4H_8O_2$.*

$$CH_3-CH_2-CH_2-\overset{\overset{\displaystyle O}{\|}}{C}-OH \qquad\qquad CH_3-\underset{\underset{\displaystyle CH_3}{|}}{CH}-\overset{\overset{\displaystyle O}{\|}}{C}-OH$$

## Electronic Structure of Atoms

**1.17**   *Write the ground-state electron configuration for each element.*

    (a) *Sodium (11)*     **Na (11 electrons): $1s^2 2s^2 2p^6 3s^1$**
    (b) *Magnesium (12)*  **Mg (12 electrons): $1s^2 2s^2 2p^6 3s^2$**
    (c) *Oxygen (8)*      **O (8 electrons):   $1s^2 2s^2 2p^4$**
    (d) *Nitrogen (7)*    **N (7 electrons):   $1s^2 2s^2 2p^3$**

**1.18**   *Write the ground-state electron configuration for each element.*

    (a) *Potassium*     **K (19 electrons):   $1s^2 2s^2 2p^6 3s^2 3p^6 4s^1$**
    (b) *Aluminum*    **Al (13 electrons): $1s^2 2s^2 2p^6 3s^2 3p^1$**
    (c) *Phosphorus*   **P (15 electrons):   $1s^2 2s^2 2p^6 3s^2 3p^3$**
    (d) *Argon*       **Ar (18 electrons): $1s^2 2s^2 2p^6 3s^2 3p^6$**

**1.19**   *Which element has the ground-state electron configuration of*

    (a) *$1s^2 2s^2 2p^6 3s^2 3p^4$*   **16 electrons = ground state sulfur**
    (b) *$1s^2 2s^2 2p^4$*         **8 electrons = ground state oxygen**

**1.20**   *Which element or ion does not have the ground-state electron configuration*
*$1s^2 2s^2 2p^6 3s^2 3p^6$?*

    (a) *$S^{2-}$*      (b) *$Cl^-$*      (c) *Ar*      (d) *$Ca^{2+}$*      ***(e) K***

**Potassium (K) has one more electron than all of the other atoms. Its configuration is**
**$1s^2 2s^2 2p^6 3s^2 3p^6 4s^1$. The potassium ion ($K^+$) has the same electron configuration as**
**the other atoms.**

**1.21**  *Define valence shell and valence electron.*

**The valence shell of an atom is the outermost shell occupied by electrons, which is the highest numbered shell.  A valence electron is an electron that occupies the valence shell.**

**1.22**  *How many electrons are in the valence shell of each element?*

    *(a) Carbon*      $1s^2 2s^2 2p^2$            **Four valence electrons**
    *(b) Nitrogen*    $1s^2 2s^2 2p^3$            **Five valence electrons**
    *(c) Chlorine*    $1s^2 2s^2 2p^6 3s^2 3p^5$    **Seven valence electrons**
    *(d) Aluminum*  $1s^2 2s^2 2p^6 3s^2 3p^1$    **Three valence electrons**
    *(e) Oxygen*      $1s^2 2s^2 2p^4$            **Six valence electrons**

**1.23**  *How many electrons are in the valence shell of each ion?*

    (a) $H^+$    **$1s^0$**        **zero valence electrons**
    (b) $H^-$    **$1s^2$**        **two valence electrons**

## Lewis Structures

**1.24**  *Judging from their relative positions in the Periodic Table, which element in each set is more electronegative?*

**Electronegativity generally increases from left to right across a row and from bottom to top in a column in the Periodic Table.**

    *(a) Carbon or **Nitrogen***    **Nitrogen is to the right of carbon in the Periodic Table.**
    *(b) **Chlorine** or Bromine*    **Chlorine is above bromine in the Periodic Table.**
    *(c) **Oxygen** or Sulfur*      **Oxygen is above sulfur in the Periodic Table.**

**1.25**  *Which compounds have nonpolar covalent bonds, which have polar covalent bonds, and which have ionic bonds?*
    *(a) LiF*      *(b) $CH_3F$*      *(c) $MgCl_2$*      *(d) HCl*

| Bond | Difference in Electronegativity | Type of Bond |
|---|---|---|
| *(a)  Li-F* | 4.0 - 1.0 = 3.0 | ionic |
| *(b)  C-F* | 4.0 - 2.5 = 1.5 | polar covalent |
| C-H | 2.5 - 2.1 = 0.4 | non-polar covalent |
| *(c)  Mg-Cl* | 3.0 - 1.2 = 1.8 | polar covalent |
| *(d)  H-Cl* | 3.0 - 2.1 = 0.9 | polar covalent |

**1.26**    *Using the symbols δ- and δ+, indicate the direction of polarity, if any, in each covalent bond.*

(a) *C-Cl*
$$\overset{\delta+ \quad \delta-}{\text{C-Cl}}$$
**Chlorine is more electronegative than carbon.**

(b) *S-H*
$$\overset{\delta- \quad \delta+}{\text{S-H}}$$
**Sulfur is more electronegative than hydrogen.**

(c) *C-S*
    C-S
**Carbon and sulfur have the same electronegativity, so there is no polarity in a C-S bond.**

(d) *P-H*
    P-H
**Phosphorus and hydrogen have the same electro negativities, so there is no polarity in a P-H bond.**

**1.27**    *Write Lewis structures for these compounds. Show all valence electrons. None of them contain a ring of atoms.*

(a) *H₂O₂*
   *Hydrogen peroxide*

H—O̤—O̤—H

(b) *N₂H₄*
   *Hydrazine*

H—N̈—N̈—H
       |    |
       H    H

(c) *CH₃OH*
   *Methanol*

      H
      |
H—C—O̤—H
      |
      H

(d) *CH₃SH*
   *Methanethiol*

      H
      |
H—C—S̈—H
      |
      H

(e) *CH₃NH₂*
   *Methanamine*

      H
      |
H—C—N̈—H
      |    |
      H    H

(f) *CH₃Cl*
   *Chloromethane*

      H
      |
H—C—C̈l:
      |
      H

(g) *CH₃OCH₃*
   *Dimethyl ether*

      H        H
      |         |
H—C—O̤—C—H
      |         |
      H        H

(h) *CH₃CH₃*
   *Ethane*

      H    H
      |    |
H—C—C—H
      |    |
      H    H

(i) *CH₂CH₂*
   *Ethylene*

   H        H
    \      /
     C=C
    /      \
   H        H

(j) *C₂H₂*
   *Acetylene*

H—C≡C—H

(k) *CO₂*
   *Carbon dioxide*

Ö=C=Ö

(l) *CH₂O*
   *Formaldehyde*

   H
    \
     C=Ö
    /
   H

(m) *CH₃COCH₃*
   *Acetone*

      H  :O:  H
      |   ‖    |
H—C—C—C—H
      |         |
      H        H

(n) *H₂CO₃*
   *Carbonic acid*

         :O:
          ‖
H—O̤—C—O̤—H

(o) *CH₃COOH*
   *Acetic acid*

      H  :O:
      |   ‖
H—C—C—O̤—H
      |
      H

**1.28**    *Write Lewis structures for these ions.*

(a) $HCO_3^-$          (b) $CO_3^{2-}$          (c) $CH_3COO^-$          (d) $Cl^-$

Bicarbonate ion          Carbonate ion          Acetate ion          Chloride ion

**1.29**    *Why are the following molecular formulas impossible?*

(a) $CH_5$   **Carbon atoms can only bond with a maximum of four hydrogen atoms. Each hydrogen atom can only bond with one other atom. Thus there is no possible stable bonding arrangement for $CH_5$ that will utilize all of the atoms.**

(b) $C_2H_7$   **Hydrogen atoms can accommodate only one bond each, so a single hydrogen atom cannot make stable bonds to both carbon atoms. Thus, the two carbon atoms must be bonded to each other. This means that each of the bonded carbon atoms can accommodate only three more bonds, for a total of six available bonding sites. Therefore, only six of the seven hydrogen atoms can be bonded to the two carbon atoms giving a stable structure.**

**1.30**    *Following the rule that each atom of carbon, oxygen, and nitrogen reacts to achieve a complete outer shell of eight valence electrons, add unshared pairs of electrons as necessary to complete the valence shell of each atom in these ions. Then, assign formal charges as appropriate.*

**1.31**    *The following Lewis structures show all valence electrons. Assign formal charges in each structure as appropriate.*

(a)
```
    H  :O:
    |   ||  ..
H – C – C – C – H
    |       ..|
    H       H
```
(b)
```
         :O:
          |
    ..
H – N – C = C – H
    |       |
    H       H
```
(c)
```
    H
    |   ..
H – C – O – H
    |   |
    H   H
```
(d)
```
    H
    |
H – C :
    |
    H
```

(a)
```
    H  :O:
    |   ||  ..  _
H – C – C – C – H
    |       |
    H       H
```
(b)
```
         :O:  ‾
          |
    ..
H – N – C = C – H
    |       |
    H       H
```
(c)
```
    H   +
    |
H – C – O – H
    |   |
    H   H
```
(d)
```
    H
    |
H – C : ‾
    |
    H
```

**1.32**    *Each compound contains both ionic and covalent bonds. Draw a Lewis structure for each, and show by charges which bonds are ionic and by dashes which bonds are covalent.*

(a) *NaOH*

```
       +   ‾ ..
   Na     :O – H
            ..
```

(b) *NaHCO$_3$*

```
              :O:
               ||
          ..       ..  ‾      +
     H – O – C – O:    Na
          ..       ..
```

(c) *NH$_4$Cl*

```
          H
          |  +               ..  ‾
     H – N – H      :Cl:
          |               ..
          H
```

(d) *CH$_3$COONa*

```
       H  :O:
       |   ||   ..  ‾      +
   H – C – C – O:    Na
       |        ..
       H
```

(e) *CH$_3$ONa*

```
        H
        |   ..  ‾      +
   H – C – O:    Na
        |   ..
        H
```

**1.33**    *Silver and oxygen can form a stable compound. Predict the formula of this compound and determine whether the compound consists of ionic or covalent bonds.*

**Silver and oxygen combine to form Ag$_2$O. The electronegativity difference between silver and oxygen is 1.6, indicating that the Ag-O bonds have polar covalent nature.**

## Polarity of Covalent Bonds
**1.34**    *Which statement is true about electronegativity?*

<u>**(a) Electronegativity increases from left to right in a period of the Periodic Table.**</u>
(b) *Electronegativity increases from top to bottom in a column of the Periodic Table.*
(c) *Hydrogen, the element with the lowest atomic number, has the smallest electronegativity.*
(d) *The higher the atomic number of an element, the greater its electronegativity.*

**Electronegativity generally increases going from left to right across a row and from bottom to top of a column in the Periodic Table, thus statement (a) is correct and statements (b), (c), and (d) are incorrect.**

**1.35**    *Why does fluorine, the element in the upper right corner of the Periodic Table, have the largest electronegativity of any element?*

**The two parameters that lead to maximum electronegativity are increasing positive charge on the nucleus and decreasing atomic radii (distance between the nucleus and the electrons in the valence shell). Fluorine is the element for which these two parameters lead to maximum electronegativity.**

**1.36**    *Arrange the single covalent bonds within each set in order of increasing polarity.*
*(a) C-H, O-H, N-H     (b) C-H, C-Cl, C-I    (c) C-C, C-O, C-N     (d) C-Li, C-Hg, C-Mg*

**C-H < N-H < O-H     C-I < C-H < C-Cl    C-C < C-N < C-O     C-Hg < C-Mg < C-Li**
**(0.4)   (0.9)   (1.4)     (0)    (0.4)    (0.5)     (0)    (0.5)   (1.0)     (0.6)   (1.3)    (1.5)**

**1.37**    *Using the values of electronegativity given in Table 1.5, predict which indicated bond in each set is more polar and, using the symbols δ+ and δ–, show the direction of the polarity*

(a) *CH₃-OH or CH₃O-H*                              (b) *H-NH₂ or CH₃-NH₂*
      δ- δ+                                                      δ+ δ-
   CH₃O-H                                                      H-NH₂

(c) *CH₃-SH or CH₃S-H*                              (d) *CH₃-F or H-F*
      δ- δ+                                                      δ+ δ-
   CH₃S-H                                                      H-F

**1.38**    *Identify the most polar bond in each molecule.*
(a) *HSCH₂CH₂OH*              (b) *CHCl₂F*                   (c) *HOCH₂CH₂NH₂*
      δ- δ+                          δ+ δ-                         δ+ δ-
   HSCH₂CH₂O-H                  Cl₂HC-F                     H-OCH₂CH₂NH₂

**In molecules (a) and (c), the most polar bond is the –OH bond, due to the large electronegativity difference between oxygen and hydrogen. The C-F bond in (c) is the most polar due to the large electronegativity difference between carbon and fluorine.**

**1.39** *Predict whether the carbon-metal bond in these organometallic compounds is nonpolar covalent, polar covalent, or ionic. For each polar covalent bond, show its direction of polarity using the symbols δ+ and δ-.*

(a) Tetraethyllead      (b) Methylmagnesium chloride      (c) Dimethylmercury

**All above metal-carbon and metal-halogen bonds are polar covalent according to Table 1.5.**

## Bond Angles and Shapes of Molecules

**1.40** *Use the VSEPR model to predict bond angles about each highlighted atom.*

**1.41** *Use the VSEPR model to predict bond angles about each atom of carbon, nitrogen, and oxygen in these molecules. Hint: First add unshared pairs of electrons as necessary to complete the valence shell of each atom and then make your predictions of bond angles.*

(d) 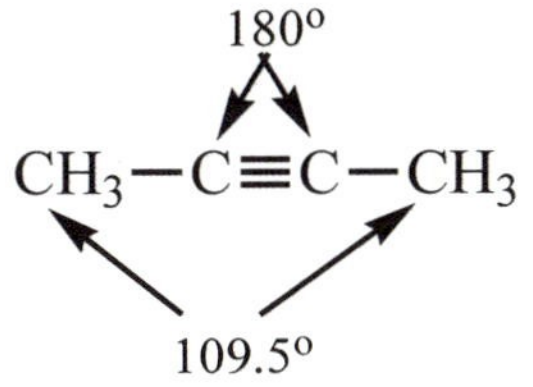 $CH_3-C\equiv C-CH_3$ $180°$ $109.5°$

(e) 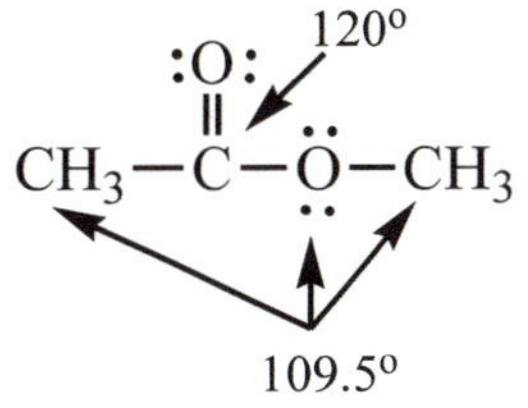 $CH_3-\overset{\overset{\displaystyle :O:}{\|}}{C}-\overset{..}{\overset{..}{O}}-CH_3$ $120°$ $109.5°$

(f) 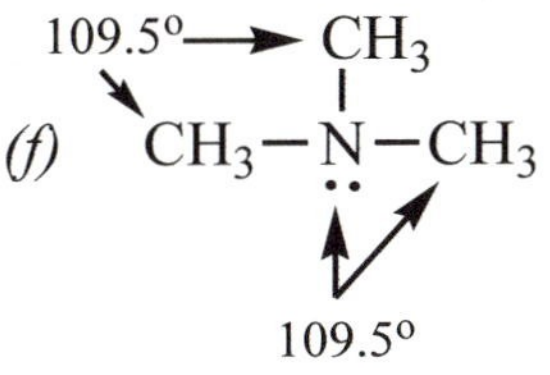 $CH_3-\overset{\overset{\displaystyle CH_3}{|}}{\underset{..}{N}}-CH_3$ $109.5°$ $109.5°$

**1.42** *Silicon is immediately below carbon in the Periodic Table. Predict the C-Si-C bond angle in tetramethylsilane, $(CH_3)_4Si$.*

**Silicon is in Group 4 of the Periodic Table and, like carbon, can be surrounded by up to four regions of electron density. Therefore, with the four regions of electron density from the Si-C bonds, the predicted C-Si-C bond angles would be 109.5° with a tetrahedral geometry around Si.**

## Polar and Nonpolar Molecules

**1.43** *Draw a three-dimensional representation for each molecule. Indicate which molecules are polar and the direction of its polarity.*

(a) $CH_3F$

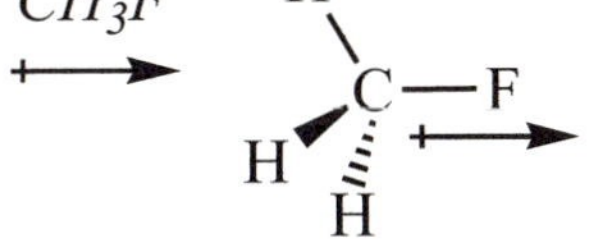

(b) $CH_2Cl_2$

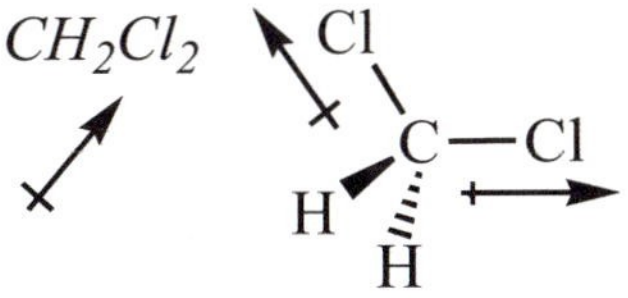

(c) $CHCl_3$

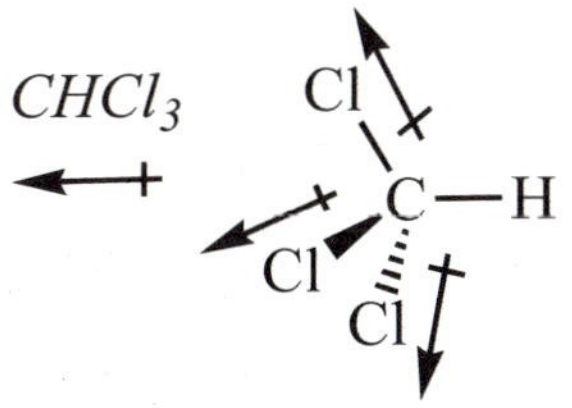

(d) $CCl_4$ 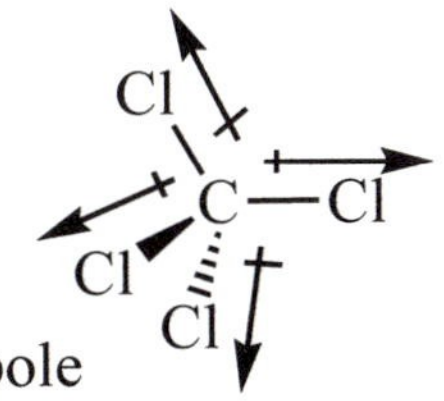 no dipole

(e) $CH_2=CCl_2$

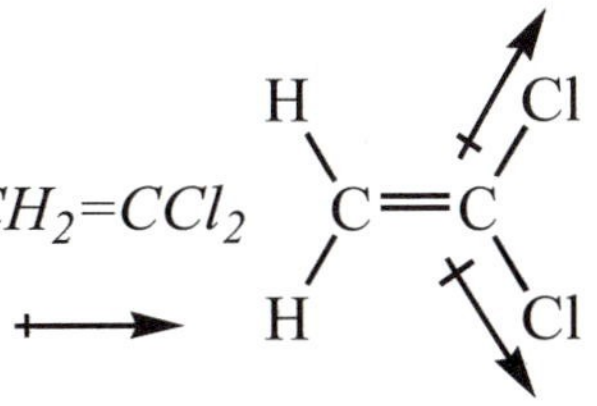

(f) $CH_2=CHCl$

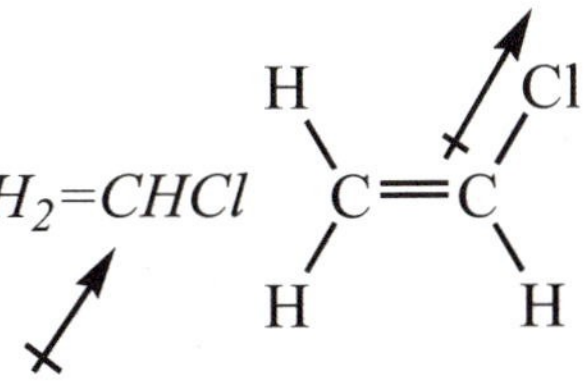

(g) $CH_3CN$     $CH_3-C\equiv N$

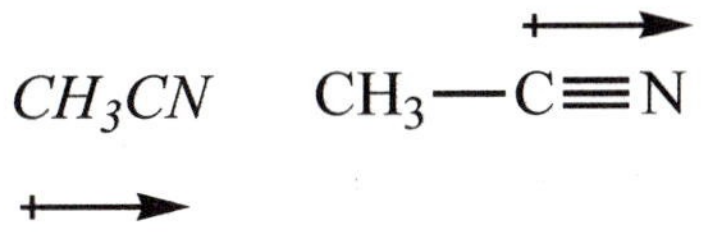

(h) $(CH_3)_2C=O$

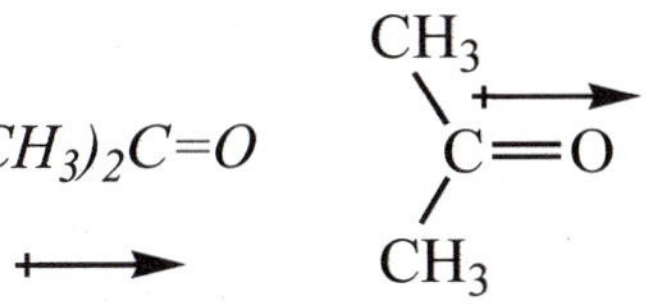

**1.44**    *Tetrafluoroethylene, $C_2F_4$, is the starting material for the synthesis of the polymer poly(tetrafluoroethylene), commonly known as Teflon . Molecules of tetrafluoroethylene are nonpolar. Propose a structural formula for this compound.*

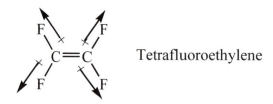

Tetrafluoroethylene

**Fluorine can only form one bond to another atom and carbon requires four bonds. Therefore, each fluorine atom in tetrafluoroethylene must be bonded to a carbon atom, and the carbon atoms are joined by a carbon-carbon double bond.  Even though each C-F bond is highly polar, the individual C-F dipoles cancel each other with a molecular dipole of zero as a net result.**

**1.45**    *Until several years ago, two chlorofluorocarbons (CFCs) most widely used as heat transfer media for refrigeration systems were Freon-11 (trichlorofluoromethane, $CCl_3F$) and Freon-12 (dichlorodifluoromethane, $CCl_2F_2$).  Draw a three-dimensional representation of each molecule, and indicate the direction of its polarity.*

### Resonance and Contributing Structures
**1.46**    *Which of these statements are true about resonance contributing structures?*

    *(a)  All resonance contributing structures must have the same number of valence electrons.*

    *(b)  All resonance contributing structures must have the same arrangement of atoms.*

    *(c)  All atoms in a resonance contributing structure must have complete valence shells.*

    *(d)  All bond angles in sets of resonance contributing structures must be the same.*

**For sets of contributing structures, electrons (usually π electrons or lone pairs) move, but atomic nuclei do not move, thus maintain the same position.  Therefore, statements (b) and (d) are true.  In addition, the total number of electrons in the inner and valence shells of the contributing structures remains the same; therefore statement (a) is also correct.  However, the movement of electrons often leaves one**

**or more atoms without a filled valence shell in a given structure, so statement (c) is false.**

**1.47** *Draw the resonance contributing structure indicated by the curved arrow(s) and assign formal charges as appropriate.*

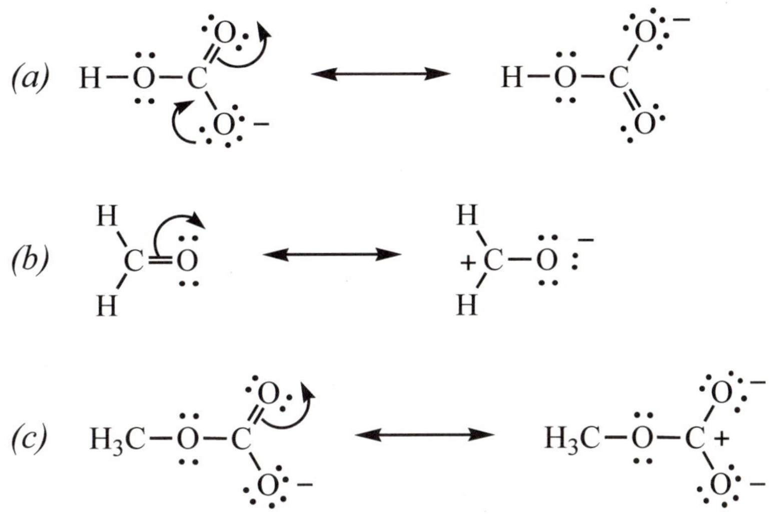

(a)

(b)

(c)

**1.48** *Using the VSEPR model, predict the bond angles about the carbon atom in each pair of resonance contributing structures in Problem 1.47. In what way do the bond angles change from one contributing structure to the other?*

**Atomic nuclei do not move between resonance contributing structures, therefore molecular geometry (bond angles and bond lengths) remains the same between all resonance contributing structures. Carbon nuclei that are carbocations or have just one π bond are *sp²* hybridized, have a planar geometry, and 120° bond angles.**

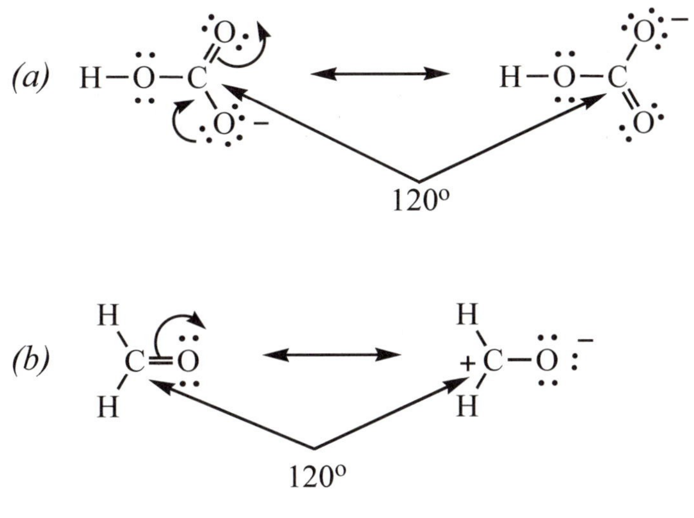

(a)

120°

(b)

120°

$109.5°$

(c)

$109.5°$        $120°$

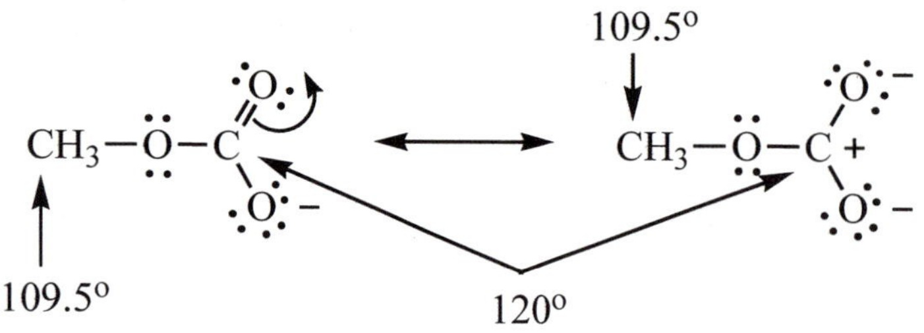

## Hybridization of Atomic Orbitals

**1.49**   *State the hybridization of each highlighted atom.*

(a) $sp^3$     (b) $sp^2$     (c) $sp$

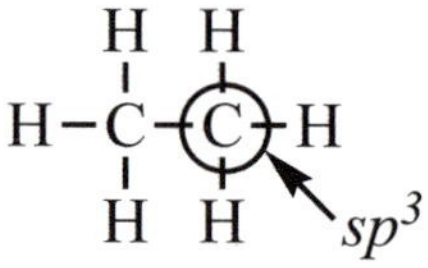

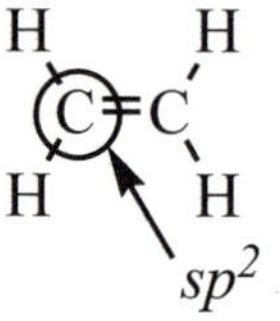

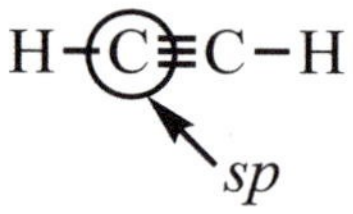

(d) $sp^3$     (e) $sp^2$ / $sp^3$     (f) $sp^2$

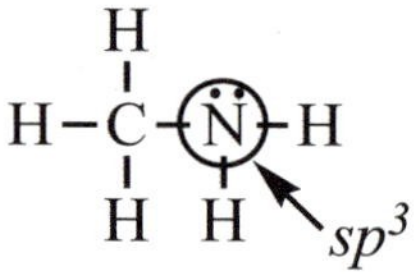

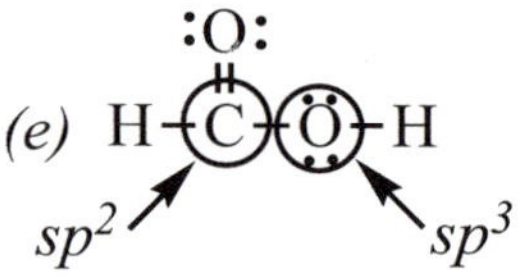

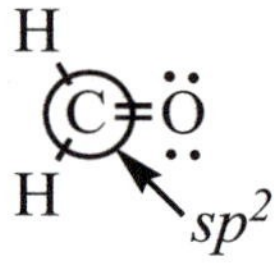

**1.50**   *Describe each highlighted bond in terms of the overlap of hybrid orbitals.*

(a) $\sigma_{sp^3\text{-}sp^3}$     (b) $\sigma_{sp^3\text{-}sp^3}$     (c) $\sigma_{sp^3\text{-}sp^3}$

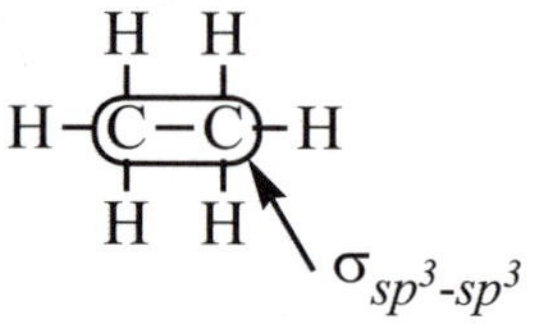

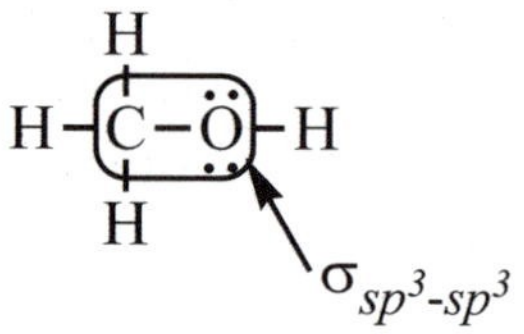

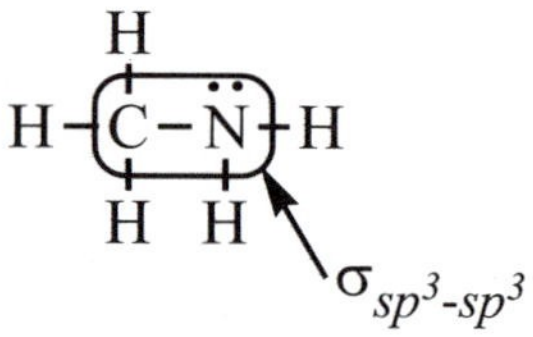

(d) $\sigma_{sp^2\text{-}sp^2}$ $\pi_{2p\text{-}2p}$     (e) $\sigma_{sp^2\text{-}sp^2}$ $\pi_{2p\text{-}2p}$     (f) $\sigma_{sp^2\text{-}sp^3}$

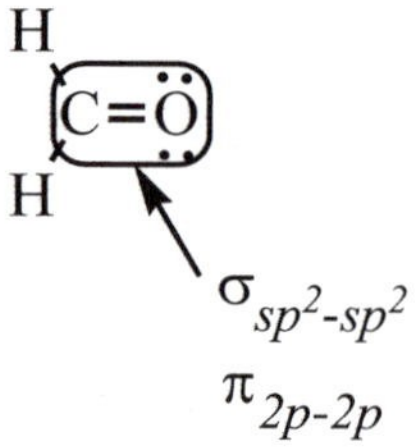

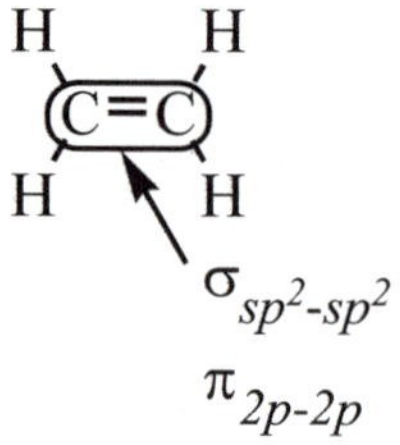

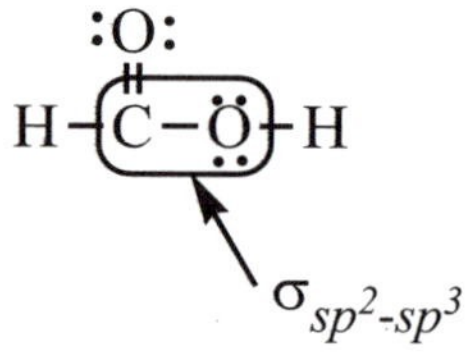

**Functional Groups**

**1.51** *Draw Lewis structures for these functional groups. Be certain to show all valence electrons on each.*

(a) *Carbonyl*     (b) *Carboxyl*     (c) *Hydroxyl*     (d) *Primary amino*

$$\underset{\displaystyle -\overset{\displaystyle :O:}{\overset{\|}{C}}-}{}\qquad \underset{\displaystyle -\overset{\displaystyle :O:}{\overset{\|}{C}}-\ddot{O}-H}{}\qquad -\ddot{O}-H \qquad -\overset{\displaystyle ..}{\underset{\displaystyle H}{N}}-H$$

**1.52** *Draw the structure for a compound of molecular formula:*

(a) $C_2H_6O$ *that is an alcohol*

$$H-\overset{\displaystyle H}{\underset{\displaystyle H}{C}}-\overset{\displaystyle H}{\underset{\displaystyle H}{C}}-\ddot{O}-H$$

(b) $C_3H_6O$ *that is an aldehyde*

$$H-\overset{\displaystyle H}{\underset{\displaystyle H}{C}}-\overset{\displaystyle H}{\underset{\displaystyle H}{C}}-\overset{\displaystyle :O:}{\overset{\|}{C}}-H$$

(c) $C_3H_6O$ *that is a ketone*

$$H-\overset{\displaystyle H}{\underset{\displaystyle H}{C}}-\overset{\displaystyle :O:}{\overset{\|}{C}}-\overset{\displaystyle H}{\underset{\displaystyle H}{C}}-H$$

(d) $C_3H_6O_2$ *that is a carboxylic acid*

$$H-\overset{\displaystyle H}{\underset{\displaystyle H}{C}}-\overset{\displaystyle H}{\underset{\displaystyle H}{C}}-\overset{\displaystyle :O:}{\overset{\|}{C}}-\ddot{O}-H$$

(e) $C_4H_{11}N$ *that is a tertiary amine.*

$$\begin{array}{c} H-C{\overset{H\;H}{\diagdown\!/}} \\ \;\;\;\;\;\;\;\;:N-\overset{H\;H}{\underset{H\;H}{C-C}}-H \\ H{\diagup}C{\diagdown} \\ \;\;H\;\;H \end{array}$$

**Functional Groups**

**1.53** *Draw condensed structural formulas for all compounds of molecular formula $C_4H_8O$ that contain:*

(a) *A carbonyl group (there are two aldehydes and one ketone).*

$$\overset{\displaystyle O}{\overset{\|}{CH_3-C-CH_2-CH_3}} \qquad \overset{\displaystyle O}{\overset{\|}{CH_3-CH_2-CH_2-C-H}} \qquad \overset{\displaystyle CH_3}{\underset{\displaystyle \underset{O}{\|}}{CH_3-CH-C-H}}$$

(b) *A carbon-carbon double bond and a hydroxyl group (there are eight).*

$$CH_2=CH-CH_2-CH_2-OH \qquad CH_3-CH=CH-CH_2-OH \qquad CH_3-CH_2-CH=CH-OH$$

$$\underset{\text{CH}_2\text{=CH-}\overset{\displaystyle\text{OH}}{\underset{|}{\text{CH}}}\text{-CH}_3}{}$$   $$\underset{\text{CH}_2\text{=}\overset{\displaystyle\text{OH}}{\underset{|}{\text{C}}}\text{-CH}_2\text{-CH}_3}{}$$   $$\underset{\text{CH}_3\text{-}\overset{\displaystyle\text{OH}}{\underset{|}{\text{C}}}\text{=CH-CH}_3}{}$$

$$\underset{\text{CH}_3\text{-}\overset{\displaystyle\text{CH}_3}{\underset{|}{\text{C}}}\text{=CH-OH}}{}$$   $$\underset{\text{CH}_2\text{=}\overset{\displaystyle\text{CH}_3}{\underset{|}{\text{C}}}\text{-CH-OH}}{}$$

**1.54**   *Draw structural formulas for:*

*(a) The eight alcohols of molecular formula $C_5H_{12}O$.*

CH₃CH₂CH₂CH₂CH₂OH   $$\underset{\text{CH}_3\text{CH}_2\text{CH}_2\overset{\displaystyle\text{OH}}{\underset{|}{\text{CH}}}\text{CH}_3}{}$$   $$\underset{\text{CH}_3\text{CH}_2\overset{\displaystyle\text{OH}}{\underset{|}{\text{CH}}}\text{CH}_2\text{CH}_3}{}$$

$$\underset{\text{CH}_3\overset{\displaystyle\text{CH}_3}{\underset{|}{\text{CH}}}\text{CH}_2\text{CH}_2\text{OH}}{}$$   $$\overset{\displaystyle\text{CH}_3}{\underset{\displaystyle\text{OH}}{\text{CH}_3\underset{|}{\overset{|}{\text{C}}}\text{CH}_2\text{CH}_3}}$$   $$\overset{\displaystyle\text{CH}_3}{\underset{\displaystyle\text{OH}}{\text{CH}_3\underset{|}{\overset{|}{\text{CH}}}\text{CHCH}_3}}$$

$$\underset{\text{CH}_3\overset{\displaystyle\text{CH}_2\text{OH}}{\underset{|}{\text{CH}}}\text{CH}_2\text{CH}_3}{}$$   $$\overset{\displaystyle\text{CH}_3}{\underset{\displaystyle\text{CH}_3}{\text{CH}_3\underset{|}{\overset{|}{\text{C}}}\text{CH}_2\text{OH}}}$$

*(b) The eight aldehydes of molecular formula $C_6H_{12}O$.*

$$\underset{\text{CH}_3\text{CH}_2\text{CH}_2\text{CH}_2\text{CH}_2\overset{\displaystyle\text{O}}{\overset{\|}{\text{CH}}}}{}$$   $$\overset{\displaystyle\text{CH}_3}{\text{CH}_3\underset{|}{\text{CH}}\text{CH}_2\text{CH}_2\overset{\displaystyle\text{O}}{\overset{\|}{\text{CH}}}}$$   $$\overset{\displaystyle\text{CH}_3}{\text{CH}_3\text{CH}_2\underset{|}{\text{CH}}\text{CH}_2\overset{\displaystyle\text{O}}{\overset{\|}{\text{CH}}}}$$

$$\underset{\displaystyle\text{CH}_3}{\text{CH}_3\text{CH}_2\text{CH}_2\underset{|}{\text{CH}}\overset{\displaystyle\overset{\text{O}}{\|}}{\text{CH}}}$$   $$\overset{\displaystyle\text{CH}_3}{\underset{\displaystyle\text{CH}_3}{\text{CH}_3\underset{|}{\overset{|}{\text{C}}}\text{CH}_2\overset{\displaystyle\overset{\text{O}}{\|}}{\text{CH}}}}$$   $$\overset{\displaystyle\text{CH}_3}{\underset{\displaystyle\text{CH}_3}{\text{CH}_3\underset{|}{\overset{|}{\text{CH}}}\text{CH}\overset{\displaystyle\overset{\text{O}}{\|}}{\text{CH}}}}$$

$$\underset{\displaystyle\text{CH}_3}{\text{CH}_3\text{CH}_2\underset{|}{\overset{\displaystyle\text{CH}_3}{\text{C}}}\overset{\displaystyle\overset{\text{O}}{\|}}{\text{—CH}}}$$   $$\underset{\displaystyle\text{CH}_2\text{CH}_3}{\text{CH}_3\text{CH}_2\underset{|}{\text{CH}}\overset{\displaystyle\overset{\text{O}}{\|}}{\text{CH}}}$$

(c) *The six ketones of molecular formula C₆H₁₂O.*

O
‖
CH₃CH₂CH₂CH₂CCH₃

O
‖
CH₃CH₂CH₂CCH₂CH₃

CH₃  O
|    ‖
CH₃CHCH₂CCH₃

CH₃
|
CH₃CHCCH₂CH₃
‖
O

CH₃ O
|   ‖
CH₃C——CCH₃
|
CH₃

O
‖
CH₃CH₂CHCCH₃
|
CH₃

(d) *The eight carboxylic acids of molecular formula C₆H₁₂O₂.*

O
‖
CH₃CH₂CH₂CH₂CH₂COH

CH₃     O
|       ‖
CH₃CHCH₂CH₂COH

CH₃   O
|     ‖
CH₃CH₂CHCH₂COH

O
‖
CH₃CH₂CH₂CHCOH
|
CH₃

CH₃ O
|   ‖
CH₃CCH₂COH
|
CH₃

CH₃  O
|    ‖
CH₃CHCHCOH
|
CH₃

CH₃ O
|   ‖
CH₃CH₂C——COH
|
CH₃

O
‖
CH₃CH₂CHCOH
|
CH₂CH₃

(e) *The three tertiary amines of molecular formula C₅H₁₃N.*

CH₃
|
CH₃NCH₂CH₂CH₃

CH₃
|
CH₃CH₂NCH₂CH₃

CH₃
|
CH₃NCHCH₃
|
CH₃

**1.55**  *Identify the functional groups in each compound. We study each compound in more detail in the indicated section.*

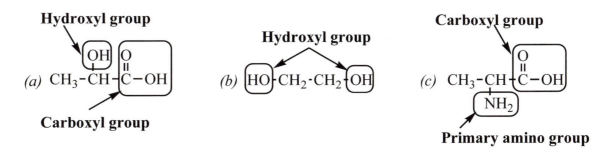

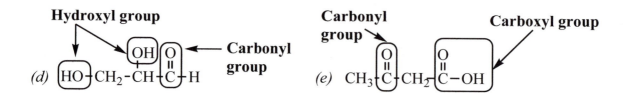

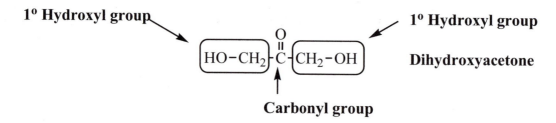

**1.56**  *Dihydroxyacetone, C₃H₆O₃, the active ingredient in many sunless tanning lotions, contains two 1° hydroxyl groups, each on a different carbon, and one ketone group. Draw a structural formula for dihydroxyacetone.*

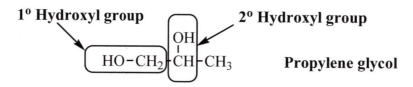

**1.57**  *Propylene glycol, C₃H₈O₂, commonly used in airplane deicers, contains a 1° alcohol and a 2° alcohol. Draw a structural formula for propylene glycol.*

**1° Hydroxyl group**     **2° Hydroxyl group**

HO–CH₂–CH–CH₃          **Propylene glycol**
        |
       OH

**1.58**  *Ephedrine is a molecule found in the dietary supplement ephedra, which has been linked to adverse health reactions such as heart attacks, strokes, and heart palpitations. The use of ephedra as dietary supplements is now banned by the FDA.*

*(a) Identify at least two functional groups in ephedrine.*

**Ephedrine**

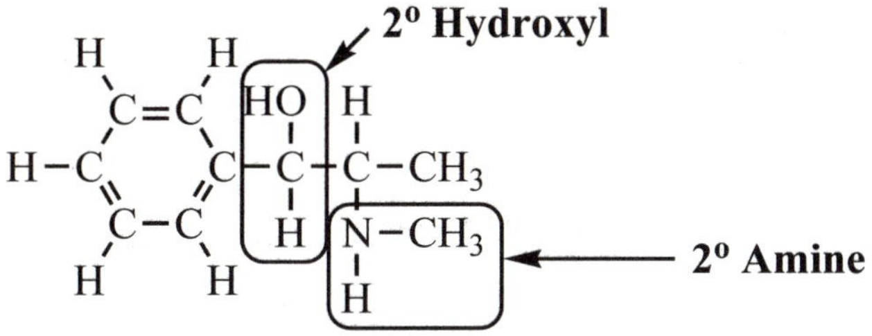

2° Hydroxyl

2° Amine

*(b) Would you predict ephedrine to be polar or nonpolar?*

**Ephedrine is expected to be polar due to the polar C-O and O-H bonds of the alcohol and the polar N-H bond of the amine.**

**1.59**   *Ozone, $O_3$, and carbon dioxide, $CO_2$, are both known as greenhouse gases. Compare and contrast their shapes and indicate the hybridization of each atom in the two molecules.*

*Carbon dioxide:* $sp$

$O=C=O$

**Linear**

$sp^2$

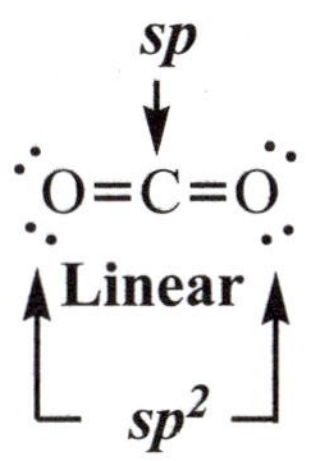

*Ozone:*

**Bent**          **Bent**

**All oxygen atoms $sp^2$ hybridized**

**According to the VSEPR model, carbon dioxide will adopt a linear geometry and ozone will be bent. A molecule of $CO_2$ will be non-polar, where ozone will be polar due to its bent geometry and distribution of charge. Note that two resonance contributing structures can be drawn for ozone.**

**Looking Ahead**

**1.60**   *Allene, $C_3H_4$, has the structural formula $H_2C=C=CH_2$. Determine the hybridization of each carbon in allene and predict the shape of the molecule.*

*allene:*   $120°$   $\longrightarrow$   $C=C=C$   $\longleftarrow$ $120°$

$180°$   $sp$

$sp^2$

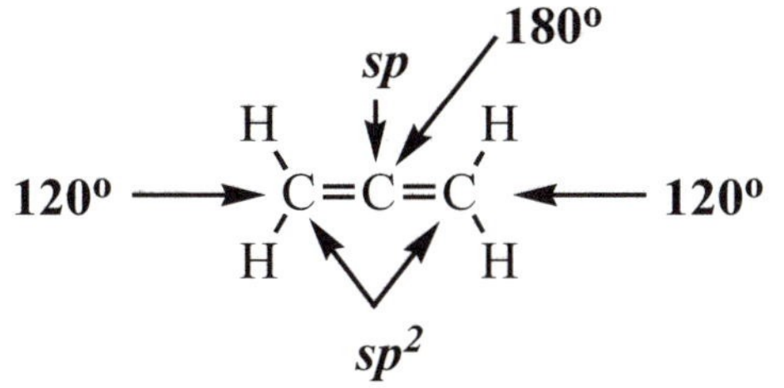

**The geometry of the carbon chain is linear.**

**1.61**   *Dimethyl sulfoxide, $(CH_3)_2SO$, is a common solvent used in organic chemistry.*

(a) *Write a Lewis structure for dimethyl sulfoxide.*

$$\begin{array}{ccc}
 & H \;\; :\overset{\displaystyle ..}{\underset{\displaystyle }{O}}:^{-} H & \\
 & | \quad | \quad | & \\
H- & C-S-C & -H \\
 & | \quad | \quad | & \\
 & H \;\; \overset{+}{\underset{}{}} \;\; H &
\end{array}$$

$sp^3$                          ↑ dipole
109.5°          $sp^3$          $sp^3$
                109.5°          109.5°

(b) *Predict the hybridization of the sulfur atom in the molecule.*

**The sulfur atom is $sp^3$ hybridized.**

(c) *Predict the geometry of dimethyl sulfoxide.*

**The geometry about the carbon and sulfur atoms will be tetrahedral.**

(d) *Is dimethyl sulfoxide a polar or a nonpolar molecule?*

**Dimethyl sulfoxide is a polar molecule due to the S=O bond.**

**1.62**   *In Chapter 5, we study a group of organic cations called carbocations. Following is the structure of one such carbocation, the tert-butyl cation.*

*tert*-**Butyl cation**
$$\begin{array}{c}
H_3C \\
\quad \diagdown \; + \\
\qquad C-CH_3 \\
\quad \diagup \\
H_3C \\
\end{array}$$
120° ($sp^2$ hybridized)

(a) *How many electrons are in the valence shell of the carbon bearing the positive charge?*

**There are six electrons in the valence shell of the carbon bearing the positive charge.**

(b) *Predict the bond angles about this carbon.*

**The carbon bearing the positive charge has three regions of electron density around it. The VSEPR model predicts bond angles of 120° around the carbon.**

(c) *Given the bond angles you predicted in (b), what hybridization do you predict for this carbon?*

**The carbon bearing the positive charge is $sp^2$ hybridized (a combination of the 2s and the $2p_x$ and $2p_y$ atomic orbitals).  The area above and below the central carbon is occupied by the unhybridized $2p_z$ orbital.**

**1.63**   *We also study the isopropyl cation, $(CH_3)_2CH^+$, in Chapter 5.*

(a) *Write a Lewis structure for this cation.  Use a plus sign to show the location of the positive charge.*

**Isopropyl cation:**

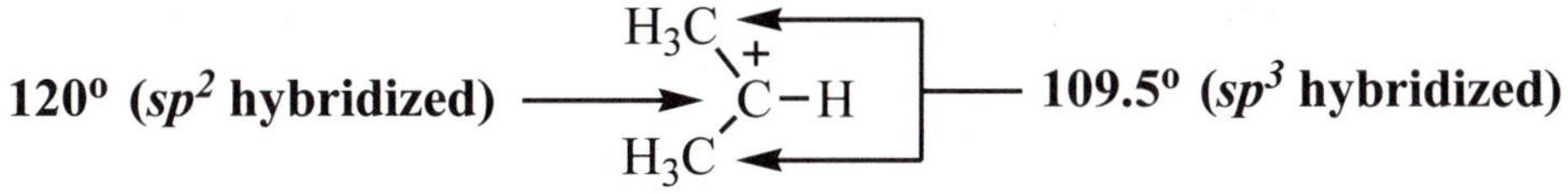

**120° ($sp^2$ hybridized)**  $\longrightarrow$     — **109.5° ($sp^3$ hybridized)**

(b) *How many electrons are in the valence shell of the carbon bearing the positive charge?*

**There are six electrons in the valence shell of the carbon bearing the positive charge.**

(c) *Use the VSEPR model to predict all bond angles about the carbon bearing the positive charge.*

**The carbon bearing the positive charge has three regions of electron density around it.  The VSEPR model predicts bond angles of 120° around the carbon.**

(d) *Describe the hybridization of each carbon in this cation.*

**The carbon bearing the positive charge is $sp^2$ hybridized (a combination of the 2s and the $2p_x$ and $2p_y$ atomic orbitals).  The –CH₃ groups are $sp^3$ hybridized (a combination of the 2s and $2p_x$, $2p_y$, and $2p_z$ atomic orbitals).**

**1.64**   *In Chapter 9, we study benzene, $C_6H_6$, and its derivatives.*

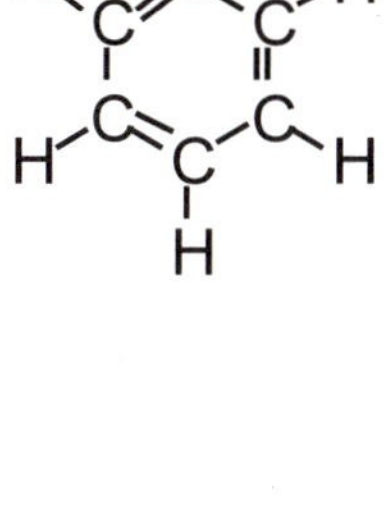

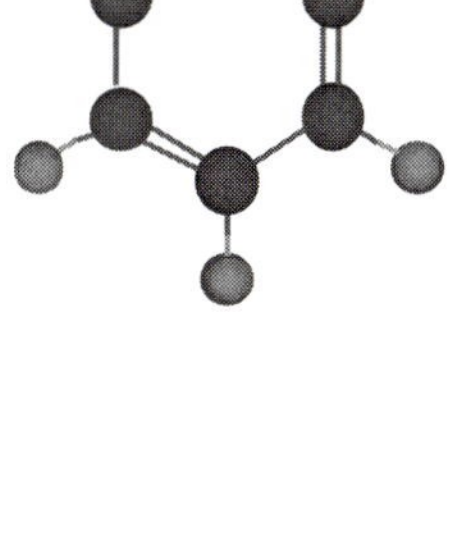

*(a) Predict each H-C-C and each C-C-C bond angle on benzene.*

**Each benzene carbon has three regions of electron density around it, so according to the VSEPR model, the carbons are trigonal planar. All H-C-C and C-C-C bond angles are 120°.**

*(b) State the hybridization of each carbon in benzene.*

**The carbons in benzene are $sp^2$ hybridized because each one makes three σ bonds and one π bond. A C-C π bond requires the overlap of an unhybridized $2p_z$ orbital from each carbon. The σ bonds result from the overlap of an H(1s) and C($sp^2$) for the C-H bonds and C($sp^2$)-C($sp^2$) for the C-C bonds.**

*(c) Predict the shape of a benzene molecule.*

**All 12 atoms lie in the same plane, therefore the molecule is planar.**

**1.65**  *Explain why <u>all</u> the carbon-carbon bonds in benzene are equal in length.*

$1.39 \times 10^{-10}$ m

**The observed structure of benzene is a hybrid of two equivalent resonance contributing structures. Because the equivalent contributing structures have identical patterns of covalent bonding and are of equal energy, the result is a resonance hybrid that has equivalent C-C bonds that have a length between a C-C bond ($1.54 \times 10^{-10}$ m) and C=C bond ($1.33 \times 10^{-10}$).**

## CHAPTER 2
### *Solutions to Problems*

**2.1**   *Write each acid-base reaction as a proton-transfer reaction.  Label which reactant is the acid and which the base; which product is the conjugate base of the original acid and which is the conjugate acid of the original base.  Use curved arrows to show the flow of electrons in each reaction.*

**Hint:  For acid-base curved arrow mechanisms, the arrows are drawn from the base non-bonded electron pair to the acid proton, forming the new bond, B-H.  The previous H-A electron bonding pair shifts to become an electron lone pair on the conjugate base, A⁻.**

$$B^- + H-A \longrightarrow B-H + {}^-A$$

| Base | Acid | Conjugate acid | Conjugate base |

*(a)* $CH_3SH + OH^- \longrightarrow CH_3S^- + H_2O$

$$CH_3S-H + {}^-O-H \longrightarrow CH_3S^- + H-O-H$$

**Acid          Base          Conjugate base          Conjugate acid**

*(b)* $CH_3OH + NH_2^- \longrightarrow CH_3O^- + NH_3$

$$CH_3O-H + {}^-N-H \longrightarrow CH_3O^- + H-N-H$$

**Acid          Base          Conjugate base          Conjugate acid**

**2.2**   *For each value of $K_a$, calculate the corresponding value of $pK_a$.  Which compound is the stronger acid?*

$$pK_a = -\log K_a$$

*(a) Acetic acid, $K_a = 1.74 \times 10^{-5}$*          **$pK_a$ (acetic acid) = 4.76**
*(b) Water, $K_a = 2.00 \times 10^{-16}$*          **$pK_a$ (water) = 15.7**

**As the $pK_a$ values decrease, acidity increases.  Therefore acetic acid, having the lower $pK_a$, is the stronger acid.**

**2.3**     *For each acid-base equilibrium, label the stronger acid, the stronger base, the weaker acid, and the weaker base. Then predict whether the position of equilibrium lies toward the right or the left.*

**In acid-base reactions, the equilibria favor the side with the weaker acid/weaker base.**

(a) $CH_3NH_2$  +  $CH_3COOH$  $\rightleftharpoons$  $CH_3NH_3^+$  +  $CH_3COO^-$
   Methylamine    Acetic acid        Methylammonium   Acetate
                                       ion          ion

> **Acetic acid is a stronger acid than the methylammonium ion; therefore the equilibrium lies far to the right.**

$CH_3NH_2$  +  $CH_3COOH$  $\longrightarrow$  $CH_3NH_3^+$  +  $CH_3COO^-$
**(Stronger base)  (Stronger acid)**      **(Weaker acid)   (Weaker base)**
             **$pK_a = 4.76$**                 **$pK_a = 10.6$**

(b) $CH_3CH_2O^-$  +  $NH_3$  $\rightleftharpoons$  $CH_3CH_2OH$  +  $NH_2^-$
   Ethoxide     Ammonia             Ethanol     Amide
    ion                                     anion

> **Ethanol is a stronger acid than the ammonia; therefore the equilibrium lies far to the left.**

$CH_3CH_2O^-$  +  $NH_3$  $\longleftarrow$  $CH_3CH_2OH$  +  $NH_2^-$
**(Weaker base)  (Weaker acid)**     **(Stronger acid)   (Stronger base)**
           **$pK_a = 38$**               **$pK_a = 15.9$**

**2.4**     *Complete this acid-base reaction. First add unshared pairs of electrons on the reacting atoms to give each a complete octet. Use curved arrows to show the redistribution of electrons in the reaction. In addition, predict whether the position of this equilibrium lies toward the left or the right.*

$$CH_3-O^- \quad + \quad CH_3-\overset{\overset{H}{|}}{\underset{\underset{CH_3}{|}}{N^+}}-CH_3 \quad \rightleftharpoons$$

**The trimethylammonium ion is a stronger acid than methanol; therefore the equilibrium lies far to the right.**

CH$_3$–Ö: $^-$  +  H—N$^+$–CH$_3$  $\rightleftharpoons$  CH$_3$–Ö–H  +  :N–CH$_3$

(with CH$_3$ groups on nitrogen)

(Stronger base)        (Stronger acid)                    (Weaker acid)       (Weaker base)
                            p$K_a$ = 10                          p$K_a$ = 16

**2.5**   *Write an equation for the reaction between each Lewis acid-base pair, showing electron flow by means of curved arrows. Hint: Aluminum is in Group 3A of the Periodic Table, just under boron.  Aluminum in AlCl$_3$ has only six electrons in its valence shell and has an incomplete octet.*

**For Lewis acid-Lewis base curved arrow mechanisms, the arrows are drawn from the Lewis base non-bonded electron pair to the Lewis acid, forming the new bond.**

*(a)  Cl$^-$   +   AlCl$_3$  $\longrightarrow$*

Lewis base
(electron pair donor)

Lewis acid
(electron pair acceptor)

*(b)  CH$_3$Cl   +   AlCl$_3$  $\longrightarrow$*

Lewis base
(electron pair donor)

Lewis acid
(electron pair acceptor)

## Arrhenius Acids and Bases

**2.6**   *Complete the net ionic equation for each acid when placed in water. Use curved arrows to show the flow of electron pairs in each reaction.  Also for each reaction, determine the direction of equilibrium using Table 2.2 as a reference for the p$K_a$ values of proton acids.*

**In acid-base reactions, the equilibria favor the side with the weaker acid/weaker base.**

(a) $NH_4^+$  +  $H_2O$  ⇌

**The hydronium ion is a stronger acid than the ammonium ion; therefore the equilibrium lies far to the left.**

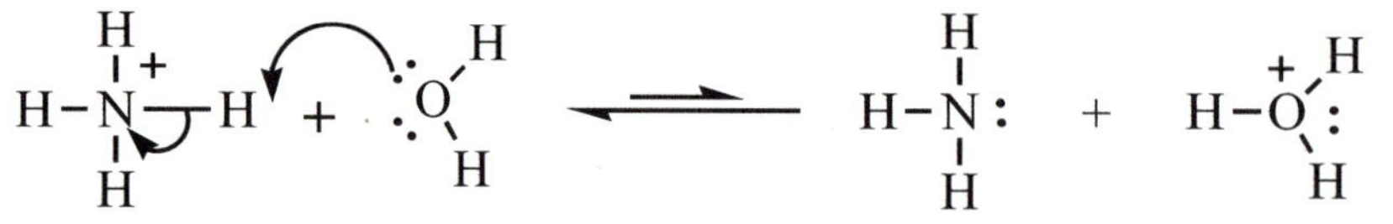

Weaker acid     Weaker base                Stronger base     Stronger acid
$pK_a = 9.24$                                                      $pK_a = -1.74$


(b) $HCO_3^-$  +  $H_2O$  ⇌

**The hydronium ion is a stronger acid than the bicarbonate ion; therefore the equilibrium lies far to the left.**

Weaker acid     Weaker base                Stronger base          Stronger acid
$pK_a = 10.33$                                                      $pK_a = -1.74$


(c) $CH_3COOH$  +  $H_2O$  ⇌

**The hydronium ion is a stronger acid than acetic acid; therefore the equilibrium lies far to the left.**

Weaker acid     Weaker base                Stronger base     Stronger acid
$pK_a = 4.76$                                                       $pK_a = -1.74$

**2.7**    *Complete the net ionic equation for each base when placed in water. Use curved arrows to show the flow of electron pairs in each reaction. Also for each reaction, determine the direction of equilibrium using Table 2.2 as a reference for the $pK_a$ values of proton acids formed.*

**In acid-base reactions, the equilibria favor the side with the weaker acid/weaker base.**

(a)  $CH_3NH_2$   +   $H_2O$   ⇌

**The methylammonium ion is a stronger acid than water; therefore the equilibrium lies far to the left.**

| Weaker base | Weaker acid | Stronger acid | Stronger base |
|---|---|---|---|
| | $pK_a = 15.7$ | $pK_a = 10.64$ | |

(b)  $HSO_4^-$   +   $H_2O$   ⇌

**Sulfuric acid is a stronger acid than water; therefore the equilibrium lies far to the left.**

| Weaker base | Weaker acid | Stronger acid | Stronger base |
|---|---|---|---|
| | $pK_a = 15.7$ | $pK_a = -5.2$ | |

(c)  $Br^-$   +   $H_2O$   ⇌

**Hydrogen bromide is a stronger acid than water; therefore the equilibrium lies far to the left.**

| Weaker base | Weaker acid | Stronger acid | Stronger base |
|---|---|---|---|
| | $pK_a = 15.7$ | $pK_a = -8$ | |

*(d)* $CO_3{}^{2-}$ + $H_2O$  ⇌

**Bicarbonate ion is a stronger acid than water; therefore the equilibrium lies far to the left.**

Weaker base        Weaker acid                    Stronger acid        Stronger base
                   $pK_a = 15.7$                   $pK_a = 10.33$

## Brønsted-Lowry Acids and Bases

**2.8**   *Complete a net ionic equation for each proton-transfer reaction using curved arrows to show the flow of electron pairs in each reaction. In addition, write Lewis structures for all starting materials and products. Label the original acid and its conjugate base; label the original base and its conjugate acid. If you are uncertain about which substance in each equation is the proton donor, refer to Table 2.2 for the pK$_a$ values of proton acids.*

*(a)* $NH_3$ + $HCl$ ⟶

Base           Acid                        Conjugate acid    Conjugate base

*(b)* $CH_3CH_2O^-$ + $HCl$ ⟶

Base                Acid                  Conjugate acid        Conjugate base

*(c)* $HCO_3^-$ + $OH^-$ ⟶

Acid                Base                  Conjugate base        Conjugate acid

(d) $CH_3COO^-$ + $NH_4^+$ $\longrightarrow$

**Base**          **Acid**          **Conjugate acid**     **Conjugate base**

(e) $NH_4^+$ + $OH^-$ $\longrightarrow$

**Base**          **Acid**          **Conjugate acid**     **Conjugate base**

(f) $CH_3COO^-$ + $CH_3NH_3^+$ $\longrightarrow$

**Base**          **Acid**          **Conjugate acid**     **Conjugate base**

(g) $CH_3CH_2O^-$ + $NH_4^+$ $\longrightarrow$

**Base**          **Acid**          **Conjugate acid**     **Conjugate base**

(h) $CH_3NH_2^+$ + $OH^-$ $\longrightarrow$

**Base**          **Acid**          **Conjugate acid**     **Conjugate base**

**2.9**    *Each of these molecules and ions can function as a base. Complete the Lewis structure of each base, and write the structural formula of the conjugate acid formed by its reaction with HCl.*

*(a)* $CH_3CH_2OH$

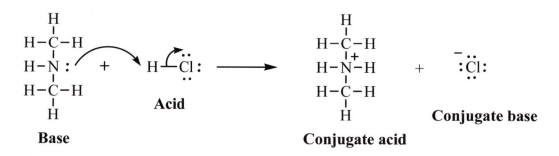

*(b)* $\overset{\overset{O}{\|}}{H}CH$

*(c)* $(CH_3)_2NH$

*(d)* $HCO_3^-$

**2.10**  *Offer an explanation for the following observations.*

(a) *$H_3O^+$ is a stronger acid than $NH_4^+$.*

**Oxygen is more electronegative than nitrogen; therefore oxygen is less able to bear a formal positive charge.**

(b) *Nitric acid, $HNO_3$, is a stronger acid than nitrous acid, $HNO_2$ ($pK_a$ 3.7).*

**The negative charge in nitrate ($NO_3^-$), the conjugate base for nitric acid, is delocalized onto three electronegative oxygen atoms through three resonance contributing structures.  Nitrite ($NO_2^-$), the conjugate base for nitrous acid, delocalizes the negative charge onto only two electronegative oxygen atoms through two resonance contributing structures.  Thus the negative charge is more delocalized in the nitrate anion, making it a more stable anion and a weaker conjugate base.  As the conjugate base strength decreases, its acid strength increases.**

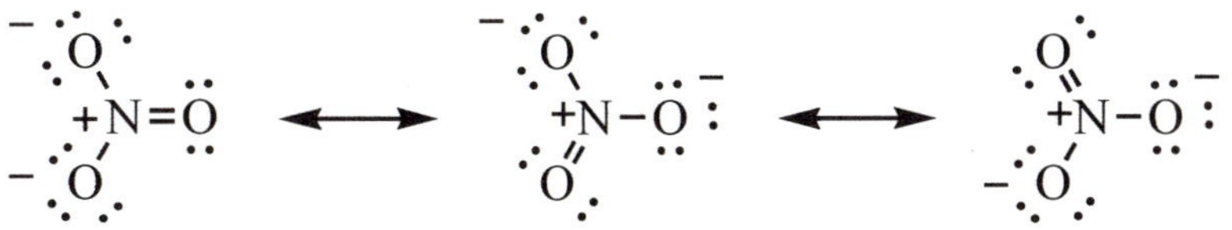

(c) *Ethanol, $CH_3CH_2OH$, and water have approximately the same acidity.*

**Deprotonation of ethanol and water gives ethoxide and hydroxide, respectively. In both cases, the conjugate base bears the negative charge on the oxygen and is stabilized only by the oxygen's electronegativity, giving the conjugate bases roughly the same stability and the acids nearly the same strength.**

(d) *Trichloroacetic acid, $CCl_3COOH$ ($pK_a$ 0.64), is a stronger acid than acetic acid, $CH_3COOH$ ($pK_a$ 4.74).*

**Charge delocalization by resonance:**

**Charge delocalization by inductive electron withdrawal:**

**R = H or Cl**

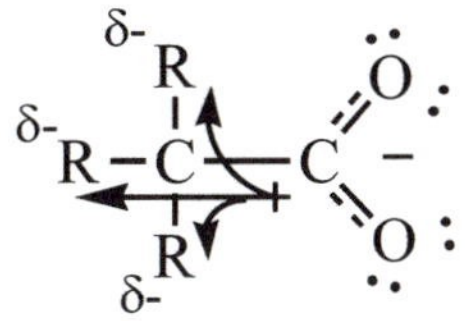

Both conjugate bases involve a stabilization by the delocalization of a negative charge onto two oxygen atoms through <u>resonance</u>, but the trichloroacetate anion is further stabilized by charge delocalization by the powerful <u>inductive</u> electron withdrawing effects exerted by the highly electronegative chlorines on the $-CCl_3$ group. The electronegative chlorines are much more effective at stabilizing a developing negative charge relative to the much less electronegative hydrogen atoms of the methyl group. Therefore, trichloroacetate anion is more stabilized and thus a weaker conjugate base than acetate, making trichloroacetic acid a stronger acid relative to acetic acid.

(e) *Trifluoroacetic acid, $CF_3COOH$ (pK$_a$ 0.23), is a stronger acid than trichloroacetic acid, $CCl_3COOH$ (pK$_a$ 0.64).*

The resonance and inductive effects on the conjugate bases are the same as in (d), but the more electronegative fluorines of the $CF_3$ substituent stabilizes the anionic conjugate base more than the less electronegative chlorines of the $CCl_3$ substituents.

**2.11** *Select the most acidic proton in the following compounds.*

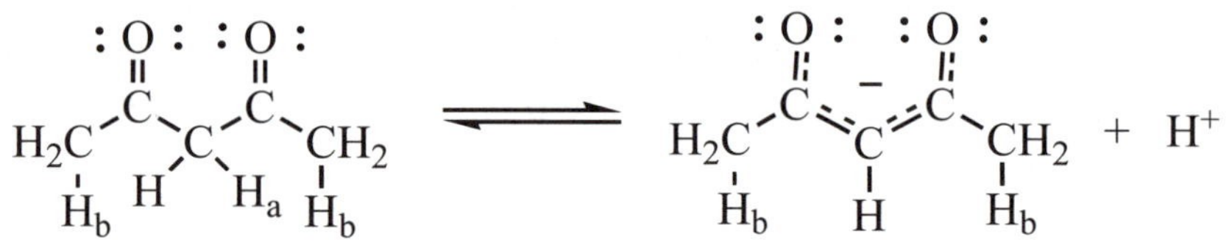

Deprotonation of H$_a$ results in a more stable conjugate base that is stabilized by resonance through three resonance contributing structures, two involving the delocalization of a negative charge placed on two electronegative oxygen atoms and a carbon atom. The dashed-line structure is a description of the resonance-stabilized hybrid structure that is a combination of all three resonance contributing structures below and shows the negative charge delocalized on the two oxygen atoms and the central carbon atom.

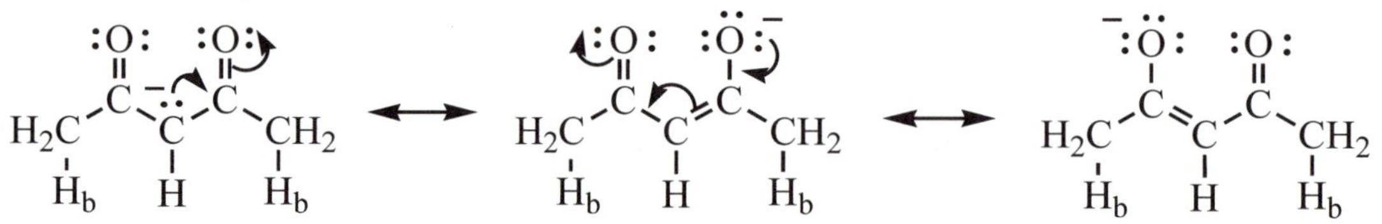

Deprotonation of H$_b$ results in a less stable conjugate base with a negative charge delocalized through only two resonance structures.

(b)

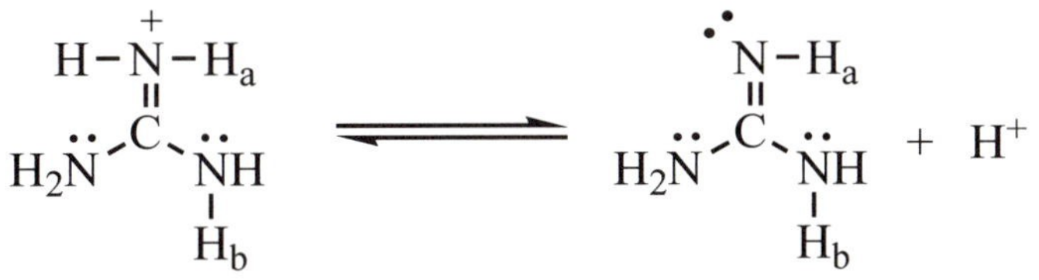

$H_a$ is the most acidic proton

**Deprotonation of $H_a$ results in the neutral conjugate base that is resonance stabilized.**

## Quantitative Measure of Acid Strength

**2.12** *Which has the larger numerical value*

(a) *The $pK_a$ of a strong acid or the $pK_a$ of a weak acid?*

**As acid strengths decrease, the $pK_a$'s increase, therefore, a weak acid has the larger $pK_a$.**

(b) *The $K_a$ of a strong acid or the $K_a$ of a weak acid?*

**As acid strengths increase, so does the $K_a$, therefore, a strong acid has the larger $K_a$.**

**2.13** *In each pair, select the stronger acid:*

**Remember, acid strength is inversely proportional to $pK_a$ and proportional to $K_a$; therefore the stronger acid (underlined and bolded in each answer) is the one with the lower $pK_a$ and the larger $K_a$.**

(a) ***Pyruvic acid ($pK_a$ 2.49)*** *or lactic acid ($pK_a$ 3.08)*
(b) *Citric acid ($pK_{a1}$ 3.08) or **phosphoric acid ($pK_{a1}$ 2.10)***
(c) *Nicotinic acid (niacin, $K_a$ 1.4 x $10^{-5}$) **or acetylsalicylic acid (aspirin, $K_a$ 3.3 x$10^{-4}$)***
(d) *Phenol ($K_a$ 1.12 x $10^{-10}$) or **acetic acid ($K_a$ 1.74 x $10^{-5}$)***

**2.14**   *Arrange the compounds in each set in order of increasing acid strength.  Consult Table 2.2 for pKa values of each acid.*

| | | |
|---|---|---|
| (a)   CH₃CH₂OH | HOCO⁻ | C₆H₅COH |
| Ethanol | Bicarbonate ion | Benzioc acid |
| pKₐ:   **15.9** | **10.33** | **4.19** |
| **Weakest** | | **Strongest** |

| | | |
|---|---|---|
| (b)   HOCOH | HOCCH₃ | HCl |
| Carbonic acid | Acetic acid | Hydrogen chloride |
| pKₐ:   **6.36** | **4.76** | **-7** |
| **Weakest** | | **Strongest** |

**2.15**   *Arrange the compounds in each set in order of increasing base strength. Consult Table 2.2 for pKa values of the conjugate acid of each base. (Hint: The stronger the acid, the weaker its conjugate base, and vice versa.)*

(a)   NH₃          HOCO⁻          CH₃CH₂O⁻
       9.24          6.34            15.9          pKₐ of conjugate acid

**Base strength increases in the order:**   HOCO⁻  <  NH₃  <  CH₃CH₂O⁻

(b)   OH⁻          HOCO⁻          CH₃CO⁻
       15.7          6.34            4.76          pKₐ of conjugate acid

**Base strength increases in the order:**   CH₃CO⁻  <  HOCO⁻  <  ⁻OH

(c)   H₂O          NH₃          CH₃CO⁻
       -1.74          9.24          4.76          pKₐ of conjugate acid

**Base strength increases in the order:**   H₂O  <  CH₃CO⁻  <  NH₃

*(d)*    $NH_2^-$           $\overset{\overset{\displaystyle O}{\displaystyle \|}}{CH_3C}O^-$           $OH^-$

      **38**               **4.76**             **15.7**       **p$K_a$ of conjugate acid**

**Base strength increases in the order:**    $\overset{\overset{\displaystyle O}{\displaystyle \|}}{CH_3C}O^-$   <   $OH^-$   <   $NH_2^-$

## Position of Equilibrium in Acid-Base Reactions

**2.16**    *Unless under pressure, carbonic acid in aqueous solution breaks down into carbon dioxide and water, and carbon dioxide is evolved as bubbles of gas. Write an equation for the conversion of carbonic acid to carbon dioxide and water.*

$$\overset{\overset{\displaystyle O}{\displaystyle \|}}{HOC}OH \quad\longrightarrow\quad CO_2 \;+\; H_2O$$

**2.17**    *Will carbon dioxide be evolved when sodium bicarbonate is added to an aqueous solution of each compound?*

*(a) $H_2SO_4$ [p$K_a$ -5.2]*       *(b) $CH_3CH_2OH$ [p$K_a$ 15.9]*       *(c) $NH_4Cl$ [p$K_a$ 9.24]*

**In order for carbon dioxide to be evolved, the bicarbonate ion must be protonated to give carbonic acid, which then breaks down to carbon dioxide and water (see Problem 2.16). Acid-base equilibria favor the side with the weaker acid and base. Therefore, the only acids that will protonate sodium bicarbonate to produce carbonic acid are the acids that are stronger than carbonic acid (the acids with lower p$K_a$'s than carbonic acid [p$K_a$ 6.36]). Of the three choices, only sulfuric acid satisfies this requirement.**

**2.18**    *Acetic acid, $CH_3COOH$, is a weak organic acid, p$K_a$ 4.76. Write equations for the equilibrium reactions of acetic acid with each base. Which equilibria lie considerably toward the left? Which lie considerably toward the right?*

**Acid-base equilibria favor the side with the weaker acid (the acid with the larger p$K_a$.**

*(a)*   $CH_3COOH \;+\; HCO_3^- \;\rightleftharpoons\; CH_3COO^- \;+\; H_2CO_3$     **Right side favored**
      **p$K_a$ 4.76**                                   **p$K_a$ 6.36**

*(b)*   $CH_3COOH \;+\; NH_3 \;\rightleftharpoons\; CH_3COO^- \;+\; NH_4^+$      **Right side favored**
      **p$K_a$ 4.76**                                   **p$K_a$ 9.24**

(c)  $CH_3COOH$ + $H_2O$ ⇌ $CH_3COO^-$ + $H_3O^+$          **Left side favored**

    **p$K_a$ 4.76**                                    **p$K_a$ -1.74**

(d)  $CH_3COOH$ + $HO^-$ ⇌ $CH_3COO^-$ + $H_2O$          **Right side favored**

    **p$K_a$ 4.76**                                    **p$K_a$ 15.7**

**2.19**   *For an acid-base reaction, one way to indicate the predominant species at equilibrium is to say that the reaction arrow points to the acid with the higher value of p$K_a$. For example*

$$NH_4^+ + H_2O \longleftarrow NH_3 + H_3O^+$$
$$pK_a\ 9.24 \qquad\qquad pK_a\ -1.74$$

$$NH_4^+ + OH^- \longrightarrow NH_3 + H_2O$$
$$pK_a\ 9.24 \qquad\qquad pK_a\ 15.7$$

*Explain why this rule works.*

**In acid-base equilibria, the position of the equilibrium favors the reaction of the stronger acid and stronger base to give the weaker acid and weaker base. The acid with the higher p$K_a$ is the weaker acid, so the arrow will point toward it.**

## Lewis Acids and Bases

**2.20**   *Complete the following acid-base reactions using curved arrow notation to show the flow of electron pairs.*

**The convention used for curved arrow notations shows the transfer of electron pairs from electron donors to electron acceptors.**

(b)

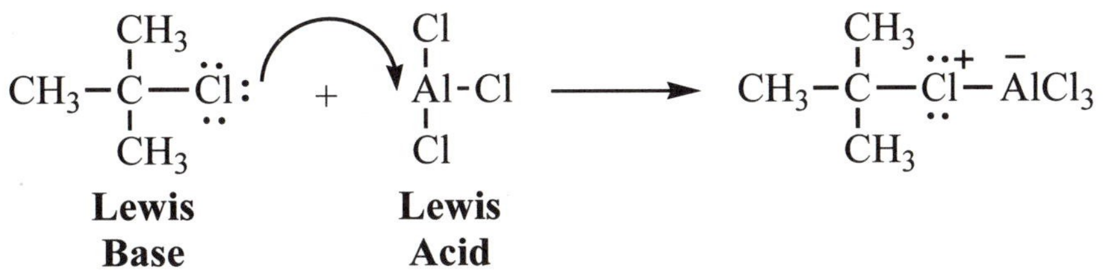

**2.21**  *Complete equations for these reactions between Lewis acid-Lewis base pairs. Label which starting material is the Lewis acid and which is the Lewis base, and use a curved arrow to show the flow of the electron pair in each reaction. In solving these problems, it is essential that you show all valence electrons for the atoms participating directly in each reaction.*

(a)

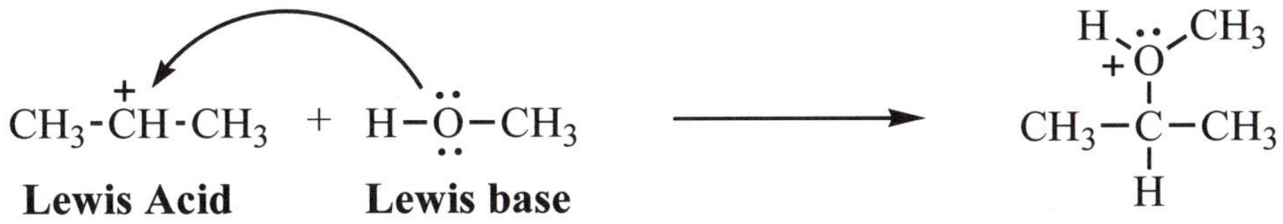

(b)

(c)

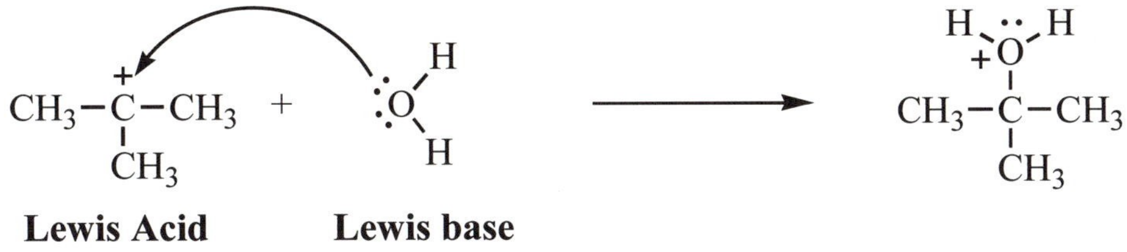

**2.22**  *Use the curved arrow notation to show the flow of electron pairs in each Lewis acid-base reaction. Be certain to show all valence electron pairs on each atom participating in the reaction.*

(a)

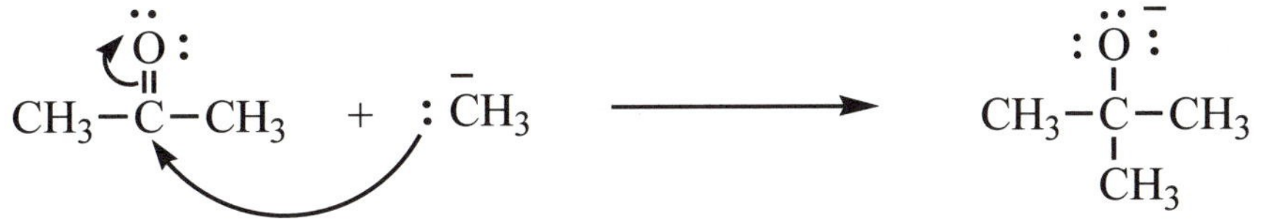

(b)

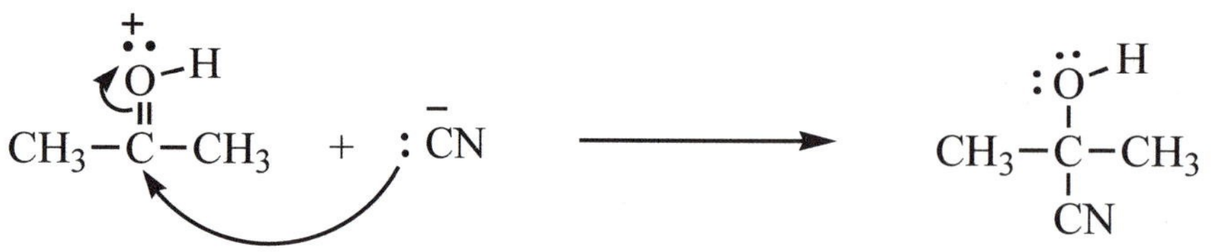

(c)

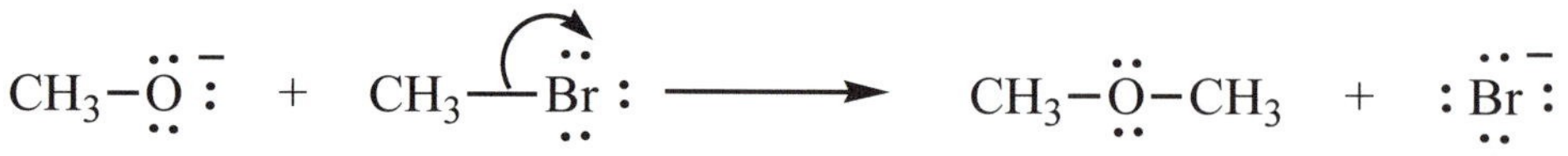

## Looking Ahead

**2.23**   *Alcohols (Chapter 8) are weak organic acids, $pK_a$ 15 - 18.  The $pK_a$ of ethanol,
$CH_3CH_2OH$, is 15.9.  Write equations for the equilibrium reactions of ethanol with each
base.  Which equilibria lie considerably toward the right?  Which lie considerably
toward the left?*

(a)   $CH_3CH_2OH$  +  $HCO_3^-$  ⇌  $CH_3CH_2O^-$  +  $H_2CO_3$         **Left side favored**
        $pK_a$ **15.9**                                  $pK_a$ **6.36**

(b)   $CH_3CH_2OH$  +  $OH^-$  ⇌  $CH_3CH_2O^-$  +  $H_2O$         **Left side slightly**
        $pK_a$ **15.9**                               $pK_a$ **15.7**              **favored**

(c)   $CH_3CH_2OH$  +  $NH_2^-$  ⇌  $CH_3CH_2O^-$  +  $NH_3$         **Right side favored**
        $pK_a$ **15.9**                               $pK_a$ **38**

(d)   $CH_3CH_2OH$  +  $NH_3$  ⇌  $CH_3CH_2O^-$  +  $NH_4^+$         **Left side favored**
        $pK_a$ **15.9**                               $pK_a$ **9.24**

**2.24**   *Phenols (Chapter 9) are weak acids and most are insoluble in water.  Phenol, $C_6H_5OH$
($pK_a$ 9.95), for example, is only slightly soluble in water but its sodium salt, $C_6H_5O^-Na^+$,
is quite soluble in water.  Will phenol dissolve in:*

**Phenol must be converted to a phenoxide anion ($C_6H_5O^-$) for it to dissolve; therefore
the acid-base equilibrium must favor the right side.  The bases <u>NaOH</u> and <u>Na$_2$CO$_3$</u>
satisfy this criterion and will dissolve phenol.**

(a)   $C_6H_5OH$  +  $OH^-$  ⇌  $C_6H_5O^-$  +  $H_2O$         **Right side favored**
        $pK_a$ **9.95**                          $pK_a$ **15.7**

(b) $C_6H_5OH$ + $HCO_3^-$ $\rightleftharpoons$ $C_6H_5O^-$ + $H_2CO_3$          **Left side favored**

$pK_a$ **9.95**                                      $pK_a$ **6.36**

(c) $C_6H_5OH$ + $CO_3^{2-}$ $\rightleftharpoons$ $C_6H_5O^-$ + $HCO_3^-$          **Right side slightly**

$pK_a$ **9.95**                                      $pK_a$ **10.33**          **favored**

**2.25**  *Carboxylic acids (Chapter 14) of six or more carbons are insoluble in water, but their sodium salts are very soluble in water. Benzoic acid, $C_6H_5COOH$ ($pK_a$ 4.19), for example, is insoluble in water, but its sodium salt, $C_6H_5COO^-Na^+$, is quite soluble in water. Will benzoic acid dissolve in:*

(a) $C_6H_5COOH$ + $OH^-$ $\rightleftharpoons$ $C_6H_5COO^-$ + $H_2O$          **Right side favored**

$pK_a$ **4.19**                                      $pK_a$ **15.7**

(b) $C_6H_5COOH$ + $HCO_3^-$ $\rightleftharpoons$ $C_6H_5COO^-$ + $H_2CO_3$          **Right side favored**

$pK_a$ **4.19**                                      $pK_a$ **6.36**

(c) $C_6H_5COOH$ + $CO_3^{2-}$ $\rightleftharpoons$ $C_6H_5COO^-$ + $HCO_3^-$          **Right side favored**

$pK_a$ **4.09**                                      $pK_a$ **10.33**

**Benzoic acid must be converted to benzoate anion ($C_6H_5COO^-$) for it to dissolve; therefore the acid-base equilibrium must favor the right side. All three bases NaOH, NaHCO$_3$, and Na$_2$CO$_3$ satisfy this criterion and will dissolve benzoic acid.**

**2.26**  *As we shall see in Chapter 16, hydrogens on a carbon adjacent to a carbonyl group are far more acidic than those not adjacent to a carbonyl group. The highlighted H in propanone, for example, is more acidic than the highlighted H in ethane. Account for the greater acidity of propanone in terms of (a) the inductive effect and (b) the resonance effect.*

$CH_3CH_2{-}H$ $\rightleftharpoons$ $CH_3\overset{..}{C}H_2$ + $H^+$          **No stabilization of the conjugate base from either resonance or inductive effects.**

(a)  $CH_3\overset{\overset{O}{\|}}{C}{-}CH_2{-}\boxed{H}$ $\rightleftharpoons$ $CH_3{-}\overset{\overset{\delta-}{\underset{\|}{O}}}{C}{-}\overset{\delta-}{CH_2}$ + $H^+$

**Propanone**          **Electron withdrawing inductive**

$pK_a$ **22**          **effect by the carbonyl group.**

(b) $CH_3\overset{\overset{O}{\|}}{C}-CH_2-\boxed{H}$ $\rightleftharpoons$ $\left[ CH_3\overset{\overset{:O:}{\|}}{C}-\overset{-}{C}H_2 \longleftrightarrow CH_3-\overset{\overset{:\overset{..}{O}:^-}{|}}{C}=CH_2 \right]$ $+ H^+$

**Resonance stabilization**

**2.27** *Explain why the protons in dimethyl ether, $CH_3\text{-}O\text{-}CH_3$, are not very acidic.*

$CH_3OCH_2-H$ $\rightleftharpoons$ $CH_3O\overset{-}{\underset{..}{C}}H_2$ $+$ $H^+$

**The conjugate base of dimethyl ether is not stabilized by resonance and has only a weak inductive effect from the electronegative oxygen.  This makes its conjugate base a strong base and thus, dimethyl ether a weak acid.**

**2.28** *Predict whether sodium hydride, NaH, will act as a base or an acid and provide a rationale for your decision.*

**NaH would act as a powerful base.  The H:⁻ anion has a lone pair of electrons to donate as a Lewis base and has a lone pair of electrons to accept a proton as a Brønsted base.**

**2.29** *Alanine is one of the 20 amino acids (it contains both an amino and a carboxyl group) found in proteins (Chapter 19).  Is alanine better represented by the structural formula A or B?  Explain.*

$$CH_3-\underset{\underset{NH_2}{|}}{CH}-\overset{\overset{O}{\|}}{C}-OH \longrightarrow CH_3-\underset{\underset{\overset{+}{N}H_3}{|}}{CH}-\overset{\overset{O}{\|}}{C}-O^-$$

(A)                                    (B)

**Structural formula (B) is a better representation of Alanine.  Both carboxylic acid ($pK_a \sim 3\text{-}4$) and basic amino groups (when protonated, $pK_a \sim 9\text{-}10$) are present on the molecule.  The acid-base equilibria would favor the formation of the more weakly acidic species.**

**2.30**    *Glutamic acid is another of the amino acids found in proteins (Chapter 19).  Glutamic acid has two carboxyl groups, one with pK$_a$ 2.10, the other pK$_a$ 4.07.*

*(a) Which carboxyl group has which pK$_a$?*

$$\text{p}K_a\ 4.07 \longrightarrow {}^2\text{HO-}\overset{\displaystyle O}{\overset{\|}{\text{C}}}\text{·CH}_2\text{CH}_2\text{·}\underset{\underset{\displaystyle +\text{NH}_3}{|}}{\overset{\displaystyle H}{\overset{|}{\text{C}}}}\text{-}\overset{\displaystyle O}{\overset{\|}{\text{C}}}\text{·OH}^1 \longleftarrow \text{p}K_a\ 2.10$$

*(b) Account for the fact that one carboxyl group is a considerably stronger acid than the other.*

**The conjugate base that results from the deprotonation of H$^1$ is stabilized by the strong electron withdrawing inductive effects of the nearby –NH$_3^+$ group. Inductive effects act by the polarization of the electrons in sigma (single) bonds and diminish with increasing distance.  The conjugate base that results from the deprotonation of H$^2$ has the negative charge further away from the –NH$_3^+$ and is less stabilized by it.**

## CHAPTER 3
### *Solutions to Problems*

**3.1**   *Do the structural formulas in each pair represent the same compound or constitutional isomers?*

**First find the longest chain of carbon atoms.  Second, number the longest chain from the end nearest the first branch. Third, compare the lengths of each chain and the size and locations of any branches.**

*(a)*

**A**                                    **B**

**Both structures contain a longest chain of six carbons.  Structure A has branching on carbons #3 and #4.  Structure B on carbons #2 and #4, therefore this pair of structural formulas represents <u>constitutional isomers</u>.**

*(b)*

**A**                                    **B**

**Both structures contain a longest chain of five carbons.  Structures A and B have identical branching and substituents, therefore this pair of structural formulas represents the <u>same compound</u>.**

**3.2**   *Draw structural formulas for the three constitutional isomers of molecular formula $C_5H_{12}$.*

**First, draw a line-angle formula for the constitutional isomer with all carbons in an unbranched chain.  Then, draw line-angle formulas for all constitutional isomers with a main chain with one less carbon and all of the possible branching combinations.  Repeat process until no more unique structures can be drawn.**

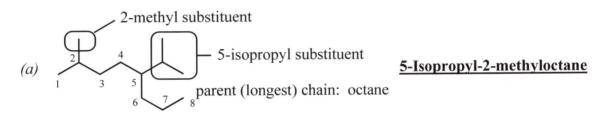

$$CH_3CH_2CH_2CH_2CH_3$$

**Pentane**

CH₃
$$CH_3CHCH_2CH_3$$

**2-Methylbutane**

CH₃
$$CH_3-C-CH_3$$
CH₃

**2,2-Dimethylpropane**

**3.3**    *Write IUPAC names for these alkanes.*

**The longest chain in each structure is an octane and numbered from the end bearing the nearest substituent. The substituents are identified and placed in front of the parent chain name alphabetically**

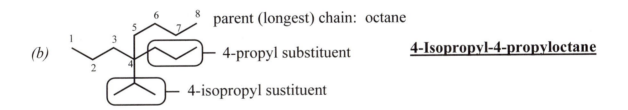

*(a)*    2-methyl substituent — 5-isopropyl substituent — parent (longest) chain: octane    **5-Isopropyl-2-methyloctane**

*(b)*    parent (longest) chain: octane — 4-propyl substituent — 4-isopropyl sustituent    **4-Isopropyl-4-propyloctane**

**3.4**    *Write the molecular formula and IUPAC name for each cycloalkane.*

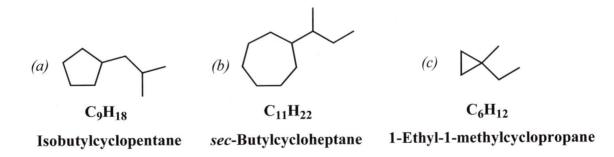

*(a)*                          *(b)*                          *(c)*

$C_9H_{18}$                   $C_{11}H_{22}$                 $C_6H_{12}$

**Isobutylcyclopentane**   ***sec*-Butylcycloheptane**   **1-Ethyl-1-methylcyclopropane**

**3.5**    *Combine the proper prefix, infix, and suffix and write the IUPAC name for each compound.*

(a)    $CH_3\overset{\overset{O}{\|}}{C}CH_3$

**Propanone**

(b)    $CH_3CH_2CH_2CH_2\overset{\overset{O}{\|}}{C}H$

**Pentanal**

(c)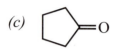

**Cyclopentanone**

(d)

**Cycloheptene**

**3.6**    Draw Newman projections for two staggered and two eclipsed conformations of 1,2-dichloroethane.

**Staggered:**

**Newman Projection**

**Line-angle**

**Eclipsed:**

**Newman Projection**

**Line-angle**

**3.7**    *Following is a chair conformation of cyclohexane with carbon atoms numbered 1 through 6.*

   *(a) Draw hydrogen atoms that are above the plane of the ring on carbons 1 and 2 and below the plane of the ring on carbon 4.*

*(b) Which of these hydrogens are equatorial? Which are axial?*

**The hydrogens on carbons #1 and #4 are axial and the hydrogen on carbon #2 is equatorial.**

*(c) Draw the other chair conformation.  Now, which hydrogens are equatorial? Which are axial? Which are above the plane of the ring, and which are below it?*

**Flipping the ring does change a group's position relative to the ring from axial to equatorial and vise versa.  But ring flipping does not change up or down orientations.  Substituents that were oriented up will remain up and those that were oriented down will remain oriented down after a ring flip.  After ring flipping the structure on the right, the resulting structure on the left has the hydrogen atoms equatorial on carbons #1 (remains oriented up) and #4 (remains oriented down) and the hydrogen atom on carbon #2 is axial (remains oriented up).**

**3.8**   *The conformational equilibria for methyl, ethyl, and isopropylcyclohexane are all about 95% in favor of the equatorial conformation, but the conformational equilibrium for tert-butylcyclohexane is almost completely on the equatorial side.  Explain why the conformational equilibria for the first three compounds are comparable but that for tert-butylcyclohexane lies considerably farther toward the equatorial conformation.*

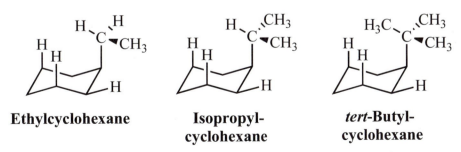

**Ethylcyclohexane**         **Isopropyl-**              ***tert*-Butyl-**
                            **cyclohexane**            **cyclohexane**

**Substituents in the axial position undergo 1,3-diaxial repulsive non-bonded interactions with other axial groups or atoms on the same side of the ring, which increases the energy of the structure. Equatorial substituents lack 1,3-diaxial repulsive non-bonded interactions. Therefore, conformations placing substituents in equatorial positions minimize these unfavorable steric interactions. Axial ethyl and isopropyl substituents can always adopt a conformation where a less bulky hydrogen atom on the substituent can be directed at the other axial substituents, minimizing 1,3-diaxial repulsions. An axial *tert*-butyl substituent will always have a methyl group directed at the other axial positions, raising the energy of that conformation. As the size of the axial groups becomes larger, the 1,3-diaxial repulsions increase, thus increasing the preference for the conformation where 1,3-diaxial repulsive forces between axial substituents are minimized or eliminated through a ring flip that places the bulky groups in an equatorial position.**

**3.9**  *Which cycloalkanes show cis-trans isomerism? For each that does, draw both isomers.*
*(a) 1,3-Dimethylcyclopentane*: **Cis and trans isomers are possible.**

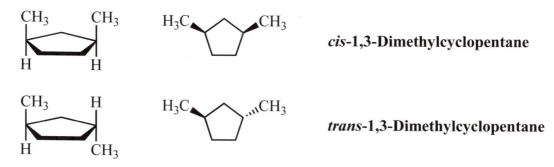

*cis*-**1,3-Dimethylcyclopentane**

*trans*-**1,3-Dimethylcyclopentane**

*(b) Ethylcyclopentane*:  **Does not show cis-trans isomerism (same compounds).**

*(c) 1-Ethyl-2-methylcyclobutane*:  **Cis and trans isomers are possible**

*cis*-**1-Ethyl-2-methylcyclobutane**

*trans*-**1-Ethyl-2-methylcyclobutane**

**3.10**  *Following is a planar hexagon representation for one isomer of 1,2,4-trimethylcyclohexane.  Draw alternative chair conformations of this compound, and state which chair conformation is the more stable.*

**More stable**
**Fewer 1,3-diaxial repulsions**

**The conformation on the right (the more stable conformation) has two $CH_3$-H 1,3-diaxial repulsions.  The conformation on the left is destabilized by one $CH_3$-$CH_3$ 1,3-diaxial repulsion and two $CH_3$-H 1,3-diaxial repulsions.**

**3.11**  *Arrange the alkanes in each set in order of increasing boiling point.*

**The only intermolecular forces acting upon alkanes are dispersion forces, which increase with increasing size and surface area of alkanes.  As branching increases within an alkane, the surface area of the molecule decreases, therefore, dispersion forces decrease.  Boiling points of isomeric alkanes decrease with increasing branching.**

*(a) 2-Methylbutane, 2,2-dimethylpropane, and pentane*

**2,2-dimethylpropane  <  2-methylbutane  <  pentane**
**(bp 9.5 °C)                    (bp 28 °C)            (bp 36 °C)**

*(b) 3,3-Dimethylheptane,  2,2,4-trimethylhexane, and nonane*

**2,2,4-trimethylhexane  <  3,3-dimethylheptane  <  nonane**
**(bp 112 °C)                       (bp 137 °C)                (bp 151 °C)**

**Structure of Alkanes**

<u>**3.12**</u>   *For each condensed structural formula, write a line-angle formula.*

(a)  $CH_3CH_2\overset{\overset{\displaystyle CH_2CH_3}{|}}{C}H\overset{}{C}HCH_2\overset{\overset{\displaystyle CH_3}{|}}{C}HCH_3$
$\qquad\qquad \underset{|}{}$
$\qquad\quad CH(CH_3)_2$

(b)  $CH_3\overset{\overset{\displaystyle CH_3}{|}}{\underset{\underset{\displaystyle CH_3}{|}}{C}}CH_3$

(c)  $(CH_3)_2CHCH(CH_3)_2$

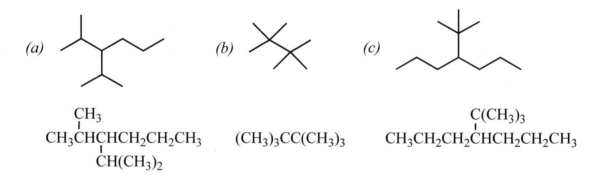

(d)  $CH_3CH_2\overset{\overset{\displaystyle CH_2CH_3}{|}}{\underset{\underset{\displaystyle CH_2CH_3}{|}}{C}}CH_2CH_3$

(e)  $(CH_3)_3CH$

(f)  $CH_3(CH_2)_3CH(CH_3)_2$

<u>**3.13**</u>   *Write a condensed structural formula and the molecular formula of each alkane.*

(a)

(b)

(c)

$CH_3\overset{\overset{\displaystyle CH_3}{|}}{\underset{\underset{\displaystyle CH(CH_3)_2}{|}}{C}}HCHCH_2CH_2CH_3$

$(CH_3)_3CC(CH_3)_3$

$CH_3CH_2CH_2\overset{\overset{\displaystyle C(CH_3)_3}{|}}{C}HCH_2CH_2CH_3$

<u>**3.14**</u>   *For each of the following condensed structural formulas, provide an even more abbreviated formula using parentheses and subscripts.*

(a)  $CH_3CH_2CH_2CH_2CH_2\overset{\overset{\displaystyle CH_3}{|}}{C}HCH_3$          $CH_3(CH_2)_4CH(CH_3)_2$

(b)  $H\overset{\overset{\displaystyle CH_2CH_2CH_3}{|}}{\underset{\underset{\displaystyle CH_2CH_2CH_3}{|}}{C}}CH_2CH_2CH_3$          $H\overset{\overset{\displaystyle (CH_2)_2CH_3}{|}}{\underset{\underset{\displaystyle (CH_2)_2CH_3}{|}}{C}}(CH_2)_2CH_3$

$$CH_2CH_2CH_3$$
$$\quad\quad\quad |$$
(c)   $CH_3CCH_2CH_2CH_2CH_2CH_3$
$$\quad\quad\quad |$$
$$CH_2CH_2CH_3$$

$$(CH_2)_2CH_3$$
$$\quad\quad |$$
$CH_3C(CH_2)_4CH_3$
$$\quad\quad |$$
$$(CH_2)_2CH_3$$

## Constitutional Isomerism

**3.15**   *Which statements are true about constitutional isomers?*

(a) *They have the same molecular formula.*   **True**
(b) *They have the same molecular weight.*   **True**
(c) *They have the same order of attachment of atoms.*   **False**
(d) *They have the same physical properties.*   **False**

**3.16**   *Each member of the following set of compounds is an alcohol; that is, each contains an -
OH (hydroxyl group, Section 1.8A).  Which structural formulas represent (1) the same
compound, (2) different compounds that are constitutional isomers, or (3) different
compounds that are not constitutional isomers?*

**The following compounds are named for your reference.  At this point, you may not
be expected to name these compounds yet.**

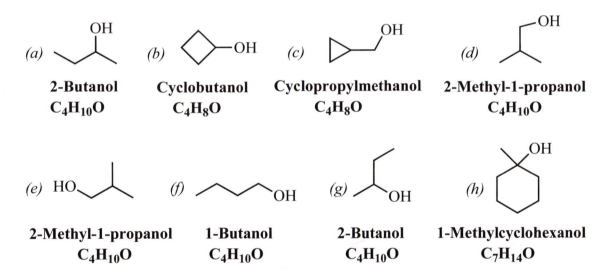

| (a) 2-Butanol $C_4H_{10}O$ | (b) Cyclobutanol $C_4H_8O$ | (c) Cyclopropylmethanol $C_4H_8O$ | (d) 2-Methyl-1-propanol $C_4H_{10}O$ |

| (e) 2-Methyl-1-propanol $C_4H_{10}O$ | (f) 1-Butanol $C_4H_{10}O$ | (g) 2-Butanol $C_4H_{10}O$ | (h) 1-Methylcyclohexanol $C_7H_{14}O$ |

(1) **Compounds a and g represent the same compound (2-butanol).**
     **Compounds d and e represent the same compound (2-methyl-1-propanol).**

(2) **Compounds a(g), d(e), and f represent constitutional isomers of $C_4H_{10}O$.**
     **Compounds b and c represent constitutional isomers of $C_4H_8O$.**

(3) **Isomers of $C_4H_{10}O$ [a(g) and d(e)] are different compounds from the isomers of
     $C_4H_8O$ [b and c], and all different compounds from compound h.**

**3.17**   *Each member of the following set of compounds is an amine; that is, each contains a nitrogen bonded to one, two or three carbon groups (Section 1.8B).  Which structural formulas represent (1) the same compound, (2) different compounds that are constitutional isomers, or (3) different compounds that are not constitutional isomers?*

**The following compounds are named for your reference.  At this point, you may not be expected to name these compounds yet.**

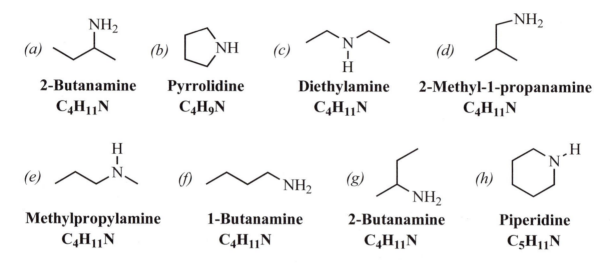

| (a) 2-Butanamine $C_4H_{11}N$ | (b) Pyrrolidine $C_4H_9N$ | (c) Diethylamine $C_4H_{11}N$ | (d) 2-Methyl-1-propanamine $C_4H_{11}N$ |

| (e) Methylpropylamine $C_4H_{11}N$ | (f) 1-Butanamine $C_4H_{11}N$ | (g) 2-Butanamine $C_4H_{11}N$ | (h) Piperidine $C_5H_{11}N$ |

**(1)  Compounds a and g represent the same compound (2-butanamine).**

**(2)  Compounds a(g), c, d, e, and f represent constitutional isomers of $C_4H_{11}N$.**

**(3)  Compounds b and h represent different compounds and also differ from the isomers of $C_4H_{11}N$.**

**3.18**   *Each member of the following set of compounds is either an aldehyde or ketone (Section 1.8C).  Which structural formulas represent (1) the same compound, (2) different compounds that are constitutional isomers, or (3) different compounds that are not constitutional isomers?*

**The following compounds are named for your reference.  At this point, you may not be expected to name these compounds yet.**

| (a) 2-Butanone $C_4H_8O$ | (b) Cyclopentanone $C_5H_8O$ | (c) 3-Pentanone $C_5H_{10}O$ | (d) 2-Methylpropanal $C_4H_8O$ |

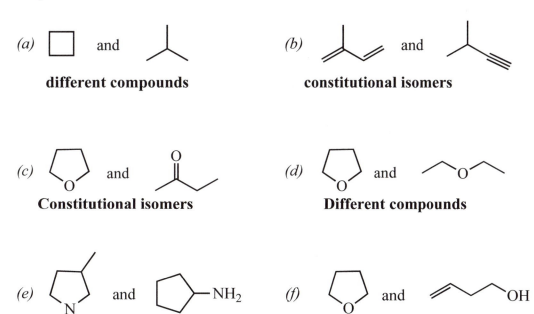

Butanal          2-Pentanone      Cyclohexanone    Cyclopentanecarbaldeyde
$C_4H_8O$        $C_5H_{10}O$     $C_6H_{10}O$     $C_6H_{10}O$

(1) None of the above compounds are the same.

(2) Compounds a, d, and e represent constitutional isomers of $C_4H_8O$.

Compounds c and f represent constitutional isomers of $C_5H_{10}O$.

Compounds g and h represent constitutional isomers of $C_6H_{10}O$

(3) Compound b is a different compound than the isomers of $C_4H_8O$, $C_5H_{10}O$, and $C_6H_{10}O$.

**3.19**    *For each pair of compounds, tell whether the structural formulas shown represent (1) the same compound, (2) different compounds that are constitutional isomers, or (3) different compounds that are not constitutional isomers.*

(a) □ and ⋏

**different compounds**

(b) ⋏═ and ⋏≡

**constitutional isomers**

(c) ⟨O⟩ and ⋎(O)

**Constitutional isomers**

(d) ⟨O⟩ and ⌒O⌒

**Different compounds**

(e) ⟨N−H⟩ and ⟨⟩−NH₂

**Constitutional isomers**

(f) ⟨O⟩ and ═⌒OH

**Constitutional isomers**

**3.20**   *Name and draw line-angle formulas for the nine constitutional isomers with molecular formula C₇H₁₆.*

**Heptane**
**(bp 94.8 °C)**

**2-Methylhexane**
**(bp 90.0 °C)**

**3-Methylhexane**
**(bp 92.0 °C)**

**2,2-Dimethylpentane**
**(bp 79.3 °C)**

**2,3-Dimethylpentane**
**(bp 89.8 °C)**

**2,4-Dimethylpentane**
**(bp 80.5 °C)**

**3-Ethylpentane**
**(bp 93.5 °C)**

**3,3-Dimethylpentane**
**(bp 86.1 °C)**

**2,2,3-Trimethylbutane**
**(bp 80.9 °C)**

**3.21**   *Tell whether the compounds in each set are constitutional isomers.*

(a)  $CH_3CH_2OH$   and   $CH_3OCH_3$
**Constitutional isomers**

(b)  $CH_3\overset{\overset{\displaystyle O}{\|}}{C}CH_3$   and   $CH_3CH_2\overset{\overset{\displaystyle O}{\|}}{C}H$
**Constitutional isomers**

(c) $CH_3\overset{\overset{\displaystyle O}{\|}}{C}OCH_3$ and  $CH_3CH_2\overset{\overset{\displaystyle O}{\|}}{C}OH$
**Constitutional isomers**

(d) $CH_3\overset{\overset{\displaystyle OH}{|}}{C}HCH_2CH_3$  and  $CH_3\overset{\overset{\displaystyle O}{\|}}{C}CH_2CH_3$
**Different compounds**

(e)    and   $CH_3CH_2CH_2CH_2CH_3$
**Different compounds**

(f)   and  $CH_2{=}CHCH_2CH_2CH_3$
**Constitutional isomers**

**3.22**   *Draw line-angle formulas for*

(a) *The four alcohols with molecular formula C₄H₁₀O.*

*(b) The two aldehydes with molecular formula C₄H₈O.*

*(c) The one ketone with molecular formula C₄H₈O.*

*(d) The three ketones with molecular formula C₅H₁₀O.*

*(e) The four carboxylic acids with molecular formula C₅H₁₀O₂.*

**Nomenclature of Alkanes and Cycloalkanes**
**3.23**   *Write IUPAC names for these alkanes and cycloalkanes.*

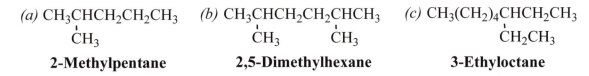

*(a)* CH₃CHCH₂CH₂CH₃
        CH₃

**2-Methylpentane**

*(b)* CH₃CHCH₂CH₂CHCH₃
        CH₃        CH₃

**2,5-Dimethylhexane**

*(c)* CH₃(CH₂)₄CHCH₂CH₃
                CH₂CH₃

**3-Ethyloctane**

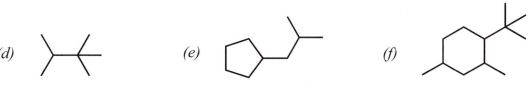

*(d)*

**2,2,3-Trimethylbutane**

*(e)*

**Isobutylcyclopentane**

*(f)*

**1-*tert*-Butyl-2,4-dimethyl-
cyclohexane**

**3.24**    *Write line-angle formulas for these alkanes.*

*(a) 2,2,4-Trimethylhexane*

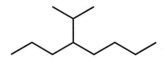

*(b) 2,2-Dimethylpropane*

*(c) 3-Ethyl-2,4,5-trimethyloctane*

*(d) 5-Butyl-2,2-dimethylnonane*

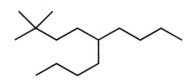

*(e) 4-Isopropyloctane*

*(f) 3,3-Dimethylpentane*

*(g) trans-1,3-Dimethylcyclopentane*

*(h) cis-1,2-Diethylcyclobutane*

**3.25**    *Explain why each is an incorrect IUPAC name.  Write the correct IUPAC name for the intended compound.*

*(a) 1,3-Dimethylbutane*

CH₃
|
CH₃CHCH₂CH₂CH₃

**The longest chain is pentane.
The IUPAC name is 2-methylpentane.**

*(b) 4-Methylpentane*

CH₃
|
CH₃CHCH₂CH₂CH₃

**The pentane chain is numbered incorrectly.
The IUPAC name is 2-methylpentane.**

*(c) 2,2-Diethylbutane*

$$CH_3$$
$$|$$
$$CH_3CH_2CCH_2CH_3$$
$$|$$
$$CH_2CH_3$$

**The longest chain is pentane.**
**The IUPAC name is 3-ethyl-3-methylpentane.**

*(d) 2-Ethyl-3-methylpentane*

$$H_3C \quad CH_3$$
$$|\quad\quad|$$
$$CH_3CH_2CHCHCH_2CH_3$$

**The longest chain is hexane.**
**The IUPAC name is 3,4-dimethylhexane.**

*(e) 2-Propylpentane*

$$CH_3$$
$$|$$
$$CH_3CH_2CH_2CHCH_2CH_2CH_3$$

**The longest chain is heptane.**
**The IUPAC name is 4-methylheptane.**

*(f) 2,2-Diethylheptane*

$$CH_3$$
$$|$$
$$CH_3CH_2CCH_2CH_2CH_2CH_2CH_3$$
$$|$$
$$CH_2CH_3$$

**The longest chain is octane.**
**The IUPAC name is 3-ethyl-3methyloctane.**

*(g) 2,2-Dimethylcyclopropane*

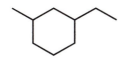

**The ring is numbered incorrectly.**
**The IUPAC name is 1,1-dimethylcyclopropane.**

*(h) 1-Ethyl-5-methylcyclohexane*

**The ring is numbered incorrectly.**
**The IUPAC name is**
**1-ethyl-3-methylcyclohexane.**

**3.26**    *Draw a structural formula for each compound.*

| *(a) Ethanol* | *(b) Ethanal* | *(c) Ethanoic acid* |
|---|---|---|
| | $$O$$ | $$O$$ |
| | $$\parallel$$ | $$\parallel$$ |
| **CH_3CH_2OH** | **CH_3CH** | **CH_3C·OH** |

(d) *Butanone*

$$CH_3\overset{\overset{\displaystyle O}{\|}}{C}CH_2CH_3$$

(e) *Butanal*

$$CH_3CH_2CH_2\overset{\overset{\displaystyle O}{\|}}{C}H$$

(f) *Butanoic acid*

$$CH_3CH_2CH_2\overset{\overset{\displaystyle O}{\|}}{C}\cdot OH$$

(g) *Propanal*

$$CH_3CH_2\overset{\overset{\displaystyle O}{\|}}{C}H$$

(h) *Cyclopropanol*

(i) *Cyclopentanol*

⬠—OH

(j) *Cyclopentene*

(k) *Cyclopentanone*

⬠=O

**3.27**  *Write the IUPAC name for each compound.*

(a) $CH_3\overset{\overset{\displaystyle O}{\|}}{C}CH_3$

**Propanone**

(b) $CH_3(CH_2)_3\overset{\overset{\displaystyle O}{\|}}{C}H$

**Pentanal**

(c) $CH_3(CH_2)_8\overset{\overset{\displaystyle O}{\|}}{C}OH$

**Decanoic acid**

(d)

**Cyclohexene**

(e)  =O

**Cyclohexanone**

(f)  —OH

**Cyclobutanol**

## Conformations of Alkanes and Cycloalkanes

**3.28**  *How many different staggered conformations are there for 2-methylpropane? How many different eclipsed conformations are there?*

**Looking down any of the carbon-carbon bonds, there is one staggered and one eclipsed conformation.**

**Staggered**              **Eclipsed**

**3.29**    *Looking along the bond between carbons 2 and 3 of butane, there are two different staggered conformations and two different eclipsed conformations. Draw Newman projections of each and arrange them in order from most stable conformation to least stable conformation.*

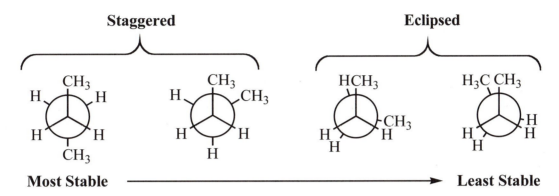

**3.30**    *Explain why each of the following Newman projections might not represent the most stable conformation of that molecule.*

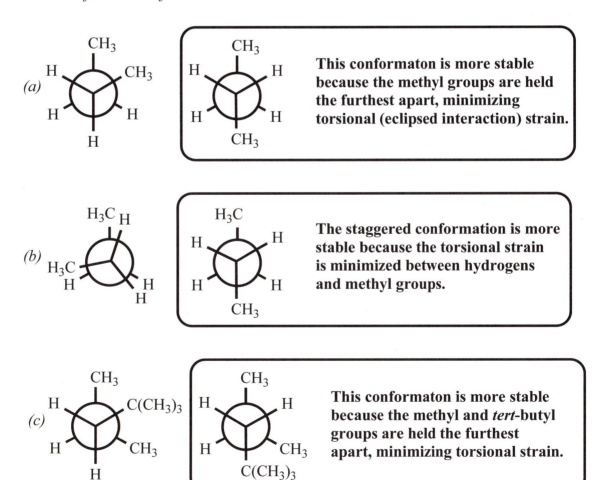

**3.31**   *Explain why the following are not different conformations of 3-hexene.*

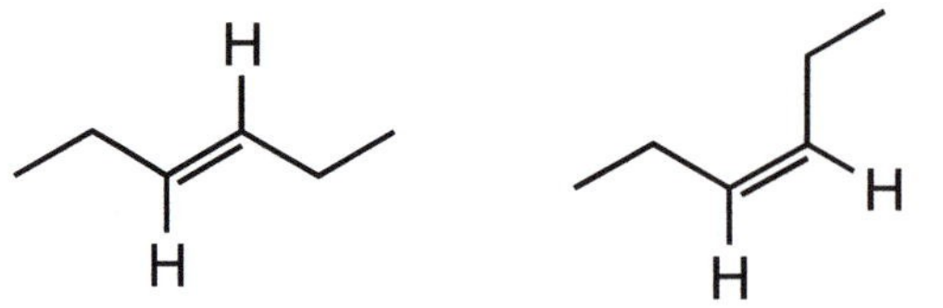

*trans*-3-Hexene          *cis*-3-Hexene

**Different conformations result from the rotation about a single bond.  Rotation about a C-C double bond is restricted, therefore does not occur under normal conditions.  *Cis*- and *trans*-3-hexene differ by the spatial orientation of the ethyl groups attached to the C-C double bond and cannot be interconverted by rotation about the C-C double bond.**

**3.32**   *Which of the following two conformations is the more stable?  Hint: use molecular models to compare structures.*

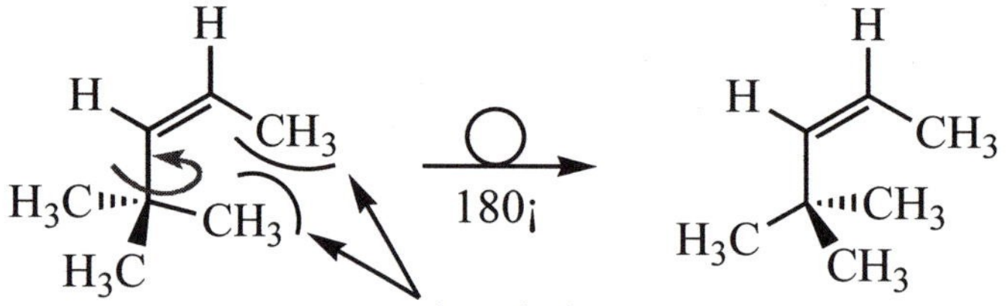

**Non-bonded repulsions between the methyl groups in A destabilize the conformation relative to conformation B.**

**3.33**   *Determine whether the following pairs of structures in each set represent the same molecule or constitutional isomers. If they are the same molecule, determine whether they are in the same or different conformations.*

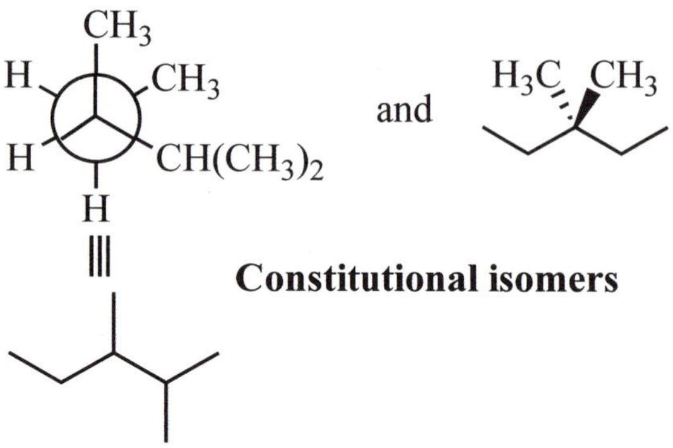

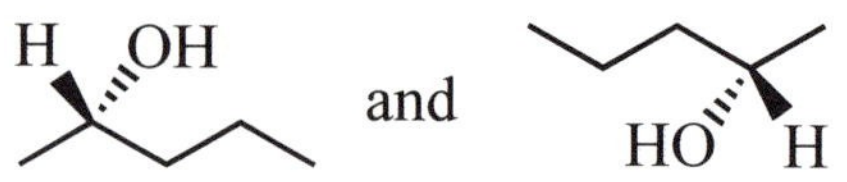

**In (a), by naming the compounds according to IUPAC rules, the structure on the left is 2,3-dimethylpentane and the structure on the right is 3,3-dimethylpentane, which**

**makes them constitutional isomers. In (b), the structures represent the same compound. By flipping the structure on the right by 180° in the plane of the paper, it can be to fit exactly on the structure on the left. You can also accomplish this operation with molecular models.**

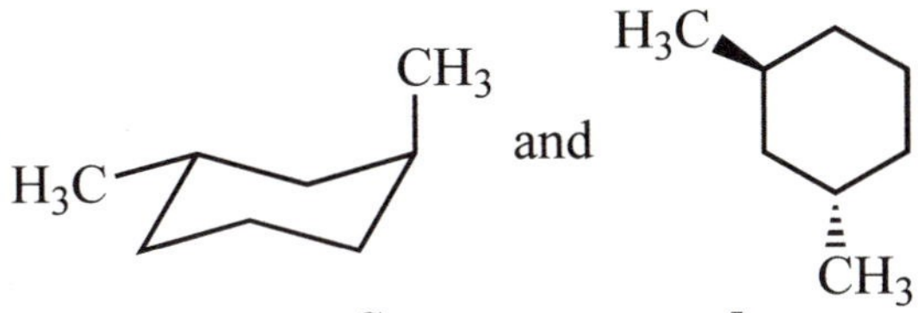

**Same compound**
**(Same conformation)**

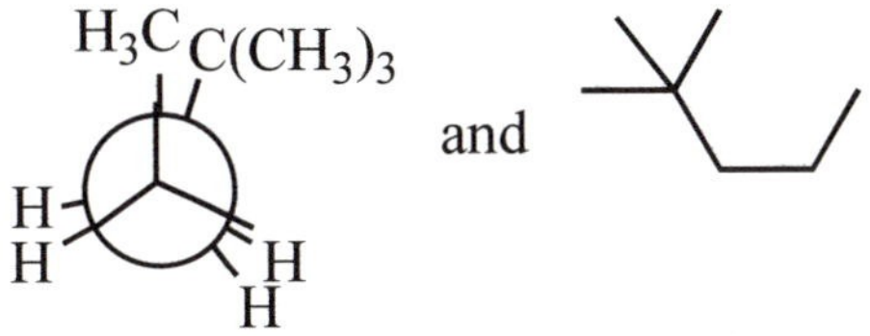

**Same compound**
**(Same conformation)**

**In (c), the both structures represent *trans*-1,3-dimethylcyclohexane. In (d), by redrawing the Newman projection on the left to match the line-angle structure on the right, it is easier to see that both structures represent 2,2-dimethylpentane.**

### Cis-trans Isomerism in Cycloalkanes

**3.34**  *What structural feature of cycloalkanes makes cis-trans isomerism in them possible?*

**Rings prevent full 360° rotation about the C-C single bonds, allowing two possible spatial orientations of the attached substituents.**

**3.35**  *Is cis-trans isomerism possible in alkanes?*

**No, alkanes do not have rings or C-C double bonds, thus all conformations are interconvertable by rotation about the C-C single bond.**

**3.36**  *Name and draw structural formulas for the cis and trans isomers of 1,2-dimethylcyclopropane.*

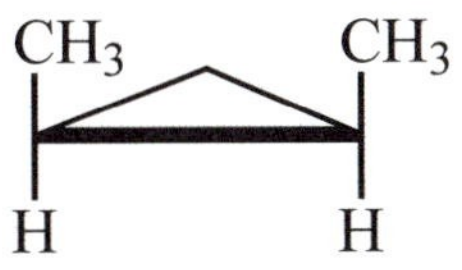

*cis*-1,2-Dimethylcyclopropane 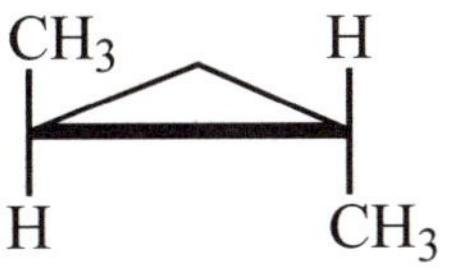            *trans*-1,2-Dimethylcyclopropane

**3.37**  *Name and draw structural formulas for all cycloalkanes with molecular formula $C_5H_{10}$. Be certain to include cis-trans isomers as well as constitutional isomers.*

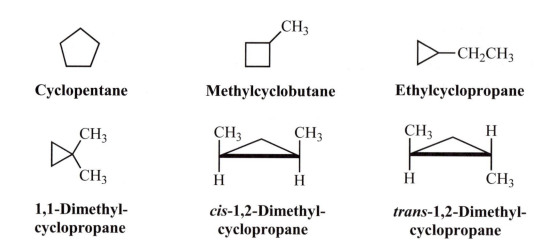

**Cyclopentane**          **Methylcyclobutane**          **Ethylcyclopropane**

**1,1-Dimethyl-cyclopropane**          *cis*-**1,2-Dimethyl-cyclopropane**          *trans*-**1,2-Dimethyl-cyclopropane**

**3.38** *Using a planar pentagon representation for the cyclopentane ring, draw structural formulas for the cis and trans isomers of:*

  (a)  *1,2-Dimethylcyclopentane*          (b)  *1,3-Dimethylcyclopentane*

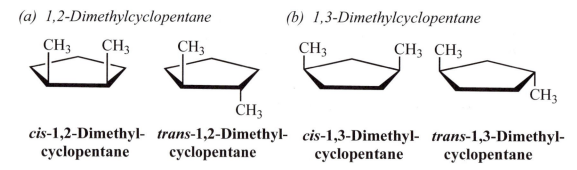

*cis*-**1,2-Dimethyl-cyclopentane**    *trans*-**1,2-Dimethyl-cyclopentane**    *cis*-**1,3-Dimethyl-cyclopentane**    *trans*-**1,3-Dimethyl-cyclopentane**

**3.39** *Draw the alternative chair conformations for the cis and trans isomers of 1,2-dimethylcyclohexane, 1,3-dimethylcyclohexane, and 1,4-dimethylcyclohexane.*
*(a) Indicate by a label whether each methyl group is axial or equatorial.*
*(b) For which isomer(s) are the alternative chair conformations of equal stability?*
*(c) For which isomer(s) is one chair conformation more stable than the other?*

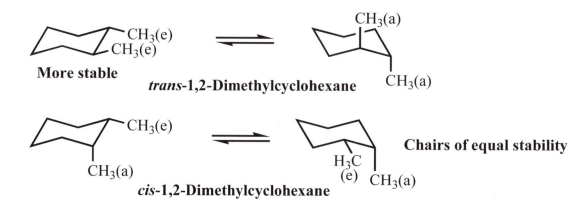

**More stable**                    $CH_3(a)$

*trans*-**1,2-Dimethylcyclohexane**          $CH_3(a)$

**Chairs of equal stability**

*cis*-**1,2-Dimethylcyclohexane**

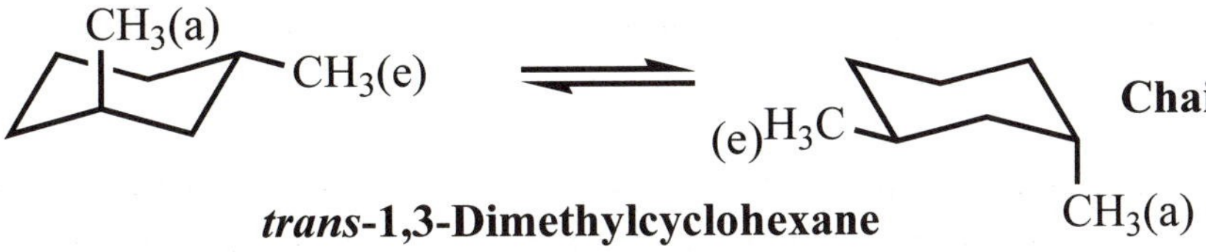

trans-1,3-Dimethylcyclohexane          **Chairs of equal stability**

CH₃(a)  CH₃(e)  ⇌  (e)H₃C  CH₃(a)

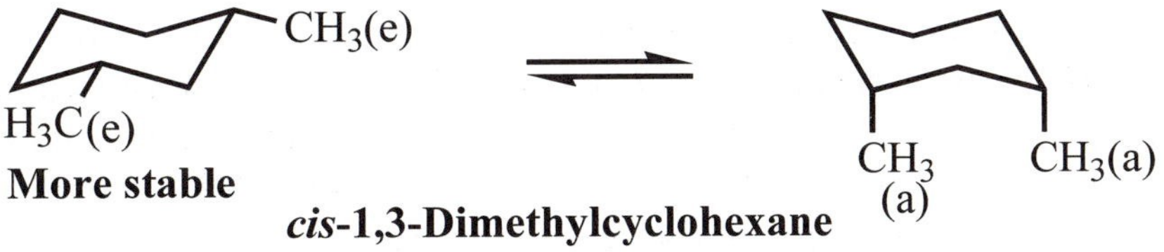

CH₃(e)  ⇌
H₃C(e)
**More stable**
cis-1,3-Dimethylcyclohexane 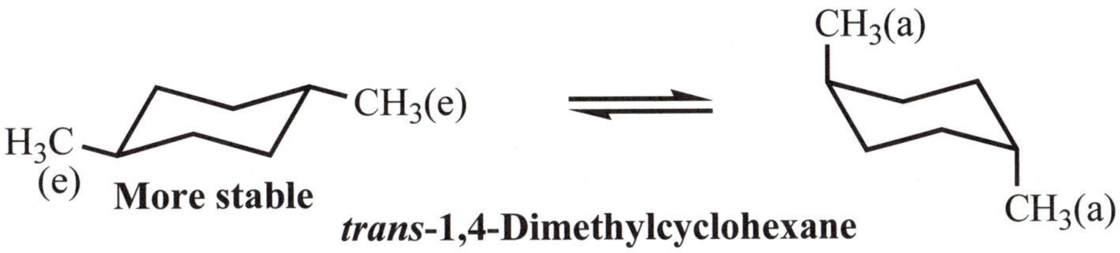
CH₃ CH₃(a)
(a)

CH₃(e)  ⇌  CH₃(a)
H₃C
(e) **More stable**
trans-1,4-Dimethylcyclohexane          CH₃(a)

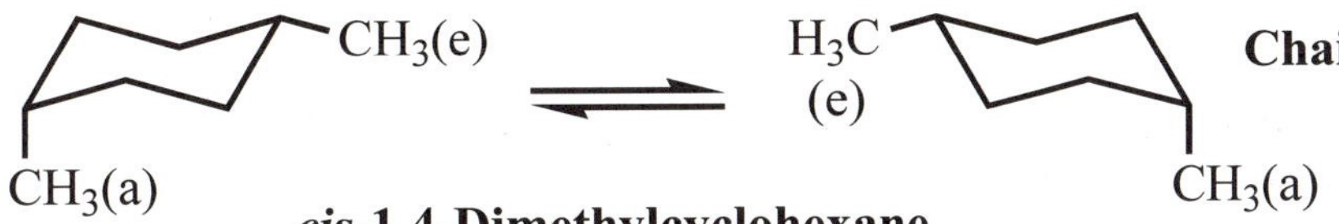

CH₃(e)  ⇌  H₃C          **Chairs of equal stability**
                    (e)
CH₃(a)                    CH₃(a)
cis-1,4-Dimethylcyclohexane

**3.40**  *Use your answers from Problem 3.39 to complete the table showing correlations between cis, trans and axial, equatorial for disubstituted derivatives of cyclohexane.*

| Position of Substitution | cis | trans |
|---|---|---|
| 1,4- | a,e  or  e,a | e,e  or  a,a |
| 1,3- | **e,e**  or  **a,a** | **a,e**  or  **e,a** |
| 1,2- | **a,e**  or  **e,a** | **e,e**  or  **a,a** |

**3.41**  *There are four cis-trans isomers of 2-isopropyl-5-methylcyclohexanol.*

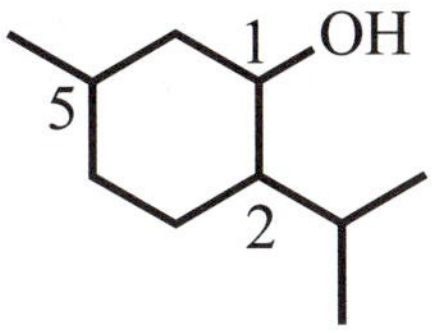

2-Isopropyl-5-methylcyclohexanol

*(a) Using a planar hexagon representation for the cyclohexane ring, draw structural formulas for these four isomers.*

**The following structures are planar hexagon representations for the four cis-trans isomers. In each, the abbreviation iPr represents the isopropyl group. An effective strategy used to arrive at these structural formulas is to take one group as a reference and then arrange the other two groups relative to it. In these drawings, –OH is the reference substituent and placed above the plane of the ring. Once the –OH is fixed, there are only two possible arrangements for the isopropyl substituent on carbon 2; either cis or trans to –OH. Similarly, there are only two possible arrangements for the methyl group on carbon 5; either cis or trans to –OH. Note that, even if you use another group as a reference, and even if you put the reference below the plane of the ring, there are still only four cis-trans isomers for this compound.**

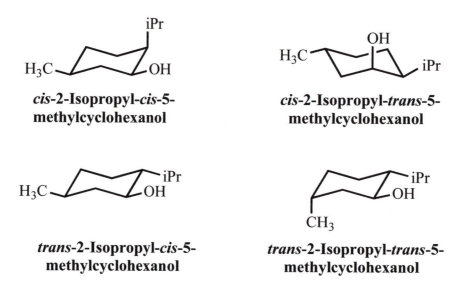

| *cis*-2-Isopropyl-<br>*cis*-5-methyl-<br>cyclohexanol | *cis*-2-Isopropyl-<br>*trans*-5-methyl-<br>cyclohexanol | *trans*-2-Isopropyl-<br>*cis*-5-methyl-<br>cyclohexanol | *trans*-2-Isopropyl-<br>*trans*-5-methyl-<br>cyclohexanol |
|---|---|---|---|

*(b) Draw the more stable chair conformation for each of your answers in part (a).*

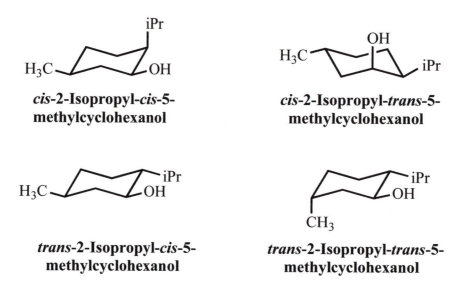

cis-2-Isopropyl-*cis*-5-
methylcyclohexanol

cis-2-Isopropyl-*trans*-5-
methylcyclohexanol

trans-2-Isopropyl-*cis*-5-
methylcyclohexanol

trans-2-Isopropyl-*trans*-5-
methylcyclohexanol

*(c) Of the four cis-trans isomers, which is the most stable? If you answered this part correctly, you picked the isomer found in nature and given the name menthol.*

**Of the four cis-trans isomers, *trans*-2-isopropyl-*cis*-5-methyl-cyclohexanol is the most stable because all of the substituents are in equatorial positions, thus minimizing diaxial non-bonding repulsions.**

**3.42**   *Draw alternative chair conformations for each substituted cyclohexane and state which chair is the more stable.*

(a) **Chairs of equal stability**

(b) **More stable**

(c) **More stable**

(d) **More stable**

**3.43**   *What kinds of conformations do the six-membered rings exhibit in adamantane?*

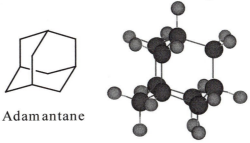

Adamantane

**In adamantine, all of the cyclohexane rings have the chair conformation.**

**Physical Properties of Alkanes and Cycloalkanes**

**3.44**   *In Problem 3.20 you drew structural formulas for all constitutional isomers of molecular formula C₇H₁₆. Predict which isomer has the lowest boiling point and which has the highest boiling point.*

**Names and boiling points for the isomers of $C_7H_{16}$ are given in the solution to problem 3.20. The isomer with the lowest boiling point is 2,2-dimethylpentane (bp 79.2 °C). The isomer with the highest boiling point is heptane (94.8 °C).**

**3.45**   *What generalizations can you make about the densities of alkanes relative to that of water?*

**(1)  All alkanes are less dense than water;  (2)  As alkane molecular weight increases, density increases;  (3)  Constitutional isomers have similar densities.**

**3.46**   *What unbranched alkane has about the same boiling point as water (see Table 3.4)? Calculate the molecular weight of this alkane, and compare it with that of water.*

**According to Table 3.4, heptane has a boiling point of 98 °C and a molecular weight of 100 g/mol, which is approximately 5.5 times that of water.  Although considerably lower in molecular weight, water molecules are held together in the liquid state by the relatively strong hydrogen bonding forces, while the much heavier heptane molecules are held together by the relatively weaker dispersion forces.**

**3.47**   *As you can see from Table 3.4, each $CH_2$ group added to the carbon chain of an alkane increases its boiling point. This increase is greater going from $CH_4$ to $C_2H_6$ and from $C_2H_6$ to $C_3H_8$ than it is from $C_8H_{18}$ to $C_9H_{20}$ or from $C_9H_{20}$ to $C_{10}H_{22}$. What do you think is the reason for this?*

**Boiling points of unbranched alkanes are related to their surface area; the larger the surface area, the greater the strength of the dispersion forces and the higher the boiling point.  The relative increase in size per $CH_2$ group is greatest between $CH_4$ and $CH_3CH_3$, and becomes progressively smaller as molecular weight increases. Therefore, the increase in boiling point per $CH_2$ group is greatest between $CH_4$ and $CH_3CH_3$, and becomes progressively smaller for the higher alkanes.**

**3.48**   *Dodecane, $C_{12}H_{26}$, is an unbranched alkane. Predict the following:*

*(a)* Will it dissolve in water?

**No, alkanes are too non-polar to dissolve in a very polar solvent such as water.**

*(b) Will it dissolve in hexane?*

**Yes, dodecane is non-polar and will dissolve in a non-polar solvent such as hexane.**

*(c) Will it burn when ignited?*

**Yes, alkanes burn when ignited in the presence of oxygen to produce water and carbon dioxide.**

*(d) Is it a liquid, solid, or gas at room temperature and atmospheric pressure?*

**It is a liquid.  Alkanes containing 5 to 17 carbons are colorless liquids.**

*(e) Is it more or less dense than water?*

**It is less dense than water. All liquid and solid alkanes are less dense than water (1.0 g/mL)**

**3.49**   *As stated in Section 3.9A, the wax found in apple skins is an unbranched alkane of molecular formula $C_{27}H_{56}$. Explain how the presence of this alkane prevents the loss of moisture from within an apple.*

**The large alkane wax in the apple skins forms a non-polar hydrophobic (water resistive) barrier that a polar compound such as water cannot pass through.**

## Reactions of Alkanes

**3.50**   *Write balanced equations for combustion of each hydrocarbon. Assume that each is converted completely to carbon dioxide and water.*

*(a) Hexane*

$$2\ CH_3(CH_2)_4CH_3\ +\ 19O_2\ \longrightarrow\ 12CO_2\ +\ 14H_2O$$

*(b) Cyclohexane*

$$+\ 9O_2\ \longrightarrow\ 6CO_2\ +\ 6H_2O$$

*(c) 2-Methylpentane*

$$\overset{\displaystyle CH_3}{2\ CH_3CHCH_2CH_2CH_3}\ +\ 19O_2\ \longrightarrow\ 12CO_2\ +\ 14H_2O$$

**3.51**  *Following are heats of combustion of methane and propane. On a gram-for-gram basis, which of these hydrocarbons is the best source of heat energy?*

| Hydrocarbon | Component of | $\Delta H_i$ [kcal/mol (kJ/mol)] |
|---|---|---|
| $CH_4$ | natural gas | -212 (-886)) |
| $CH_3CH_2CH_3$ | LPG | -530 (-2220) |

$$\Delta H_{comb}\ CH_4\ (kcal/g) = \left(\frac{-212\ kcal}{1\ mol\ CH_4}\right)\left(\frac{1\ mol\ CH_4}{16.0\ g\ CH_4}\right) = -13.2\ kcal/g$$

$$\Delta H_{comb}\ C_3H_8\ (kcal/g) = \left(\frac{-530\ kcal}{1\ mol\ C_3H_8}\right)\left(\frac{1\ mol\ C_3H_8}{44.1\ g\ C_3H_8}\right) = -12.0\ kcal/g$$

**Methane has higher heat of combustion per gram (-13.2 kcal/g) than propane (−12.0 kcal/g), therefore on a gram-for-gram basis, it is a better source of heat energy.**

**3.52**  *When ethanol is added to gasoline to produce gasohol, it promotes more complete combustion of the gasoline and is an octane booster (Section 3.11B). Compare the heats of combustion of 2,2,4-trimethylpentane (1304 kcal/mol) and ethanol (327 kcal/mol). Which has the higher heat of combustion in kcal/mol? In kcal/g?*

$$\Delta H_{comb}\ ethanol\ (kcal/g) = \left(\frac{-327\ kcal}{1\ mol\ ethanol}\right)\left(\frac{1\ mol\ ethanol}{46.1\ g\ ethanol}\right) = -7.09\ kcal/g$$

$$\Delta H_{comb}\ C_8H_{18}\ (kcal/g) = \left(\frac{-1304\ kcal}{1\ mol\ C_8H_{18}}\right)\left(\frac{1\ mol\ C_8H_{18}}{114.2\ g\ C_8H_{18}}\right) = -11.42\ kcal/g$$

**2,2,4-Trimethylpentane has both the greater heat of combustion per mole (-1304 kcal/mol) and the greater heat of combustion per gram (-11.42 kcal/mol) than ethanol (327 kcal/mol and −7.09 kcal/g, respectively).**

**Looking Ahead**

**3.53**  *Explain why 1,2-dimethylcyclohexane can exist as cis-trans isomers, while 1,2-dimethylcyclododecane cannot.*

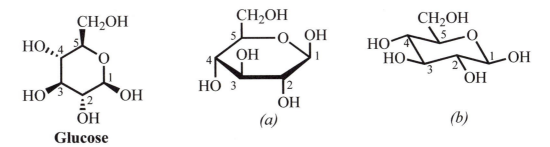

CH$_3$

CH$_3$

*cis*-**1,2-dimethyl-**
**cyclohexane**

CH$_3$
CH$_3$

*trans*-**1,2-dimethyl-**
**cyclohexane**

CH$_3$

CH$_3$

**1,2-dimethyl-**
**cyclododecane**

**Using a molecular model kit, the cyclohexane ring shows hindered rotation about the C-C single bonds, allowing the possibility of cis-trans isomers in disubstituted six-membered rings.  The much larger cyclododecane is flexible enough to allow unrestricted rotation about the C-C single bonds, thus no cis-trans isomerization can exist.**

**3.54**  *On the left is a representation of the glucose molecule. We discuss the structure and chemistry of glucose in Chapter 18.*

*(a) Convert this representation to a planar hexagon representation.*

*(b) Convert this representation to a chair conformation. Which substituent groups in the chair conformation are equatorial? Which are axial?*

**All non-hydrogen groups are equatorial and all of the hydrogens are axial, giving the most stable structure for this isomer of glucose.**

CH$_2$OH

HO,,4  5

O

HO  3

1

OH

2

OH

**Glucose**

CH$_2$OH

5

O

OH

4

OH

1

HO  3

2

OH

*(a)*

CH$_2$OH

5

O

HO  4

HO

1

OH

3    2

OH

*(b)*

**3.55**    *Following is the structural formula of cholic acid (Section 21.5A), a component of human bile whose function is to aid in the absorption and digestion of dietary fats.*

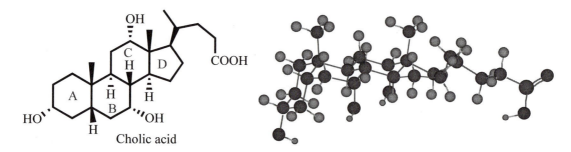

Cholic acid

*(a) What is the conformation of ring A? Of ring B? Of ring C? Of ring D?*

**Rings A, B, and C are in the chair conformation.  The five-membered ring D is in an envelope conformation**

*(b) There are hydroxyl groups on ring A, B, and C. Tell whether each is axial or equatorial.*

**The hydroxyl on ring A is equatorial.  The hydroxyls on rings B and C are axial.**

*(c) Is the methyl group at the junction of rings A/B axial or equatorial to ring A? Is it axial or equatorial to ring B?*

**With respect to A, this methyl group is an equatorial substituent.  However, it is axial to ring B.**

*(d) Is the methyl group at the junction of rings C/D axial or equatorial to ring C?*

**The methyl group at the junction of rings C and D is axial to ring C.**

**3.56**    *Following is the structural formula and ball-and-stick model of cholestanol. The only difference between this compound and cholesterol (Section 21.5A) is that cholesterol has a carbon-carbon double bond in ring B.*

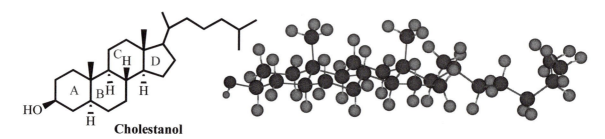

**Cholestanol**

*(a) Describe the conformation of rings A, B, C, and D in cholestanol.*

**Six-membered rings A, B, and C are all in chair conformations. The five-membered ring D is in an envelope conformation.**

*(b) Is the hydroxyl group on ring A axial or equatorial?*

**The hydroxyl group is equatorial.**

*(c) Consider the methyl group at the junction of rings A/B. Is it axial or equatorial to ring A? Is it axial or equatorial to ring B?*

**The methyl group at the A/B junction is axial with respect to both rings.**

*(d) Is the methyl group at the junction of rings C/D axial or equatorial to ring C?*

**The methyl group is axial to ring C.**

**3.57**   *As we have seen in Section 3.5, the IUPAC system divides the name of a compound into a prefix (showing the number of carbon atoms, an infix (showing the presence of carbon-carbon single, double, or triple bonds), and a suffix (showing the presence of an alcohol, amine, aldehyde, ketone, or carboxylic acid). Assume for the purposes of this problem that, to be alcohol (-ol) or amine (-amine), the hydroxyl or amino group must be bonded to a tetrahedral (sp$^3$ hybridized) carbon atom.*

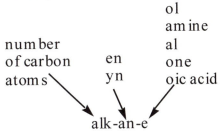

*Given this information, write the structural formula of a compound with an unbranched chain of four carbon atoms that is an:*

*(a)* $CH_3CH_2CH_2CH_3$            *(b)* $CH_2=CHCH_2CH_3$            *(c)* $HC\equiv CCH_2CH_3$

*(d)* $CH_3CH_2CH_2CH_2OH$
*(e)* $CH_2=CH\overset{\displaystyle OH}{\overset{|}{C}}HCH_3$
*(f)* $HC\equiv C\overset{\displaystyle OH}{\overset{|}{C}}HCH_3$

*(g)* $CH_3CH_2\overset{\displaystyle NH_2}{\overset{|}{C}}HCH_3$
*(h)* $CH_2=CH\overset{\displaystyle NH_2}{\overset{|}{C}}HCH_3$
*(i)* $HC\equiv C\overset{\displaystyle NH_2}{\overset{|}{C}}HCH_3$

$$\text{(j) } CH_3CH_2CH_2\overset{\overset{\displaystyle O}{\|}}{C}H$$

$$\text{(k) } CH_2{=}CHCH_2\overset{\overset{\displaystyle O}{\|}}{C}H$$

$$\text{(l) } HC{\equiv}CCH_2\overset{\overset{\displaystyle O}{\|}}{C}H$$

$$\text{(m) } CH_3CH_2\overset{\overset{\displaystyle O}{\|}}{C}CH_3$$

$$\text{(n) } CH_2{=}CH\overset{\overset{\displaystyle O}{\|}}{C}CH_3$$

$$\text{(o) } HC{\equiv}C\overset{\overset{\displaystyle O}{\|}}{C}CH_3$$

$$\text{(p) } CH_3CH_2CH_2\overset{\overset{\displaystyle O}{\|}}{C}OH$$

$$\text{(q) } CH_2{=}CHCH_2\overset{\overset{\displaystyle O}{\|}}{C}OH$$

$$\text{(r) } HC{\equiv}CCH_2\overset{\overset{\displaystyle O}{\|}}{C}OH$$

*(Note: There is only one structural formula possible for some parts of this problem.  For other parts, two or more structural formulas are possible.  Where two are more are possible, we will deal with how the IUPAC system distinguishes among them when we come to the chapters on those particular functional groups.)*

# CHAPTER 4
## *Solutions to Problems*

**4.1**    *Write the IUPAC name of each unsaturated hydrocarbon.*

(a)

(b)

(c)

**3,3-Dimethyl-1-pentene**      **2,3-Dimethyl-2-butene**      **3,3-Dimethyl-1-butyne**

**4.2**    *Name each alkene and specify its configuration using the cis-trans system.*

(a)

(b)

***cis*-4-Methyl-2-pentene**      ***trans*-2,2-Dimethyl-3-hexene**

**4.3**    *Name each alkene and specify its configuration by the E,Z system.*

(a)

(b)

**(*E*)-1-Chloro-2,3-dimethyl-2-pentene**      **(*Z*)-1-Bromo-1-chloropropene**

(c)

**(*E*)-2,3,4-Trimethyl-3-heptene**

**4.4**    *Write the IUPAC name for each cycloalkene.*

(a)

(b)

(c)

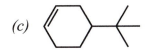

**1-Isopropyl-4-methyl-**
**cyclohexene**       **Cyclooctene**       **4-*tert*-Butylcyclohexene**

**4.5**   *Draw structural formulas for the other two cis-trans isomers of 2,4-heptadiene.*

**cis,trans-2,4-Heptadiene**          **cis,cis-2,4-Heptadiene**

**4.6**   *How many cis-trans isomers are possible for the following unsaturated alcohol?*

$$CH_3C{=}CHCH_2CH_2C{=}CHCH_2CH_2C{=}CHCH_2OH$$

(with $CH_3$ groups above each first carbon, and asterisks * below the second and third double bonds)

**The unsaturated alcohol has two double bonds that have the possibility of cis or trans configurations (indicated by an asterisk). Using the equation $2^n$ = maximum number of stereoisomers possible, where n = the number of double bonds that show cis or trans configuration, $2^2$ = four possible cis-trans stereoisomers.**

**4.7**   *Each carbon atom in ethane and in ethylene is surrounded by eight valence electrons and has four bonds to it. Explain how the VSEPR model (Section 1.4) predicts a bond angle of 109.5° about each carbon in ethane but an angle of 120° about each carbon in ethylene.*

**Both carbons in ethane have four electron pairs (all bonding pairs) surrounding the carbon. A tetrahedral arrangement of bonding pairs with bond angles of 109.5° minimizes electron pair repulsions. In ethylene, both carbons have three regions of electron density (remember, in the VSEPR model, we treat multiple bonds as a single region of electron density) surrounding the carbon. A trigonal planar arrangement of three regions of electron density with bond angles of 120° minimizes repulsions.**

**4.8**   *Predict all bond angles about each highlighted carbon atom. To make these predictions, use the valence-shell electron-pair repulsion (VSEPR) model.*

**4.9**     *For each highlighted carbon atom in Problem 4.8, identify which orbitals are used to form each sigma bond and which are used to form each pi bond.*

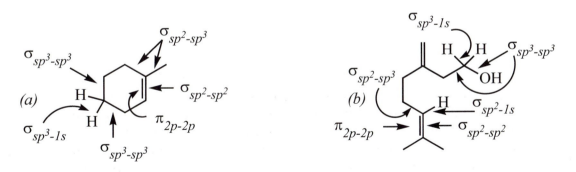

**4.10**    *Predict all bond angles about each highlighted carbon atom.*

**4.11**    *For each highlighted carbon atom in Problem 4.10, identify which orbitals are used to form each sigma bond and which are used to form each pi bond.*

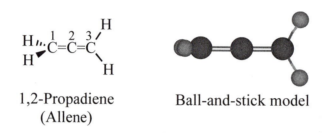

**4.12** *Following is the structure of 1,2-propadiene (allene). In it, the plane created by H-C-H of carbon 1 is perpendicular to that created by H-C-H of carbon 3.*

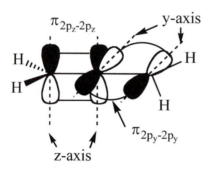

1,2-Propadiene          Ball-and-stick model
(Allene)

*(a) State the orbital hybridization of each carbon in allene.*

**Carbons 1 and 3 are $sp^2$ hybridized. Carbon 2 is $sp$ hybridized.**

*(b) Account for the molecular geometry of allene in terms of the orbital overlap model. Specifically, explain why all four hydrogen atoms are not in the same plane.*

**Each of the terminal carbons is $sp^2$ hybridized and thus has three $sp^2$ orbitals oriented 120° apart. Two of these $sp^2$ hybrid orbitals form sigma bonds with hydrogen atoms while the third forms a sigma bond with the central carbon. Each terminal carbon has one $2p$ orbital (C1 has a $2p_z$ orbital and C3 has a $2p_y$ orbital) that each forms a π bond with the corresponding $2p$ in the same axis of the central carbon's perpendicular unhybridized $2p$ orbitals ($2p_z$ and $2p_y$ orbitals). Because these two $2p$ orbitals are at right angles to each other, the $2p$ orbitals on the terminal carbons must be at right angles to each other to permit full orbital overlap.**

## Nomenclature of Alkenes and Alkynes
**4.13**   *Draw a structural formula for each compound.*

*(a) trans-2-Methyl-3-hexene*     *(b) 2-Methyl-3-hexyne*     *(c) 2-Methyl-1-butene*

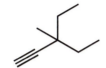

*(d) 3-Ethyl-3-methyl-1-pentyne*     *(e) 2,3-Dimethyl-2-butene*     *(f) cis-2-Pentene*

*(g) (Z)-1-Chloropropene*     *(h) 3-Methylcyclohexene*

**4.14**   *Draw a structural formula for each compound.*

*(a) 1-Isopropyl-4-methylcyclohexene*     *(b) (6E)-2,6-Dimethyl-2,6-octadiene*

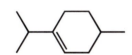
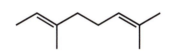

*(c) trans-1,2-Diisopropylcyclopropane*     *(d) 2-Methyl-3-hexyne*

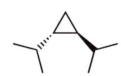
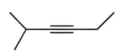

*(e) 2-Chloropropene*     *(f) Tetrachloroethylene*

**4.15**   *Write the IUPAC name for each compound.*

*(a)*          *(b)*          *(c)*

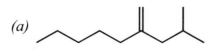

**2-Isobutyl-1-heptene          1,4,4-Trimethylcyclopentene     1,3-Cyclopentadiene**

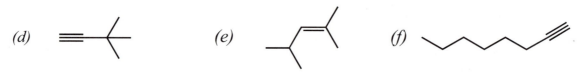

*(d)*   **3,3-Dimethyl-1-butyne**          *(e)*   **2,4-Dimethyl-2-pentene**          *(f)*   **1-Octyne**

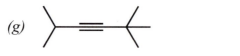

*(g)*   **2,2,5-Dimethyl-3-hexyne**          *(h)*   **3-Methyl-1-pentyne**

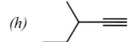

**4.16**   *Explain why each name is incorrect, and then write a correct name.*

*(a) 1-Methylpropene*

$$\overset{4}{CH_3}\overset{3}{CH}=\overset{2}{C}\overset{\overset{1}{CH_3}}{\underset{H}{}}$$

**Wrong parent chain and
need an *E* or *Z* designation
Correct name:  2-Butene**

*(b) 3-Pentene*

$$\overset{5}{CH_3}\overset{4}{CH_2}\overset{3}{CH}=\overset{2}{CH}\overset{1}{CH_3}$$

**Numbered parent chain
from the wrong end and
need an *E* or *Z* designation
Correct name:  2-Pentene**

*(c) 2-Methylcyclohexene*

**Incorrect numbering of ring
Correct name:  1-Methylcyclohexene**

*(d) 3,3-Dimethylpentene*

**Need to indicate position of double bond
Correct name:  3,3-Dimethyl-1-pentene**

*(e) 4-Hexyne*

$$\overset{7}{CH_3}\overset{6}{CH_2}\overset{5}{CH_2}\overset{4}{C}\equiv\overset{3}{C}\overset{2}{CH_2}\overset{1}{CH_3}$$

**Numbered parent chain
from the wrong end
Correct name:  3-Hexyne**

*(f) 2-Isopropyl-2-butene*

**Wrong parent chain and
need an *E* or *Z* designation
Correct name:  3,4-Dimethyl-2-pentene**

**4.17**   *Explain why each name is incorrect, and then write a correct name.*

(a) *2-Ethyl-1-propene*

**Wrong parent chain (butene)**
**Correct name:  2-Methyl-1-butene**

(b) *5-Isopropylcyclohexene*

**Incorrect numbering of the ring.**
**Correct name:  4-Isopropylcyclohexene**

(c) *4-Methyl-4-hexene*

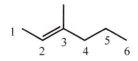

**Parent chain numbered**
**from the wrong end and**
**need *E* or *Z* designation.**
**Correct name:  3-Methyl-2-hexene**

(d) *2-sec-Butyl-1-butene*

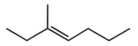

**Wrong parent chain (1-pentene)**
**Correct name:  2-Ethyl-3-methyl-1-pentene**

(e) *6,6-Dimethylcyclohexene*

**Incorrect numbering of the ring.**
**Correct name:  3,3-Dimethylcyclohexene**

(f) *2-Ethyl-2-hexene*

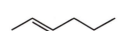

**Wrong parent chain and**
**need *E* or *Z* designation.**
**Correct name:  3-Methyl-3-heptene**

## Cis-Trans Isomerism in Alkenes and Cycloalkenes

**4.18**   *Which of these alkenes show cis-trans isomerism? For each that does, draw structural formulas for both isomers.*

(a) *1-Hexene*                      (b) *2-Hexene*
(c) *3-Hexene*                      (d) *2-Methyl-2-hexene*
(e) *3-Methyl-2-hexene*            (f) *2,3-Dimethyl-2-hexene*

**For alkenes to exist as a pair of cis-trans isomers, both carbons of the double bond must have two different substituents.  Thus (b), (c), and (e) can exist as a cis or trans isomer.**

(b)

*cis*-**2-Hexene**                    *trans*-**2-Hexene**

(c)

**cis-3-Hexene**                    **trans-3-Hexene**

(e)

**cis-3-Methyl-2-hexene**          **trans-3-Methyl-2-hexene**

**4.19**  *Which of these alkenes shows cis-trans isomerism? For each that does, draw structural formulas for both isomers.*
  (a) 1-Pentene                    (b) 2-Pentene
  (c) 3-Ethyl-2-pentene            (d) 2,3-Dimethyl-2-pentene
  (e) 2-Methyl-2-pentene           (f) 2,4-Dimethyl-2-pentene

**For alkenes to exist as a pair of cis-trans isomers, both carbons of the double bond must have two different substituents.  Thus only (b) can exist as a cis or trans isomer.**

$H_3C$      $CH_2CH_3$          $H_3C$      $H$

$H$         $H$                 $H$         $CH_2CH_3$

**cis-2-Pentene**                 **trans-2-Pentene**

**4.20**  *Which alkenes can exist as pairs of cis-trans isomers? For each alkene that does, draw the trans isomer.*

  (a) $CH_2=CHBr$                  (b) $CH_3CH=CHBr$
  (c) $(CH_3)_2C=CHCH_3$          (d) $(CH_3)_2CHCH=CHCH_3$

**For alkenes to exist as a pair of cis-trans isomers, both carbons of the double bond must have two different substituents.  Thus only (b) and (d) can exist as a cis or trans isomer.**

(b)

$H_3C$      $Br$                $H_3C$      $H$

$H$         $H$                 $H$         $Br$

**cis-1-Bromopropene**            **trans-1-Bromopropene**

*(d)*

**cis-4-Methyl-2-pentene**    **trans-4-Methyl-2-pentene**

**4.21**    *There are three compounds of molecular formula C₂H₂Br₂. Two of these compounds have a dipole greater than zero, and one has no dipole. Draw structural formulas for these three compounds and explain why two have dipole moments but the third one has none.*

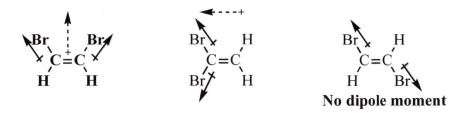

**No dipole moment**

**4.22**    *Name and draw structural formulas for all alkenes of molecular formula C₅H₁₀. As you draw these alkenes, remember that cis and trans isomers are different compounds and must be counted separately.*

**1-Pentene**    ***trans*-2-Pentene**    ***cis*-2-Pentene**

**2-Methyl-2-butene**    **3-Methyl-1-butene**    **2-Methyl-1-butene**

**4.23**    *Name and draw structural formulas for all alkenes of molecular formula C₆H₁₂ that have these carbon skeletons. (Remember cis and trans isomers.)*

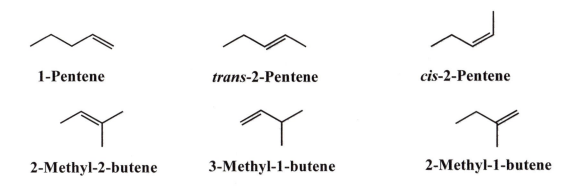

*(a)*  C-C-C-C-C

$CH_2{=}CHCH_2CH_2CH_3$

**2-Methyl-1-pentene**    **2-Methyl-2-pentene**

$$CH_3-\overset{\overset{\displaystyle CH_3}{|}}{CH} \quad\quad \overset{CH_3}{\underset{H}{C}}=\overset{\underset{H}{C}}{}$$

**cis-4-Methyl-2-pentene**

$$CH_3-\overset{\overset{\displaystyle CH_3}{|}}{CH} \quad\quad \overset{H}{\underset{H}{C}}=\overset{\underset{CH_3}{C}}{}$$

**trans-4-Methyl-2-pentene**

$$\overset{CH_3}{\underset{}{CH_3CHCH_2CH=CH_2}}$$

**4-Methyl-1-pentene**

(b)   C-C-C-C (with C, C substituents)

$$H_2C=\overset{\overset{\displaystyle CH_3}{|}}{\underset{\underset{\displaystyle CH_3}{|}}{C}}CHCH_3$$

**2,3-Dimethyl-1-butene**

$$\overset{H_3C}{\underset{H_3C}{}}C=C\overset{CH_3}{\underset{CH_3}{}}$$

**2,3-Dimethyl-2-butene**

(c)   C-C-C-C (with C above and C below center)

$$\overset{\overset{\displaystyle CH_3}{|}}{\underset{\underset{\displaystyle CH_3}{|}}{CH_3CCH=CH_2}}$$

**3,3-Dimethyl-1-butene**

**4.24**   *Arrange the groups in each set in order of increasing priority.*

    *(a) -CH₃, -Br, -CH₂CH₃*           *(b) -OCH₃, -CH(CH₃)₂, -CH₂CH₂NH₂*

     **-CH₃ < -CH₂CH₃ < -Br**         **-CH₂CH₂NH₂ < -CH(CH₃)₂ < -OCH₃**

    *(c) -CH₂OH, -COOH, -OH*          *(d) –CH=CH₂, -CH=O, -CH(CH₃)₂*

     **-CH₂OH < -COOH < -OH**         **-CH(CH₃)₂ <–CH=CH₂ < -CH=O**

**According to the Cahn-Ingold-Prelog convention, the priority of each group increases as the atomic number increases.**

**4.25**   *Draw the structural formula for at least one bromoalkene of molecular formula $C_5H_9Br$ that:*

    *(a) Shows E,Z isomerism*

       **Draw any structural formula where each carbon of the double bond has different substituents. Two examples of bromoalkenes that fit the criteria include:**

**(*E*)-2-Bromo-2-pentene**                    **(*Z*)-1-Bromo-1-pentene**

*(b) Does not show E,Z isomerism.*

**Draw any structural formula where one carbon of the double bond has two identical substituents. Two examples of bromoalkenes that fit the criteria include:**

**5-Bromo-1-pentene**                    **1-Bromo-3-methyl-2-butene**

**4.26**   *For each molecule that shows cis-trans isomerism, draw the cis isomer.*

**Molecules (a) and (c) do not show cis-trans isomerism.**

**4.27**   *Explain why each name is incorrect or incomplete, and then write a correct name.*

**Cis-trans isomerism is not possible in the given examples because there are identical substituents on at least one carbon in each of the double bonds. For the correct name, delete all designations of configuration.**

*(a) (Z)-2-Methyl-1-pentene*                    *(b) (E)-3,4-Diethyl-3-hexene*

**Correct name: 2-Methyl-1-pentene**          **Correct name: 3,4-Diethyl-3-hexene**

*(c) trans-2,3-Dimethyl-2-hexene*          *(d) (1Z,3Z)-2,3-Dimethyl-1,3-butadiene*

**Correct name: 2,3-Dimethyl-2-hexene**     **Correct name: 2,3-Dimethyl-1,3-butadiene**

**4.28**   *Draw structural formulas for all compounds of molecular formula $C_5H_{10}$ that are*

*(a) Alkenes that do not show cis-trans isomerism.*

**1-Pentene     2-Methyl-2-butene     3-Methyl-1-butene     2-Methyl-1-Butene**

*(b) Alkenes that do show cis-trans isomerism.*

***trans*-2-Pentene                    *cis*-2-Pentene**

*(c) Cycloalkanes that do not show cis-trans isomerism.*

CH$_3$                     CH$_2$CH$_3$              H$_3$C      CH$_3$

**Cyclopentane     Methylcyclobutane     Ethylcyclopropane     1,1-Dimethyl
                                                                 cyclopropane**

*(d) Cycloalkanes that do show cis-trans isomerism.*

H$_3$C      CH$_3$                        H$_3$C      CH$_3$
  H        H                                H        H
***cis*-1,2-Dimethylcyclopropane          *trans*-1,2-Dimethylcyclopropane**

**4.29**   *β-Ocimene, a triene found in the fragrance of cotton blossoms and several essential oils, has the IUPAC name (3Z)-3,7-dimethyl-1,3,6-octatriene.   Draw a structural formula for β-ocimene.*

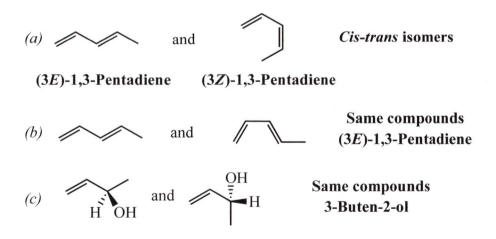

**(3Z)-3,7-Dimethyl-1,3,6-octatriene**

**4.30**   *Oleic acid and elaidic acid are the cis and trans isomers of 9-octadecenoic acid.  One of these fatty acids, a colorless liquid that solidifies at 4°C, is a major component of butter fat. The other, a white solid with a melting point of 44-45°C, is a major component of partially hydrogenated vegetable oils. Which of these two fatty acids is the cis isomer and which is the trans isomer?*

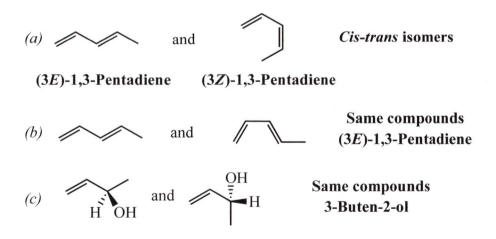

**Oleic Acid**                              **Elaidic acid**

**Most naturally occurring unsaturated fatty acids contain cis double bonds.  When cis unsaturated fatty acids are partially hydrogenated, these cis double bonds isomerize to trans double bonds.**

**4.31**   *Determine whether the following pairs of structures represent the same molecule, cis-trans isomers, or constitutional isomers.  If they are the same molecule, determine whether they are in the same or different conformations.*

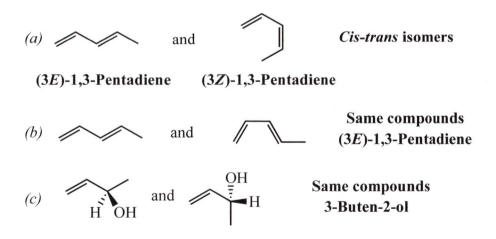

(a)  _____  and  _____          ***Cis-trans* isomers**

**(3E)-1,3-Pentadiene       (3Z)-1,3-Pentadiene**

(b)  _____  and  _____          **Same compounds**
                                    **(3E)-1,3-Pentadiene**

(c)  _____  and  _____          **Same compounds**
                                    **3-Buten-2-ol**

*(d)* and    **Same compounds
(2E,4E,6E)-2,4,6-Octatriene**

## Terpenes

**4.32**   *Show that the structural formula of vitamin A (Section 4.3G) can be divided into four isoprene units joined by head-to-tail linkages and cross-linked at one point to form the six-membered ring.*

**Isoprene chain cross-linked here**

**Vitamin A**

**4.33**   *Following is the structural formula of lycopene, a deep-red compound that is partially responsible for the red color of ripe fruits, especially tomatoes. Approximately 20 mg of lycopene can be isolated from 1 kg of fresh ripe tomatoes.*

(a) *Show that lycopene is a terpene; that is, show that its carbon skeleton can be divided into two sets of four isoprene units with the units in each set joined head-to-tail.*

isoprene unit
(bolded)                            **Lycopene**              **\* indicates cis-trans
isomerism possible**

(b) *How many of the carbon-carbon double bonds in lycopene have the possibility for cis-trans isomerism? Lycopene is the all trans isomer.*

**There are 11 carbon-carbon double bonds that have the potential for cis-trans isomerism.**

**4.34**   *As you might suspect, β-carotene, a precursor to vitamin A, was first isolated from carrots. Dilute solutions of β-carotene are yellow, hence its use as a food coloring. In plants, it is almost always present in combination with chlorophyll to assist in the harvesting of the energy of sunlight. As tree leaves die in the fall, the green of their chlorophyll molecules is replaced by the yellow and reds of carotene and carotene-related molecules.*

Head-to-head bond
joining two units
containing four isoprenes

Isoprene chain cross-linked here

**β-Carotene**

Isoprene chain cross-linked here

*(a) Compare the carbon skeletons of β-carotene and lycopene. What are the similarities? What are the differences?*

**The main structural difference between β-carotene and lycopene is that β-carotene has six-membered rings on the ends, not an open chain as in lycopene. On the other hand, both β-carotene and lycopene can be divided into two sets of four isoprene units as shown above. All of the double bonds that can show cis-trans isomerism are trans in both molecules.**

*(b) Show that β-carotene is a terpene.*

**β-Carotene has eight isoprene units indicated above in bolded bonds.**

**4.35**  *α-Santonin, isolated from the flower heads of certain species of Artemisia, is an anthelmintic; that is, a drug used to rid the body of worms (helminths). It has been estimated that over one third of the world's population is infested with these parasites.*

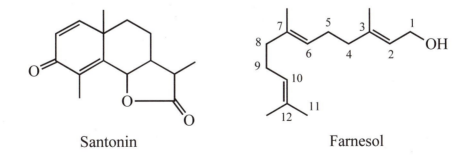

Santonin                          Farnesol

*Locate the three isoprene units in santonin and show how the carbon skeleton of farnesol might be coiled and then cross-linked to give santonin. Two different coiling patterns of the carbon skeleton of farnesol can lead to santonin. Try to find them both.*

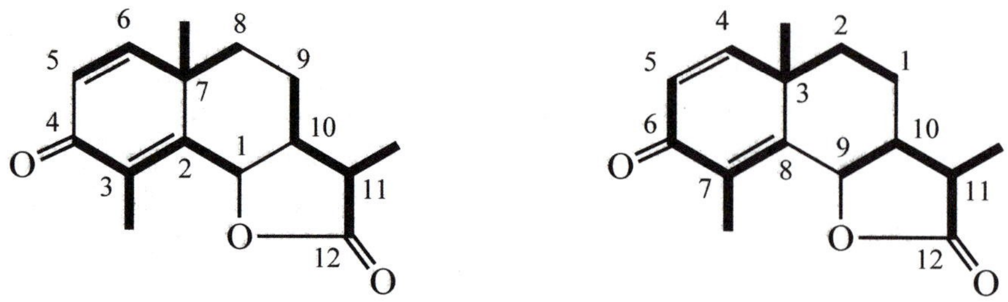

**4.36** *Periplanone is a pheromone (a chemical sex attractant) isolated from a species of cockroach. Show that the carbon skeleton of periplanone classifies it as a terpene.*

**Periplanone has four isoprene units, highlighted in bold, in its structure.**

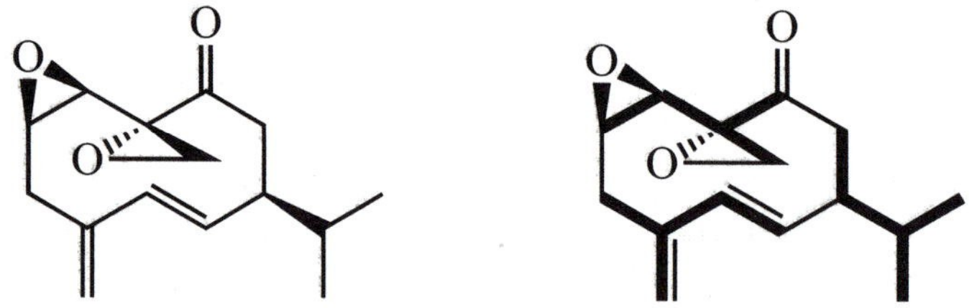

**Periplanone**

**4.37** *Gossypol, a compound found in the seeds of cotton plants, has been used as a male contraceptive in overpopulated countries such as China. Show that gossypol is a terpene.*

Gossypol

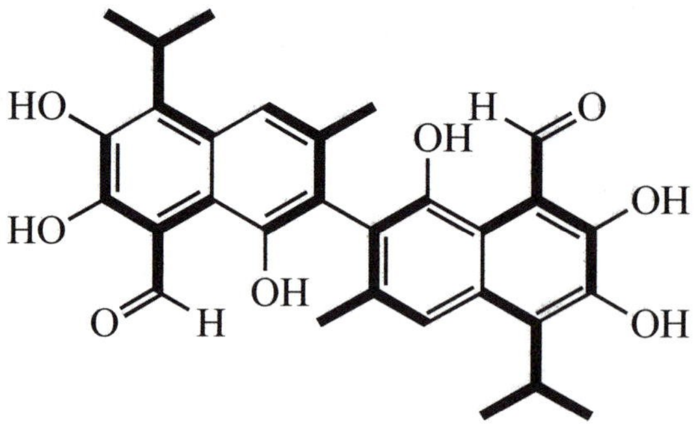

**4.38** *In many parts of South America, extracts of the leaves and twigs of Montanoa tomentosa are used as a contraceptive, to stimulate menstruation, to facilitate labor, and as an abortifacient. The compound responsible for these effects is zoapatanol.*

**Zoapatanol**

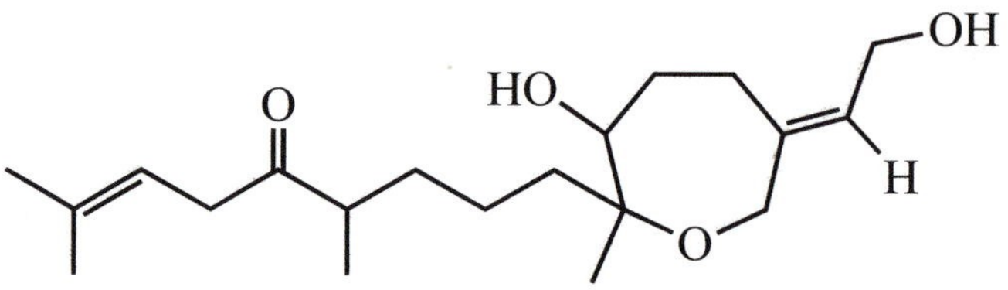

(a) *Show that the carbon skeleton of zoapatanol can be divided into four isoprene units bonded head-to-tail and then cross-linked in one point along the chain.*

**\* Indicates where *cis-trans* isomerism is possible.**

(b) *Specify the configuration about the carbon-carbon double bond to the seven-membered ring according to the E,Z system.*

**The double bond to the seven-membered ring has the *E* configuration**

(c) *How many cis-trans isomers are possible for zoapatanol?  Consider the possibilities for cis-trans isomerism in cyclic compounds and about carbon-carbon double bonds.*

**The ring substituents provide an opportunity for cis-trans isomerism, as does the double bond off of the seven-membered ring.  The two opportunities for cis-trans isomerism in the molecule give four possible cis-trans stereoisomers.**

| Ring | Double bond |
|---|---|
| Cis | Cis |
| Cis | Trans |
| Trans | Cis |
| Trans | Trans |

**4.39**  *Pyrethrin II and pyrethrosin are natural products isolated from plants of the chrysanthemum family.  Pyrethrin II is a natural insecticide and is marketed as such.*

(a) *Label all carbon-carbon double bonds in each about which cis-trans isomerism is possible.*

**Carbon-carbon double bonds where cis-trans isomerism is possible are labeled with an asterisk.**

(b) *Why are cis-trans isomers possible about the three-membered ring in pyrethrin II, but not about its five-membered ring?*

**The carbon-carbon double bond in the five-membered ring has a planar geometry; therefore, substituents attached to them can neither be on the same or opposite side of another substituent on the ring.**

*(c) Show that the ring system of pyrethrosin is composed of three isoprene units.*

Pyrethrin II                                               Pyrethrosin

**\* Indicates a double bond where *cis-trans* isomerism is possible.**

**4.40**  *Cuparene and herbertene are naturally occurring compounds isolated from various species of lichen. Determine whether one or both of these compounds can be classified as terpenes.*

**Cuparene**                                               **Herbertene**

**Both cuparene and herbertene can be classified as terpenes, with four connected isoprene units.**

**Looking Ahead**

**4.41**  *Explain why the =C-C single bond in 1,3-butadiene is slightly shorter than the =C-C single bond in 1-butene.*

1.47□                              1.51□

1,3-butadiene                     1-butene

**Partial overlap of the *p*-orbitals on carbons 2 and 3 of 1,3-butadiene form a bond with π-bond character. This delocalization of *p*-electrons is known as resonance and is a powerful stabilizing process in molecules.**

**4.42**   *What effect might the ring size in the following cycloalkenes have on the reactivity of the C=C double bond in each?*

90û          108û          111û          120û
                                          (ideal)

**Smaller rings impose non-ideal bond angles on the alkene carbons thus increasing the reactivity of the cycloalkene to relieve angle strain. Reactions can either cleave the C=C bond breaking the ring and relieving angle strain or add a reagent, thus converting from *sp²* to *sp³* hybridized carbons. For example, the addition of H₂ to cyclobutene yields cyclobutane:**

90û   $\xrightarrow[\text{Pt}]{\text{H}_2}$   90û

**120û- 90û= 30û**                **109û-90û= 19û**

**Cyclobutene imposes a 30° deviation from the ideal bond angle of 120° for an *sp²* hybridized carbon. Upon addition of hydrogen, cyclobutene is converted to cyclobutane, which has a deviation of 19° from the ideal bond angle of 109° for an *sp³* hybridized carbon. Although cyclobutane still experiences angle strain from non-ideal bond angles, it is less strained than cyclobutene.**

**4.43**   *What effect might each substituent have on the electron density surrounding the alkene C=C bond?*

*(a)* $\overset{..}{\underset{..}{O}}CH_3$  ⟷  $H_2C\overset{..}{\underset{..}{O}}CH_3$     **Resonance increases electron density surrounding the double bond.**

*(b)* C≡N:  ⟷  $H_2C=C=\overset{..}{N}:^-$     **Resonance decreases electron density surrounding the double bond.**

*(c)* $Si(CH_3)_3$     **No resonance, therefore -Si(CH₃)₃ has very little effect on the electron density of the carbon-carbon double bond.**

**4.44**   *In Section 21.1 on the biochemistry of fatty acids, we will study the three long-chain unsaturated carboxylic acids shown below. Each has 18 carbons and is a component of animal fats, vegetable oils, and biological membranes. Because of their presence in animal fats, they are called fatty acids.*

Oleic acid        $CH_3(CH_2)_7CH=CH(CH_2)_7COOH$

Linoleic acid     $CH_3(CH_2)_4CH=CHCH_2CH=CH(CH_2)_7COOH$

Linolenic acid    $CH_3CH_2CH=CHCH_2CH=CHCH_2CH=CH(CH_2)_7COOH$

*(a) How many cis-trans isomers are possible for each fatty acid?*

**Maximum number of cis-trans isomers = $2^n$ where n = number of possible cis-trans double bonds**

**Oleic acid:  one double bond           $2^1$ = two cis-trans isomers**

**Linoleic acid:  two double bonds        $2^2$ = four cis-trans isomers**

**Linolenic acid:  three double bonds     $2^3$ = eight cis-trans isomers**

*(b) These three fatty acids occur in biological membranes almost exclusively in the cis configuration. Draw line-angle formulas for each fatty acid showing the cis configuration about each carbon-carbon double bond.*

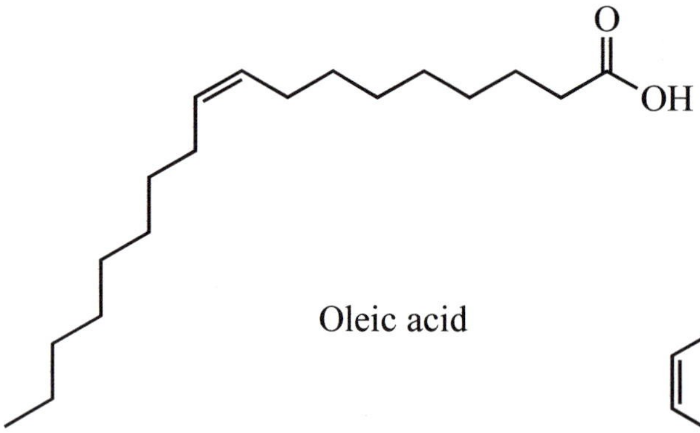

Oleic acid

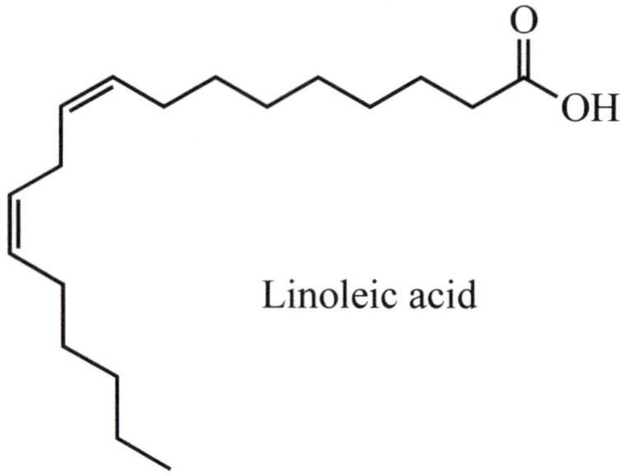

Linoleic acid

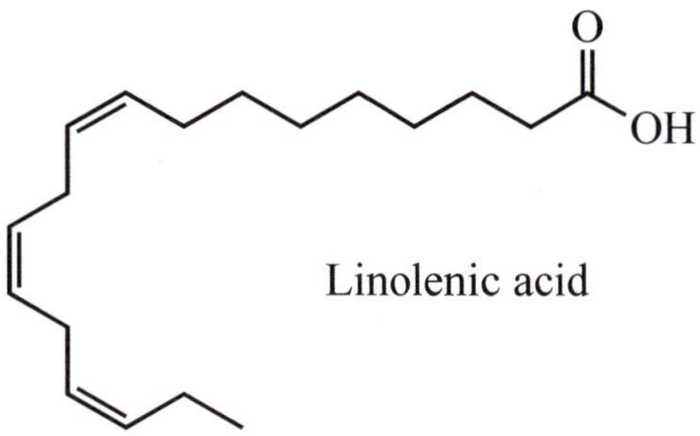

Linolenic acid

**4.45**   *Assign an E or Z configuration and a cis or trans configuration to these carboxylic acids, each of which is an intermediate in the citric acid cycle (Section 22.7).  Under each is given its common name.*

**Fumaric acid may be designated either *E* or *trans*.  Aconitic acid may be designated either *Z* or *trans*.   Note that while the two carboxyl groups in aconitic acid are *cis* or *Z* to each other, the main carbon chain is *trans*.**

## CHAPTER 5
*Solutions to Problems*

**5.1**   *In what way would the reaction energy diagram drawn in Example 5.1 change if the reaction were endothermic?*

**In endothermic reactions, the products are higher in energy than the reactants.**

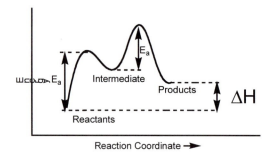

**5.2**   *Name and draw a structural formula for the major product of each alkene addition reaction:*

(a)  $CH_3CH=CH_2$ +  HI ⟶

**2-Iodopropane**

(b)  ⬡=CH₂ + HI ⟶

**1-Iodo-1-methylcyclohexane**

**5.3**   *Arrange these carbocations in order of increasing stability:*

⬡–$\overset{+}{C}H_2$   <   ⬡–$CH_3$   <   ⬡$\overset{+}{C}H_3$

**The order of increasing stability of carbocations is:**
**methyl < primary < secondary < tertiary**

**5.4**    *Propose a mechanism for the addition of HI to 1-methylcyclohexene to give 1-iodo-1-methylcyclohexane. Which step in your mechanism is rate determining?*

**Step 1: Protonation of the alkene to give the most stable 3° carbocation intermediate.**

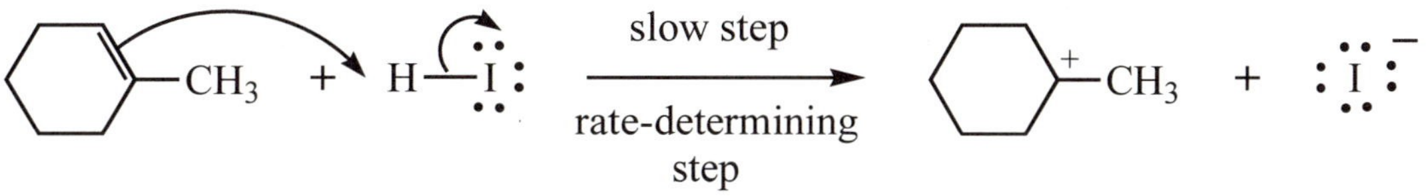

**Step 2: Nucleophilic attack of the iodide anion on the 3° carbocation intermediate to give the product.**

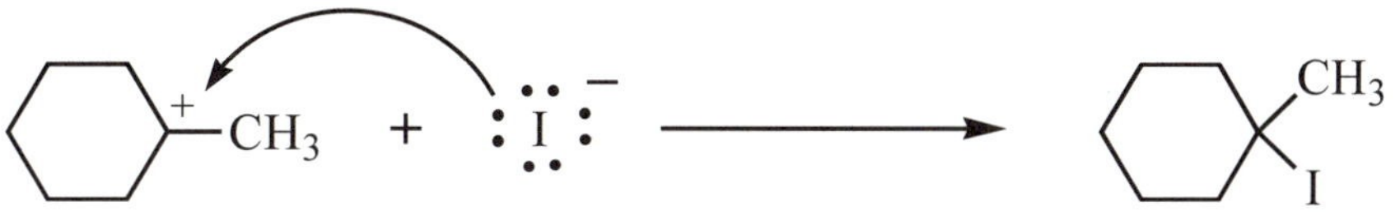

**5.5**    *Draw a structural formula for the product of each alkene hydration reaction:*

(a)

**2-Methyl-2-butanol**

(b)

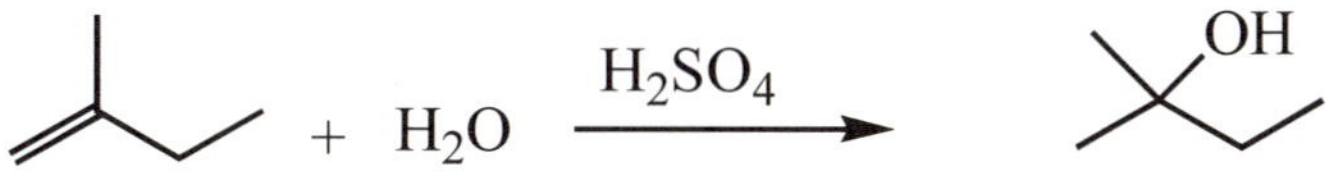

**2-Methyl-2-butanol**

**5.6**    *Propose a mechanism for the acid-catalyzed hydration of 1-methylcyclohexene to give 1-methylcyclohexanol. Which step in your mechanism is rate determining?*

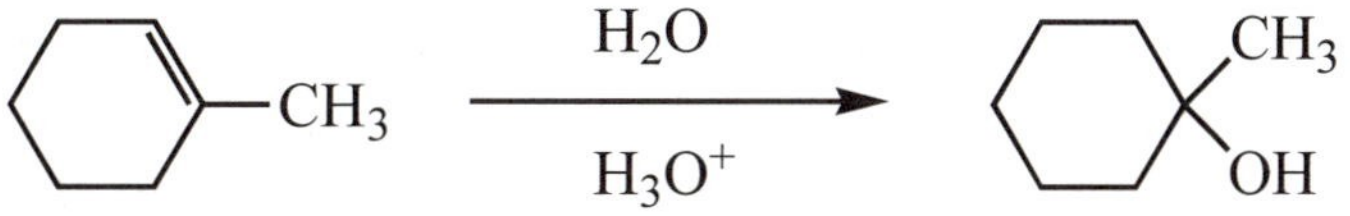

**Step 1: Protonation of the alkene to give the most stable 3° carbocation intermediate.**

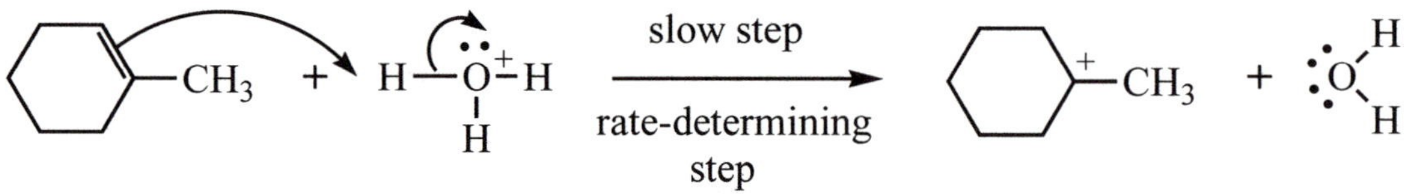

**Step 2: Nucleophilic attack of the water on the 3° carbocation intermediate to give the protonated alcohol.**

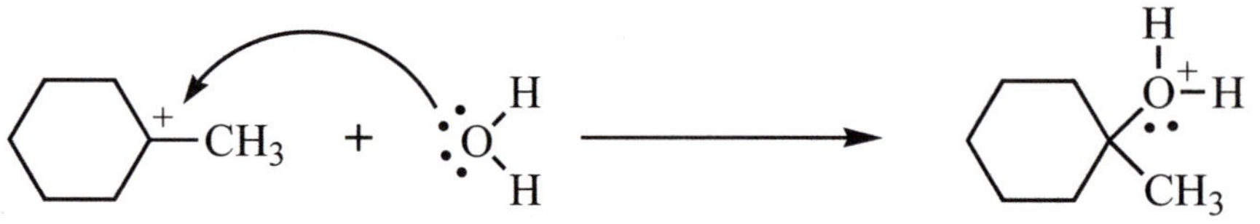

**Step 3: The protonated alcohol loses a proton to form the product.**

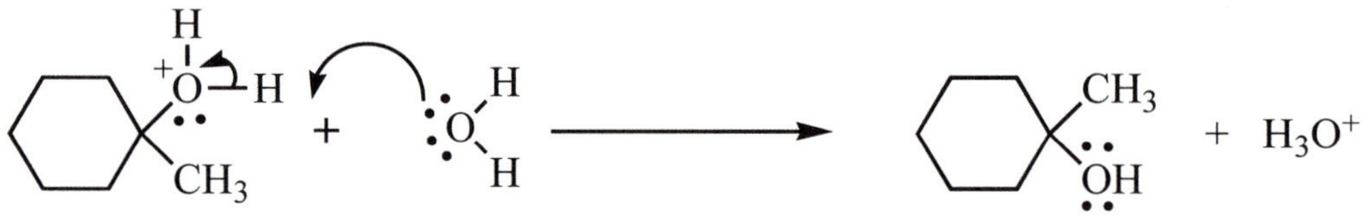

**5.7** *Complete these reactions:*

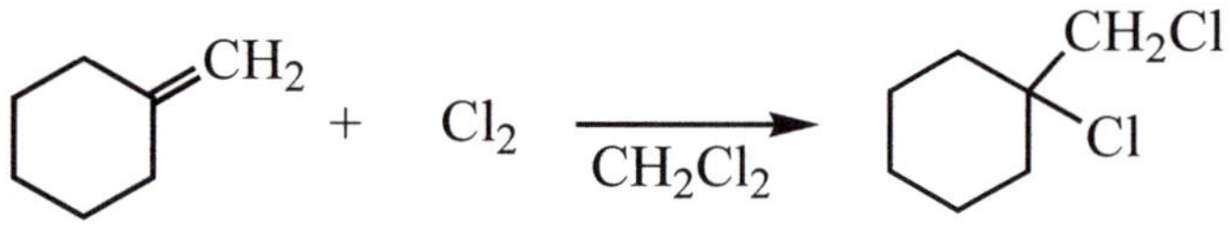

**Energy Diagrams**

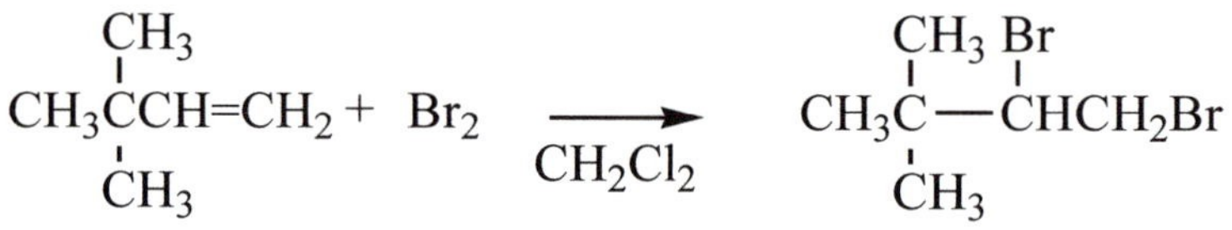

**5.8** *Describe the differences between a transition state and a reaction intermediate.*

A transition state is a point on the reaction coordinate where the energy is at a maximum. Because a transition state is at an energy maximum, it cannot be isolated and its structure cannot be determined experimentally. A reaction intermediate corresponds to an energy minimum between two transition states. Because the reaction intermediate is at an energy higher than either reactants or products, it is highly reactive and rarely, if ever, can one be isolated.

**5.9**    *Sketch an energy diagram for a one-step reaction that is very slow and only slightly exothermic. How many transition states are present in this reaction? How many intermediates are present?*

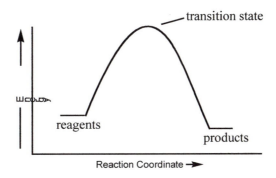

**One-step reactions have one transition state and no intermediates.**

**5.10**   *Sketch an energy diagram for a two-step reaction that is endothermic in the first step, exothermic in the second step, and exothermic overall. How many transition states are present in this two-step reaction? How many intermediates are present?*

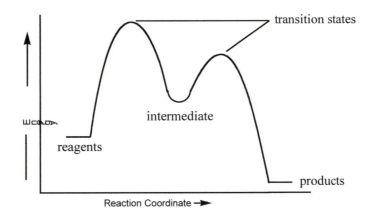

**This two-step reaction has two transition states and one intermediate.**

**5.11**   *Determine whether each of the following statements is true or false and provide a rationale for your decision:*

   *(a) A transition state can never be lower in energy than the reactants from which it was formed.*

   **True, transition states are always energy maxima on reaction energy diagrams.**

*(b) An endothermic reaction cannot have more than one intermediate.*  **False**

> **An endothermic reaction can have many steps associated with it.  For each pair of steps between reactant and product, an intermediate might be formed.**

*(c) An exothermic reaction cannot have more than one intermediate.*  **False**

> **As with an endothermic reaction, an exothermic reaction can have many steps associated with it. For each pair of steps between reactant and product, an intermediate might be formed.**

## Electrophilic Additions

**5.12**   *From each pair, select the more stable carbocation:*

*(a)*  $CH_3CH_2CH_2^+$   *or*   $\boxed{CH_3\overset{+}{C}HCH_3}$     **2û Carbocations more stable than 1û carbocations.**

      **Primary**           **Secondary**

*(b)*  $CH_3\overset{\overset{\displaystyle CH_3}{|}}{\underset{+}{C}H}CHCH_3$   *or*   $\boxed{CH_3\overset{\overset{\displaystyle CH_3}{|}}{\underset{+}{C}}CH_2CH_3}$     **3û Carbocations more stable than 2û carbocations.**

      **Secondary**           **Tertiary**

**5.13**   *From each pair, select the more stable carbocation:*

*(a)*   **3û Carbocations more stable than 2û carbocations.**

      **Secondary**           **Tertiary**

*(b)*   **2û Carbocations more stable than 1û carbocations.**

      **Secondary**           **Primary**

**5.14** *Draw structural formulas for the isomeric carbocation intermediates formed by the reaction of each alkene with HCl. Label each carbocation as primary, secondary, or tertiary, and state which, if either, of the isomeric carbocations is formed more readily.*

(a)

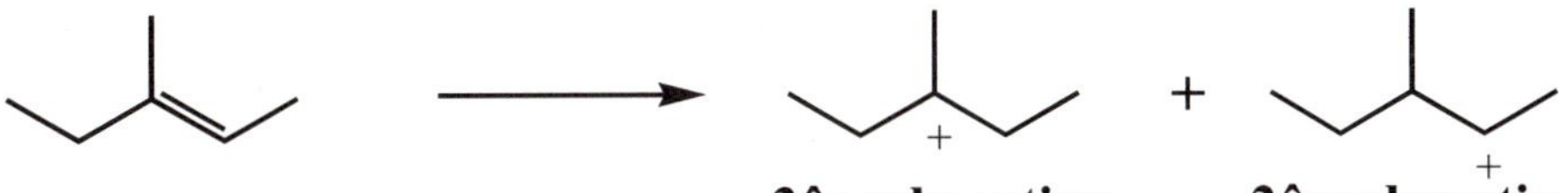

**3°carbocation
formed more
readily**         **2°carbocation**

(b)

**2°carbocation**         **2°carbocation**

**Both carbocations are of equal stability,
thus formed at approximately equal rates.**

(c)     CH₃

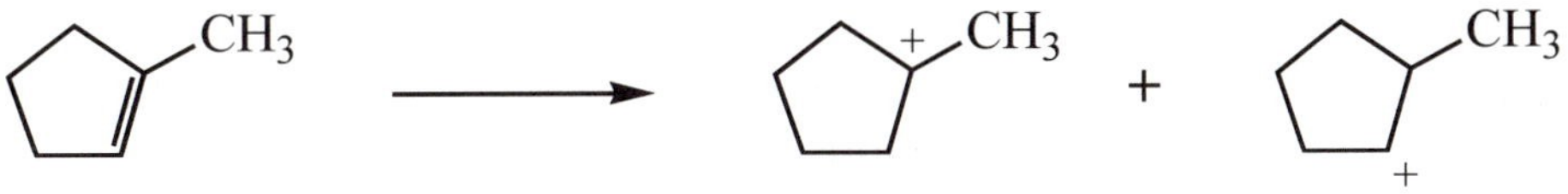

CH₃          CH₃

**3°carbocation
formed more
readily**         **2°carbocation**

(d)     =CH₂

CH₃          CH₂

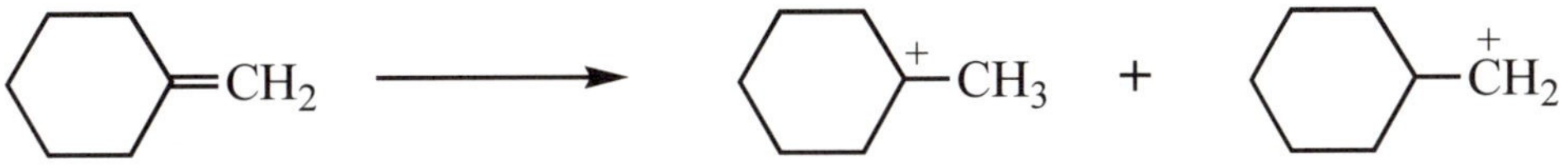

**3°carbocation
formed more
readily**         **1°carbocation**

**5.15** *From each pair of compounds, select the one that reacts more rapidly with HI, draw the structural formula of the major product formed in each case, and explain the basis for your ranking:*

(a)        and

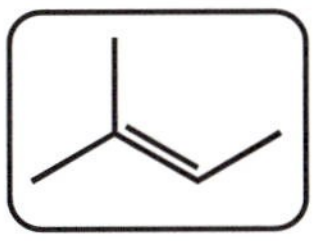

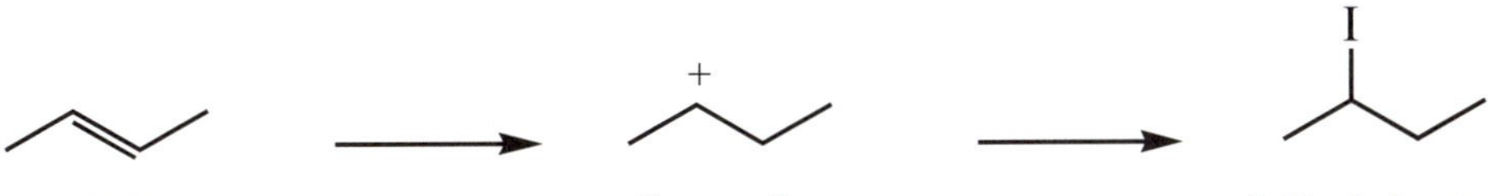

***trans*-2-Butene**         **Secondary
carbocation**         **2-Iodobutane**

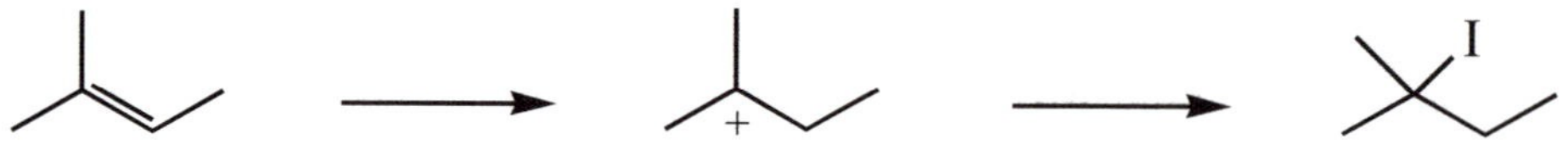

2-Methyl-2-butene            Tertiary              2-Iodo-2-methyl
                             carbocation                butane
                                                   (major product)

The formation of the more stable tertiary carbocation intermediate proceeds at a faster rate, therefore 2-methyl-2-butene will react faster than *trans*-2-butene.

*(b)*

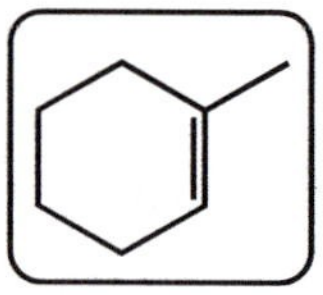

and

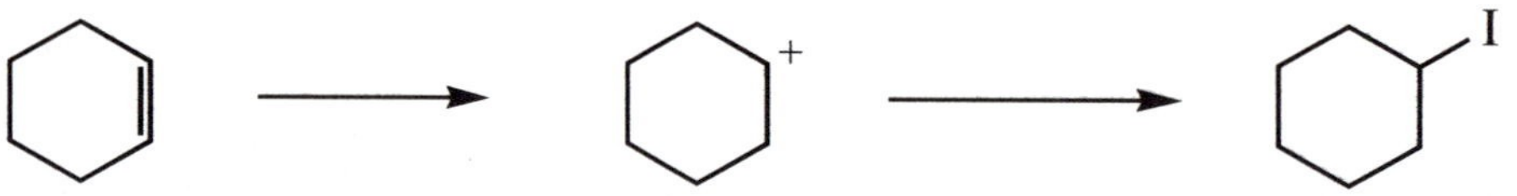

Cyclohexene                  Secondary            Iodocyclohexane
                             carbocation

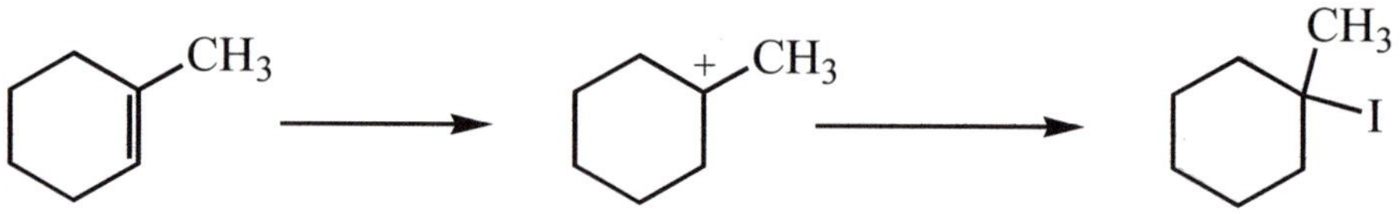

1-Methylcyclohexene          Tertiary             1-Iodo-1-methyl
                             carbocation            cyclohexane
                                                  (major product)

The formation of the more stable tertiary carbocation intermediate proceeds at a faster rate, therefore 1-methylcyclohexene will react faster than cyclohexene.

**5.16**  *Complete these equations by predicting the major product formed in each reaction:*

*(a)*

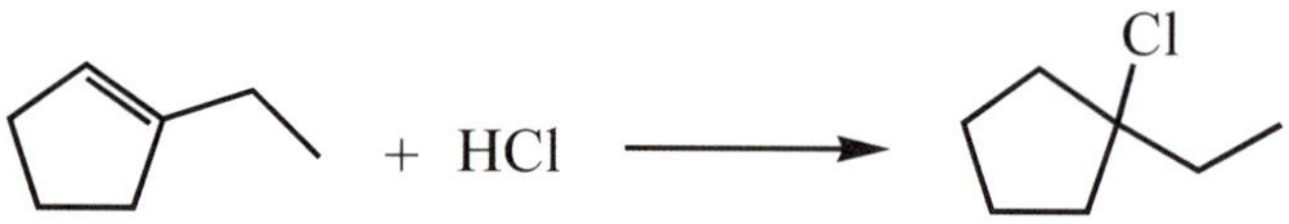

*(b)*

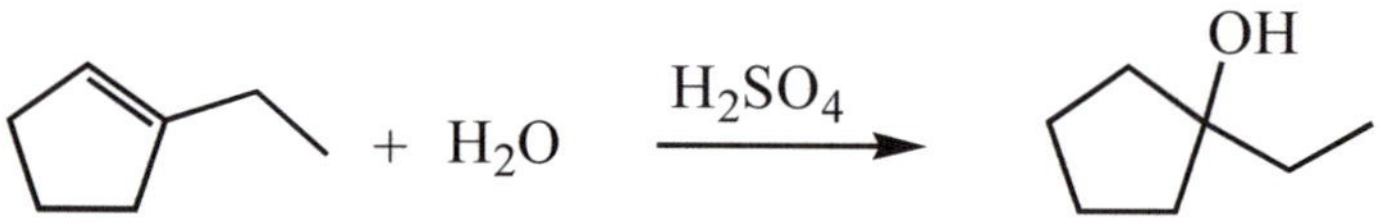

(c)

(d)

(e)

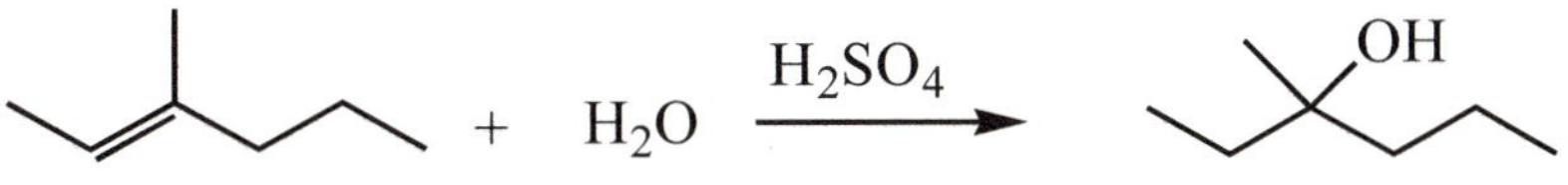

(f)

**5.17**   *The reaction of 2-methyl-2-pentene with each reagent is regioselective.   Draw a structural formula for the product of each reaction, and account for the observed regioselectivity:*

(a) 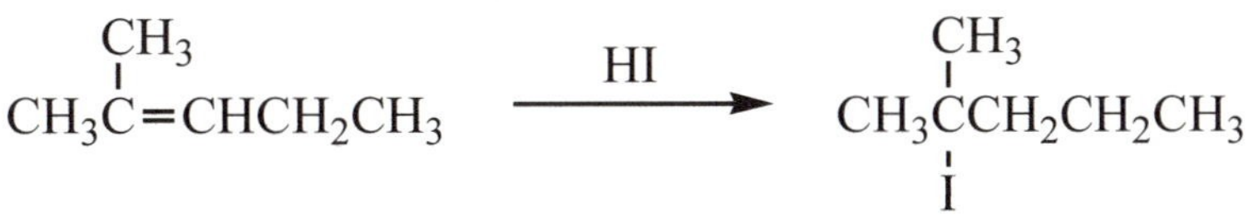   $CH_3\overset{\overset{\displaystyle CH_3}{|}}{C}=CHCH_2CH_3$   $\xrightarrow{\text{HI}}$   $CH_3\overset{\overset{\displaystyle CH_3}{|}}{\underset{\underset{\displaystyle I}{|}}{C}}CH_2CH_2CH_3$

(b) 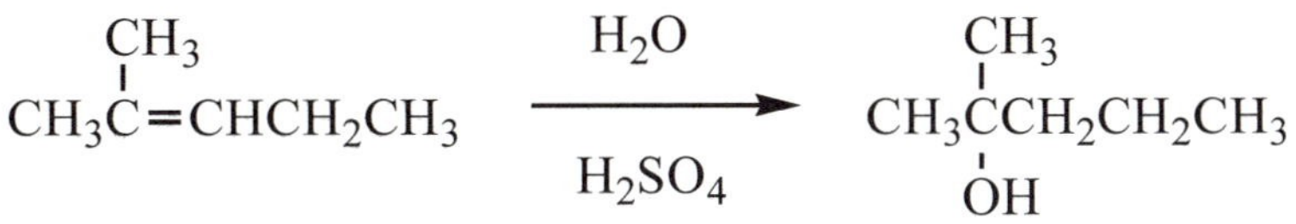   $CH_3\overset{\overset{\displaystyle CH_3}{|}}{C}=CHCH_2CH_3$   $\xrightarrow[\text{H}_2\text{SO}_4]{\text{H}_2\text{O}}$   $CH_3\overset{\overset{\displaystyle CH_3}{|}}{\underset{\underset{\displaystyle OH}{|}}{C}}CH_2CH_2CH_3$

**The first step in the reaction is the protonation of the alkene to generate a carbocation.  The reaction path that produces the more stable carbocation occurs at a faster rate, thus producing the observed regioselectivity.**

**5.18**   *The addition of bromine and chlorine to cycloalkenes is stereoselective.   Predict the stereochemistry of the product formed in each reaction, and account for your predicted stereoselectivity:*

**Halogens undergo anti addition to alkenes.  In cycloalkenes, the anti addition results in *trans* dihalides.**

(a)

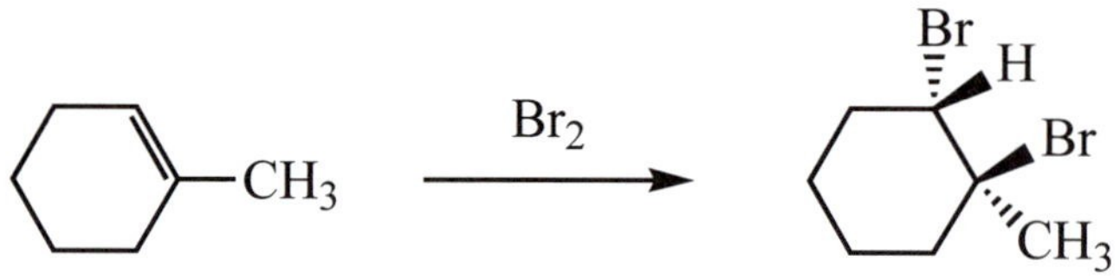

(b)

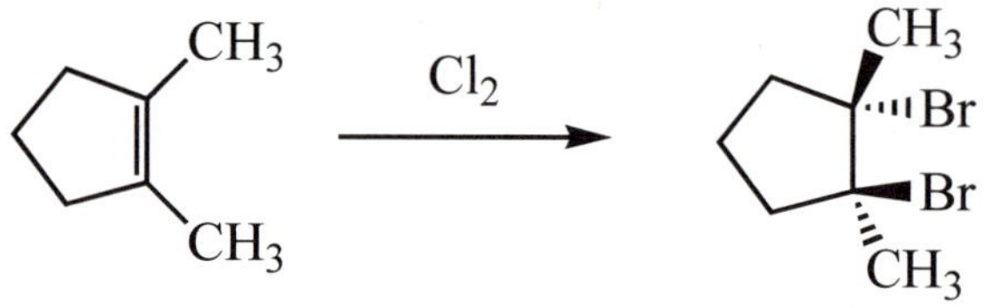

**5.19** *Draw a structural formula for an alkene with the indicated molecular formula that gives the compound shown as the major product.  Note that more than one alkene may give the same compound as the major product.*

(a) $C_5H_{10}$ + $H_2O$ $\xrightarrow{H_2SO_4}$

(b) $C_5H_{10}$ + $Br_2$ $\longrightarrow$

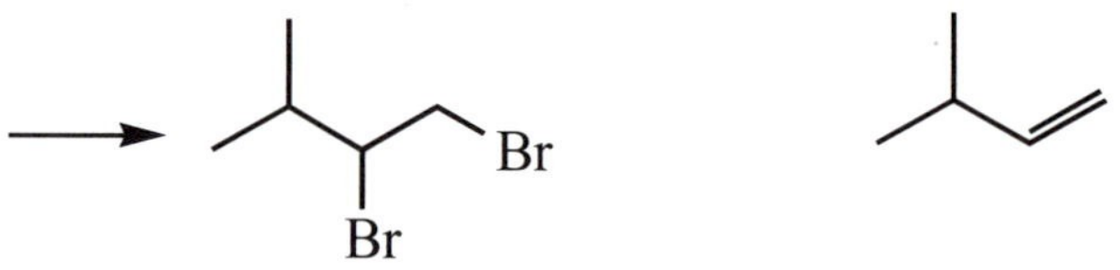

(c) $C_7H_{12}$ + HCl $\longrightarrow$

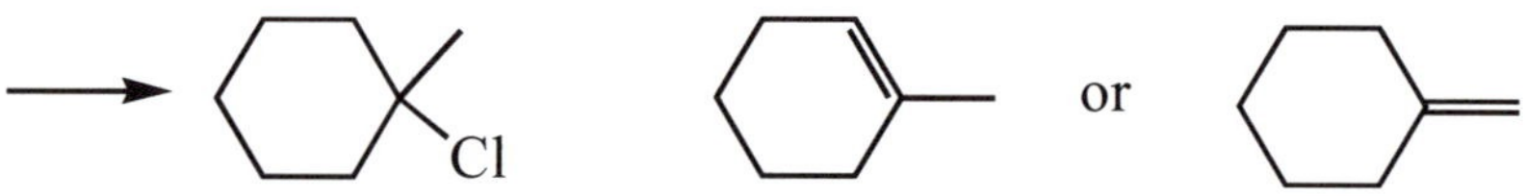

**5.20** *Draw the structural formula for an alkene with molecular formula $C_5H_{10}$ that reacts with $Br_2$ to give each product:*

(a)

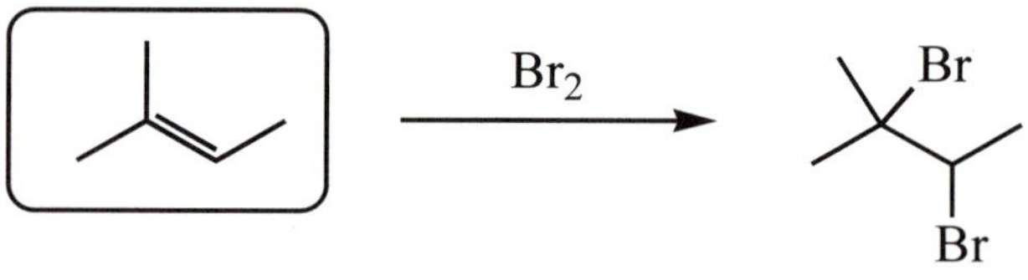

(b)

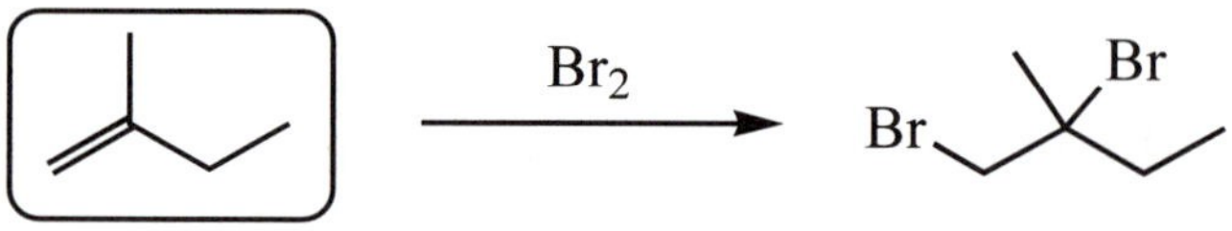

(c)

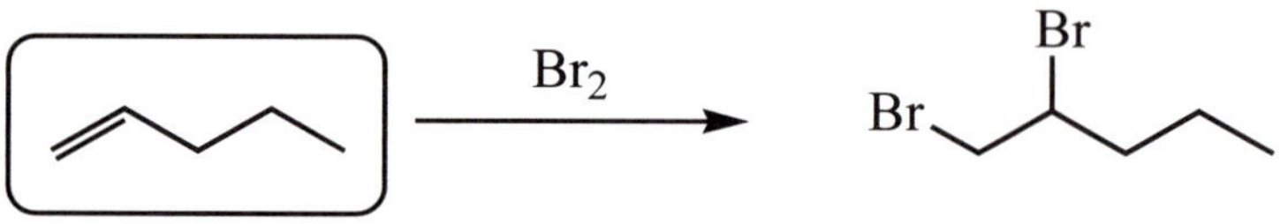

**5.21**   *Draw the structural formula for a cycloalkene with molecular formula C$_6$H$_{10}$ that reacts with Cl$_2$ to give each compound.*

**Chlorine undergoes an anti-addition to alkenes.**

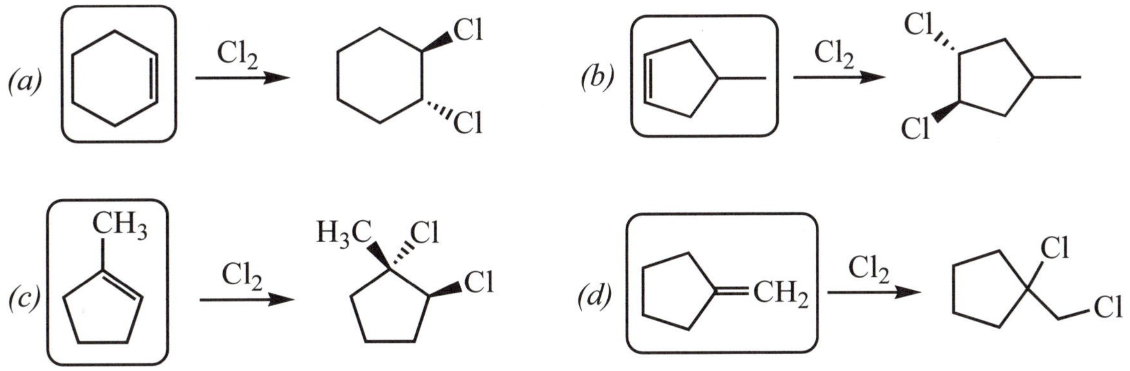

(a)

(b)

(c)

(d)

**5.22**   *Draw the structural formula for an alkene with molecular formula C$_5$H$_{10}$ that reacts with HCl to give the indicated chloroalkane as the major product:*

(a)

(b)

(c)

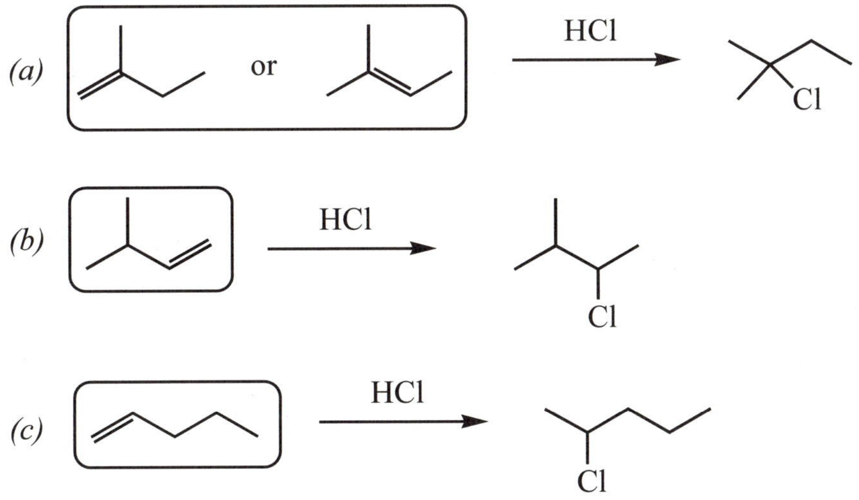

**For part (a), either alkene shown gives 2-chloro-2-methylbutane as the major product as predicted by Markovnikov's rules.  For part (b), 2-methyl-2-butene would not work because the major product predicted by Markovnikov's rules would be 2-chloro-2-methylbutane.  For part (c), *cis* or *trans*-2-butene would not be good choices because 2-chloropentane and 3-chloropentane would be formed in about the same amount.**

**5.23** *Draw the structural formula of an alkene that undergoes acid-catalyzed hydration to give the indicated alcohol as the major product. More than one alkene may give each compound as the major product.*

(a) $\boxed{CH_3CH_2CH{=}CHCH_2CH_3}$ $\xrightarrow[\text{H}_2\text{SO}_4]{\text{H}_2\text{O}}$ $CH_3CH_2\overset{\overset{\displaystyle OH}{|}}{C}HCH_2CH_2CH_3$

(b) $\xrightarrow[\text{H}_2\text{SO}_4]{\text{H}_2\text{O}}$ (cyclobutane with OH and CH$_3$)

(c) $\boxed{\ H_2C{=}\overset{\overset{\displaystyle CH_3}{|}}{C}CH_2CH_3 \quad \text{or} \quad CH_3\overset{\overset{\displaystyle CH_3}{|}}{C}{=}CHCH_3\ }$ $\xrightarrow[\text{H}_2\text{SO}_4]{\text{H}_2\text{O}}$ $CH_3\overset{\overset{\displaystyle OH}{|}}{\underset{\underset{\displaystyle CH_3}{|}}{C}}CH_2CH_3$

(d) $\boxed{CH_3CH{=}CH_2}$ $\xrightarrow[\text{H}_2\text{SO}_4]{\text{H}_2\text{O}}$ $CH_3\overset{\overset{\displaystyle OH}{|}}{C}HCH_3$

**5.24** *Draw the structural formula of an alkene that undergoes acid-catalyzed hydration to give each alcohol as the major product. More than one alkene may give each compound as the major product.*

(a) $\xrightarrow[\text{H}_3\text{O}^+]{\text{H}_2\text{O}}$ (cyclohexane ring)—OH

(b) $\xrightarrow[\text{H}_3\text{O}^+]{\text{H}_2\text{O}}$ (cyclopentane ring with CH$_3$, OH, CH$_3$)

(c) —CH$_3$ $\xrightarrow[\text{H}_3\text{O}^+]{\text{H}_2\text{O}}$ (cyclohexane ring with OH and CH$_3$)

*(d)*

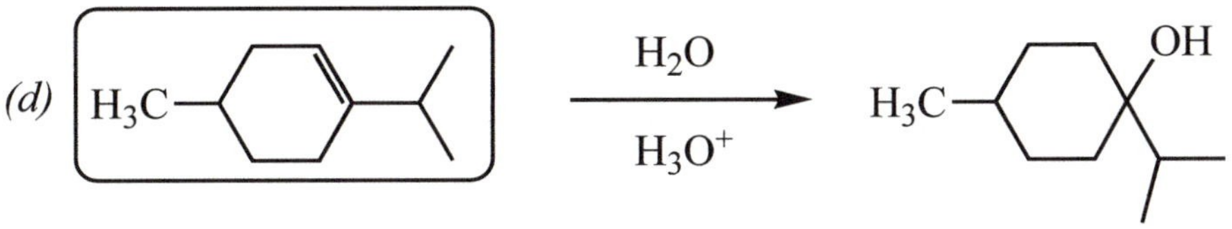

**5.25**   *Terpin is prepared commercially by the acid-catalyzed hydration of limonene:*

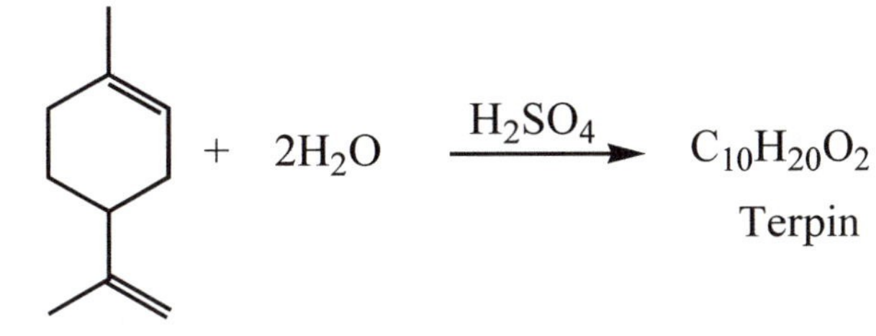

Limonene

*(a) Propose a structural formula for terpin and a mechanism for its formation.*

**Add water to each double bond by regioselective protonation of the double bond
to form a 3° carbocation.  Reaction of each carbocation with water and the loss
of a proton yields terpin.  Note that the reactions do not necessarily proceed in
the order shown.  Both carbon-carbon double bonds will react at similar rates
because in each case, a tertiary carbocation is formed.**

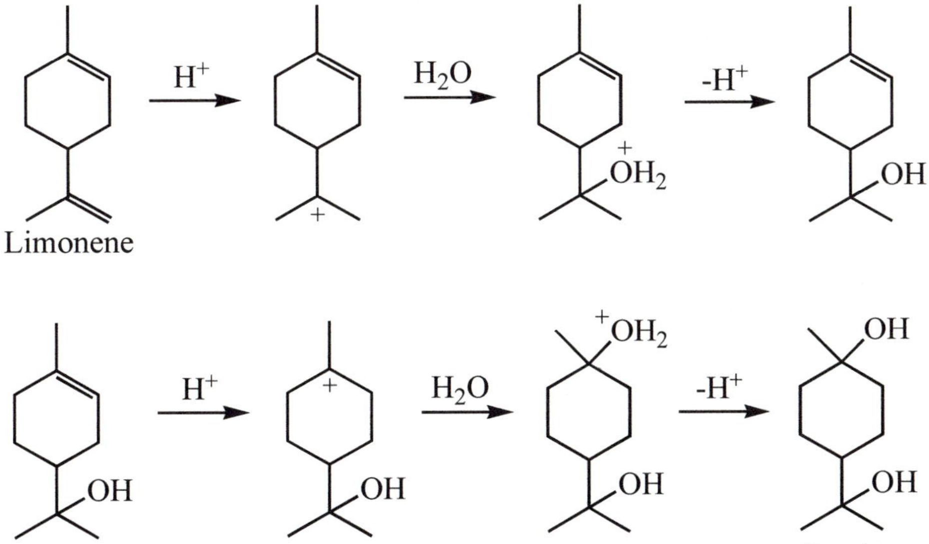

Limonene

Terpin

*(b) How many cis-trans isomers are possible for the structural formula you propose?*

**There are two cis-trans isomers possible:**

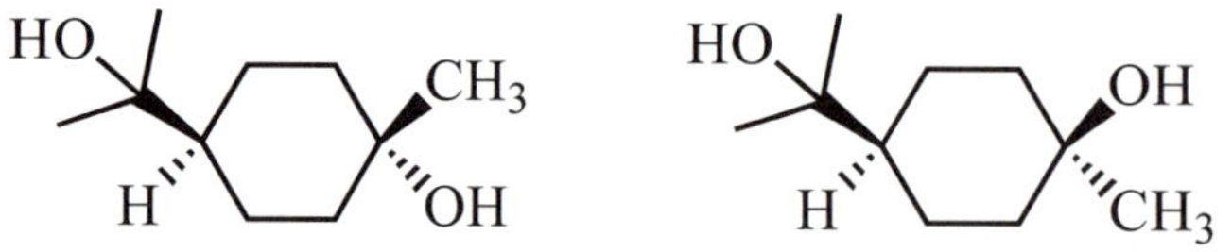

*(c) Terpin hydrate, the isomer in terpin in which the one-carbon and three-carbon substituent are cis to each other, is used as an expectorant in cough medicines. Draw alternative chair conformations for terpin hydrate, and state which of the two is the more stable.*

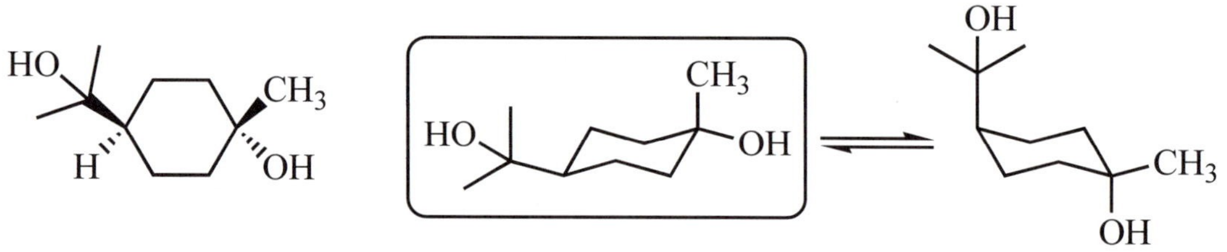

**The circled conformation is more stable because the bulky three-carbon substituent (about the same size as a *tert*-butyl group) is equatorial.**

**5.26**   *Treatment of 2-methylpropene with methanol in the presence of a sulfuric acid catalyst gives tert-butyl methyl ether:*

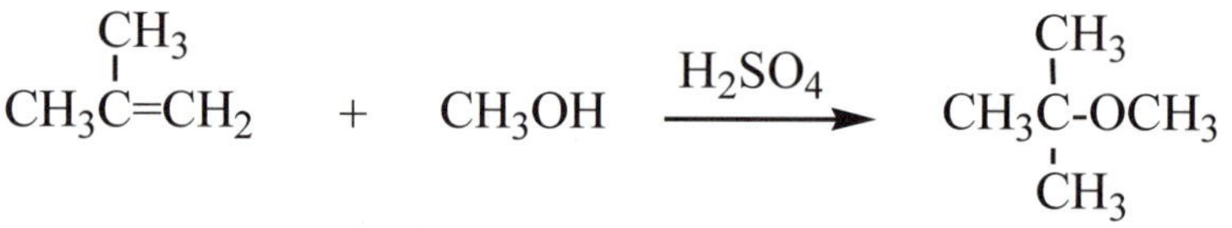

2-Methylpropene          Methanol                    *tert*-Butyl methyl
                                                                        ether

*Propose a mechanism for formation of this ether.*

**Step 1:  Protonation of the alkene gives a stable 3° carbocation intermediate.**

**Step 2: Nucleophilic attack of methanol on the carbocation intermediate yields a protonated ether intermediate.**

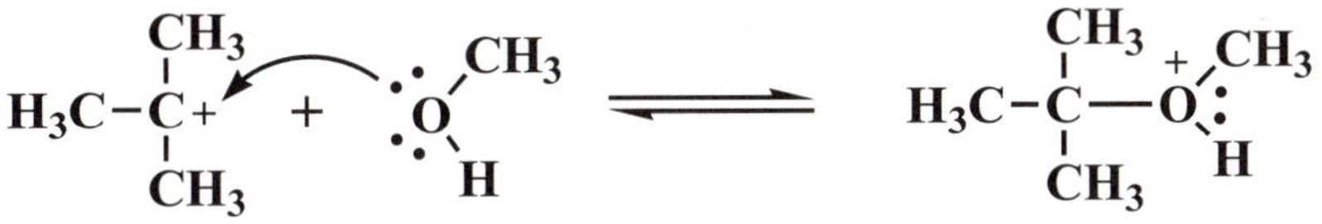

**Step 3: Proton loss by the protonated ether intermediate yields the ether and regenerates the acid catalyst.**

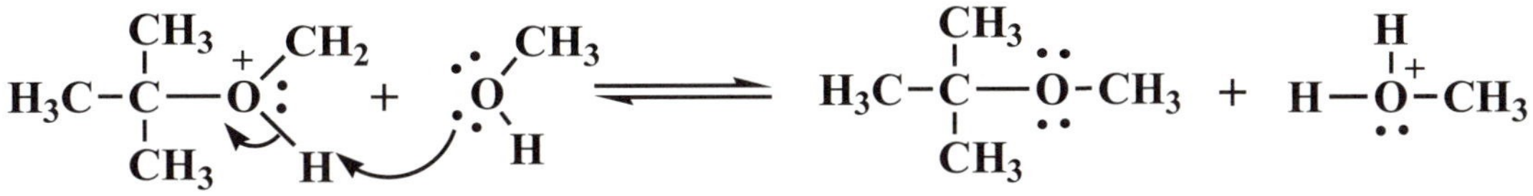

**5.27** *Treatment of 1-methylcyclohexene with methanol in the presence of a sulfuric acid catalyst gives a compound of molecular formula $C_8H_{16}O$. Propose a structural formula for this compound and a mechanism for its formation.*

**The mechanism for this reaction is similar to the one proposed in problem 5.26 except for the different starting alkenes.**

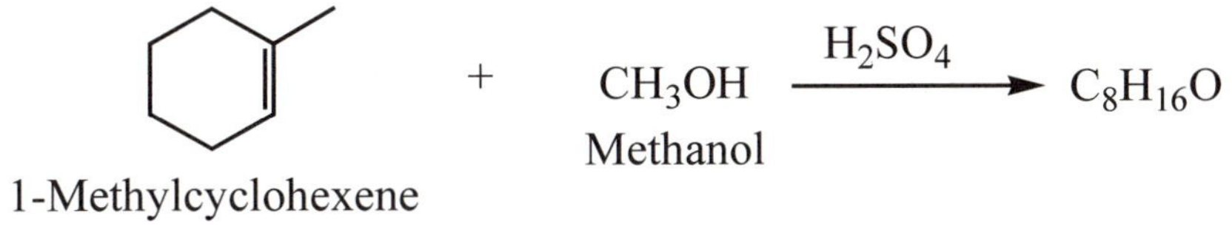

1-Methylcyclohexene + CH$_3$OH $\xrightarrow{\text{H}_2\text{SO}_4}$ C$_8$H$_{16}$O
Methanol

**Step 1:**

**Step 2:** CH$_3$OH +

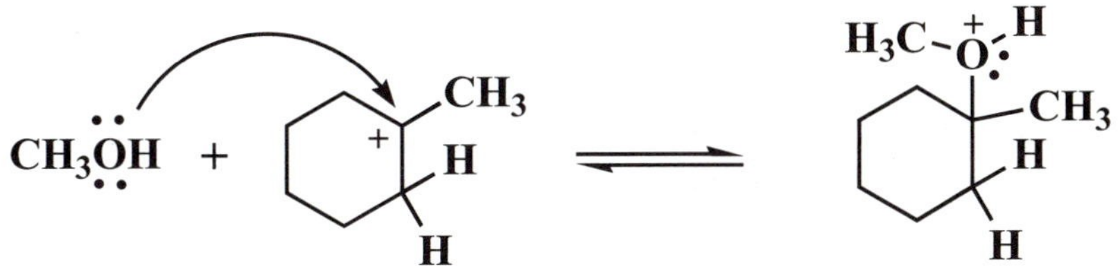

**Step 3:**

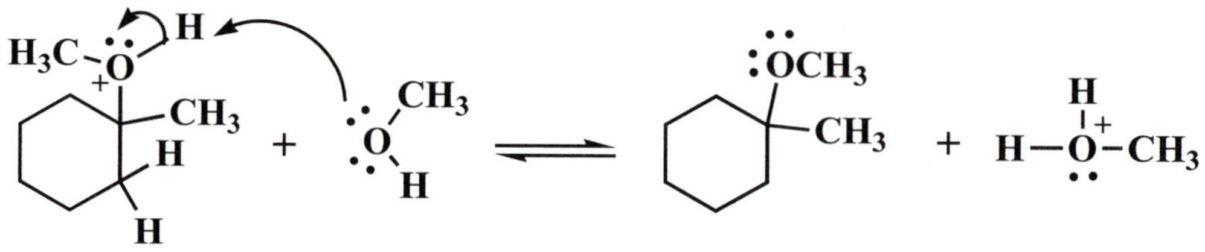

**5.28**   *cis-3-Hexene and trans-3-hexene are different compounds and have different physical and chemical properties. Yet, when treated with $H_2O/H_2SO_4$, each gives the same alcohol. What is the alcohol, and how do you account for the fact that each alkene gives the same one?*

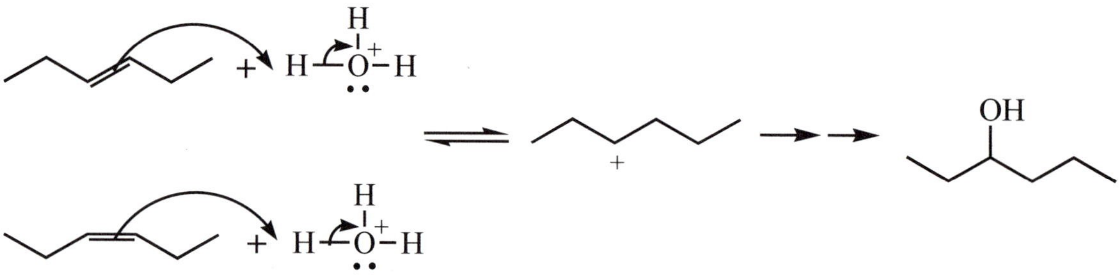

**Both *cis-* and *trans*-3-hexene yield 3-hexanol upon acid-catalyzed hydration because they both form the same carbocation intermediate that leads to 3-hexanol according to the above partial mechanism.**

## Oxidation-Reduction

**5.29**   *Which of these transformations involve oxidation, which involve reduction, and which involve neither oxidation nor reduction?*

(a)  $CH_3\overset{\underset{|}{OH}}{C}HCH_3 \longrightarrow CH_3\overset{\overset{O}{||}}{C}CH_3$   **Oxidation (loss of $H_2$)**

(b)  $CH_3\overset{\underset{|}{OH}}{C}HCH_3 \longrightarrow CH_3CH=CH_2$   **Neither oxidation or reduction**
**There is a loss of O, but gain of H.**

(c)  $CH_3CH=CH_2 \longrightarrow CH_3CH_2CH_3$   **Reduction (gain of $H_2$)**

**5.30**   *Write a balanced equation for the combustion of 2-methylpropene in air to give carbon dioxide and water.  The oxidizing agent is $O_2$, which makes up approximately 20% of air.*

$$CH_3\overset{\underset{|}{CH_3}}{C}=CH_2 \ + \ 6O_2 \longrightarrow 4CO_2 \ + \ 4H_2O$$

**5.31**    *Draw the product formed by treating each alkene with aqueous OsO₄/ROOH:*

(a)

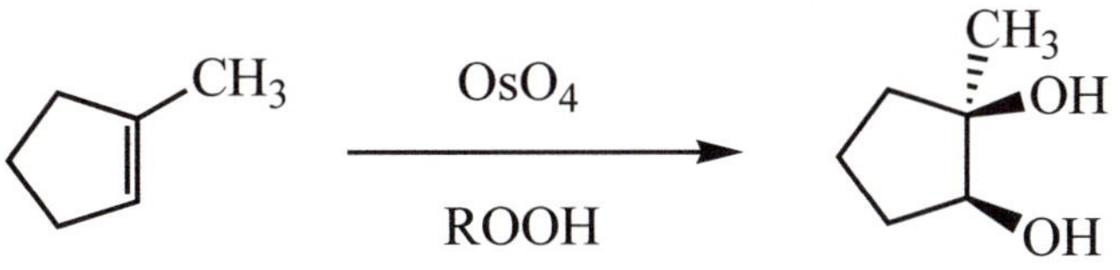

(b)

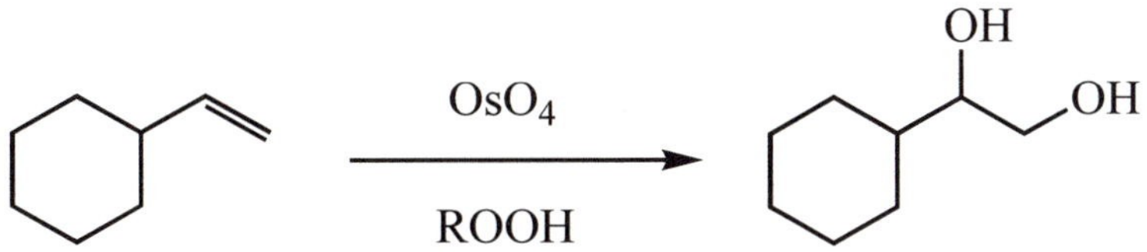

(c)

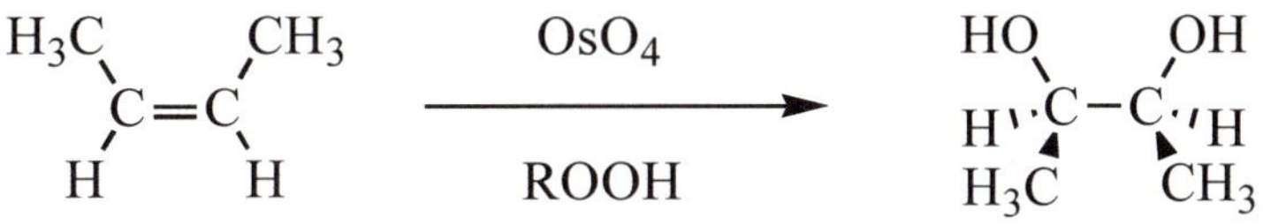

**5.32**    *What alkene, when treated with OsO₄/ROOH, gives each glycol?*

(a)

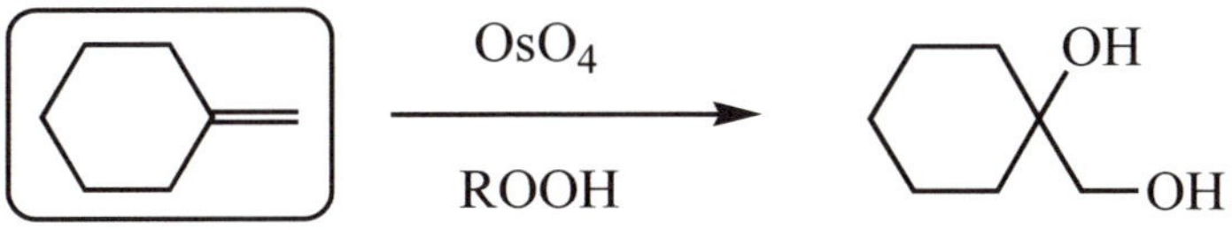

(b)

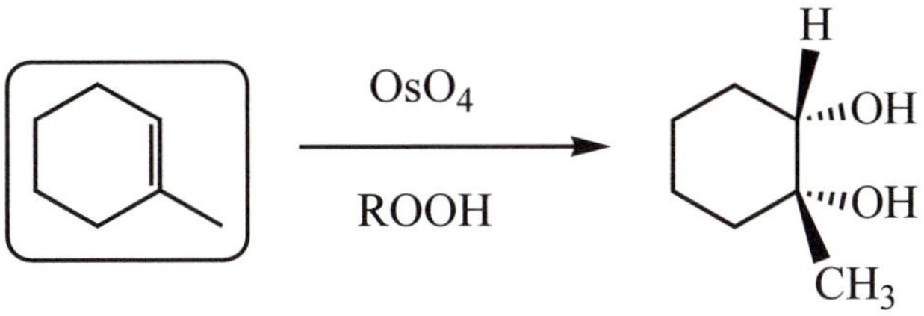

(c)

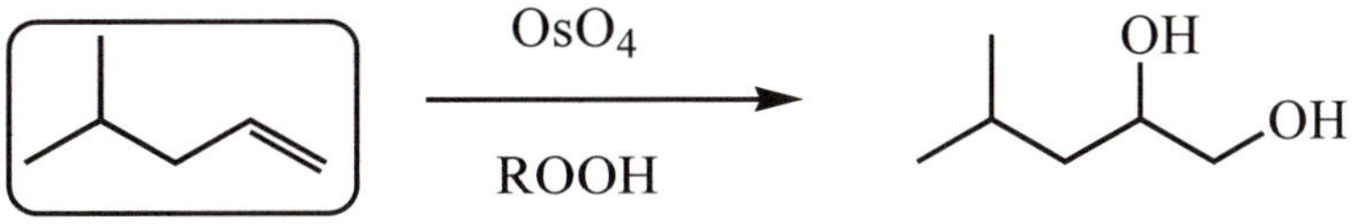

**5.33**    *Draw the product formed by treating each alkene with H₂/Ni:*

(a)

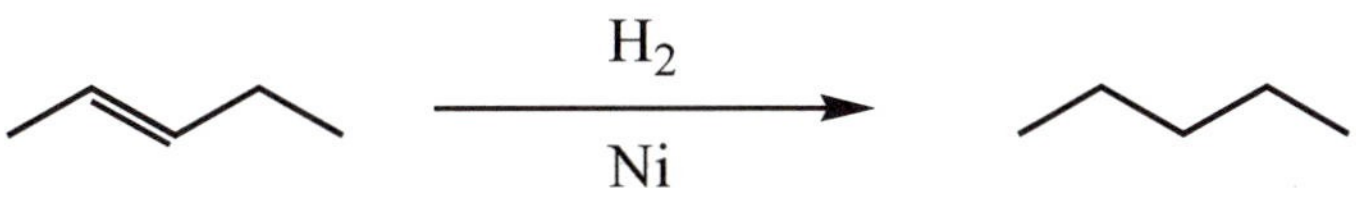

(b)

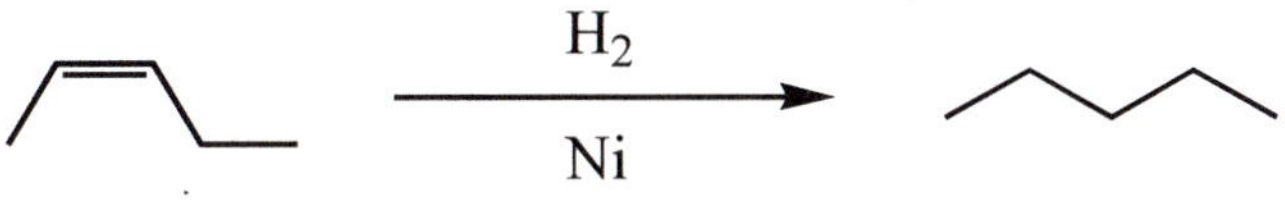

(c)

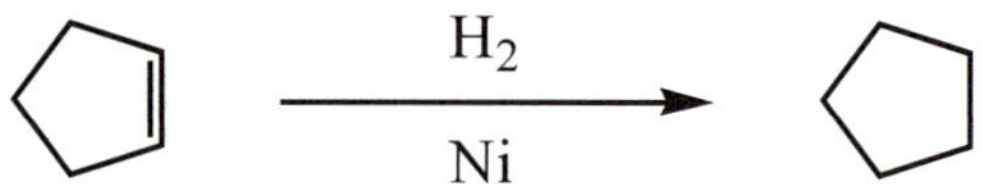

(d)

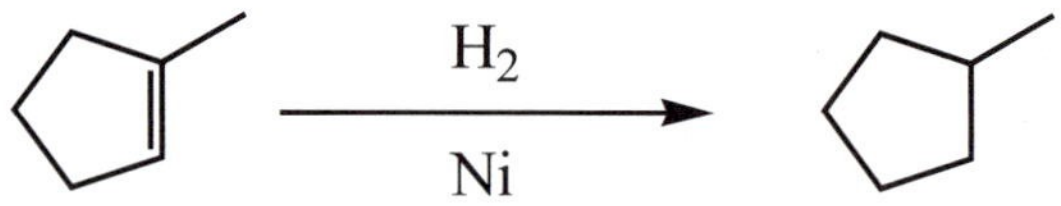

**5.34**  *Hydrocarbon A, C₅H₈, reacts with 2 moles of Br₂ to give 1,2,3,4-tetrabromo-2-methylbutane. What is the structure of hydrocarbon A?*

**The skeleton and location of the double bonds of hydrocarbon A are given through the bromination product.**

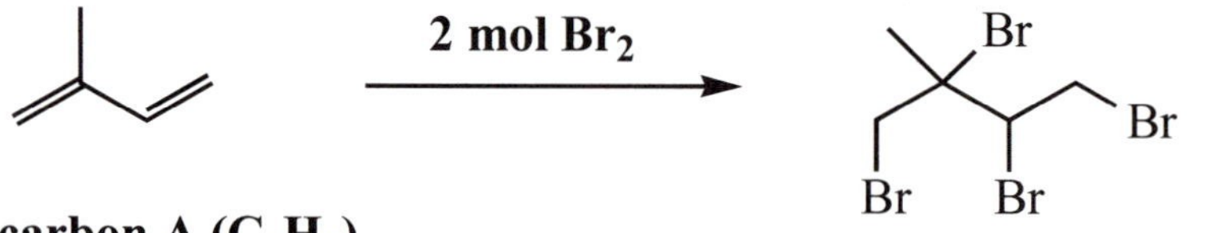

**Hydrocarbon A (C₅H₈)**
**2-Methyl-1,3-butadiene**

**1,2,3,4-Tetrabromo-2-methylbutane**

**5.35**  *Two alkenes, A and B, each have the formula C₅H₁₀. Both react with H₂/Pt and with HBr to give identical products. What are the structures of A and B?*

**The key to solving the problem is recognizing that the skeleton of the alkenes must be the same with only the position of the double bond being different.**

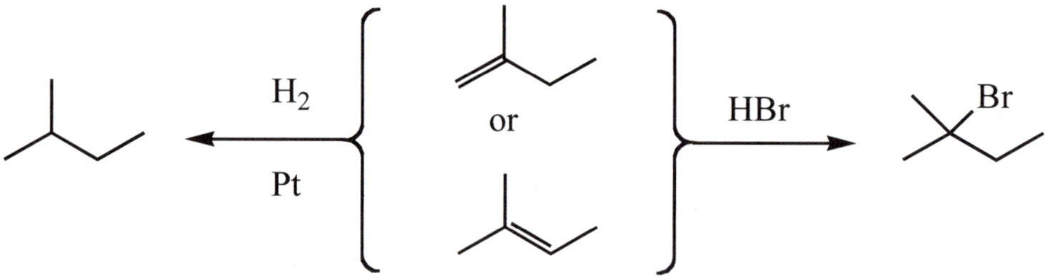

**Structures A and B**

**Synthesis**
**5.36**  *Show how to convert ethylene into these compounds:*

(a)   $H_2C=CH_2$   $\xrightarrow[\textbf{Ni}]{\textbf{H}_2}$   $CH_3CH_3$

(b)   $H_2C=CH_2$  $\xrightarrow[\text{H}_2\text{SO}_4]{\text{H}_2\text{O}}$  $CH_3CH_2OH$

(d)   $H_2C=CH_2$  $\xrightarrow{\text{Br}_2}$  $BrCH_2CH_2Br$

(e)   $H_2C=CH_2$  $\xrightarrow[\text{ROOH}]{\text{OsO}_4}$  $HOCH_2CH_2OH$

(f)   $H_2C=CH_2$  $\xrightarrow{\text{HCl}}$  $CH_3CH_2Cl$

**5.37**   *Show how to convert cyclopentene into these compounds.*

(a)  $\xrightarrow{\text{Br}_2}$

(b)  $\xrightarrow[\text{ROOH}]{\text{OsO}_4}$

(c)  $\xrightarrow[\text{H}_2\text{SO}_4]{\text{H}_2\text{O}}$

(d)  $\xrightarrow{\text{HBr}}$

(e)  $\xrightarrow[\text{Pt}]{\text{H}_2}$

**5.38**   *Show how to convert 1-butene to these compounds.*

(a)  $\xrightarrow[\text{Ni}]{\text{H}_2}$

(b)  $\xrightarrow[\text{H}_2\text{SO}_4]{\text{H}_2\text{O}}$

(c)  $\xrightarrow{\text{HBr}}$

(d) 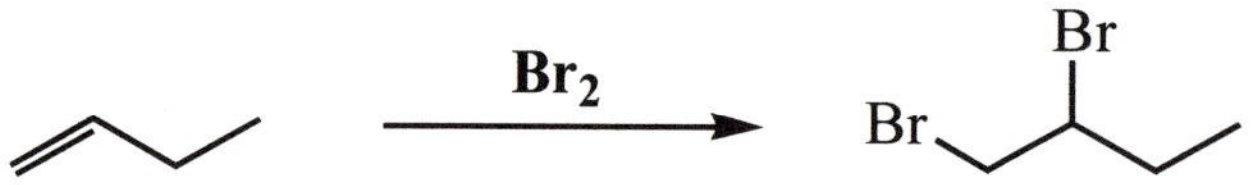    $\xrightarrow{\text{Br}_2}$

**5.39**    *Show how the following compounds can be synthesized in good yields from an alkene.*

(a) 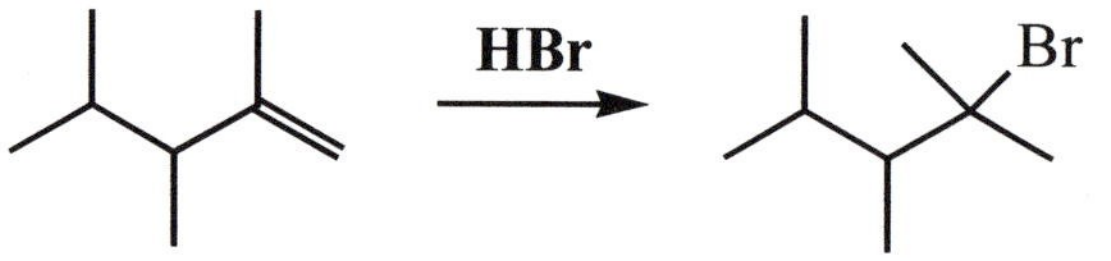    $\xrightarrow{\text{HBr}}$

(b) 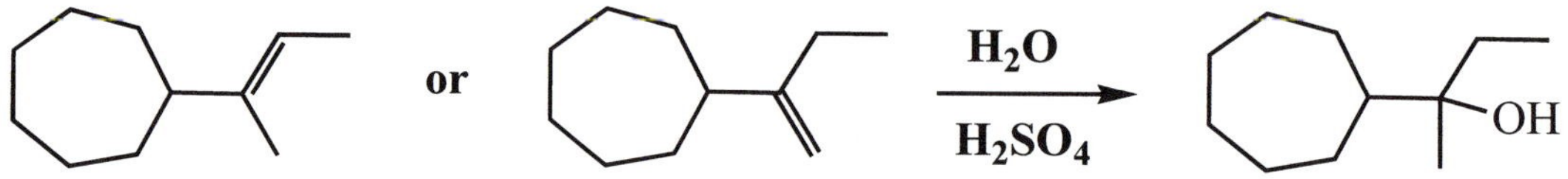       or    $\xrightarrow[\text{H}_2\text{SO}_4]{\text{H}_2\text{O}}$

(c) 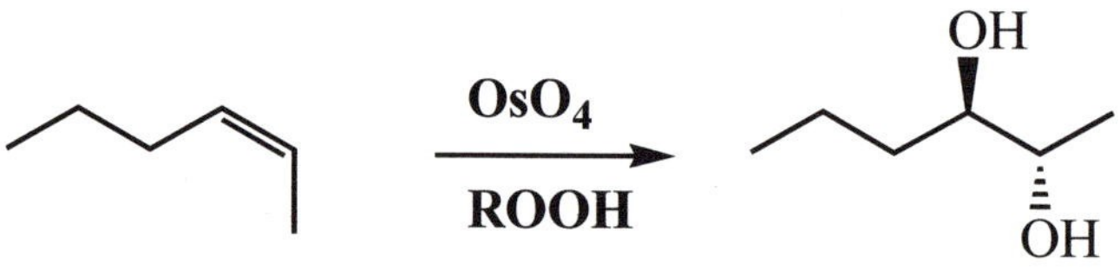    $\xrightarrow[\text{ROOH}]{\text{OsO}_4}$

(d) 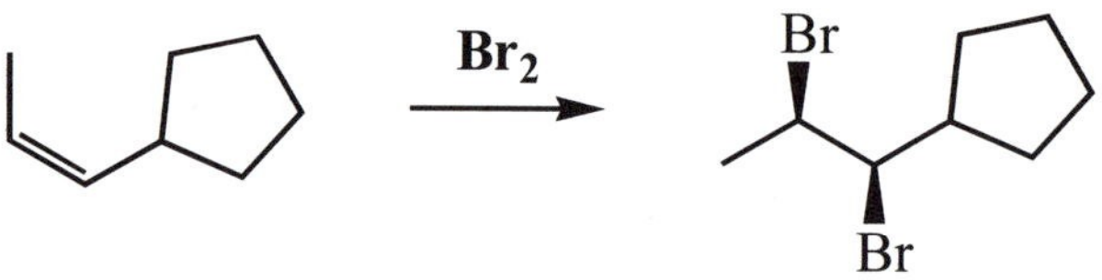    $\xrightarrow{\text{Br}_2}$

**Looking Ahead**

**5.40**    *Each of the following 2° carbocations is more stable than the tertiary carbocation shown: Provide an explanation for each cation's enhanced stability.*

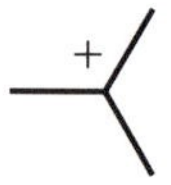

Tertiary
carbocation

(a) 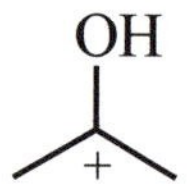          (b)           (c)

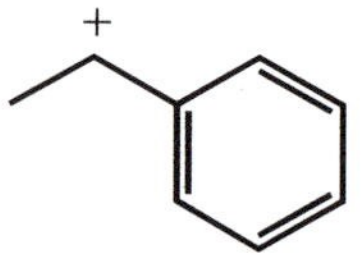

*Provide an explanation for each cation's enhanced stability.*

(a) 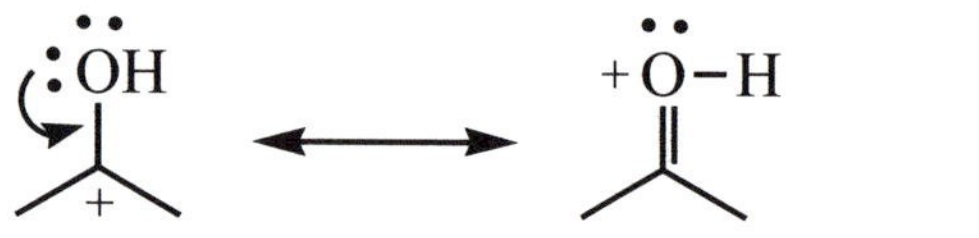    $\longleftrightarrow$
**very stable
resonance structure**

**The positive charge is delocalized between two atoms via two resonance structures. The resonance structure on the right is especially stable because each atom has a filled octet.**

(b)

**The positive charge is delocalized between two atoms via two equivalent resonance structures.**

(c)

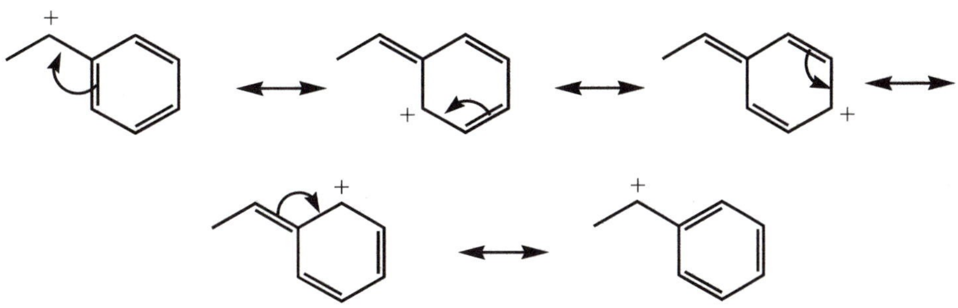

**The positive charge is delocalized over four atoms via five resonance structures.**

**5.41**  *Recall that an alkene possesses a π cloud of electrons above and below the plane of the C=C bond. Any reagent can therefore react with either face of the double bond. Determine whether the reaction of each of the given reagents with the top face of cis-2-butene will produce the same product as the reaction of the same reagent with the bottom face. (Hint: Build molecular models of the products and compare them.)*

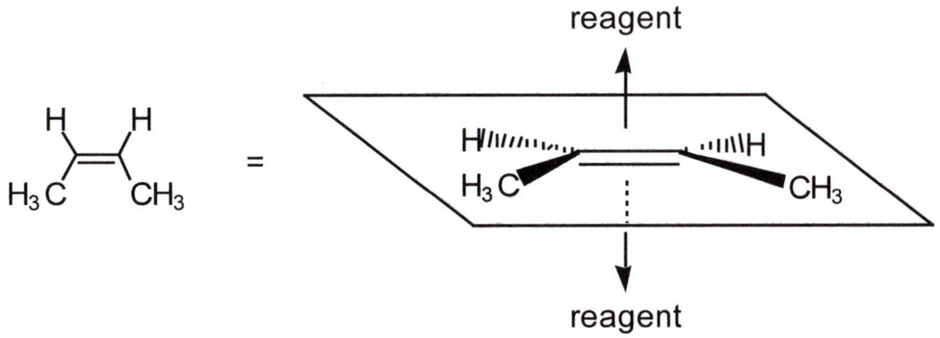

(a)

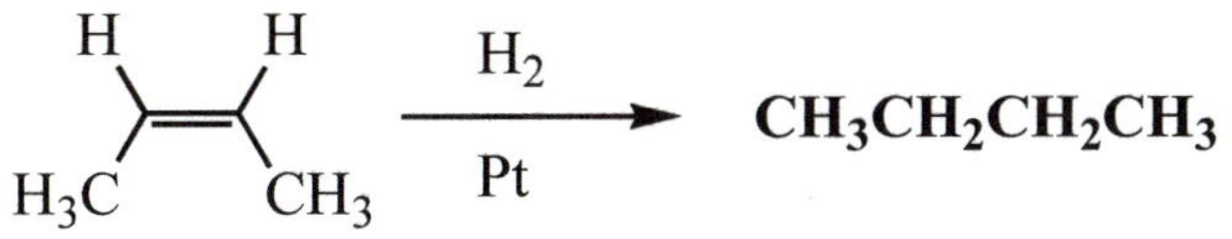

$CH_3CH_2CH_2CH_3$

**Same molecule results from either top or bottom attack**

(b)

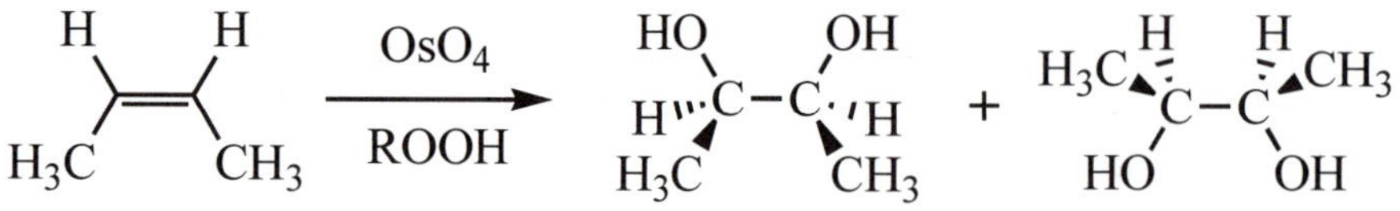

**Same molecule results from
either top or bottom attack**

(c)

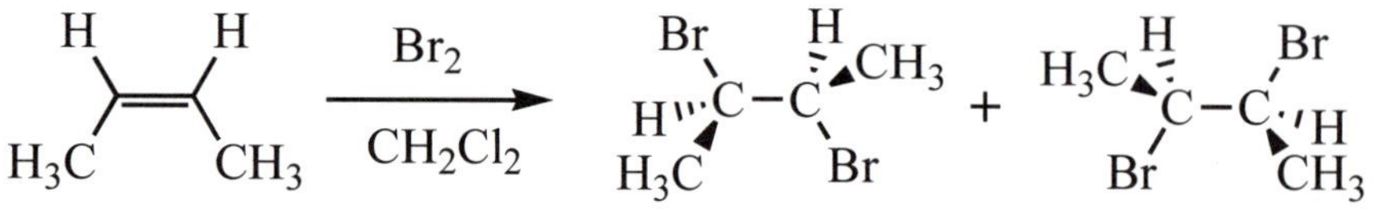

**Two different molcules formed that are non-superposable
mirror images.  These are referred to as <u>enantiomers</u>.**

**In Chapter 6, we will further define stereoisomers and explore the stereoisomers
referred to as enantiomers.  We will also develop the concept of non-
superposable mirror images.**

**5.42**   *Each of the following reactions yield two products in differing amounts:
Draw the products of each reaction and determine which product is favored.*

(a)

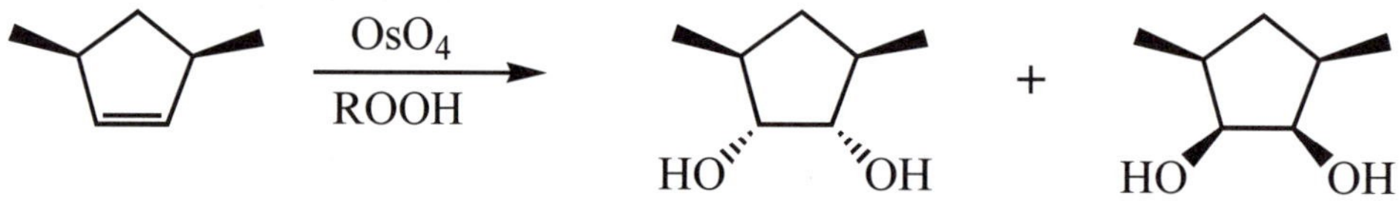

Major product          Minor product

**The major product results from the approach of the OsO$_4$ on the least hindered
side of the molecule (the side opposite the methyl groups).**

(b)

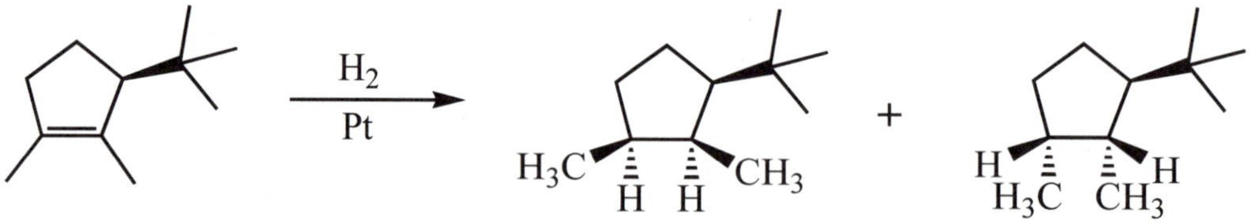

Major product          Minor product

**The major product results from the addition of the hydrogen on the least
hindered side of the molecule (the side opposite the *tert*-butyl group).**

## CHAPTER 6
### Solutions to Problems

**6.1**    *Each molecule has one stereocenter: Draw stereorepresentations of the enantiomers of each.*

(a)

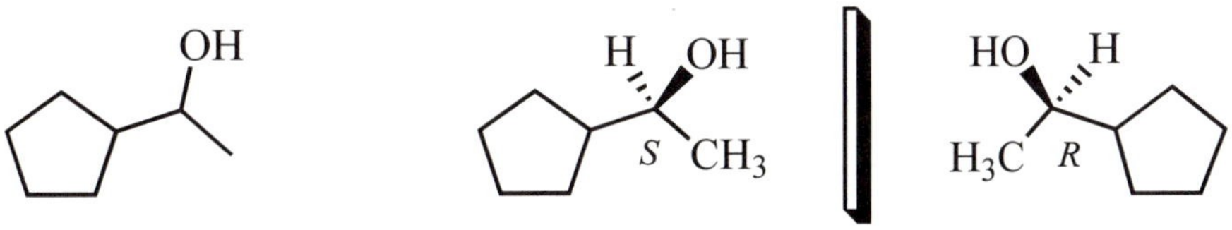

(b)

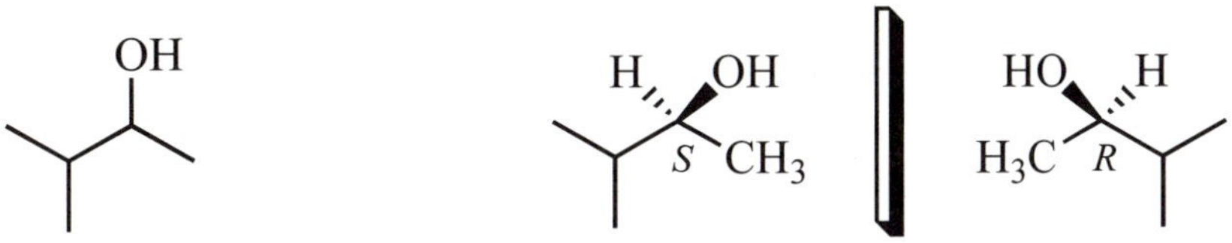

**6.2**    *Assign an R or S configuration to each stereocenter:*

(a)

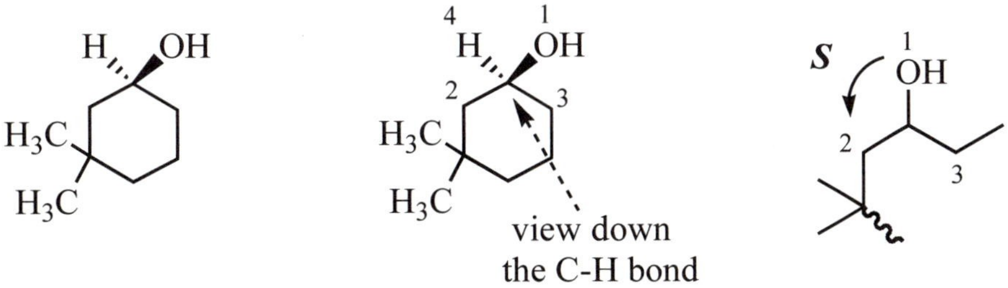

view down
the C-H bond

(b)

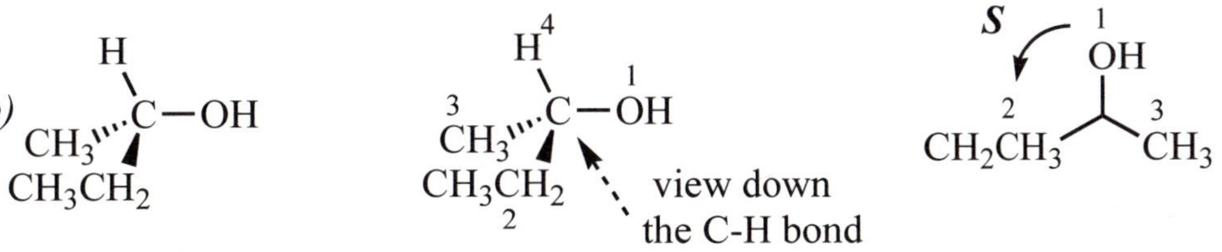

view down
the C-H bond

(c)

view down
the C-H bond

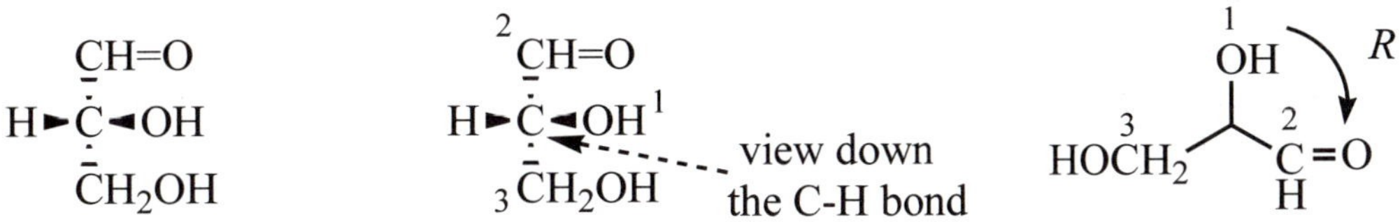

**6.3**    *Following are stereorepresentations of the four stereoisomers of 3-chloro-2-butanol:*

(1)                (2)                (3)                (4)

*(a) Which compounds are enantiomers?*

**Compounds (1) and (3) are enantiomers.**
**Compounds (2) and (4) are enantiomers.**

*(b) Which compounds are diastereomers?*

**Compounds (1) and (2) are diastereomers.**
**Compounds (1) and (4) are diastereomers.**
**Compounds (2) and (3) are diastereomers.**
**Compounds (3) and (4) are diastereomers.**

**6.4**    *Following are four Newman projection formulas for tartaric acid:*

(1)              (2)              (3)              (4)

*(a) Which represent the same compound?*

**Compounds (1) and (4) are enantiomers having the configuration of (2R,3R) for (1) and (2S,3S) for (2).**

**Compounds (2) and (3) are the same compound having the configuration of (2R,3S).**

*(b) Which represent enantiomers?*

**Enantiomers are stereoisomers that are non-superposable mirror images.**
**Compounds (1) and (4) are enantiomers and fit the above criteria.**

*(c)  Which represent(s) meso tartaric acid?*

**Compounds (2) and (3) represent meso tartaric acid.  They have the
configuration (2R,3S) and possess an internal mirror plane of symmetry with
two stereocenters that are mirror images of each other.**

**6.5**    *How many stereoisomers exist for 1,3-cyclopentanediol?*

**1,3-Cyclopentanediol has three stereoisomers possible.  The trans isomers exist as a
pair of enantiomers and the cis isomer is a meso compound.**

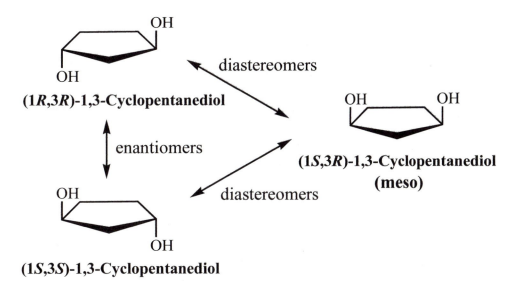

**6.6**    *How many stereoisomers exist for 1,4-cyclohexanediol?*

**1,4-Cyclohexanediol has two stereoisomers.  Although 1,4-cyclohexanediol is an
achiral molecule with no carbon stereocenters , the ring does allow for cis-trans
isomerism.  Each molecule is achiral because it possesses a mirror plane of
symmetry.  In the following figure, the mirror plane in each molecule is
perpendicular to the plane of the paper and through the hydroxyls.**

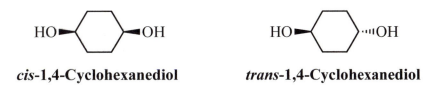

*cis*-1,4-Cyclohexanediol          *trans*-1,4-Cyclohexanediol

**6.7**    *The specific rotation of progesterone, a female sex hormone (Table 17.3), is +172°,*
*measured at 20°C. Calculate the observed rotation for a solution prepared by dissolving*
*40 mg of progesterone in 100 mL of dioxane and placing it in a sample tube 1.00 dm*
*long.*

**The concentration of progesterone, expressed in grams per milliliter is:**
**40 mg/100 mL = 0.04 g/100 mL**

$$\text{specific rotation} = \frac{\text{observed rotation (degrees)}}{\text{length (dm)} \times \text{concentration (g / 100 mL)}}$$

**Rearranging this formula to solve for observed rotation gives:**

**observed rotation (degrees) =**
**specific rotation × length (dm) × concentration (g / 100 mL)**

**Plugging in the experimental values gives the final answer:**

**observed rotation (degrees) = +172û× 1.00 dm × 0.040 = + 6.9û**

**6.8**    *Define the term stereoisomer. Name four types of stereoisomers.*

**Stereoisomers have the same molecular formula and the same order of attachment
of atoms in their molecules but different three-dimensional orientations of their
atoms in space. The four types of stereoisomers include enantiomers (chiral
stereoisomers that are not superposable mirror images), diastereomers
(stereoisomers that are not non-superposable mirror images), cis-trans isomers
(stereoisomers that differ by orientation of atoms about a ring or double bond), and
meso isomers (an achiral isomer that has two or more stereocenters).**

**6.9**    *In what way are constitutional isomers different from stereoisomers? In what way are
they the same?*

**Constitutional and stereoisomers have the <u>same</u> molecular formula but are different
in that stereoisomers have the same atom connectivity, but have different spatial
orientation of their atoms.**

**6.10**   *Which of these objects are chiral (assume that there is no label or other identifying
mark)?*

**An object, like a molecule, is chiral if it is not superposable upon its mirror image.**

*(a) A pair of scissors*

**Chiral. You can buy left- and right-handed scissors. Try cutting paper with a left-handed scissors using your right hand!**

*(b) A tennis ball*

**Not chiral. A tennis ball has two mirror planes because of the seams.**

*(c) A paper clip*

**Not chiral. A paper clip has a horizontal mirror plane.**

*(d) A beaker*

**Not chiral. A beaker has a vertical mirror plane bisecting it through the spout.**

*(e) The swirl created in water as it drains out of a sink or bathtub*

**Chiral. Spirals can be clockwise and counter clockwise. Their mirror images are not superposable and they lack mirror planes.**

**6.11**   *Think about the helical coil of a telephone cord or the spiral binding on a notebook, and suppose that you view the spiral from one end and find that it has a left-handed twist. If you view the same spiral from the other end, does it have a right-handed twist or a left-handed twist from that end as well?*

**A spiral with a left-handed twist will have a left-handed twist when viewed from either end.**

**6.12**   *Next time you have the opportunity to view a collection of augers or other seashells that have a helical twist, study the chirality of their twists. Do you find an equal number of left-handed and right-handed augers, or, for example, do they all have the same handedness? What about the handedness of augers compared with that of other spiral shells?*

**This question was just meant to encourage you to think about chirality in nature. Please share your observations with your friends and family. Share your observations with other members of the class and come to a collective conclusion.**

**6.13**   *Next time you have an opportunity to examine any of the seemingly endless varieties of spiral pasta (rotini, fusilli, radiatori, tortiglione), examine their twist. Do the twists of*

*any one kind all have a right-handed twist, do they all have a left-handed twist, or are
they a racemic mixture?*

**This question was just meant to encourage you to think about chirality in everyday
objects.  Please share your observations with your friends and family. Share your
observations with other members of the class and come to a collective conclusion.**

**6.14**   *One reason we can be sure that sp³-hybridized carbon atoms are tetrahedral is the
number of stereoisomers that can exist for different organic compounds.*

(a) *How many stereoisomers are possible for CHCl₃, CH₂Cl₂,  and CHBrClF if the four
bonds to carbon have a tetrahedral geometry?*

**The only way a tetrahedral carbon atom can have a stereocenter is if it has four
different substituents bonded to it.  Only CHBrClF has four different
substituents bonded to it, and it is the only one of the choices that has a
stereoisomer (in this case, an enantiomer).  Prove this with molecular models.**

**non-superposable mirror images (enantiomers)**

(b) *How many stereoisomers are possible for each compound if the four bonds to the
carbon have a square planar geometry?*

**A carbon atom with square planar geometry can have cis-trans isomers if it has
at least two or more different groups bonded to it.  CH₂Cl₂ and CHBrClF satisfy
this criterion.**

*cis*            *trans*            *cis (Br-Cl)*          *trans (Cl-Br)*

**6.15**   *Which of the following statements is true?*

(a) *Enantiomers are always chiral.*  **True**
(b) *A diastereomer of a chiral molecule must also be chiral.*  **False**
(c) *A molecule that possesses an internal plane of symmetry can never be chiral.*  **True**
(d) *An achiral molecule will always have an enantiomer.*  **False**
(e) *An achiral molecule will always have a diastereomer.*  **False**
(f) *A chiral molecule will always have an enantiomer.*  **True**
(g) *A chiral molecule will always have a diastereomer.*  **False**

**Enantiomers**

**6.16**  *Which compounds contain stereocenters?*

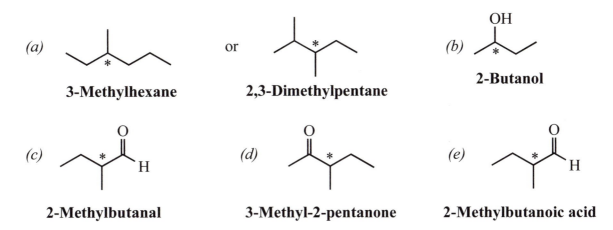

(a)  **2-Chloropentane**

(b)  **3-Chloropentane**

(c)  **3-Chloro-1-pentene**

(d)  **1,2-Dichloropropane**

**2-Chloropentane (a), 3-chloro-1-pentene (c), and 1,2-dichloropentane (d) possess one stereocenter (as indicated with an asterisk) and are therefore chiral.  3-Chloropentane (b) does not possess a stereocenter and has an internal plane of symmetry, and therefore is achiral.**

**6.17**  *Using only C, H, and O, write structural formulas for the lowest-molecular-weight chiral molecule of each of the following compounds:*

(a)  **3-Methylhexane**    or    **2,3-Dimethylpentane**

(b)  **2-Butanol**

(c)  **2-Methylbutanal**

(d)  **3-Methyl-2-pentanone**

(e)  **2-Methylbutanoic acid**

**6.18**  *Which alcohols with molecular formula $C_5H_{12}O$ are chiral?*

**There are eight alcohols with the formula $C_5H_{12}O$, of which three are chiral.  The carbon stereocenters are indicated with an asterisk.**

**Chiral**

**Chiral**

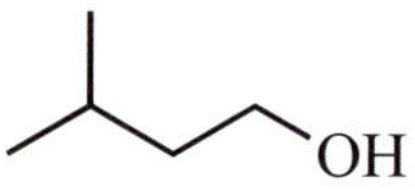

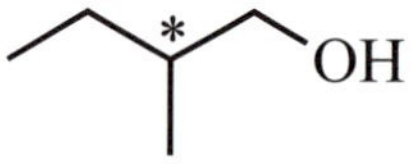

**Chiral**

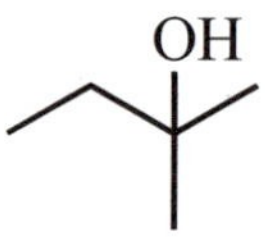

**6.19**    *Which carboxylic acids with molecular formula C₆H₁₂O₂ are chiral?*

**There are eight carboxylic acids with the formula C₆H₁₂O₂, of which three are chiral. The carbon stereocenters are indicated with an asterisk.**

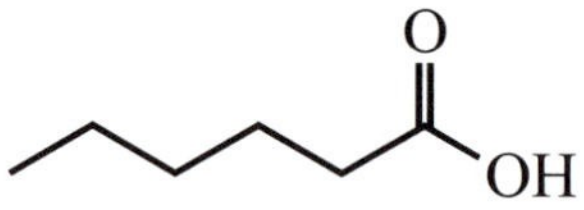

**Achiral** 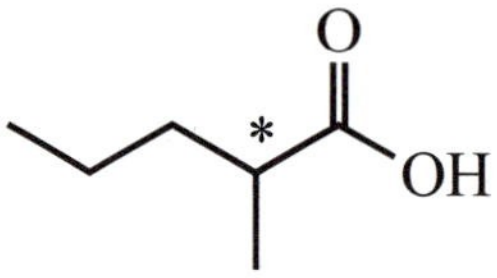          **Chiral** 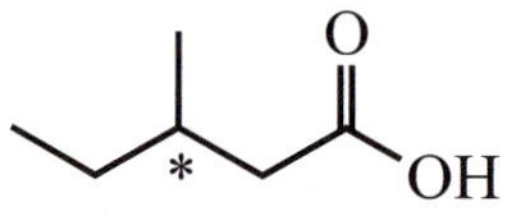          **Chiral**

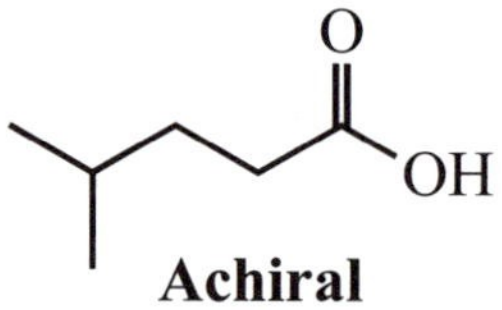

**Chiral**

**Achiral** 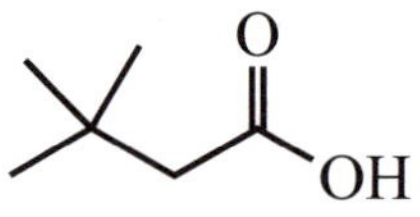          **Achiral** 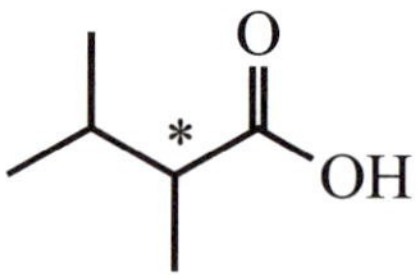          **Chiral**

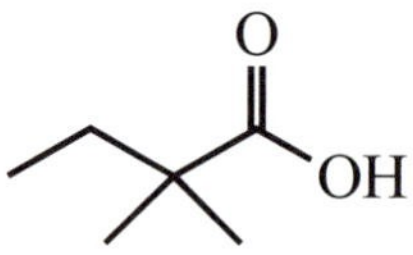

**Achiral** 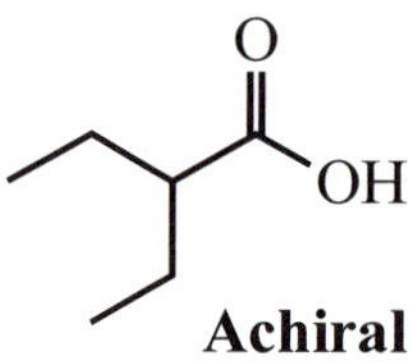          **Achiral**

**6.20**    *Draw the enantiomer for each molecule:*

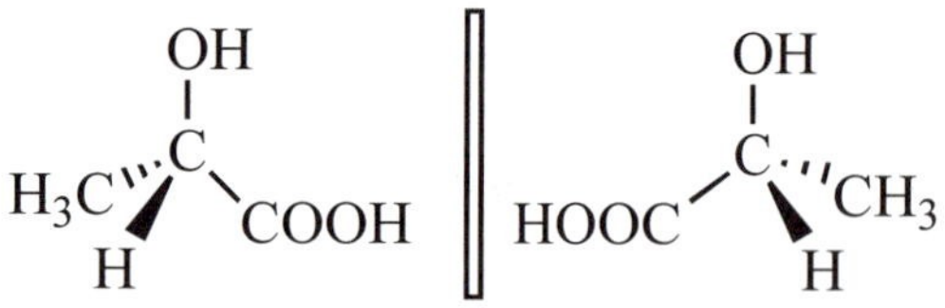

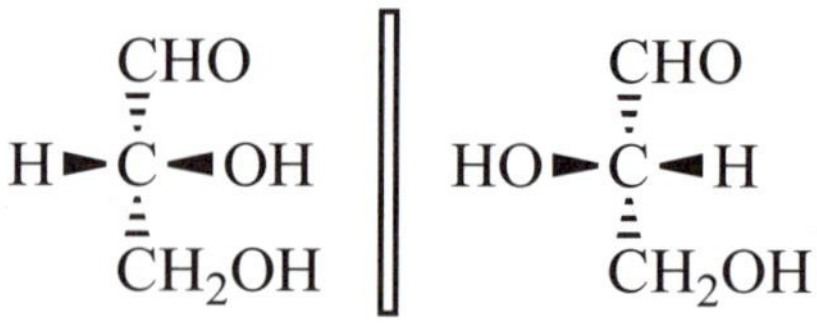

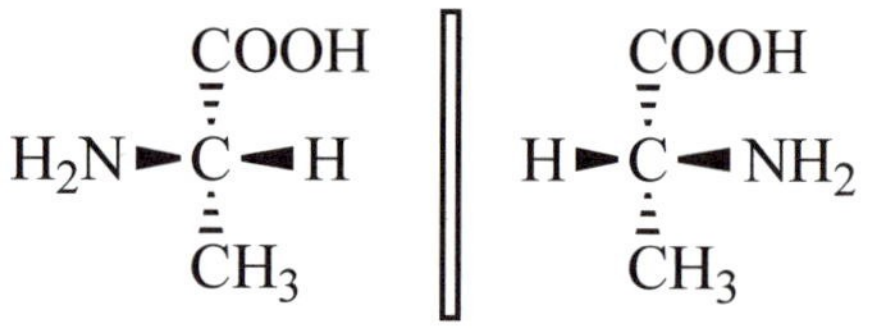

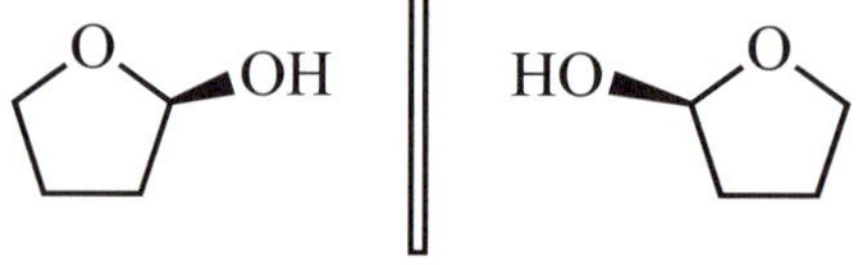

(e)

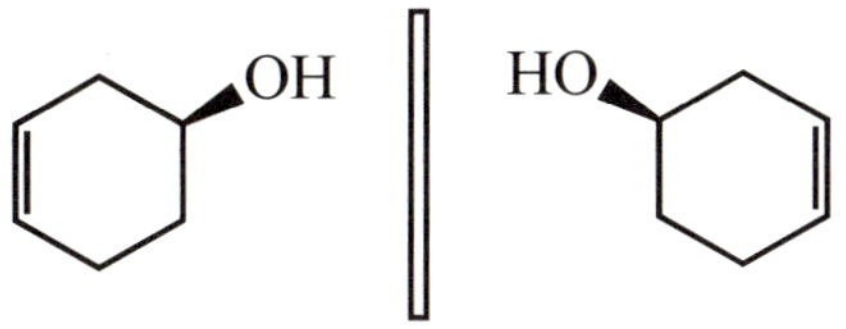

(f)

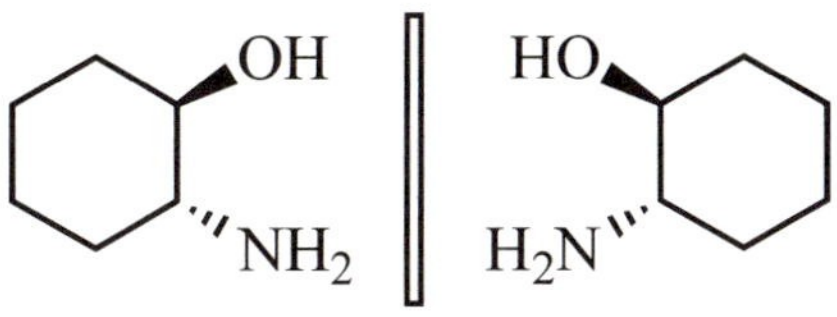

(g)

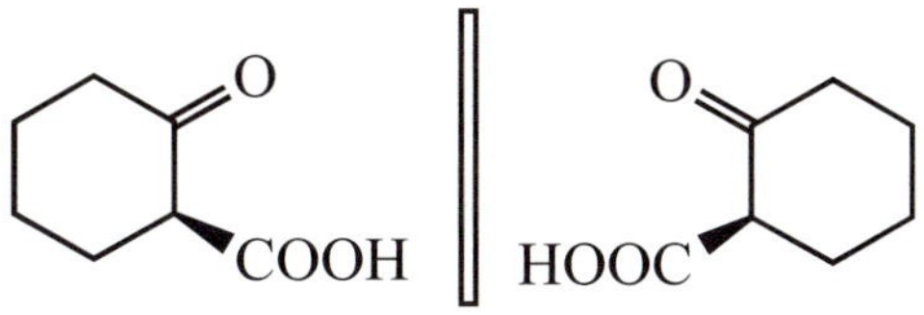

(h)

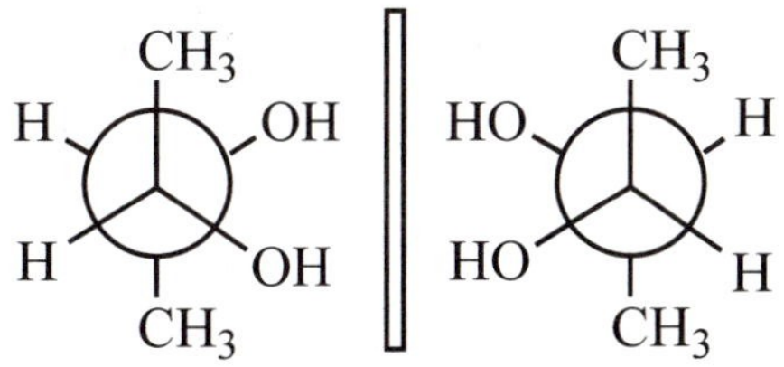

(i)

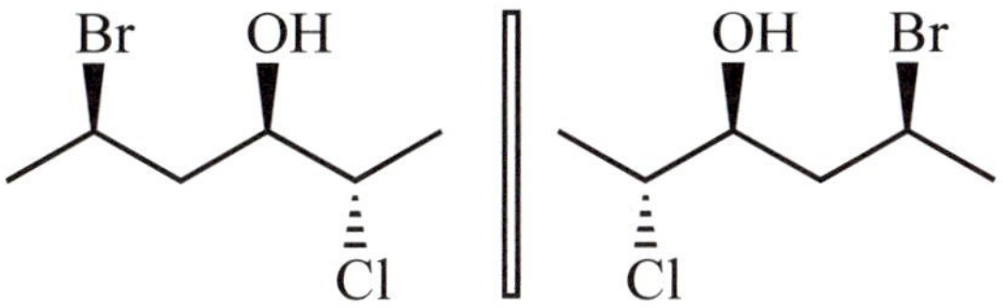

(j)

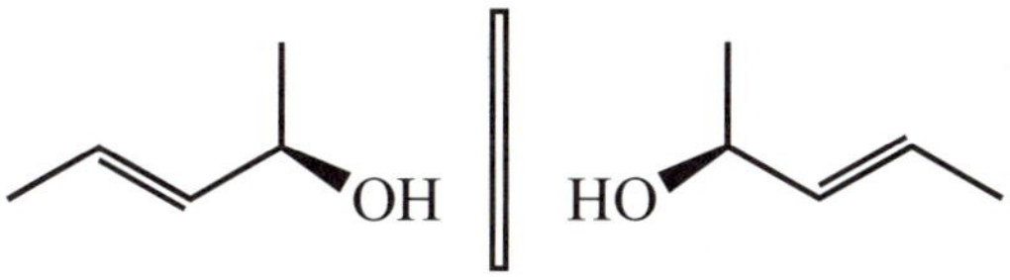

(k)

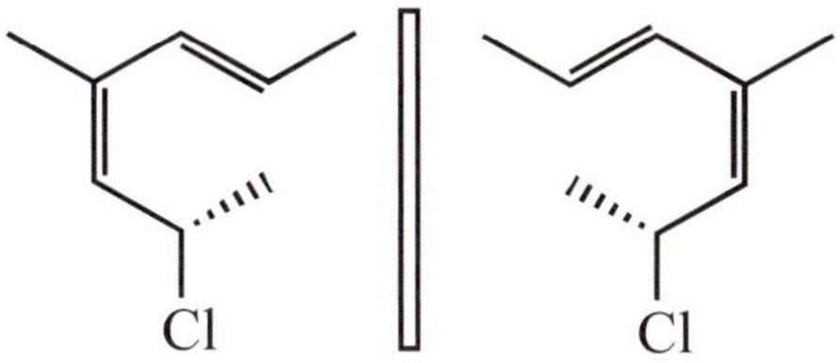

(l)

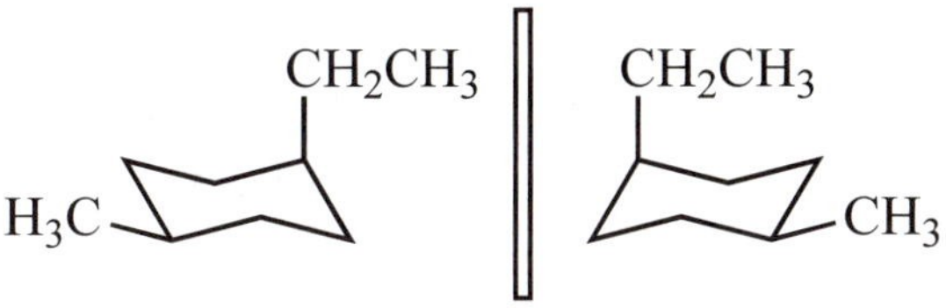

**6.21** *Mark each stereocenter in these molecules with an asterisk (note that not all contain stereocenters):*

(a)

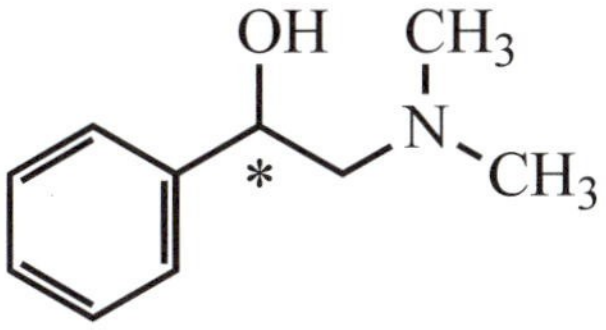

(b)

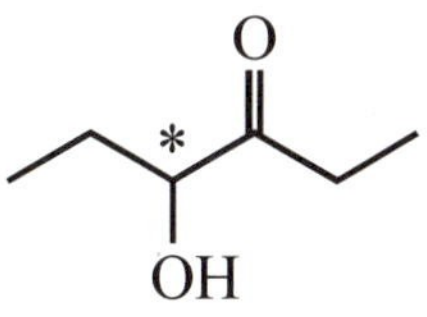

(c)

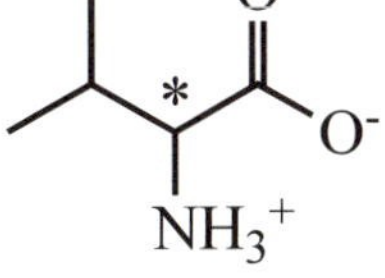

(d)

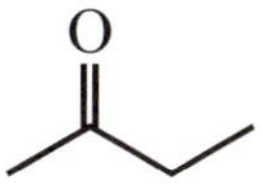

**No stereocenter**

**6.22** *Mark each stereocenter in these molecules with an asterisk (note that not all contain stereocenters):*

(a) HO⌐⌐⌐OH (with OH)
**Achiral**

(b) HO⌐⌐⌐ (with OH, *)
**Chiral**

(c) ⌐⌐⌐ (with OH, *)
**Chiral**

(d) ⌐⌐⌐ (with OH)
**Achiral**

**6.23** *Mark each stereocenter in these molecules with an asterisk (note that not all contain stereocenters):*

(a) OH
**No stereocenter**

(b) OH / *COOH

(c) *COOH / NH₂

(d) O
**No stereocenter**

(e) HO⌐⌐⌐OH / OH
**No stereocenter**

(f) * / OH

(g) HOOC⌐⌐⌐COOH / HO COOH
**No stereocenter**

**6.24** *Following are eight stereorepresentations of lactic acid:*

(a) COOH / C / H OH / CH₃ / **S**

(b) CH₃ / C / HO H / HOOC / **S**

(c) COOH / C / HO CH₃ / H / **S**

(d) CH₃ / C / H COOH / HO / **S**

(e) H►C◄OH / COOH / CH₃ **R**

(f) H►C◄OH / CH₃ / **S** COOH

(g) H₃C►C◄COOH / OH / H **R**

(h) H►C◄COOH / CH₃ / OH **R**

*Take (a) as a reference structure. Which stereorepresentations are identical with (a) and which are mirror images of (a)?*

**Structures (a) thru (d) and (f) are the stereorepresentations of (S)-lactic acid.**

**Structures (e), (g), and (h) are stereorepresentations of (R)-lactic acid and therefore, enantiomers of structure (a).**

## Designation of Configuration: The *R*,*S* Convention

**6.25**   *Assign priorities to the groups in each set:*

(a)  *-H* **(4)**    *-CH₃* **(3)**    *-OH* **(1)**    *-CH₂OH* **(2)**

(b)  *-CH₂CH=CH₂* **(3)**    *-CH=CH₂* **(1)**    *-CH₃* **(4)**   *-CH₂COOH* **(2)**

(c)  *-CH₃* **(3)**   *-H* **(4)**   *-COO⁻* **(2)**   *-NH₃⁺* **(1)**

(d)  *-CH₃* **(4)**    *-CH₂SH* **(2)**    *-NH₃⁺* **(1)**   *-COO⁻* **(3)**

**6.26**   *Which molecules have R configurations?*

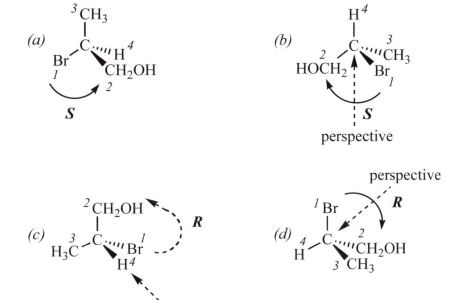

**An easy way to determine stereochemistry with non-ideal perspectives [such as in (c) where the lowest priority group is pointing towards you] is to look down the bond from #4 group to the stereocenter (opposite that of Cahn-Ingold-Prelog convention). Determine *R* or *S* as usual and then report the opposite. So the counterclockwise arrow really describes *R* stereochemistry in this case.**

**Of the four stereorepresentations, (c) and (d) have the *R* configuration.**

**6.27**    *Following are structural formulas for the enantiomers of carvone:*

(-)-Carvone
(Spearmint
oil)

(+)-Carvone
(Caraway and
dillseed oil)

**(R)-(-)-Carvone**          **(S)-(+)-Carvone**

*Each enantiomer has a distinctive odor characteristic of the source from which it can be
isolated. Assign an R or S configuration to the stereocenter in each. How can they have
such different properties when they are so similar in structure?*

**In an achiral environment, enantiomers have the same physical and chemical
properties. But, in chiral environments, enantiomers can behave very differently.
The odor receptors responsible for detecting the carvone smells must be chiral
themselves, therefore, able to physiologically differentiate the enantiomers.**

**6.28**    *Following is a staggered conformation of one of the stereoisomers of 2-butanol:*

*(a) Is this (R)-2-butanol or (S)-2-butanol?*

**The structure drawn is (S)-2-butanol.**

*(b) Draw a Newman projection for this staggered conformation, viewed along the bond
between carbons 2 and 3.*

View along
the $C_2$-$C_3$ bond.

*(c) Draw a Newman projection for one more staggered conformation of this molecule.
Which of your conformations is the more stable? Assume that -OH and -CH$_3$ are
comparable in size.*

**There are two additional staggered conformations of (*S*)-2-butanol:**

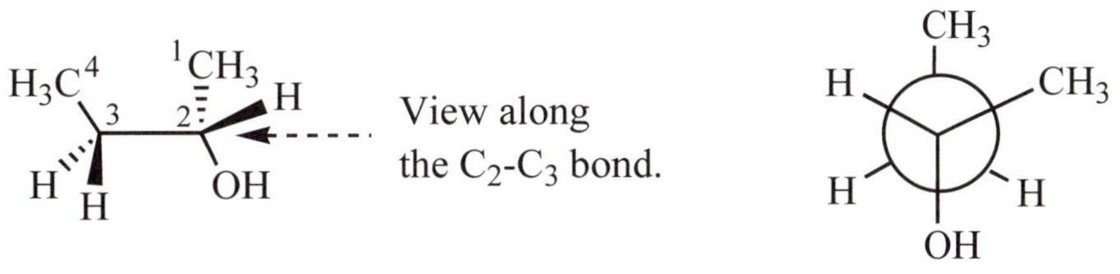

**More stable**

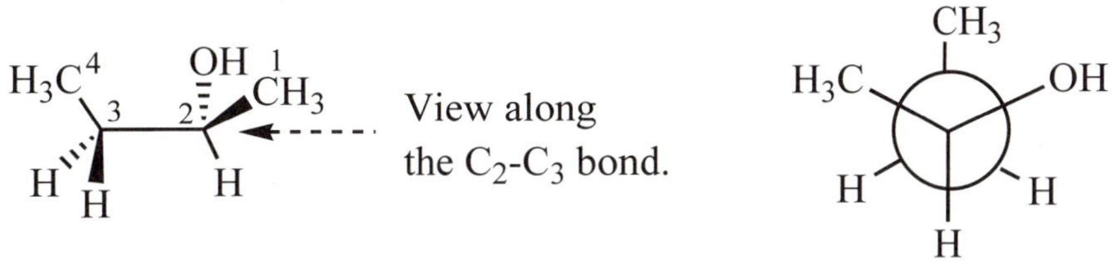

**View along the C$_2$-C$_3$ bond.**

**Less stable**

Assuming that –OH and –CH$_3$ are the same size, then the structure in part (b) and the first structure (more stable) shown in part (c) are of equal stability. These are more stable than the second structure (less stable) shown in part (c). This second structure is less stable because both the –OH and –CH$_3$ groups are adjacent to the other –CH$_3$ group, increasing the repulsive forces and thus decreasing its stability.

## Molecules with Two or More Stereocenters

**6.29**   *Write the structural formula of an alcohol with molecular formula C$_6$H$_{14}$O that contains two stereocenters.*

An asterisk indicates the stereocenters. There are four possible stereoisomers of 3-methyl-2-pentanol.

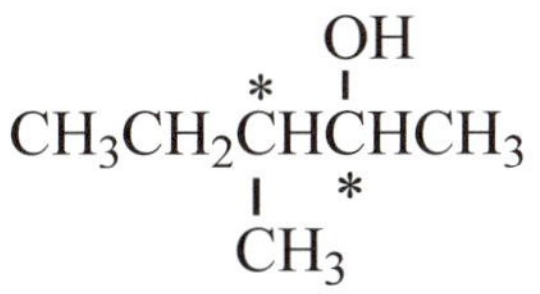

3-Methyl-2-pentanol

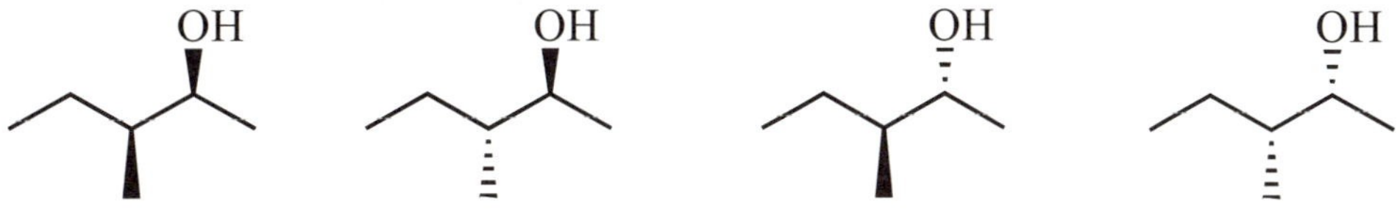

**6.30**   *For centuries, Chinese herbal medicine has used extracts of Ephedra sinica to treat asthma. Investigation of this plant resulted in the isolation of ephedrine, a potent dilator*

*of the air passages of the lungs.  The naturally occurring stereoisomer is levorotatory and has the following structure:*

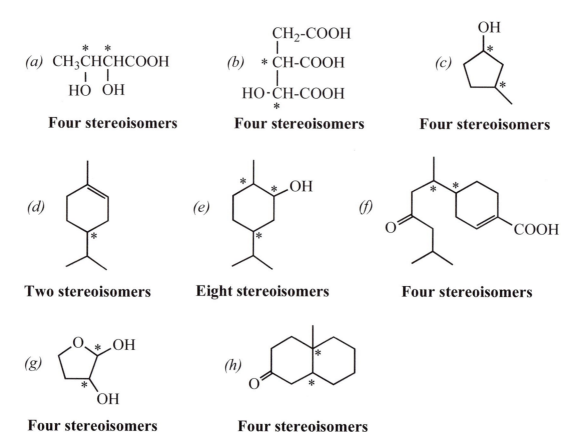

*Ephedrine*
*[α] = -41û*

*Assign R or S configuration to each stereocenter.*

**6.31**   *The specific rotation of naturally occurring ephedrine, shown in Problem 6.30, is -41°. What is the specific rotation of its enantiomer?*

**The specific rotations of enantiomers are equal in magnitude, but opposite in sign. The specific rotation of the enantiomer of naturally occurring ephedrine is +41°.**

**6.32**   *Label each stereocenter in these molecules with an asterisk.  How many stereoisomers are possible for each molecule?*

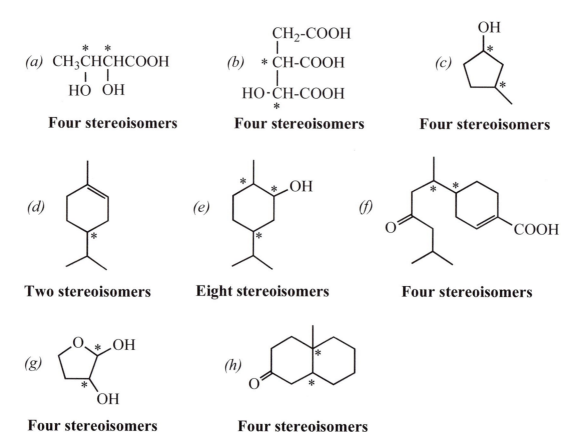

**6.33**    *Label the four stereocenters in amoxicillin, which belongs to the family of semisynthetic penicillins:*

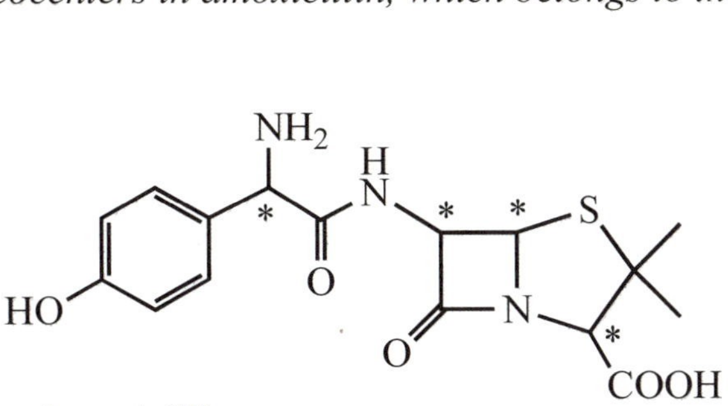

Amoxicillin

**6.34**    *Label all stereocenters in loratadine (Claritin) and fexofenadine (Allegra), now the top-selling antihistamines in the United States.  How many stereoisomers are possible for each compound?*

*(a)*

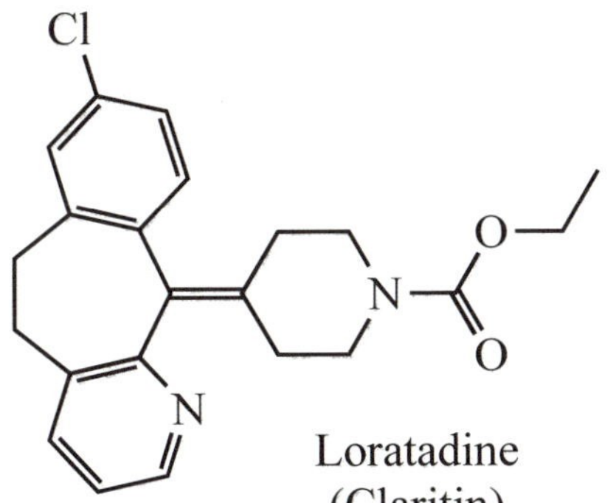

Loratadine
(Claritin)

**Loratadine has no stereocenters.
It is an achiral molecule.**

*(b)*

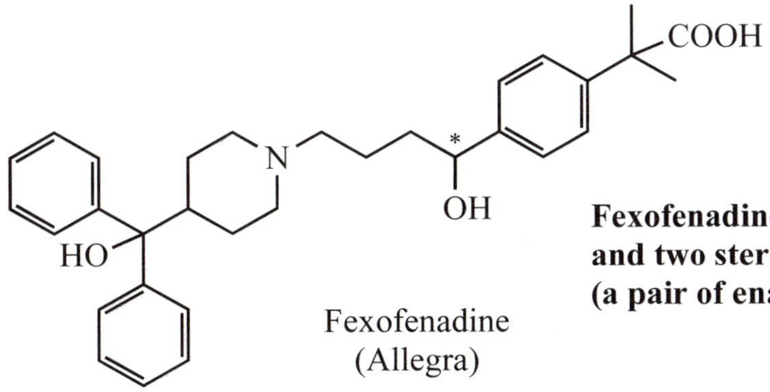

Fexofenadine
(Allegra)

**Fexofenadine has one stereocenter
and two stereoisomers.
(a pair of enantiomers)**

**6.35**  *Following are structural formulas for three of the most widely prescribed drugs used to treat depression.  Label all stereocenters in each compound, and state the number of stereoisomers possible for each.*

*(a)*

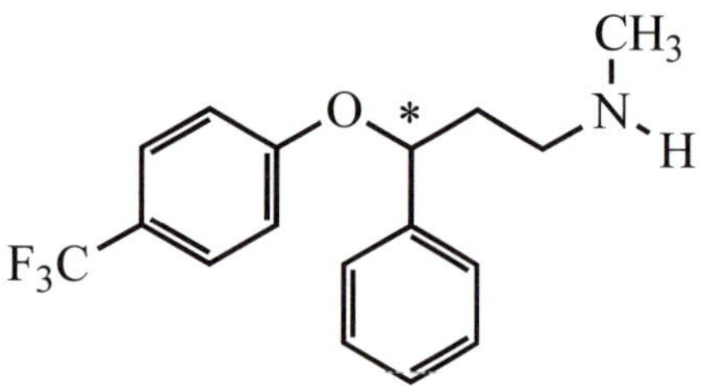

**Fluoxetine**
**(Prozac)**
**Two stereoisomers possible**

*(b)*

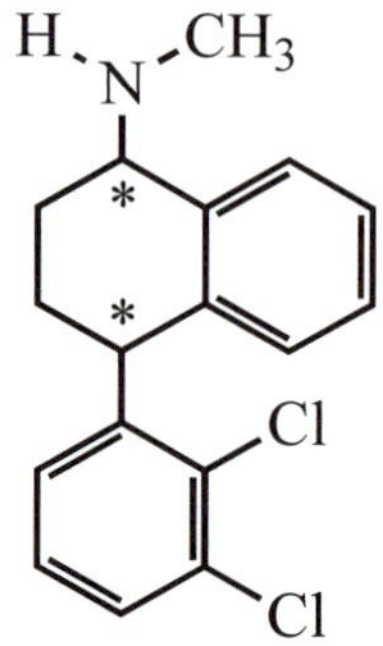

**Sertraline**
**(Zoloft)**
**Four stereoisomers possible**

*(c)*

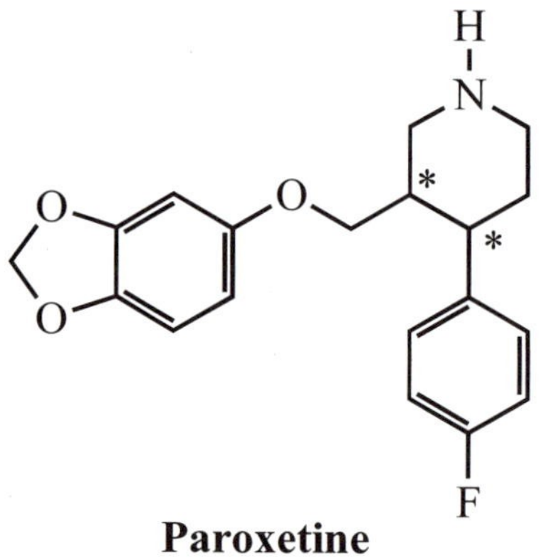

**Paroxetine**
**(Paxil)**
**Four stereoisomers possible**

**6.36**  *Triamcinolone acetonide, the active ingredient in Azmacort Inhalation Aerosol, is a steroid used to treat bronchial asthma:*

*(a) Label the eight stereocenters in this molecule.*

Triamcinolone acetonide

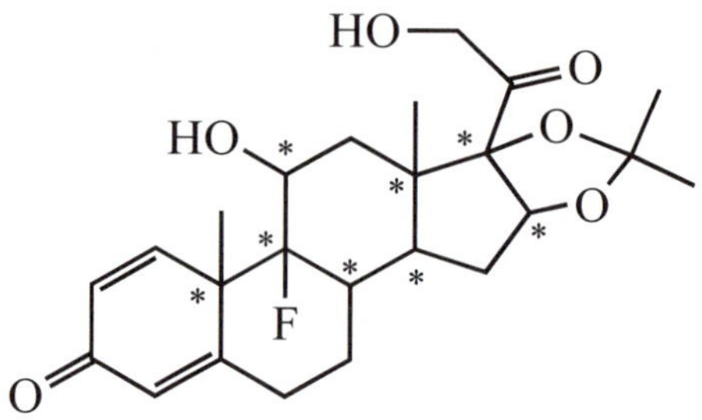

(b) *How many stereoisomers are possible for the molecule? (Of this number, only one is the active ingredient in Azmacort.)*

**$2^n$ = maximum number of stereoisomers possible where n = the number of stereocenters. Triamcinolone acetonide has $2^8 = 256$ possible stereoisomers.**

**6.37** *Which of these structural formulas represent meso compounds?*

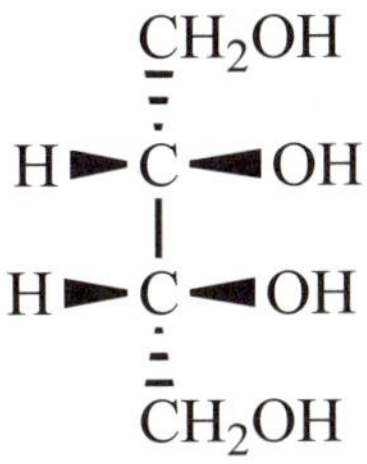

**Compounds (a), (c), (d), and (f) each have at least two stereocenters and an internal mirror plane, and therefore are meso isomers.**

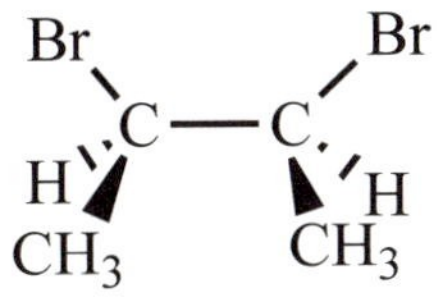

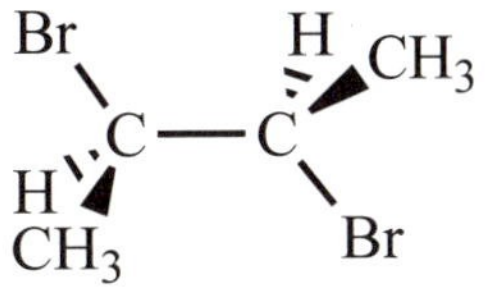

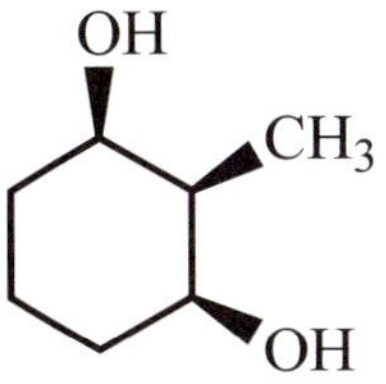

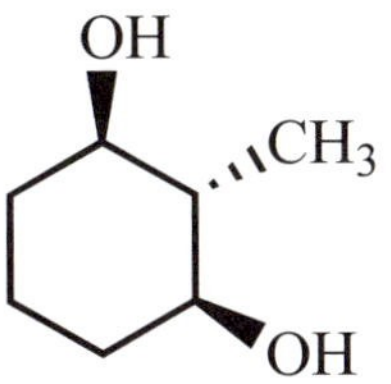

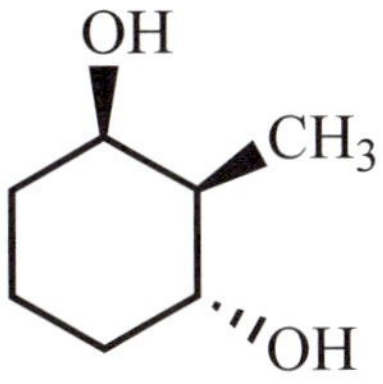

**6.38** *Draw a Newman projection, viewed along the bond between carbons 2 and 3, for both the most stable and the least stable conformations of meso-tartaric acid:*

$$\begin{array}{cc} OH & OH \\ | & | \\ \end{array}$$
$$HOOC\text{-}CH\text{-}CH\text{-}COOH$$

**The carboxylic acid groups are the largest groups in this molecule. The most stable conformer will have the carboxylic acid groups opposite (anti) to each other. In the least stable conformer, the carboxylic acid groups are eclipsed.**

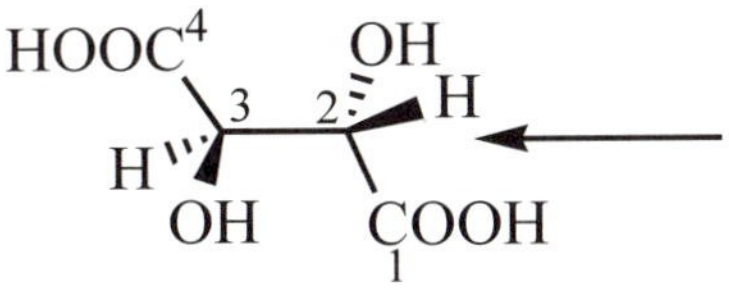

Viewed along this bond

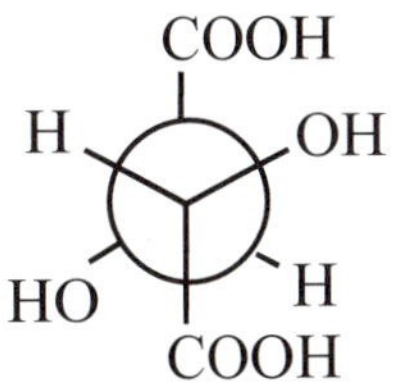

**Most stable conformer**

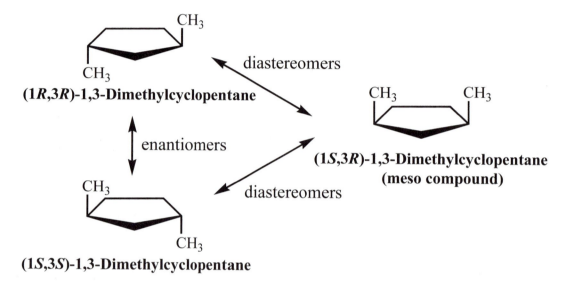

**Least stable conformer**

**6.39**   *How many stereoisomers are possible for 1,3-dimethylcyclopentane?  Which are pairs of enantiomers? Which are meso compounds?*

**(1R,3R)-1,3-Dimethylcyclopentane**

diastereomers

enantiomers

diastereomers

**(1S,3R)-1,3-Dimethylcyclopentane (meso compound)**

**(1S,3S)-1,3-Dimethylcyclopentane**

**6.40**   *In Problem 3.39, you were asked to draw the more stable chair conformation of glucose, a molecule in which all groups on the six-membered ring are equatorial.*

*(a) Identify all stereocenters in this molecule.*

*(b) How many stereoisomers are possible?*

**There are $2^5 = 32$ stereoisomers possible**

*(c) How many pairs of enantiomers are possible?*

**There is one pair of enantiomers for glucose [(+)-glucose and (-)-glucose] and 16 pairs of enantiomers for molecules with the same constitution as glucose.**

*(d) What is the configuration (R or S) at carbons 1 and 5 in the stereoisomer shown?*

**Carbon 1 has the configuration of *R* and the configuration of carbon 5 is also *R*.**

**<u>6.41</u>** *What is a racemic mixture? Is a racemic mixture optically active? That is, will it rotate the plane of polarized light?*

**A racemic mixture is a 50:50 mixture of both enantiomers. Enantiomers will rotate plane-polarized light in equal, but opposite directions. Racemic mixtures will not be optically active because one enantiomer will cancel the optical rotation of the other.**

## Looking Ahead

**<u>6.42</u>** *Predict the product(s) of the following reactions (in cases where more than one stereoisomer is possible, show each stereoisomer):*

*(a)*

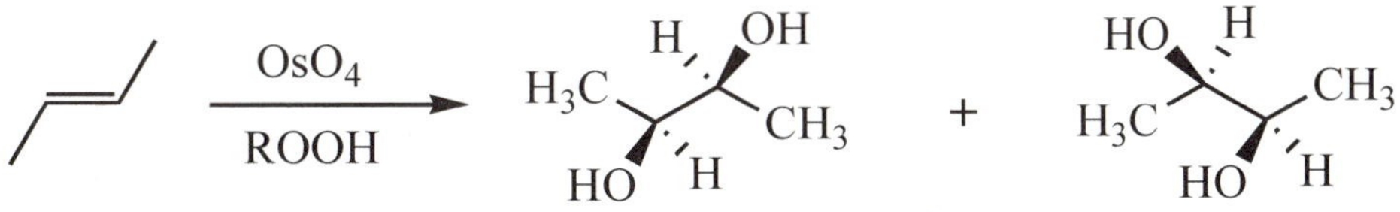


Racemic mixture

*(b)*

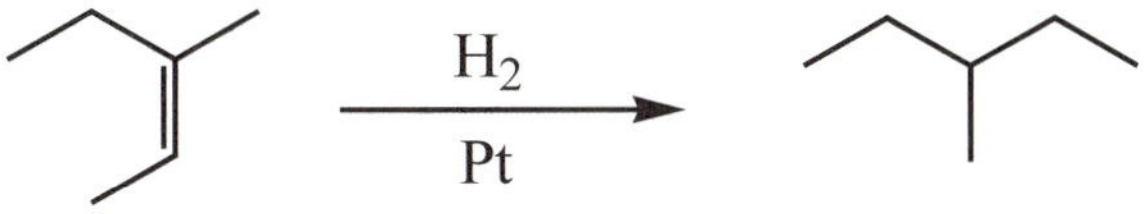


*(c)*

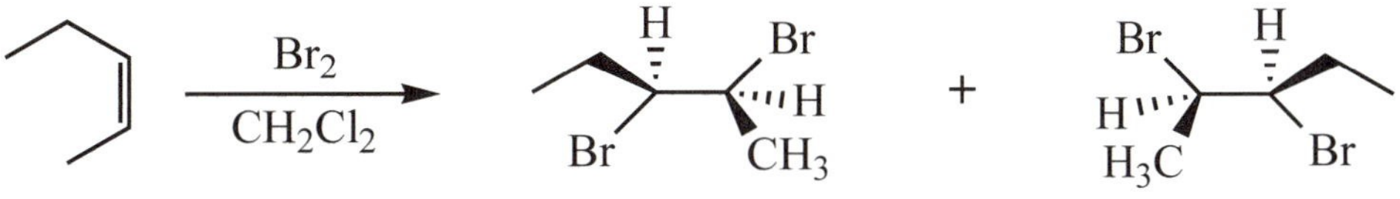


Racemic mixture

*(d)*

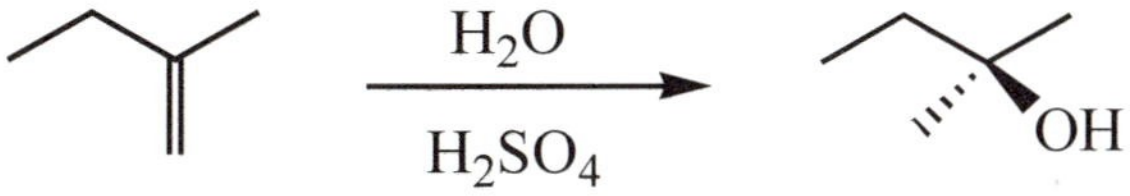

**6.43**   *What alkene, when treated with H₂/Pd, will ensure a 100% yield of the stereoisomer shown.*

(a)

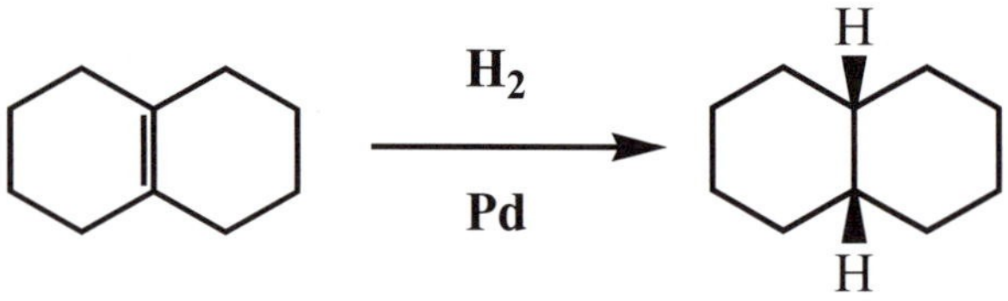

**Only product formed**

(b)

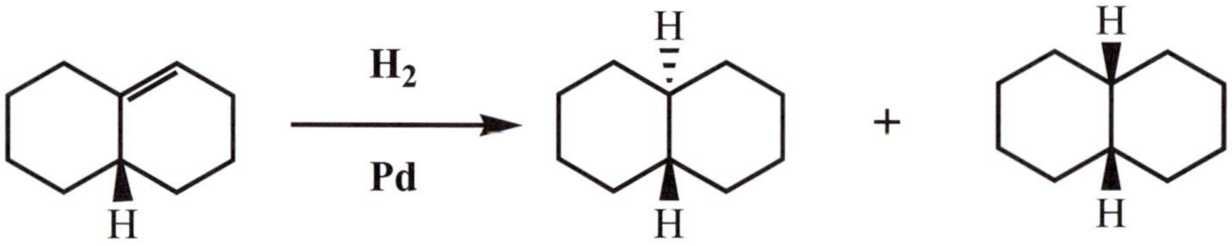

**Hydrogenation of (a) yields the same compound regardless of the side of attack of H₂ on the alkene, whereas hydrogenation of (b) gives two products, only one of which is *cis*-decalin.**

**6.44**   *Which of the following reactions will yield a racemic mixture of products?*

(a)

**Achiral product**

(b)

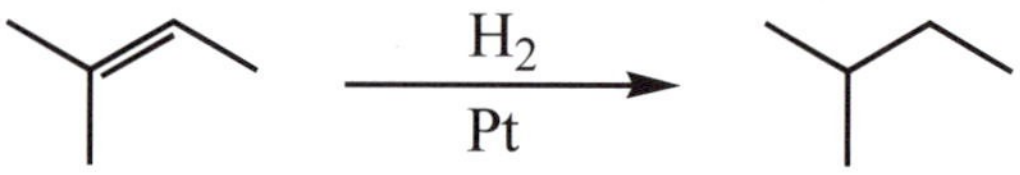

**Achiral product**

(c)

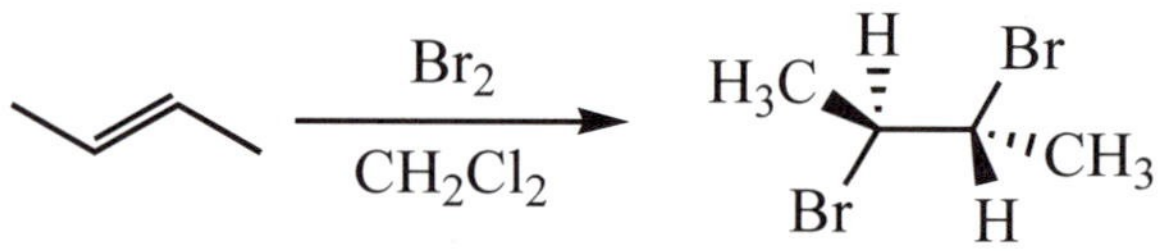

**Meso isomer product (achiral)**

(d)

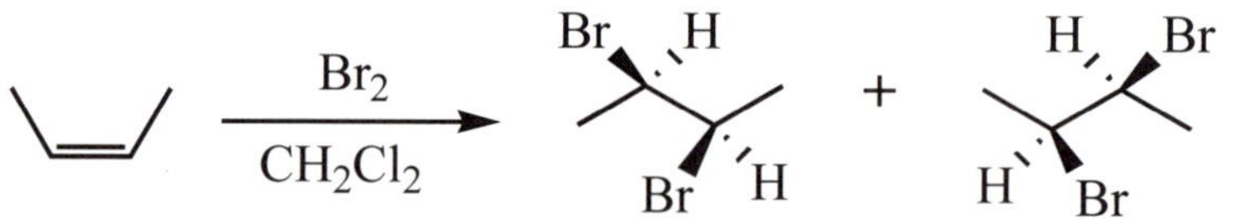

**Racemic mixture**

(e)

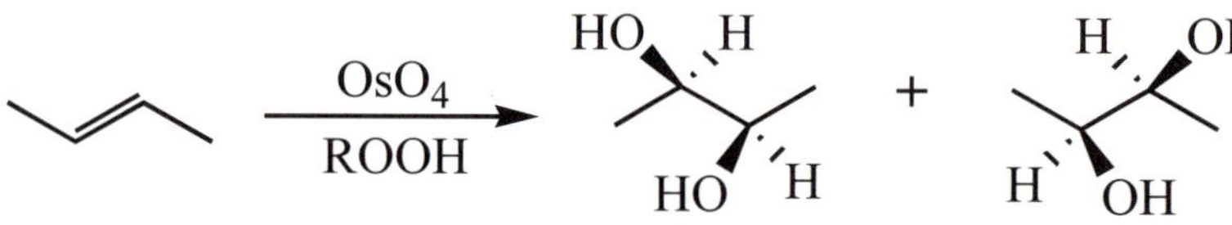

**Racemic mixture**

(f)

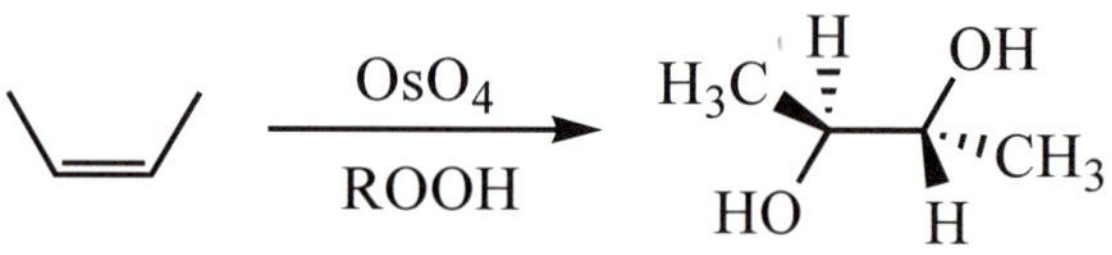

**Meso isomer product (achiral)**

**6.45**  *Draw all the stereoisomers that can be formed in the following reaction:*

HCl

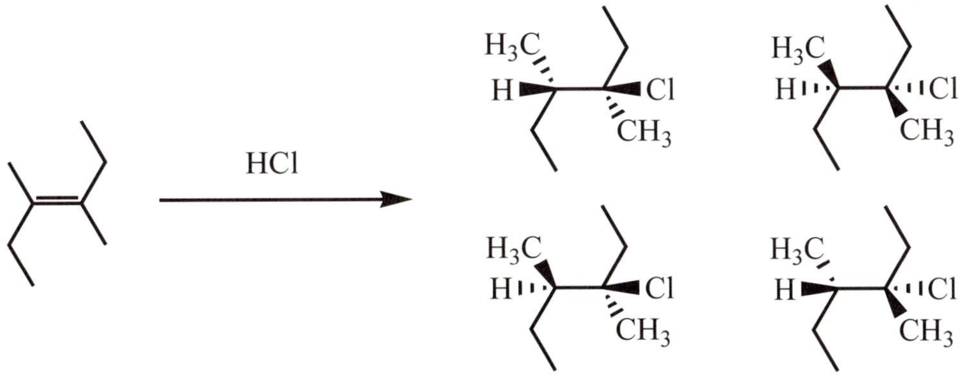

*Comment on the utility of this particular reaction as a synthetic method.*

**Both faces of the reactant molecule are the same thus equal probability of each face reacting to form syn (same side) and anti addition (opposite side) products. This situation results in a product with two stereocenters, thus setting up the possibility of producing four stereoisomers. If a selective synthesis of one stereoisomer over the other possible stereoisomers is desired, this may not be a very useful synthetic method.**

**6.46**  *Explain why the products of the following reaction do not rotate plane-polarized light:*

(a)

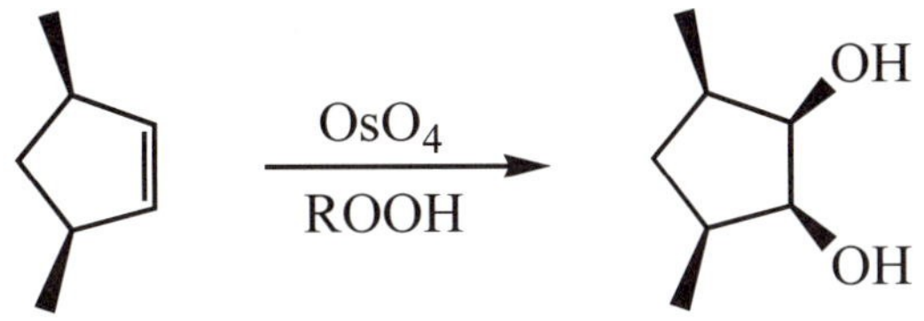

OsO₄ / ROOH

**The product is the achiral meso stereoisomer.**

(b)

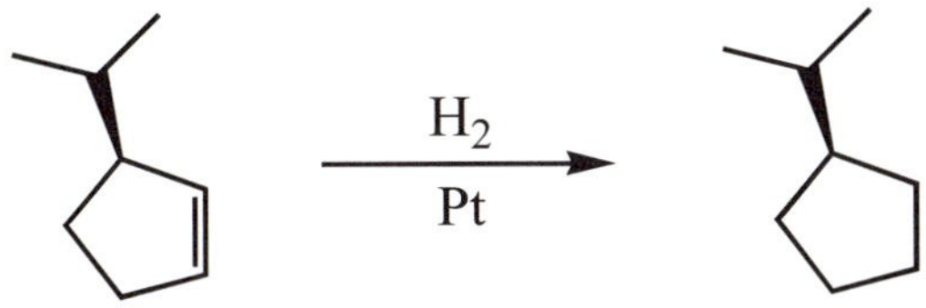

H₂ / Pt

**The product is achiral.**

## CHAPTER 7
### Solutions to Problems

**7.1**    *Write the IUPAC name for each compound:*

(a)

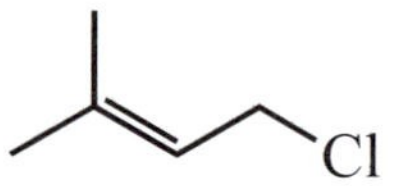

**1-Chloro-3-methyl-2-butene**

(b)

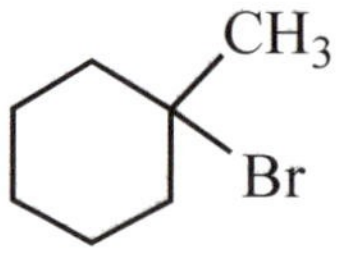

**1-Bromo-1-methylcyclohexane**

(c)     $\begin{array}{c}\text{Cl}\\|\\\text{CH}_3\text{CHCH}_2\text{Cl}\end{array}$

**1,2-Dichloropropane**

(d)

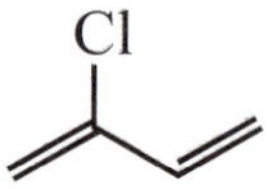

**2-Chloro-1,3-butadiene**

**7.2**    *Complete these nucleophilic substitution reactions:*

(a)      $\text{Br} + \text{CH}_3\text{CH}_2\text{S}^-\text{Na}^+ \longrightarrow$  $\text{S}-\text{CH}_2\text{CH}_3 + \text{Na}^+\text{Br}^-$

(b)      $\text{Br} + \text{CH}_3\overset{\overset{\text{O}}{||}}{\text{C}}\text{O}^-\text{Na}^+ \longrightarrow$  $\text{O}\overset{\overset{\text{O}}{||}}{\text{C}}\text{CH}_3 + \text{Na}^+\text{Br}^-$

**7.3**    *Write the expected product for each nucleophilic substitution reaction and predict the mechanism by which the product is formed:*

(a) 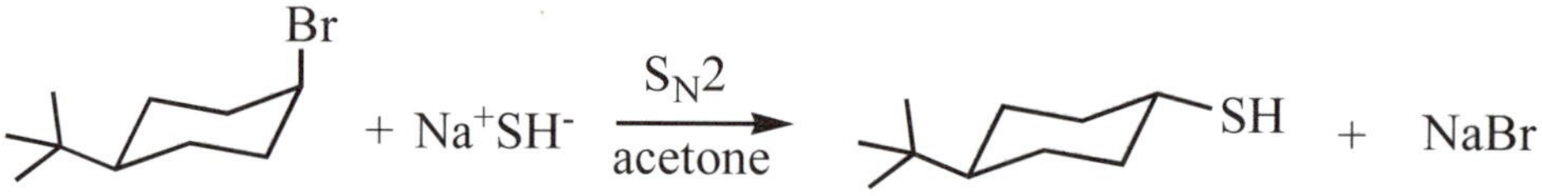    $+ \text{Na}^+\text{SH}^- \xrightarrow[\text{acetone}]{\text{S}_\text{N}2}$ $\text{SH} + \text{NaBr}$

**The SH⁻ is a good nucleophile and, because the reaction involves a secondary alkyl halide with bromide anion as good leaving group, the reaction follows an S$_\text{N}$2 mechanism. The S$_\text{N}$2 results in the inversion off stereochemistry at the carbon bearing the leaving group.**

**Mechanism:**

$\longrightarrow$ $\text{SH} + \text{Br}^-$

(b)

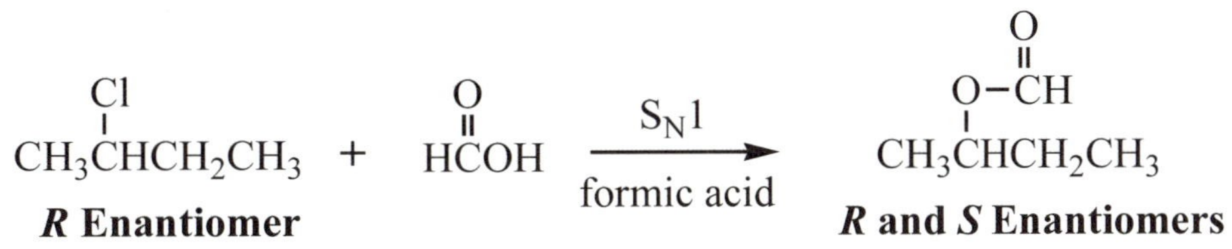

$$CH_3CHCH_2CH_3 \quad + \quad HCOH \xrightarrow[\text{formic acid}]{S_N1} CH_3CHCH_2CH_3$$

with Cl above first carbon

**R Enantiomer**

**R and S Enantiomers**

**Mechanism:**

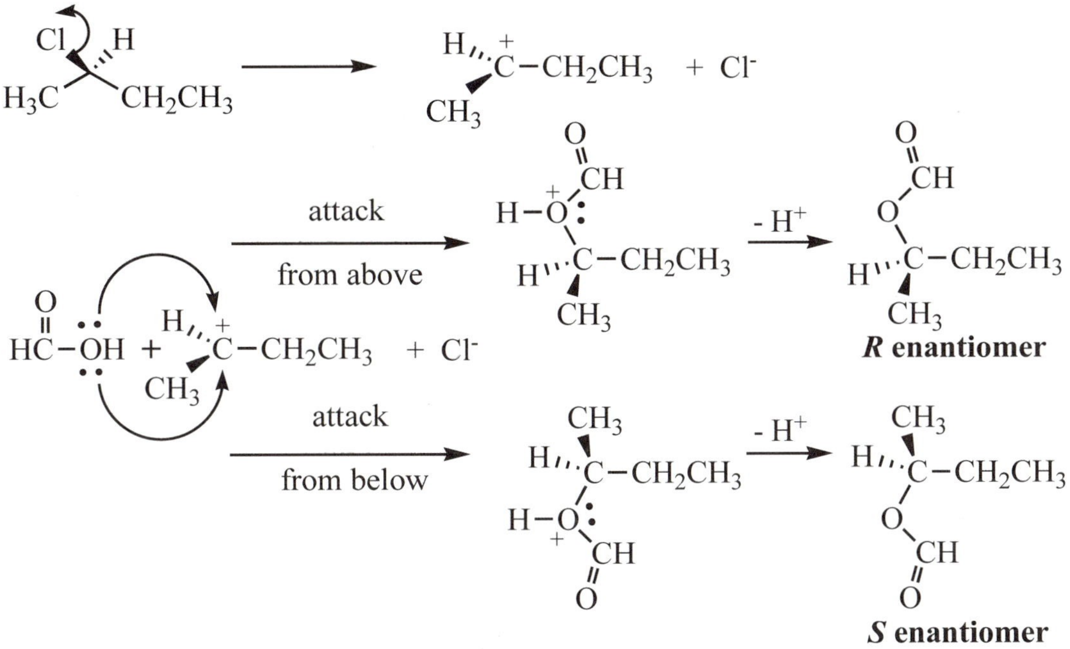

A $S_N1$ reaction is favored for a poor nucleophile (formic acid) reacting with a secondary halide. Formic acid is also an excellent ionizing solvent. $S_N1$ reactions result in racemization.

**7.4** *Predict the β-elimination product formed when each chloroalkane is treated with sodium ethoxide in ethanol (if two products might be formed, predict which is the major product):*

(a)

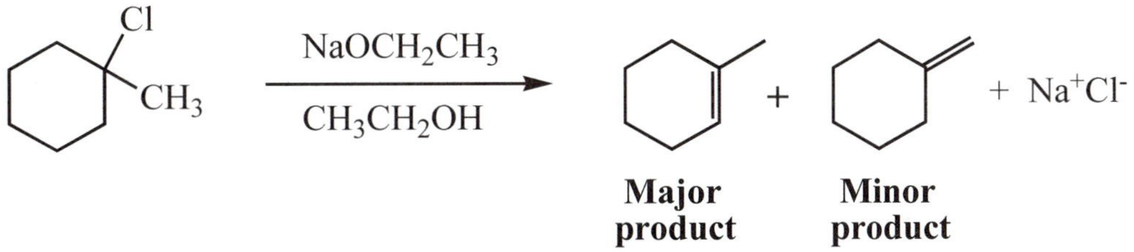

$$\xrightarrow[\text{CH}_3\text{CH}_2\text{OH}]{\text{NaOCH}_2\text{CH}_3}$$

**Major product** + **Minor product** + $Na^+Cl^-$

(b)

$$\xrightarrow[\text{CH}_3\text{CH}_2\text{OH}]{\text{NaOCH}_2\text{CH}_3}$$

+ $Na^+Cl^-$

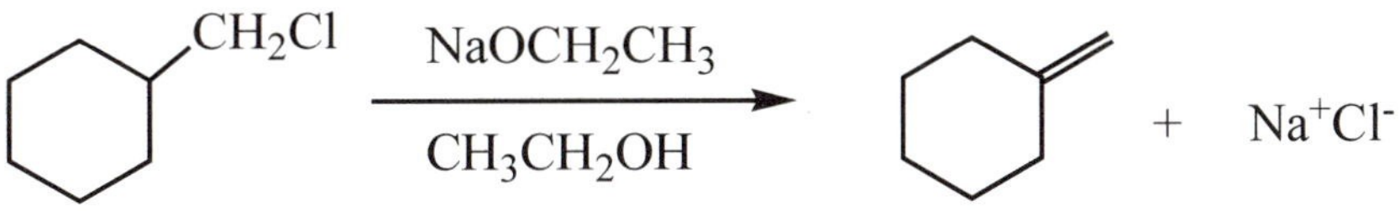

*(c)* 

Cl ⟋⟍ CH₃   NaOCH₂CH₃ / CH₃CH₂OH  →  [cyclohexene with methyl] + [cyclohexene with methyl] + Na⁺Cl⁻

**Equal amounts**

**7.5**   *Predict whether each elimination reaction proceeds predominantly by an E1 or E2 mechanism, and write a structural formula for the major organic product.*

**These reactions proceed by an E2 mechanism.  E2 mechanisms are favored when using secondary or tertiary halides with nucleophiles that are also strong bases like methoxide or ethoxide.  If two or more alkenes are possible products, the most stable product is the major product according to Zaitsev's rule.  Trans alkenes are more stable than cis alkenes**

*(a)*   Br / (structure) + CH₃O⁻Na⁺ →(methanol)

**Major product**

+ [trans alkene] + CH₃OH

[cis alkene]

**Minor product**

*(b)*   [cyclohexane with Cl and CH₃] + CH₃CH₂O⁻Na⁺ →(ethanol) [cyclohexene with CH₃] + CH₃CH₂OH

**7.6**   *Predict whether each reaction proceeds predominantly by substitution (Sₙ1 or Sₙ2) or elimination (E1 or E2) or whether the two compete, and write structural formulas for the major organic product(s):*

*(a)*   Br / (structure) + CH₃O⁻ Na⁺ →(methanol) [alkene] + [OCH₃ structure]

**Major product by E2 reacton**        **Major product by Sₙ2 reacton**

**Reaction (a) involves the methoxide anion, which is both a strong base and strong nucleophile.  Secondary halides in polar solvents can undergo either substitution or elimination reactions.  Sₙ2 reactions are favored with strong nucleophiles and E2 is favored with strong bases, therefore these two reactions are competing in (a).  The E2 reaction will favor the most stable alkene as the major elimination product, which is the trans isomer.**

*(b)* 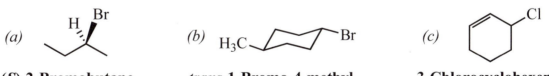 + Na⁺ I⁻ → (acetone)

**Reaction (b) involves the iodide anion, which is weak base and strong nucleophile. S$_N$2 reactions are favored when secondary halides react with strong nucleophilic anions that are also weakly basic in weakly ionizing solvents. S$_N$2 reactions proceed with complete inversion of stereocenters that are bonded to the leaving group, thus yielding the *cis*-1-iodo-4-methylcycloheane from *trans*-1-chloro-4-methylcyclohexane.**

## Nomenclature

**7.7**    *Write the IUPAC name for each compound:*

*(a)* CH$_2$=CF$_2$

**1,1-Difluoroethene**

*(b)* ⬡—Br

**3-Bromocyclopentene**

*(c)*

**2-Chloro-5-methylhexane**

*(d)* Cl(CH$_2$)$_6$Cl

**1,6-Dichlorohexane**

*(e)* CF$_2$Cl$_2$

**Dichlorodifluoroethane**

*(f)*

**3-Bromo-3-ethylpentane**

**7.8**    *Write the IUPAC name for each compound (be certain to include a designation of configuration, where appropriate, in your answer):*

*(a)*

**(S)-2-Bromobutane**

*(b)* H$_3$C—◯—Br

**trans-1-Bromo-4-methyl-cyclohexane**

*(c)*

**3-Chlorocyclohexene**

*(d)*

**(E)-1-Chloro-2-butene**

*(e)*

**(R)-2-Bromo-2-chlorobutane**

*(f)*

**meso-2,3-Dibromobutane**

**7.9**     *Draw a structural formula for each compound (given are IUPAC names):*

(a) *3-Bromopropene*          (b) *(R)-2-Chloropentane*      (c) *meso-3,4-Dibromohexane*

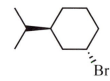     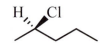     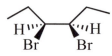

(d) *trans-1-Bromo-3-isopropylcyclohexane*     (e) *1,2-Dichloroethane*

(f)  *Bromocyclobutane*

**7.10**    *Draw a structural formula for each compound (given are common names):*

(a) *Isopropyl chloride*        (b) *sec-Butyl bromide*        (c) *Allyl iodide*

(d) *Methylene chloride*        (e) *Chloroform*               (f) *tert-Butyl chloride*

**CH₂Cl₂**                        **CHCl₃**

$CH_2Cl_2$                        $CHCl_3$

(g) *Isobutyl chloride*

**7.11**    *Which compounds are 2° alkyl halides?*

**2-Iodooctane and *trans*-1-chloro-4-methylcyclohexane are 2°alkyl halides.**

*(a) Isobutyl chloride*     *(b) 2-Iodooctane*        *(c) trans-1-Chloro-4-methylcyclohexane*

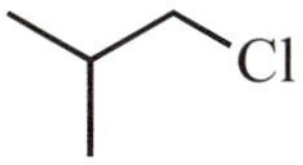

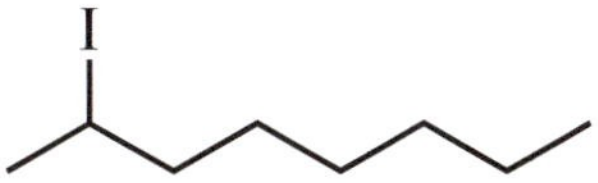

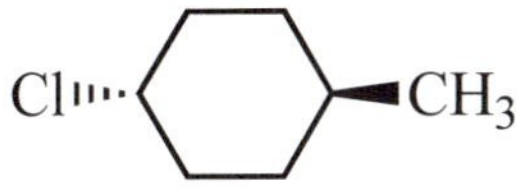

**1ûalkyl halide**          **2ûalkyl halide**                **2ûalkyl halide**

## Synthesis of Alkyl Halides

**7.12**    *What alkene or alkenes and reaction conditions give each alkyl halide in good yield?*
        *(Hint: Review Chapter 5)*

**All of the following reactions take place via the Markovnikov addition of a hydrogen halide to an alkene.**

*(a)*

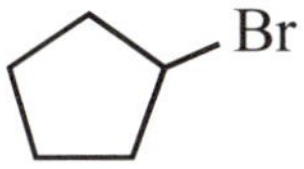

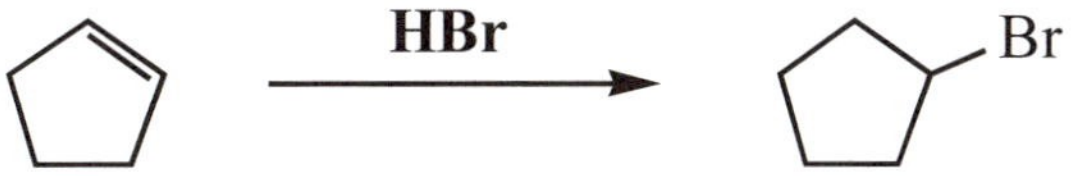

*(b)* $CH_3\overset{\overset{\displaystyle CH_3}{|}}{\underset{\underset{\displaystyle Br}{|}}{C}}CH_2CH_2CH_3$     or

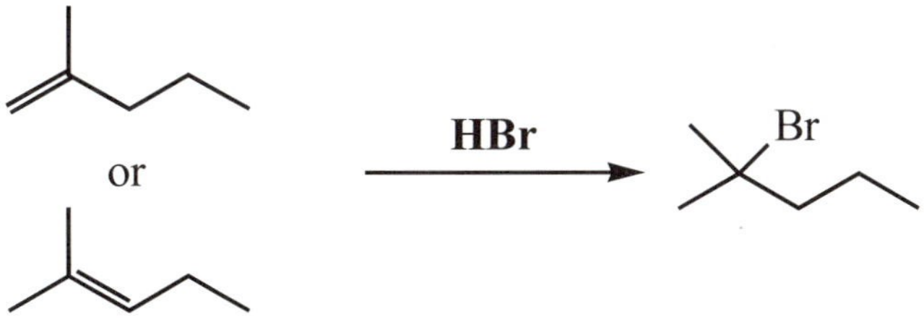

*(c)*

or

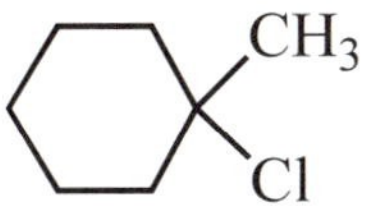

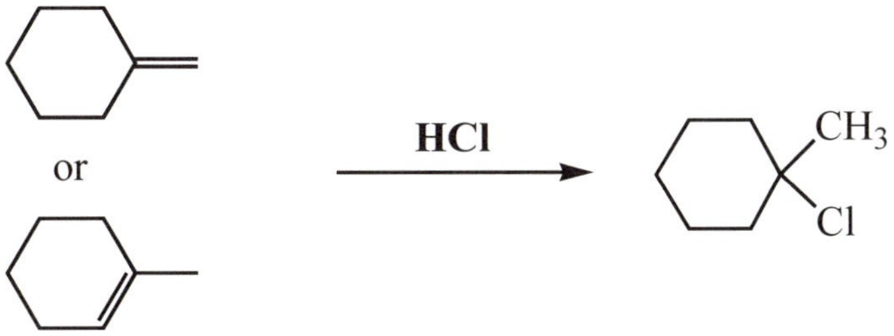

**7.13**    *Show reagents and conditions that bring about these conversions:*

*(a)*

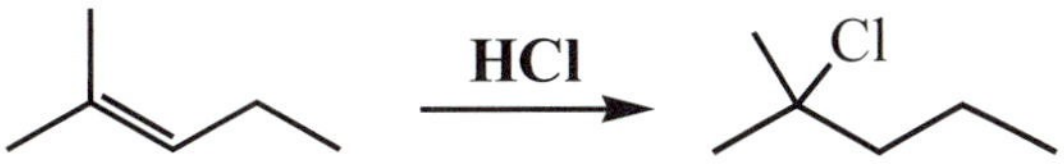

*(b)* $CH_3CH_2CH{=}CH_2$  $\xrightarrow{\text{HI}}$  $CH_3CH_2\overset{\overset{\displaystyle I}{|}}{C}HCH_3$

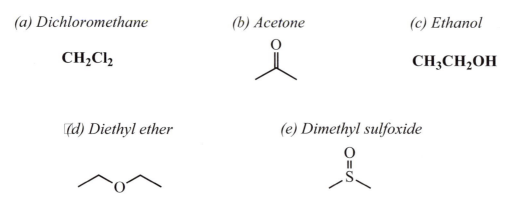

## Nucleophilic Aliphatic Substitution

7.14    *Write structural formulas for these common organic solvents:*

    *(a) Dichloromethane*         *(b) Acetone*         *(c) Ethanol*

    **$CH_2Cl_2$**                               **$CH_3CH_2OH$**

    *(d) Diethyl ether*         *(e) Dimethyl sulfoxide*

7.15    *Arrange these protic solvents in order of increasing polarity:*

    *(a)  $H_2O$*         *(b)  $CH_3CH_2OH$*    *(c)  $CH_3OH$*

**Alkyl groups decrease the polarity of solvents.  In order of increasing polarity, they are:**

$$CH_3CH_2OH  <  CH_3OH  <  H_2O$$

7.16    *Arrange these aprotic solvents in order of increasing polarity:*
    *(a) Acetone*    *(b) Pentane*      *(c) Diethyl ether*

**The carbonyl group of acetone is a polar functional group, so acetone is the most polar of the three.  The oxygen atom of diethyl ether adds polarity to this solvent compared to the hydrocarbon pentane.  In order of increasing polarity, they are:**

**Increasing polarity**

**7.17**    *From each pair, select the better nucleophile:*

(a) $H_2O$   or  $\boxed{OH^-}$    **The conjugate base is more nucleophilic than its acid.**

(b) $CH_3COO^-$   or  $\boxed{OH^-}$    **When comparing the same nucleophilic atom, nucleophilic strength is proportional to base strength.**

(c) $CH_3SH$   or  $\boxed{CH_3S^-}$    **The conjugate base is more nucleophilic than its acid.**

**7.18**    *Which statements are true for $S_N2$ reactions of haloalkanes?*

(a) *Both the haloalkane and the nucleophile are involved in the transition state.*

**True:  Both nucleophile and alkyl halide are reacting in the rate-determining step.**

(b) *The reaction proceeds with inversion of configuration at the substitution center.*

**True:  Backside attack at the substitution center results in the inversion of configuration.**

(c) *The reaction proceeds with retention of optical activity.*

**True:  $S_N2$ reactions result in an inversion of stereochemistry at the substitution center, therefore, optical activity is retained in the product.  Therefore, the product is optically active and rotates plane-polarized light with the same magnitude but in the *opposite* direction of the original.**

(d) *The order of reactivity is $3° > 2° > 1° >$ methyl.*

**False:  Steric crowding influences the order of reactivity in $S_N2$ reactions.  As the number of substituent groups on the substitution center increases, the reaction rate decreases.   For $S_N2$ reactions, the order of reactivity is: methyl $> 1° > 2° > 3°$.**

(e) *The nucleophile must have an unshared pair of electrons and bear a negative charge.*

**False:  All nucleophiles are Lewis bases and therefore must have an unshared pair of electrons.  Moderate nucleophiles such as ammonia ($NH_3$) are strong enough as nucleophiles to react in $S_N2$ reactions, even though they do not carry a negative charge.**

*(f) The greater the nucleophilicity of the nucleophile, the greater the rate of reaction.*

**True:  The nucleophile is involved in the rate-determining step; therefore, as nucleophilic strength increases, the rate also increases.**

**7.19**    *Complete these S_N2 reactions:*

(a) $Na^+I^-$  +  $CH_3CH_2CH_2Cl$  $\xrightarrow{acetone}$  $CH_3CH_2CH_2I$ + NaCl

(b) $NH_3$ + —Br $\xrightarrow{ethanol}$ —$\overset{+}{N}H_3$ $Br^-$

(c) $CH_3CH_2O^-Na^+$ + $CH_2=CHCH_2Cl$ $\xrightarrow{ethanol}$ $CH_2=CHCH_2OCH_2CH_3$ + NaCl

**7.20**    *Complete these S_N2 reactions:*

(a)

(b) $CH_3\overset{I}{\underset{|}{C}}HCH_2CH_3$ + $CH_3CH_2S^-Na^+$ $\xrightarrow{acetone}$ + $Na^+I^-$

(c) $CH_3\overset{CH_3}{\underset{|}{C}}HCH_2CH_2Br$ + $Na^+I^-$ $\xrightarrow{acetone}$ + $Na^+Br^-$

(d) $(CH_3)_3N$ + $CH_3I$ $\xrightarrow{acetone}$ $(CH_3)_4N^+$ + $I^-$

(e) —$CH_2Br$ + $CH_3O^-Na^+$ $\xrightarrow{methanol}$ —$CH_2OCH_3$ + $Na^+Br^-$

(f) $H_3C$Cl + $CH_3S^-Na^+$ $\xrightarrow{ethanol}$ $H_3C$ + $Na^+Cl^-$

(g)

(h)

**7.21**   *You were told that each reaction in Problem 7.20 proceeds by an $S_N2$ mechanism. Suppose you were not told the mechanism. Describe how you could conclude, from the structure of the haloalkane, the nucleophile, and the solvent, that each reaction is in fact an $S_N2$ reaction.*

(a) **A 2° halide, a moderate nucleophile, and a moderately ionizing solvent all favor an $S_N2$ mechanism.**

(b) **A 2° halide, with ethyl thiolate as a good nucleophile that is a weak base, and a weakly ionizing solvent all favor an $S_N2$ mechanism.**

(c) **A 2° halide, with iodide as a good nucleophile that is a weak base, and a weakly ionizing solvent all favor an $S_N2$ mechanism.**

(d) **Methyl halides can only undergo $S_N2$ reactions because their extremely unstable carbocations preclude $S_N1$ reactions. Trimethylamine is a moderate nucleophile and acetone is a weakly ionizing solvent, further supporting an $S_N2$ mechanism.**

(e) **A 1° halide, with a good nucleophile like methoxide always favors an $S_N2$ mechanism.**

(f) **A 2° halide, with methyl thiolate as a good nucleophile that is a weak base, and a moderately ionizing solvent all favor an $S_N2$ mechanism.**

(g) **Piperidine is a moderate amine nucleophile and ethanol is a moderately ionizing solvent. $S_N2$ mechanisms are favored with moderate nucleophiles reacting with 1° alkyl halides.**

(h) **Ammonia is a moderate nucleophile and ethanol is a moderately ionizing solvent. An $S_N2$ mechanism is favored in this case because the alkyl halide is primary.**

**7.22**    *In the following reactions, a haloalkane is treated with a compound that has two nucleophilic sites. Select the more nucleophilic site in each part, and show the product of each S_N2 reaction.*

**When comparing nucleophilicity of atoms in the same row of the Periodic Table, as in reactions (a) and (b), nucleophilicity increases with basicity of the atom (oxygen is less basic than nitrogen). For reaction (c), when comparing atoms in the same column of the Periodic Table, nucleophilicity increases going from top to bottom; therefore, sulfur is more nucleophilic than oxygen.**

*(a)* $HOCH_2CH_2NH_2$   + $CH_3I$   $\xrightarrow{\text{ethanol}}$     + $I^-$

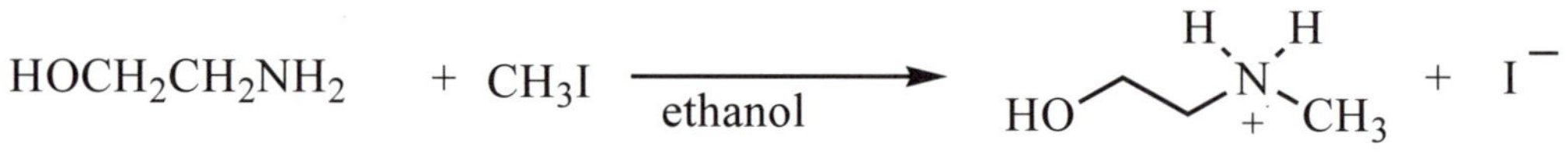

*(b)*     + $CH_3I$   $\xrightarrow{\text{ethanol}}$     + $I^-$

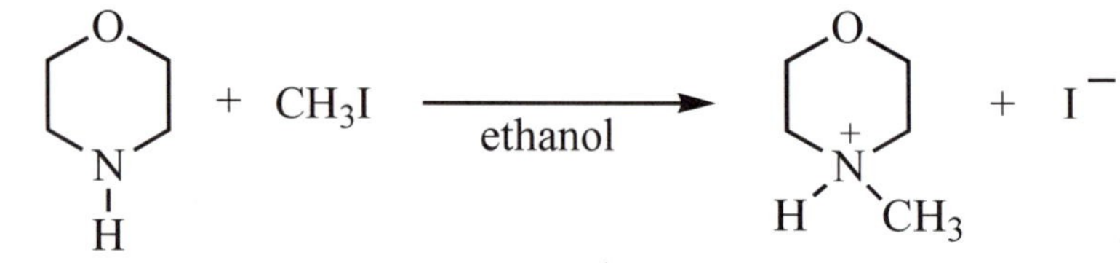

*(c)* $HOCH_2CH_2SH$   + $CH_3I$   $\xrightarrow{\text{ethanol}}$     + $HI$

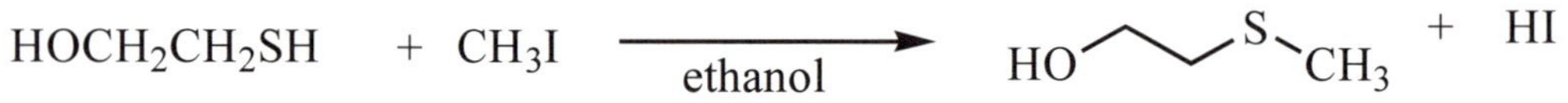

**7.23**    *Which statements are true for S_N1 reactions of haloalkanes?*

*(a) Both the haloalkane and the nucleophile are involved in the transition state of the rate-determining step*

**False:  The haloalkane ionizes to a carbocation and halide anion in the rate-determining step.**

*(b) The reaction at a stereocenter proceeds with retention of configuration.*

**False:  The carbocation carbon is $sp^2$ hybridized and has a trigonal planar geometry, which precludes it from being a stereocenter.  An S_N1 reaction at a stereocenter proceeds with racemization.**

*(c) The reaction at a stereocenter proceeds with loss of optical activity.*

**True:  The trigonal planar carbocation can be attacked by a nucleophile from either face.  Attack on one face gives one enantiomer and attack at the other face yields the other enantiomer.**

*(d) The order of reactivity is 3° > 2° > 1° > methyl.*

**True:  The order of reactivity proceeds according to the carbocation intermediate stability.  Carbocations become more stable with increasing alkyl substitution on the positively charged carbon.**

*(e) The greater the steric crowding around the reactive center, the lower the rate of reaction*

**False:  The nucleophilic attack on the carbocation is not part of the rate-determining step, therefore steric crowding at the reactive center has no effect on rate.**

*(f) The rate of reaction is greater with good nucleophiles as compared with poor nucleophiles*

**False: The nucleophilic attack on the carbocation is not part of the rate-determining step, therefore the strength of the nucleophile has no effect on reaction rate because the nucleophile is not part of the rate-determining step.**

**To summarize, statements (a), (e), and (f) are false because the rate-determining step only involves the alkyl halide.  Statement (b) is false because the optical activity is lost due to the achiral trigonal planer carbocation intermediate formed.**

**7.24**    *Draw a structural formula for the product of each S$_N$1 reaction:*

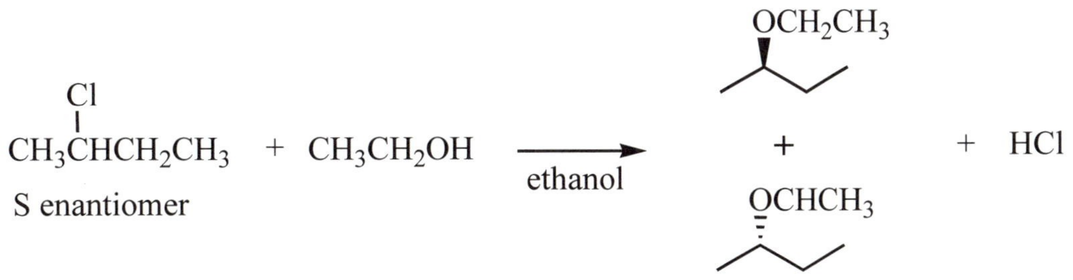

**Racemic mixture**

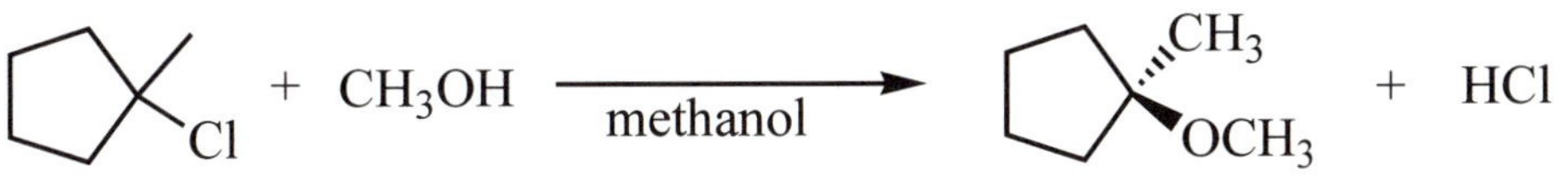

(c) 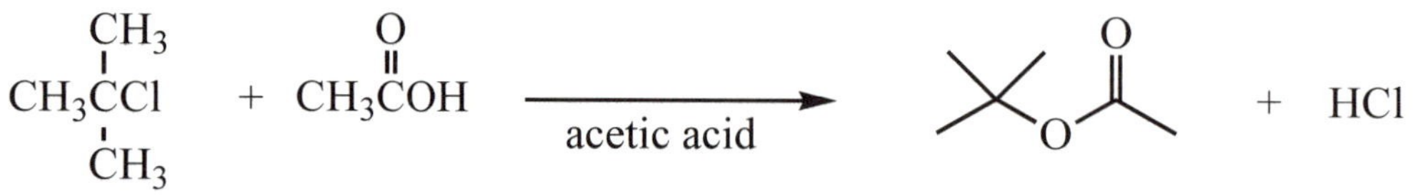  CH₃CCl  +  CH₃COH  $\xrightarrow{\text{acetic acid}}$  [structure]  +  HCl

(d) 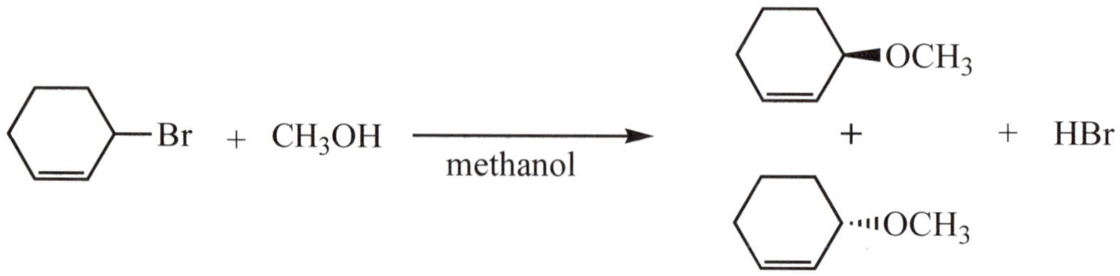  [cyclohexene]—Br  +  CH₃OH  $\xrightarrow{\text{methanol}}$  [structures with OCH₃]  +  +  HBr

**Racemic mixture**

**7.25**  *You were told that each substitution reaction in Problem 7.24 proceeds by an S$_N$1 mechanism. Suppose that you were not told the mechanism. Describe how you could conclude, from the structure of the haloalkane, the nucleophile, and the solvent, that each reaction is in fact an S$_N$1 reaction.*

   *(a)* **Chloride is a good leaving group and the resulting 2° carbocation is a relatively stable intermediate. The most important factor influencing the reaction towards S$_N$1 is ethanol, which is a poor nucleophile and a moderately ionizing solvent.**

   *(b)* **Chloride is a good leaving group and thus results in a very stable 3° carbocation intermediate. An equally important factor influencing the reaction towards S$_N$1 is methanol, which is a poor nucleophile and a good ionizing solvent. 3° Alkyl halides almost exclusively undergo S$_N$1 reactions with poor nucleophiles.**

   *(c)* **Chloride is a good leaving group and acetic acid is a strongly ionizing solvent/poor nucleophile. 3° Alkyl halides exclusively undergo S$_N$1 reactions with poor nucleophiles. All these factors strongly favor S$_N$1 reactions.**

   *(d)* **Methanol is a good ionizing solvent and a poor nucleophile. Bromide is a good leaving group resulting in a relatively stable 2° carbocation, all favoring an S$_N$1 reaction.**

**7.26**  *Select the member of each pair that undergoes nucleophilic substitution in aqueous ethanol more rapidly:*

   **The order of reactivity for alkyl halides in S$_N$1 reactions is:**
   **3° > 2° > 1° > methyl**

(a) Cl  or  (boxed structure with Cl)

(b) Br  or  (boxed) Br

(c) Br  or  (boxed) CH₃, Br

**7.27**  *Propose a mechanism for the formation of the products (but not their relative percentages) in this reaction:*

$$CH_3CCl(CH_3)(CH_3) \xrightarrow[25°C]{\substack{20\% \ H_2O \\ 80\% \ CH_3CH_2OH}} CH_3COCH_2CH_3(CH_3) + CH_3COH(CH_3) + CH_3C=CH_2(CH_3) + HCl$$

85% (for the first two products), 15% (for the alkene)

**All of the reaction products shown can be produced from the same carbocation intermediate that results from the ionization of the carbon-halogen bond. The reaction is run in an ionizing solvent, there are no good nucleophiles present, and the 3° carbocation is a very stable intermediate. All of these factors are very favorable for $S_N1$ and E1 reaction mechanisms.**

**$S_N1$ reaction mechanism with ethanol as a nucleophile:**

$$H_3C-C(CH_3)(CH_3)-Cl \xrightarrow{slow} H_3C-C^+(CH_3)(CH_3) + Cl^-$$

$$H_3C-C^+(CH_3)(CH_3) + HOCH_2CH_3 \longrightarrow H_3C-C(CH_3)(CH_3)-{}^+OCH_2CH_3 \longrightarrow CH_3COCH_2CH_3(CH_3) + H_3O^+$$

**S_N1 reaction mechanism with water as a nucleophile:**

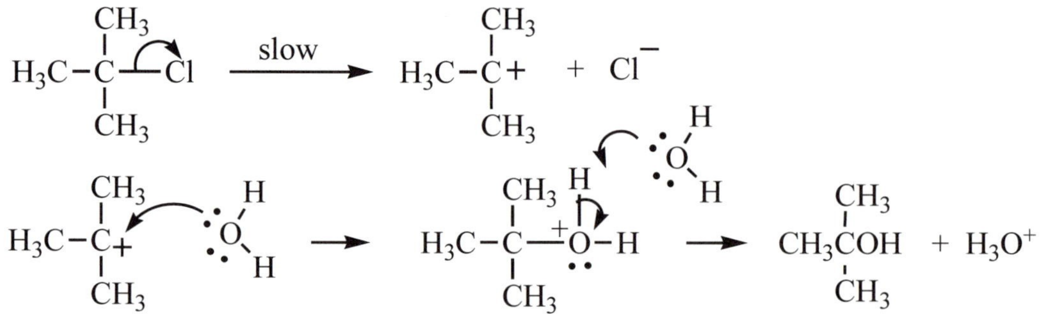

**E1 reaction mechanism:**

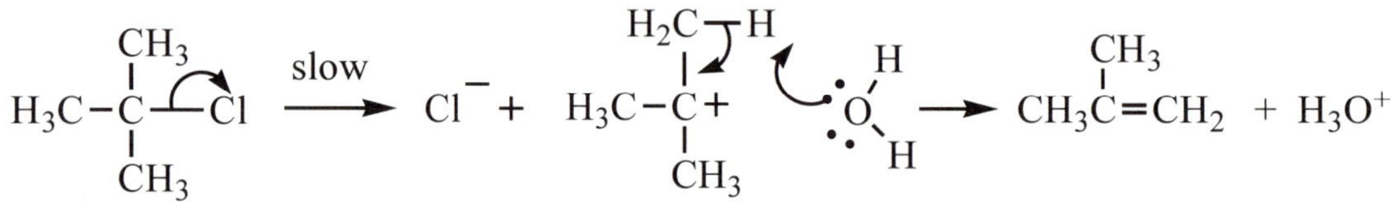

**7.28**   *The rate of reaction in Problem 7.27 increases by 140 times when carried out in 80% water to 20% ethanol, compared with 40% water to 60% ethanol. Account for this difference.*

**The reaction mechanism involves the formation of a carbocation in the rate-determining step. The addition of the more polar water to ethanol makes the solvent mixture more polar and more ionizing than pure ethanol. Thus, as the percentage of water in the ethanol increases, carbocations form by ionization of the carbon-chloride bond more easily in the water-ethanol solvent mixture relative to pure ethanol.**

**7.29**   *Select the member of each pair that shows the greater rate of S_N2 reaction with KI in acetone:*

*(a)*   or

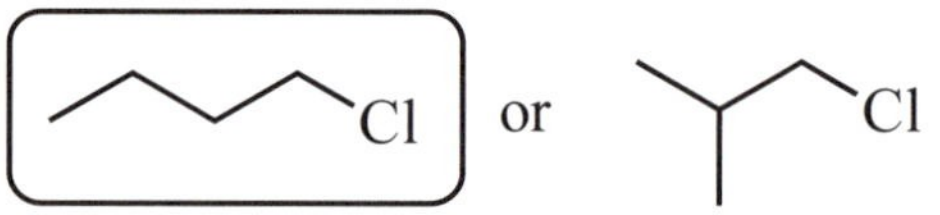

**Both compounds are 1° halides, but 1-chlorobutane is less sterically hindered than isobutyl chloride.**

*(b)* /\/\Cl   or   [ /\/\Br ]

**The alkyl halides have the same structure, but bromide is a better leaving group than chloride.**

*(c)* [ structure with Cl ]   or   [ structure with Cl ]

**Although the compounds are 1° chlorides, isopentyl chloride is less sterically hindered than neopentyl chloride.**

*(d)* [ structure with Br ]   or   [ structure with Br ]

**Although the compounds are 2° bromides, 2-bromopentane is less sterically hindered than 2-bromo-3-methylbutane.**

**7.30**  *What hybridization best describes the reacting carbon in the S$_N$2 transition state?*

**The reacting carbon in an S$_N$2 reaction transition state can be described as *sp$^2$* hybridized, with the substituents bonding to *sp$^2$* hybridized orbitals and the incoming nucleophile and leaving groups each partially bonded to an unhybridized *p* orbital. The molecular shape of this transition state can be described as trigonal bipyramidal.**

$$\left[ \text{Nu} - - \bigcirc\!\!-\overset{|}{\underset{}{C}}\!\!\bigcirc - - \text{L} \right]^{\ddagger}$$

**7.31**   *Haloalkenes such as vinyl bromide, $CH_2=CHBr$, undergo neither $S_N1$ nor $S_N2$ reactions. What factors account for this lack of reactivity?*

**Haloalkenes fail to undergo $S_N1$ reactions because the alkenyl carbocations produced through ionization of the carbon-halogen bond are too unstable.  The electron rich π-bonds of alkenyl halides repel nucleophiles in $S_N2$ reaction.**

**7.32**   *Show how you might synthesize the following compounds from a haloalkane and a nucleophile:*

**7.33**   *Show how you might synthesize each compound from a haloalkane and a nucleophile:*

(c)  + CH₃CO⁻ Na⁺ ⟶ + NaBr

(d)  + S⁻Na⁺ ⟶ + NaBr

(e)  + CH₃CO⁻ Na⁺ ⟶ + NaBr

(f)  Br + O⁻Na⁺ ⟶ + NaBr

## β-Eliminations

**7.34** *Draw structural formulas for the alkene(s) formed by treatment of each haloalkane with sodium ethoxide in ethanol. Assume that elimination is by an E2 mechanism. Where two alkenes are possible, use Zaitsev's rule to predict which alkene is the major product:*

(a) 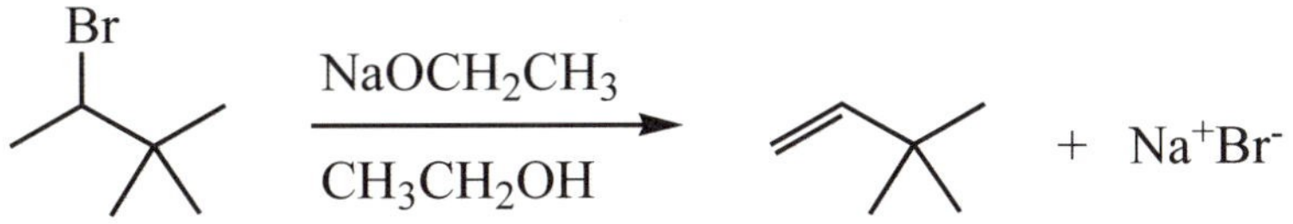 $\xrightarrow[\text{CH}_3\text{CH}_2\text{OH}]{\text{NaOCH}_2\text{CH}_3}$ + Na⁺Br⁻

(b) 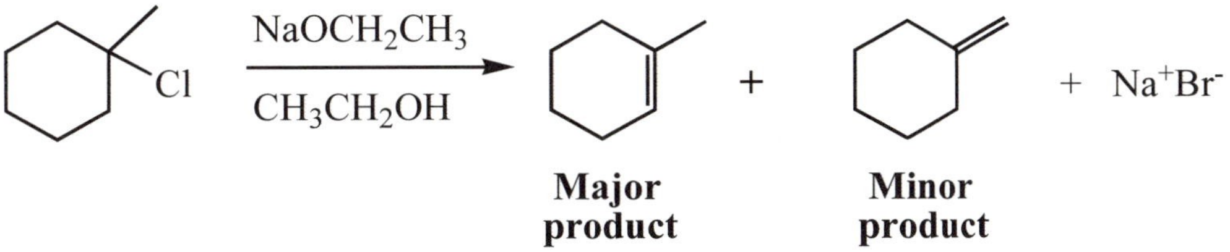 $\xrightarrow[\text{CH}_3\text{CH}_2\text{OH}]{\text{NaOCH}_2\text{CH}_3}$ **Major product** + **Minor product** + Na⁺Br⁻

(c) 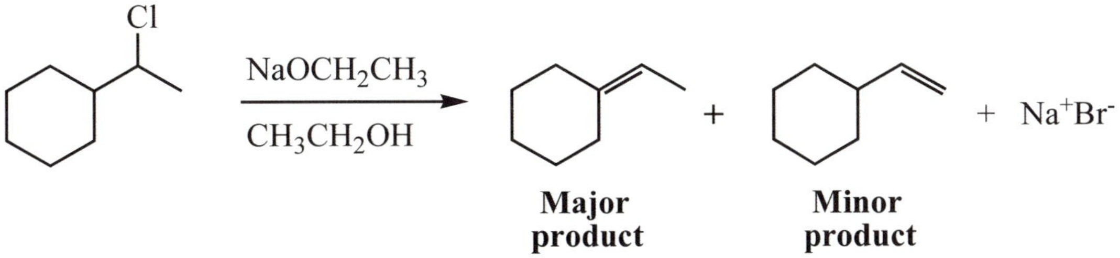 $\xrightarrow[\text{CH}_3\text{CH}_2\text{OH}]{\text{NaOCH}_2\text{CH}_3}$ **Major product** + **Minor product** + Na⁺Br⁻

(d) 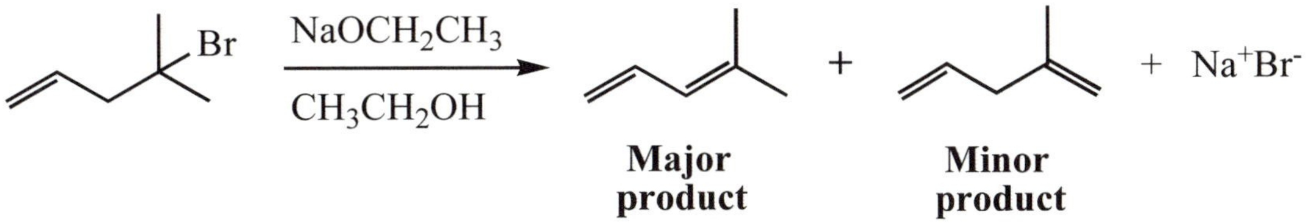 $\xrightarrow[\text{CH}_3\text{CH}_2\text{OH}]{\text{NaOCH}_2\text{CH}_3}$ **Major product** + **Minor product** + Na⁺Br⁻

**7.35**  *Which of the following haloalkanes undergo dehydrohalogenation to give alkenes that do not show cis-trans isomerism?*

(a)

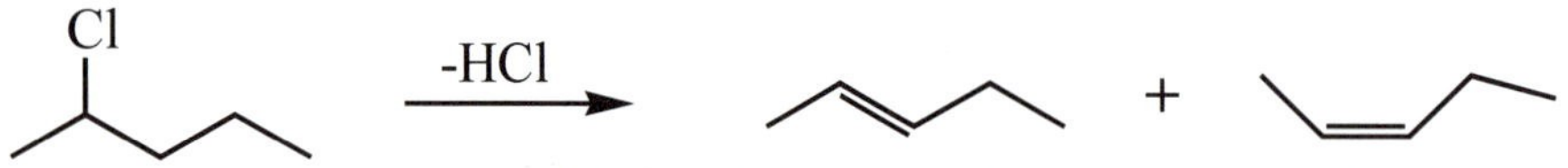

Major Zaitsev products:  cis-trans isomers are possible

(b)

Major Zaitsev products:  cis-trans isomers are possible

(c)

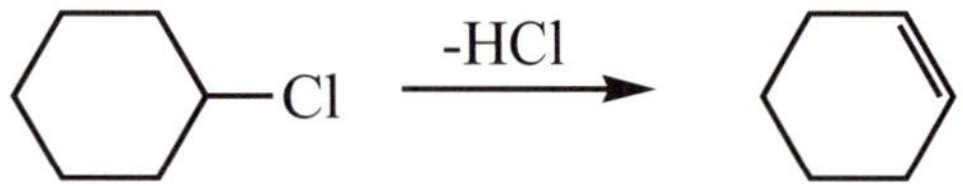

No cis-trans isomers are possible

(d)

No cis-trans isomers are possible

**7.36**  *How many isomers, including cis-trans isomers, are possible for the major product of dehydrohalogenation of each of the following haloalkanes?*

(a) 3-Chloro-3-methylhexane

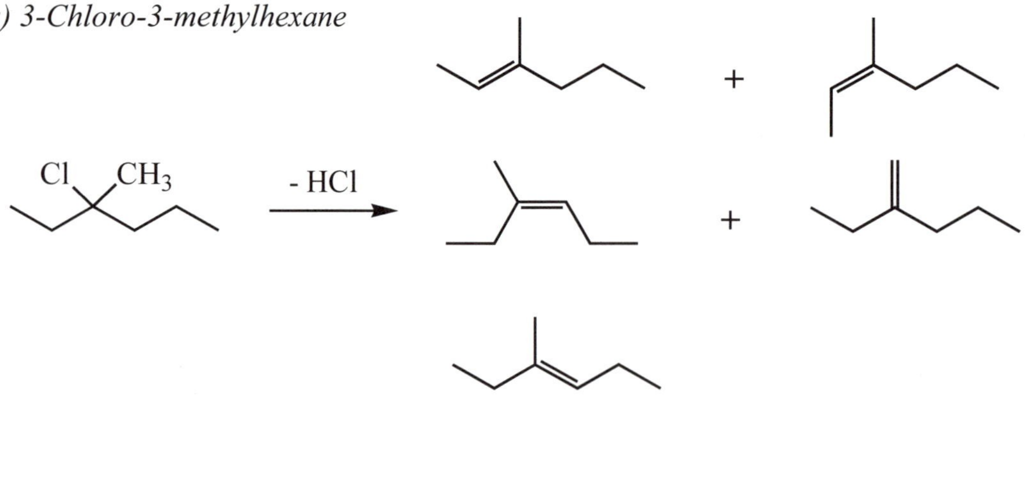

(b) 3-Bromohexane

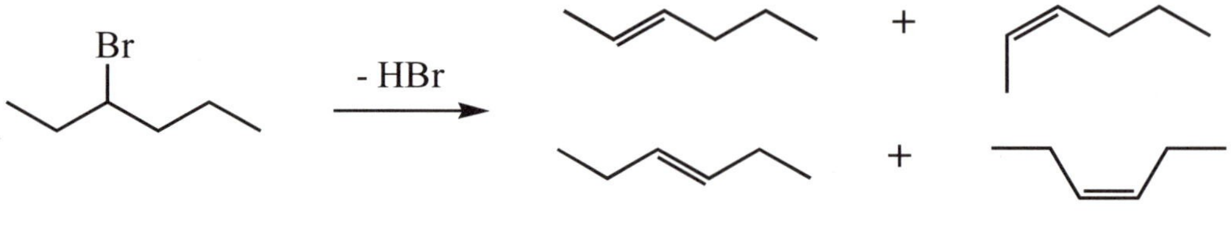

**7.37**   *What haloalkane might you use as a starting material to produce each of the following alkenes in high yield and uncontaminated by isomeric alkenes?*

(a)   [cyclohexyl]—CH$_2$Br   $\xrightarrow{\text{-HBr}}$   [cyclohexylidene]=CH$_2$

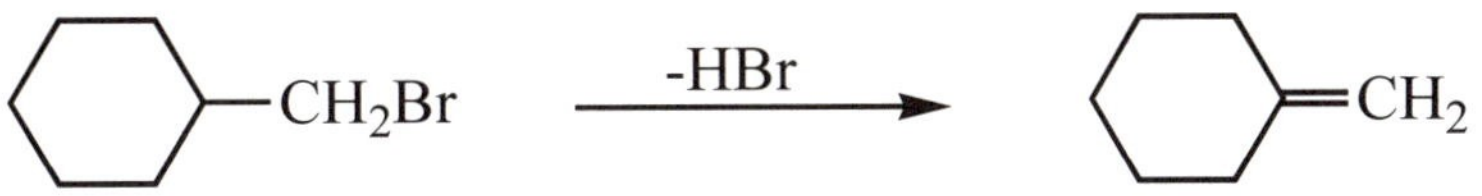

(b)   $\underset{|}{\overset{CH_3}{CH_3\overset{|}{C}HCH_2CH_2CH_2Br}}$   $\xrightarrow{\text{-HBr}}$   $\underset{|}{\overset{CH_3}{CH_3\overset{|}{C}HCH_2CH=CH_2}}$

**7.38**   *For each of the following alkenes, draw structural formulas of all chloroalkanes that undergo dehydrohalogenation when treated with KOH to give that alkene as the major product (for some parts, only one chloroalkane gives the desired alkene as the major product. For other parts, two chloroalkanes may work):*

**When elimination reactions can give two or more possible alkenes, Zaitsev's rule predicts that the most stable alkene (the most substituted alkene) will be the major product.**

(a)   [structure: 1-chloro-1-methylcyclohexane]   **or**   [structure: 2-methyl-1-chlorocyclohexane]   $\xrightarrow{\text{KOH}}$   [structure: 1-methylcyclohexene]

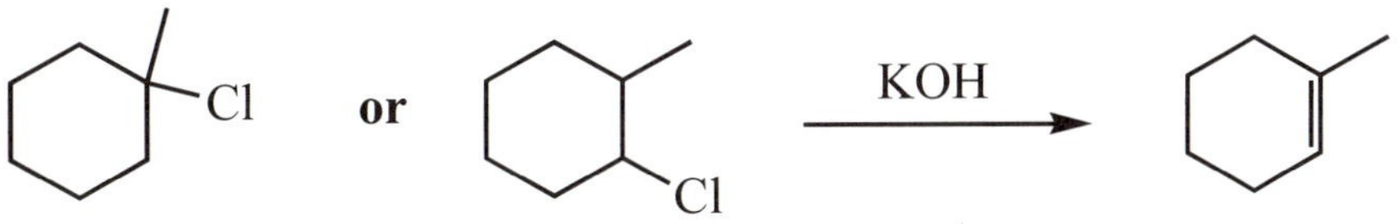

(b)   [structure: chloromethylcyclohexane]   $\xrightarrow{\text{KOH}}$   [structure: methylenecyclohexane, =CH$_2$]

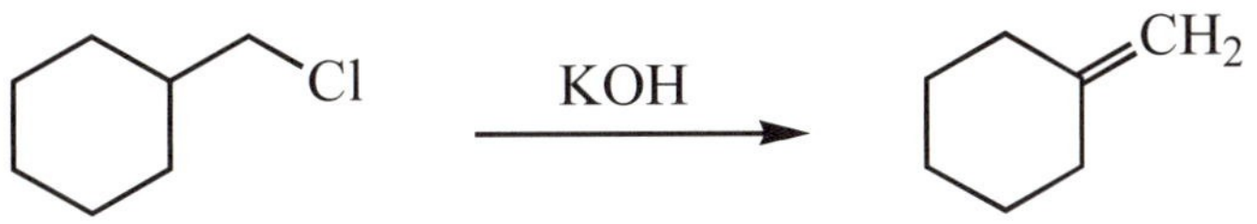

(c)   [structure]   $\xrightarrow{\text{KOH}}$   [structure]

(d)   [structure]   **or**   [structure]   $\xrightarrow{\text{KOH}}$   [structure]

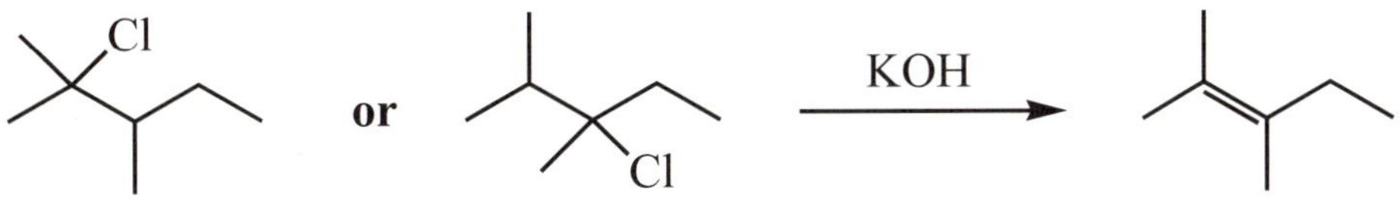

(e)   [structure]   $\xrightarrow{\text{KOH}}$   [structure]

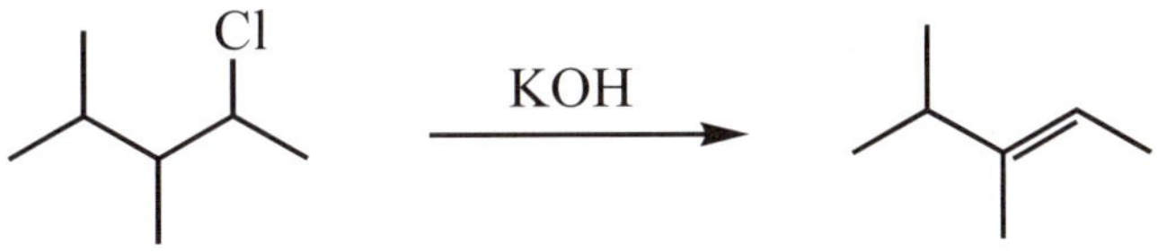

**7.39**  *When cis-4-chlorocyclohexanol is treated with sodium hydroxide in ethanol, it gives only the substitution product trans-1,4-cyclohexanediol (1). Under the same experimental conditions, trans-4-chlorocyclohexanol gives 3-cyclohexenol (2) and the bicyclic ether (3):*

*cis*-4-Chloro-          (1)          *trans*-4-Chloro-
cyclohexanol                          cyclohexanol

(a) *Propose a mechanism for the formation of product (1), and account for its configuration.*

(b) *Propose a mechanism for formation of product (2).*

(c) *Account for the fact that the bicyclic ether (3) is formed from the trans isomer, but not from the cis isomer.*

**$S_N2$ reactions occur with a backside attack by the nucleophile on the carbon bearing the leaving group. In *trans*-4-chlorocyclohexanol, the nucleophilic atom is created with the deprotonation of the hydroxyl proton. The alkoxide nucleophile is now perfectly oriented to initiate a backside attack. The alkoxide nucleophile from *cis*-4-chlorocyclohexanol is situated on the same side as the carbon-halogen bond, therefore, cannot undergo a backside attack and complete the substitution.**

## Synthesis

**7.40** *Show how to convert the given starting material into the desired product (note that some syntheses require only one step, whereas others require two or more steps):*

(a) $\xrightarrow[\text{CH}_3\text{CH}_2\text{OH}]{\text{NaOCH}_2\text{CH}_3}$

(b) $\xrightarrow{\text{HBr}}$

(c) $\xrightarrow[\text{CH}_3\text{CH}_2\text{OH}]{\text{NaOCH}_2\text{CH}_3}$ $\xrightarrow[\text{H}_2\text{SO}_4]{\text{H}_2\text{O}}$

(d) $\xrightarrow[\text{CH}_3\text{CH}_2\text{OH}]{\text{NaOCH}_2\text{CH}_3}$

(e) $\xrightarrow[\text{CH}_3\text{CH}_2\text{OH}]{\text{NaOCH}_2\text{CH}_3}$ $\xrightarrow[\text{ROOH}]{\text{OsO}_4}$

(f) $\xrightarrow[\text{CH}_3\text{CH}_2\text{OH}]{\text{NaOCH}_2\text{CH}_3}$ $\xrightarrow{\text{Br}_2}$

## Looking Ahead

**7.41** *The Williamson ether synthesis involves treating a haloalkane with a metal alkoxide. Following are two reactions intended to give benzyl tert-butyl ether. One reaction gives the ether in a good yield, the other reaction does not. Which reaction gives the ether? What is the product of the other reaction, and how do you account for its formation?*

Reaction (a) will give the best yield of ether product because it involves favorable conditions for an $S_N2$ reaction, involving a 1° halide substrate and a good nucleophile. The substrate also precludes E2 reactions with the absence of β-protons. Reaction (b) will predominately undergo an E2 reaction with a strong base deprotonating the β-proton on a sterically hindered 3° halide to produce an alkene as the major product.

**7.42** *The following ethers can, in principle, be synthesized by two different combinations of haloalkane or halocycloalkane and metal alkoxide. Show one combination that forms ether bond (1) and another that forms ether bond (2). Which combination gives the higher yield of ether?*

(a) **The formation of bond (2) gives a higher yield in the ether because an $S_N2$ reaction between a nucleophile and a primary halide yields less E2 product than the reaction between nucleophile and secondary halide, as in the formation of bond (1).**

*(b)* **The formation of bond (1) gives exclusively the ether product by an S$_N$2. No E2 reaction is possible because methyl bromide lacks the required β-proton. Attempts to form bond (2) will result predominately in an E2 reaction yielding an alkene as the major product.**

**Bond (1):** ⟩—O⁻Na⁺ + CH$_3$Br ⟶ (1) ⟶ **Only product**

**Bond (2):** ⟩—Br + CH$_3$O⁻Na⁺ ⟶ (2) + ⟩=

Minor          Major
product          product

*(c)* **The formation of bond (1) gives a higher yield in the ether because an S$_N$2 reaction between a nucleophile and a primary halide yields less E2 product than the reaction between nucleophile and secondary halide, as in the formation of bond (2).**

**Bond (1):** ⟍⟋⟍Br + ⟩—O⁻Na⁺ ⟶ (1) (2)

**Bond (2):** ⟍⟋⟍O⁻Na⁺ + Br—⟩ ⟶

**7.43**  *Propose a mechanism for this reaction:*

$$Cl-CH_2-CH_2-OH \xrightarrow{Na_2CO_3,\ H_2O} H_2C{-}CH_2\ (Ethylene\ oxide)$$

2-Chloroethanol          Ethylene oxide

$$Cl-CH_2CH_2-\overset{..}{\underset{..}{O}}-H\ +\ \overset{..}{\underset{..}{:}}OH\ \longrightarrow\ Cl-CH_2CH_2-\overset{..}{\underset{..}{O}}:^-\ \longrightarrow\ H_2C{-}CH_2$$

**7.44**  *An OH group is a poor leaving group, and yet substitution occurs readily in the following reaction. Propose a mechanism for this reaction that shows how OH overcomes its limitation of being a poor leaving group.*

**Hydroxyl is not the leaving group under acidic conditions.  Instead, the hydroxyl is protonated by the strong acid, HBr, then leaves as water, which is a better leaving group than hydroxide.**

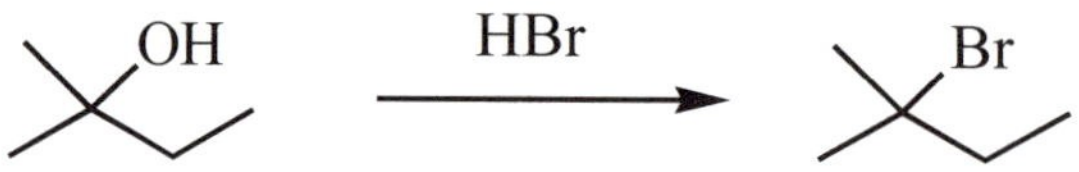

**Step 1:  Protonation of the hydroxyl by HBr.**

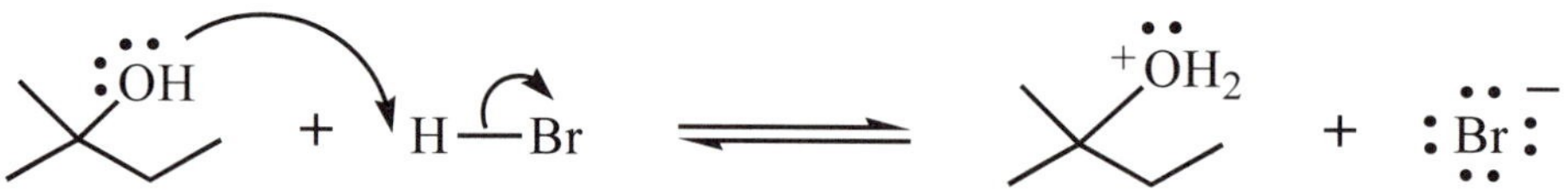

**Step 2:  Loss of water to form the 3° carbocation.**

**Step 3:  Nucleophilic attack of the bromide anion on the carbocation to form the alkyl halide.**

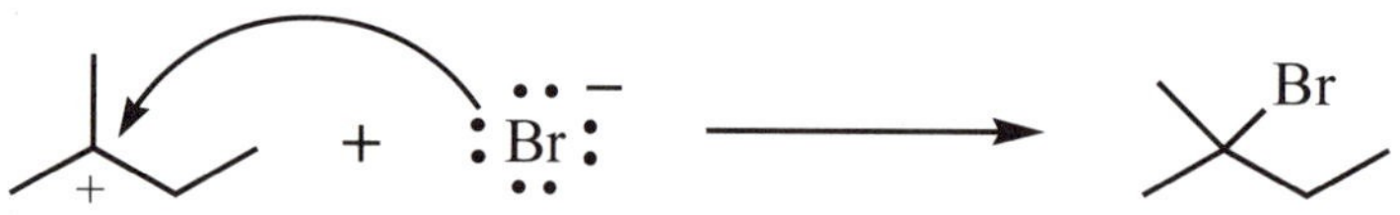

**7.45**  *Explain why (S)-2-bromobutane becomes optically inactive when treated with sodium bromide in dimethylsulfoxide (DMSO):*

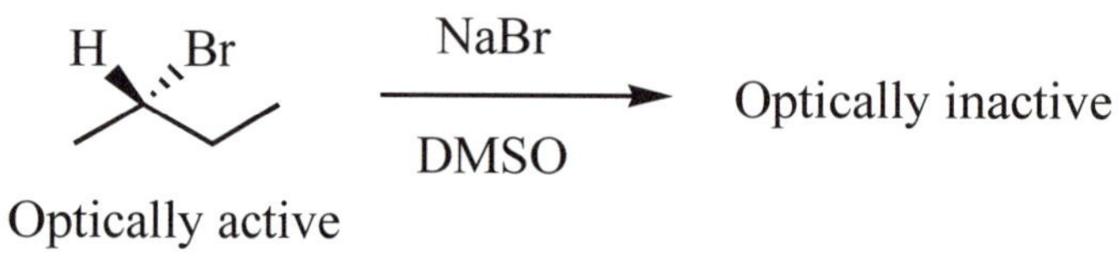

**Under these $S_N2$ conditions, the bromide nucleophile inverts the stereochemistry of the stereocenter.   After 50% of the reaction is complete, racemization is attained.**

**7.46**   *Explain why phenoxide is a much poorer nucleophile than cyclohexoxide:*

Sodium phenoxide                    Sodium cyclohexoxide

**The negative charge on phenoxide is delocalized by resonance over four different atoms and 5 resonance structures.  This leaves the electron pairs less available for nucleophilic reactions relative to cyclohexanoxide, which has no electron delocalization by resonance.**

**7.47**   *In ethers, each side of the oxygen is essentially an OR group and is thus a poor leaving group.  Epoxides are three-membered ring ethers.  Explain why an epoxide reacts readily with a nucleophile despite being an ether.*

R–O–R   +   :Nu⁻   ⟶   no reaction
An ether

An epoxide

**Although alkoxides are poor leaving groups, the release of strain in the three-membered ring is a driving force behind ring opening of epoxides by nucleophiles.**

**CHAPTER 8**
*Solutions to Problems*

**8.1**     *Write the IUPAC name for each alcohol:*

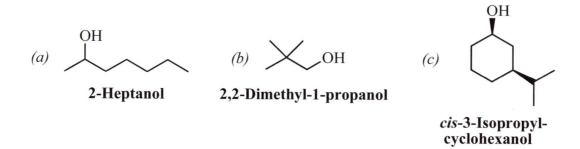

(a)     **2-Heptanol**          (b)     **2,2-Dimethyl-1-propanol**          (c)     ***cis*-3-Isopropyl-
cyclohexanol**

**8.2**     *Classify each alcohol as primary, secondary, or tertiary:*

(a)     **Primary**          (b)     **Secondary**          (c) $CH_2=CHCH_2OH$     **Primary**          (d)     **Tertiary**

**8.3**     *Write the IUPAC name for each unsaturated alcohol:*

(a)          OH          (b)          —OH

(*E*)-3-Penten-1-ol          2-Cyclopenten-1-ol

**8.4**     *Predict the position of equilibrium for this acid-base reaction. (Hint: Review Section 2.5.)*

$$CH_3CH_2O^- Na^+ \; + \; CH_3\overset{\overset{\displaystyle O}{\|}}{C}OH \; \rightleftharpoons \; CH_3CH_2OH \; + \; CH_3\overset{\overset{\displaystyle O}{\|}}{C}O^- Na^+$$

pK$_a$ = 4.76                    pK$_a$ = 15.9

**Acid-base equilibria favor the side with the weaker acid/weaker base.  Ethanol is a weaker acid; therefore, equilibrium lies to the right.**

**8.5**    *For each of the following alcohols, draw structural formulas for the alkenes that form upon acid-catalyzed dehydration, and predict which alkene is the major product from each alcohol.*

(a)    HO $\diagup$ $\xrightarrow{\text{H}_2\text{SO}_4}$   [structure]   +   [structure]

                                    **Major product**      **Minor product**

(b)    [cyclopentane with OH and CH$_3$] $\xrightarrow{\text{H}_2\text{SO}_4}$   [cyclopentene–CH$_3$]   +   [cyclopentane=CH$_2$]

                                      **Major product**      **Minor product**

**8.6**    *Draw the product formed by treating each alcohol in Example 8.6 with chromic acid.*

(a)    [hexanol chain]—OH $\xrightarrow{\text{H}_2\text{CrO}_4}$   [hexanoic acid]   +   Cr$^{3+}$

(b)    [2-pentanol with OH] $\xrightarrow{\text{H}_2\text{CrO}_4}$   [2-pentanone]   +   Cr$^{3+}$

(c)    [cyclohexane]—OH $\xrightarrow{\text{H}_2\text{CrO}_4}$   [cyclohexanone]=O   +   Cr$^{3+}$

**8.7**    *Write the IUPAC and common name for each ether:*

(a)  CH$_3$CHCH$_2$OCH$_2$CH$_3$  
      |  
    CH$_3$

(b)  [cyclopentane]—OCH$_3$

**1-Ethoxy-2-methylpropane**  
**or  Ethyl isobutyl ether**

**Methoxycyclopentane**  
**or  Cyclopentyl methyl ether**

**8.8**    *Arrange these compounds in order of increasing boiling point:*

    CH$_3$OCH$_2$CH$_2$OCH$_3$        HOCH$_2$CH$_2$OH        CH$_3$OCH$_2$CH$_2$OH

$$CH_3OCH_2CH_2OCH_3 \quad < \quad CH_3OCH_2CH_2OH \quad < \quad HOCH_2CH_2OH$$

| 1,2-Dimethoxyethane (Methyl cellosolve) bp 84 °C | 2-Methoxyethanol bp 125 °C | 1,2-Ethanediol (Ethylene glycol) bp 198 °C |

**Hydrogen bonding has a great influence on boiling points.  Although 1,2-dimethoxyethane is a polar molecule, there are only weak intermolecular forces between its molecules in the liquid state.  Both 2-methoxyethanol and 1,2-ethanediol have intermolecular hydrogen bonding between molecules.  Because 1,2-ethanediol has more sites for hydrogen bonding, it has a higher boiling point than 2-methoxyethanol.**

**8.9**     *Draw the structural formula of the epoxide formed by treating 1,2-dimethylcyclopentene with a peroxycarboxylic acid.*

**8.10**     *Show how to convert cyclohexene to cis-1,2-cyclohexanediol.*

**8.11**     *Write the IUPAC name for each thiol:*

(a)  **3-Methyl-1-butanethiol**          (b)  **3-Methyl-2-butanethiol**

**8.12**   *Which of the following are secondary alcohols?*

(a) 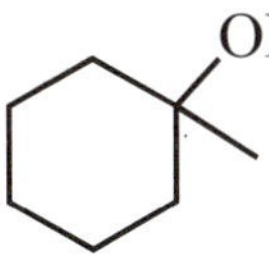    (b)  $(CH_3)_3COH$    (c) 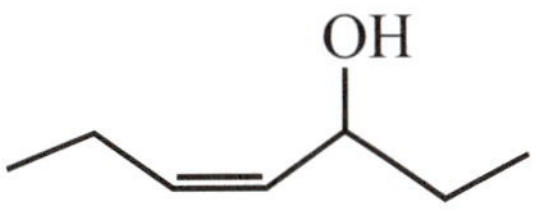    (d)

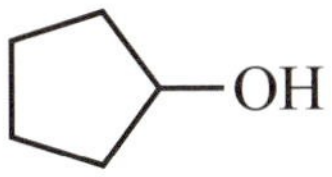

**3û̂alcohol**                    **3û̂alcohol**              **2û̂alcohol**                    **2û̂alcohol**

**Alcohols (c) and (d) are secondary alcohols.**


**8.13**   *Name these compounds:*

(a)     (b)    (c)

**1-Pentanol**                **1,3-Propanediol**           **3-Buten-1-ol**

(d) 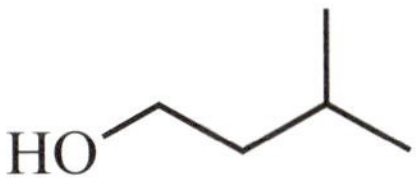    (e)    (f)

**3-Methyl-1-butanol**                                          **1-Butanethiol**

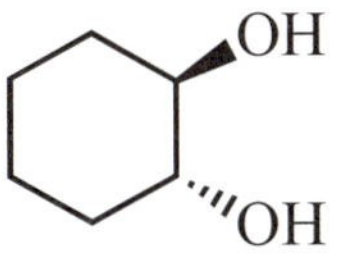

                        ***trans*-1,2-Cyclohexanediol**

**8.14**   *Draw a structural formula for each alcohol:*

(a) *Isopropyl alcohol*        (b) *Propylene glycol* 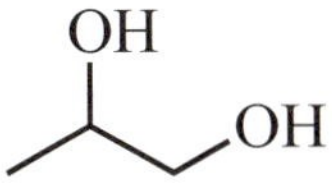        (c) *(R)-5-Methyl-2-hexanol*

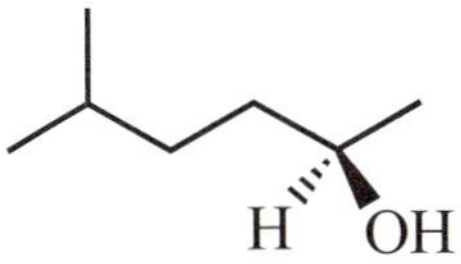

(d) *2-Methyl-2-propyl-1,3-propanediol*         (e) *2,2-Dimethyl-1-propanol*

(f) *2-Mercaptoethanol* 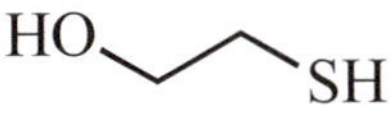        (g) *1,4-Butanediol* 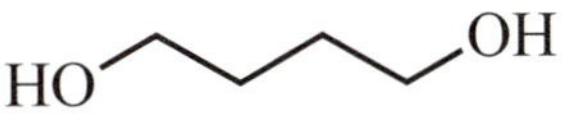        (h) *(Z)-5-Methyl-2-hexen-1-ol*

*(i) cis-3-Pentene-1-ol*          *(j) trans-1,4-Cyclohexanediol*

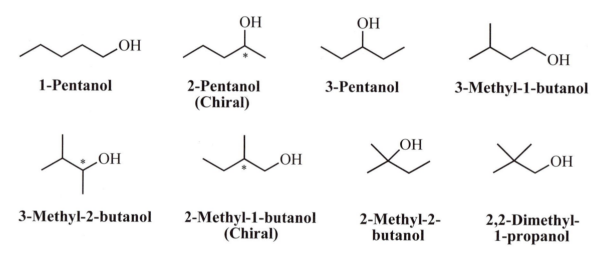

**8.15**    *Write names for these ethers:*

*(a)*                                    *(b)*                                    *(c)*

**Dicyclopentyl ether**          **Dibutyl ether**          **2-Ethoxyethanol**

**8.16**    *Name and draw structural formulas for the eight isomeric alcohols with molecular formula $C_5H_{12}O$. Which are chiral?*

**1-Pentanol**          **2-Pentanol**          **3-Pentanol**          **3-Methyl-1-butanol**
                        **(Chiral)**

**3-Methyl-2-butanol**    **2-Methyl-1-butanol**    **2-Methyl-2-**      **2,2-Dimethyl-**
                          **(Chiral)**             **butanol**         **1-propanol**

## Physical Properties

**8.17**    *Arrange these compounds in order of increasing boiling point (values in °C are -42, 78, 117, and 198):*

(a) $CH_3CH_2CH_2CH_2OH$   (b) $CH_3CH_2OH$   (c) $HOCH_2CH_2OH$   (d) $CH_3CH_2CH_3$
  **1-Butanol**              **Ethanol**          **Ethylene glycol**      **Propane**
  **(bp 117 °C)**            **(bp 78 °C)**        **(bp 198 °C)**          **(bp -42 °C)**

**Intermolecular forces and molecular size (greater attractive dispersion forces) influence boiling points.  As intermolecular forces or molecular size increase, so does the boiling point.   Intermolecular hydrogen bonding is a very strong intermolecular force and a powerful influence on boiling points.  The highest boiling point belongs to ethylene glycol with two –OH groups to engage in intermolecular hydrogen bonding.  Both ethanol and 1-butanol have one –OH group engaging in hydrogen bonding, but 1-butanol has longer hydrocarbon chain, therefore it has greater**

dispersion forces resulting in a higher boiling point than ethanol.  Propane has only dispersion forces holding it in the liquid state and no hydrogen bonding, therefore it has the lowest boiling point of the series.  The order for increasing boiling point is:

**(d)  <  (b)  <  (a)  <  (c)**

**8.18**  *Arrange these compounds in order of increasing boiling point (values in °C are -42, -24, 78, and 118):*

(a) $CH_3CH_2OH$            (b) $CH_3OCH_3$         (c) $CH_3CH_2CH_3$        (d) $CH_3COOH$

**Ethanol**            **Dimethyl ether**          **Propane**                **Acetic acid**
**bp 78 °C**              **-24 °C**                 **-42 °C**                  **118 °C**

Intermolecular forces and molecular size influence boiling points.  As intermolecular forces such as dipole interactions and hydrogen bonding or molecular size increase, so does the boiling point.  Non-polar propane has the weakest intermolecular forces with only dispersion forces.  Dimethyl ether is a polar molecule with dipole intermolecular forces, but it cannot form intermolecular hydrogen bonds with other dimethyl ether molecules.  Both acetic acid and ethanol can participate in intermolecular hydrogen bonding through their hydroxyls, but the carbonyl oxygen in acetic acid can participate in additional hydrogen bonding with the carbonyl oxygen acting as a hydrogen bond acceptor.

**8.19**  *Propanoic acid and methyl acetate are constitutional isomers, and both are liquids at room temperature:*

$$CH_3CH_2\overset{\overset{O}{\|}}{C}OH \qquad CH_3\overset{\overset{O}{\|}}{C}OCH_3$$

Propanoic acid                    Methyl acetate

*One of these compounds has a boiling point of 141°C; the other has a boiling point of 57°C. Which compound has which boiling point?*

Propanoic acid has the higher boiling point (141 °C) due to intermolecular hydrogen bonding.  The carboxyl group can function both as a hydrogen bond donor (through the –OH group) and acceptor (through the C=O and C-O groups) and is responsible for a high degree of intermolecular association between molecules in the liquid state.  Methyl acetate does not have intermolecular hydrogen bonding, thus it has the lower of the two boiling points (57 °C).

**8.20**  *Draw all possible staggered conformations of ethylene glycol (HOCH₂CH₂OH).  Can you explain why the gauche conformation is more stable than the anti conformation by approximately 1 kcal/mol?*

**Anti conformation**          **Gauche conformation**

**Usually, gauche conformations are higher in energy than the anti conformation, but in the case of ethylene glycol, intramolecular hydrogen bonding between adjacent hydroxyls stabilizes the gauche conformation.**

**8.21**  *Following are structural formulas for 1-butanol and 1-butanethiol:*

1-Butanol          1-Butanethiol

*One of these compounds has a boiling point of 98.5°C; the other has a boiling point of 117°C. Which compound has which boiling point?*

**The S-H bond of thiols is much less polar than the O-H bond of alcohols; therefore, thiols do not form significant hydrogen bonds.  Intermolecular forces must be broken before a liquid can boil.  The strong intermolecular hydrogen bonding between molecules of 1-butanol is responsible for its higher boiling point than 1-butanethiol.**

**8.22**  *From each pair of compounds, select the one that is more soluble in water:*

*(a)* $CH_2Cl_2$   *or*   $\boxed{CH_3OH}$

**Methanol ($CH_3OH$) is soluble in all proportions of water while methylene chloride ($CH_2Cl_2$) is insoluble in water.   The highly polar –OH group of methanol is capable of participating both as hydrogen bond donor and hydrogen bond acceptor with water and, therefore, methanol interacts strongly with water by intermolecular association.  No such interaction is possible with dichloromethane.**

$$(b) \quad \boxed{\underset{\substack{\|\\O}}{CH_3\overset{\displaystyle O}{C}CH_3}} \quad or \quad CH_3\overset{\displaystyle CH_2}{C}CH_3$$

**Propanone (acetone), $CH_3COCH_3$, has a large dipole moment and can function as a hydrogen bond acceptor from water through its carbonyl oxygen and is soluble in water in all proportions. 2-Methylpropene (isobutylene) has no interaction with water and is insoluble in water.**

$$(c) \quad CH_3CH_2Cl \quad or \quad \boxed{NaCl}$$

**Chloroethane is a nonpolar organic compound and insoluble is water. Sodium chloride is an ionic compound and easily dissolves in water by ionization.**

$$(d) \quad CH_3CH_2CH_2SH \quad or \quad \boxed{CH_3CH_2CH_2OH}$$

**Sulfur is less electronegative than oxygen, so a S-H bond is less polarized than an O-H bond. Hydrogen bonding is therefore weaker with thiols than with alcohols, so the alcohol is more able to interact with water molecules through hydrogen bonding. The alcohol is more soluble in water.**

$$(e) \quad \boxed{CH_3CH_2\overset{\displaystyle OH}{C}HCH_2CH_3} \quad or \quad CH_3CH_2\overset{\displaystyle O}{C}CH_2CH_3$$

**The hydroxyl group is both a hydrogen bond donor and acceptor while the ketone carbonyl oxygen is only a hydrogen bond acceptor, thus the alcohol interacts more strongly with water through hydrogen bonding, and is more soluble in water.**

**8.23**    *Arrange the compounds in each set in order of decreasing solubility in water:*

**In general, the more strongly a molecule can take part in hydrogen bonding with water, the greater the molecule will interact with water molecules and dissolve.**
*(a) Ethanol; butane; diethyl ether*

**Ethanol  >  Diethyl ether  >  Butane**

**The hydroxyl group on ethanol is both a hydrogen bond donor and acceptor, bringing strong interactions with water. The oxygen on diethyl ether can accept**

**hydrogen bonds from water, but does not act as a hydrogen bond donor; therefore ether only weakly interacts with water and is slightly soluble. Butane is a nonpolar hydrocarbon, which forms no interactions with water and is insoluble.**

*(b) 1-Hexanol; 1,2-hexanediol; hexane*

**1,2-Hexanediol > 1-Heaxanol > Hexane**

**The two hydroxyl groups on 1,2-hexanediol are both hydrogen bond donors and acceptors, allowing stronger interactions with water and taking part in more hydrogen bonding than just the one hydroxyl in 1-hexanol. Hexane is a nonpolar hydrocarbon, which forms no interactions with water and is insoluble.**

**8.24**  *Each of the following compounds is a common organic solvent. From each pair of compounds, select the solvent with the greater solubility in water:*

**The solvent with the greater solubility in water is circled. Solubility in water increases with increasing hydrogen bonding ability and decreases with increasing surface area of hydrophobic groups such as hydrocarbon chains.**

*(a)* $CH_2Cl_2$   or   $\boxed{CH_3CH_2OH}$      *(b)* $CH_3CH_2OCH_2CH_3$   or   $\boxed{CH_3CH_2OH}$

*(c)* $\boxed{CH_3\overset{\overset{\displaystyle O}{\|}}{C}CH_3}$ or $CH_3CH_2OCH_2CH_3$    *(d)* $\boxed{CH_3CH_2OCH_2CH_3}$   or   $CH_3(CH_2)_3CH_3$

**Synthesis of Alcohols**

**8.25**  *Give the structural formula of an alkene or alkenes from which each alcohol or glycol can be prepared:*

*(a)*    $\underset{H}{\overset{CH_3}{\diagdown}}C=C\underset{CH_3}{\overset{H}{\diagup}}$   **or**   $\underset{H}{\overset{CH_3}{\diagdown}}C=C\underset{H}{\overset{CH_3}{\diagup}}$   $\xrightarrow[H_2SO_4]{H_2O}$   $CH_3\overset{\overset{\displaystyle OH}{|}}{C}HCH_2CH_3$

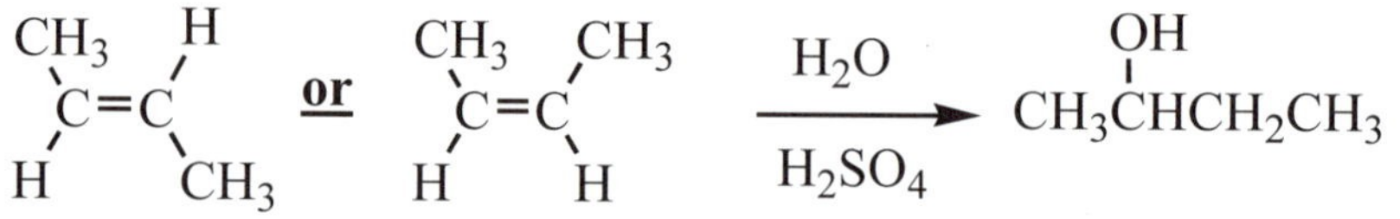

**or**   $H_2C=CHCH_2CH_3$

*(b)*   ⬡—$CH_3$   **or**   ⬡=$CH_2$   $\xrightarrow[H_2SO_4]{H_2O}$   ⬡$\langle\overset{CH_3}{OH}$

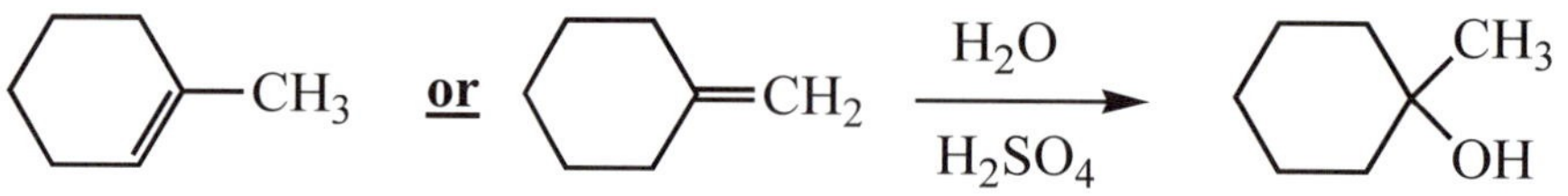

(c)

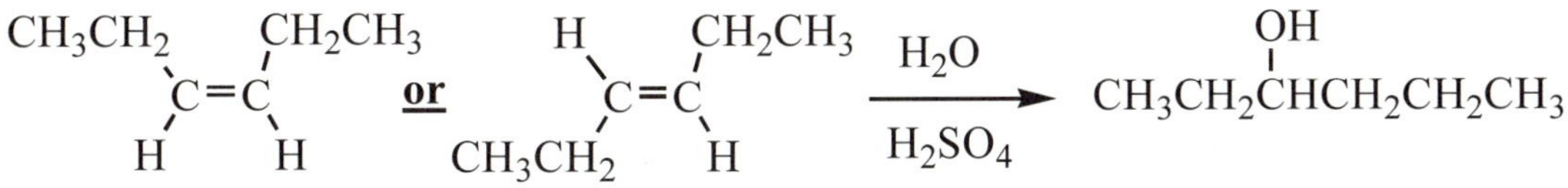

$$\underset{H}{\overset{CH_3CH_2}{>}}C=C\underset{H}{\overset{CH_2CH_3}{<}} \quad \underline{or} \quad \underset{CH_3CH_2}{\overset{H}{>}}C=C\underset{H}{\overset{CH_2CH_3}{<}} \xrightarrow[H_2SO_4]{H_2O} \overset{OH}{\underset{|}{CH_3CH_2CHCH_2CH_2CH_3}}$$

(d) $\overset{CH_3}{\underset{|}{CH_3C}}=CHCH_2CH_3$ <u>or</u> $H_2C=\overset{CH_3}{\underset{|}{CCH_2CH_2CH_3}}$ $\xrightarrow[H_2SO_4]{H_2O}$ $\overset{OH}{\underset{|}{CH_3\overset{|}{C}CH_2CH_2CH_3}}$ $\underset{CH_3}{|}$

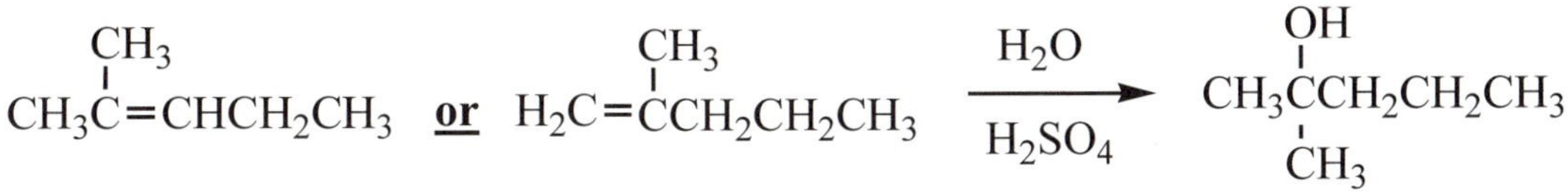

(e) ⬠ $\xrightarrow[H_2SO_4]{H_2O}$ ⬠—OH

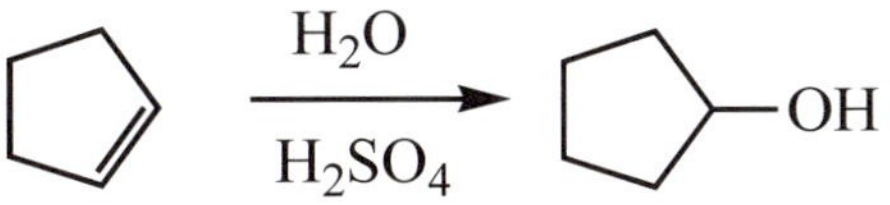

(f) $CH_3CH=CH_2$ $\xrightarrow[H_2O_2]{OsO_4}$ $\overset{HO\ \ OH}{\underset{|\ \ \ |}{CH_3CHCH_3}}$

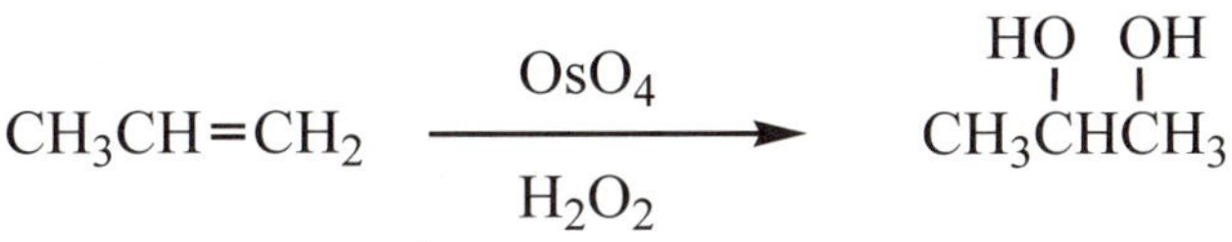

**8.26**   *The addition of bromine to cyclopentene and the acid-catalyzed hydrolysis of cyclopentene oxide are both stereoselective; each gives a trans product.  Compare the mechanisms of these two reactions, and show how each mechanism accounts for the formation of the trans product.*

**Both mechanisms involve an attack by the nucleophile on the three-membered ring intermediate from the opposite side as shown, resulting in an anti addition.**

**Acid-catalyzed ring opening of an epoxide:**

**Step 1:  Protonation of epoxide.**

**Step 2:  Ring opening of epoxide.**

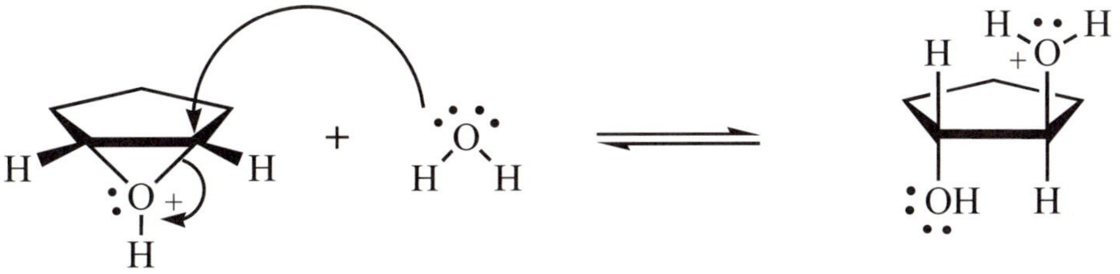

**Step 3:  Loss of proton from the protonated diol intermediate forming *trans*-1,2-cyclopentanediol.**

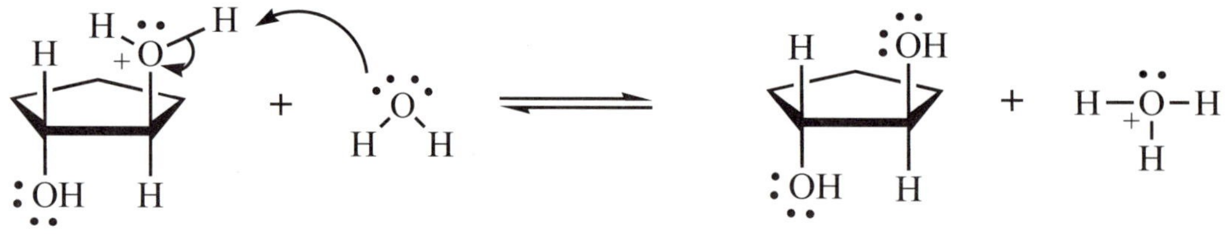

**Reaction of an alkene with bromine:**

**Step 1:  Bromine attack on the double bond to form the bromonium cation.**

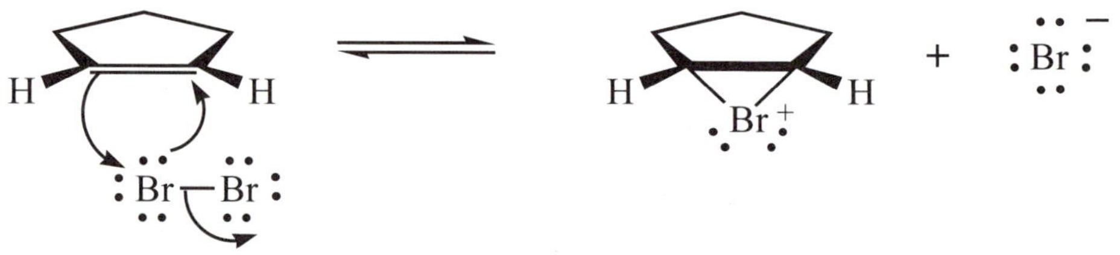

**Step 2:  Bromide attack on the bromonium cation to form *trans*-1,2-cyclopentane.**

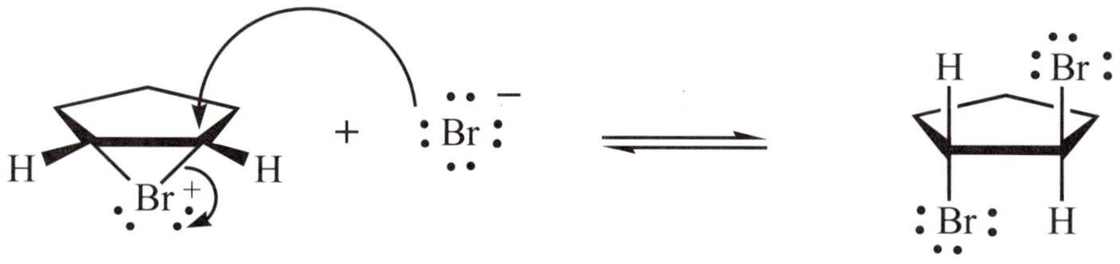

## Acidity of Alcohols and Thiols

**8.27**  *From each pair of compounds, select the stronger acid and, for each stronger acid, write a structural formula for its conjugate base:*

(a) $H_2O$  or  $\boxed{H_2CO_3}$    (b) $CH_3OH$  or  $\boxed{CH_3COOH}$   (c) $\boxed{CH_3COOH}$  or  $CH_3CH_2SH$

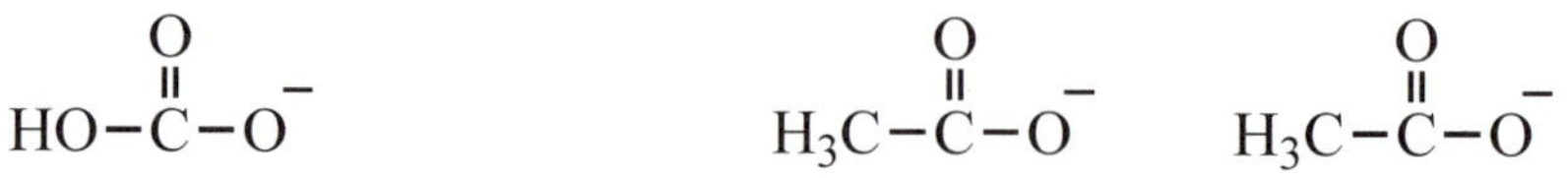

$$HO-\overset{\overset{\displaystyle O}{\|}}{C}-O^-$$    $$H_3C-\overset{\overset{\displaystyle O}{\|}}{C}-O^-$$    $$H_3C-\overset{\overset{\displaystyle O}{\|}}{C}-O^-$$

**Use Table 2.2 for the p$K_a$ values for common acids.  Alkyl thiols are more acidic than their alcohol analogs and have p$K_a$ values around 8-9.**

**8.28**    *Arrange these compounds in order of increasing acidity (from weakest to strongest):*

$$CH_3CH_2CH_2OH \quad < \quad CH_3CH_2CH_2SH \quad < \quad CH_3CH_2\overset{\displaystyle O}{\overset{\|}{C}}OH$$

**Use Table 2.2 to approximate the p$K_a$ values for common acids.  Alkyl thiols (p$K_a$ values between 8 and 9) are more acidic than their alcohol analogs (p$K_a$ values between 16-18).  Carboxylic acids are more acidic than alcohol and thiols and have p$K_a$ values between 4 and 5.**

**8.29**    *From each pair of compounds, select the stronger base and, for each stronger base, write the structural formula of its conjugate acid:*

(a)  OH⁻ or $\boxed{CH_3O^-}$      (b)  $CH_3CH_2S^-$  or  $\boxed{CH_3CH_2O^-}$      (c)  $CH_3CH_2O^-$  or  $\boxed{NH_2^-}$
        **CH₃OH**                                    **CH₃CH₂OH**                                  **NH₃**

**8.30**    *Label the stronger acid, stronger base, weaker acid, and weaker base in each of the following equilibria, and predict the position of each equilibrium; that is, does each lie considerably to the left, does it lie considerably to the right, or are the concentrations evenly balanced?*

*(a)* **The equilibrium lies to the right.**

$$CH_3CH_2O^- \quad + \quad HCl \quad \rightleftharpoons \quad CH_3CH_2OH \quad + \quad Cl^-$$

| Stronger base | Stronger acid | Weaker acid | Weaker base |
|---|---|---|---|
| | p$K_a$ = -7 | p$K_a$ = 15.9 | |

*(b)* **The equilibrium lies to the right.**

$$CH_3\overset{\displaystyle O}{\overset{\|}{C}}OH \quad + \quad CH_3CH_2O^- \quad \rightleftharpoons \quad CH_3\overset{\displaystyle O}{\overset{\|}{C}}O^- \quad + \quad CH_3CH_2OH$$

| Stronger acid | Stronger base | Weaker base | Weaker acid |
|---|---|---|---|
| p$K_a$ = 4.74 | | | p$K_a$ = 15.9 |

**8.31**    *Predict the position of equilibrium for each acid-base reaction; that is, does each lie considerably to the left, does it lie considerably to the right, or are the concentrations evenly balanced?*

*(a)* **Concentrations are evenly balanced.**

$CH_3CH_2OH$ + $Na^+OH^-$ ⇌ $CH_3CH_2O^-Na^+$ + $H_2O$
$pK_a$ 15.9          **Stronger base**                    **Weaker base**        $pK_a$ 15.7
**Weaker acid**                                                                                      **Stronger acid**

*(b)* **Equilibrium lies to the right.**

$CH_3CH_2SH$ + $Na^+OH^-$ → $CH_3CH_2S^-Na^+$ + $H_2O$
$pK_a$ 8.5          **Stronger base**                    **Weaker base**        $pK_a$ 15.7
**Stronger acid**                                                                                  **Weaker acid**

*(c)* **Equilibrium lies to the left.**

$CH_3CH_2OH$ + $CH_3CH_2S^-Na^+$ ← $CH_3CH_2O^-Na^+$ + $CH_3CH_2SH$
$pK_a$ 15.9          **Weaker base**                    **Stronger base**        $pK_a$ 8.5
**Weaker acid**                                                                                    **Stronger acid**

*(d)* **Equilibrium lies to the right.**

$$CH_3CH_2S^-Na^+ + CH_3\overset{O}{\overset{\|}{C}}OH \rightarrow CH_3CH_2SH + CH_3\overset{O}{\overset{\|}{C}}O^-Na^+$$

**Stronger base**          $pK_a$ 4.76                    $pK_a$ 8.5              **Weaker base**
                                   **Stronger acid**              **Weaker acid**

## Reactions of Alcohols

**8.32**   *Show how to distinguish between cyclohexanol and cyclohexene by a simple chemical test. (Hint: Treat each with Br₂ in CCl₄ and watch what happens.)*

**Cyclohexanol is unreactive toward bromine in carbon tetrachloride. Cyclohexene will decolorize a deeply red solution of bromine in carbon tetrachloride according to the following reaction:**

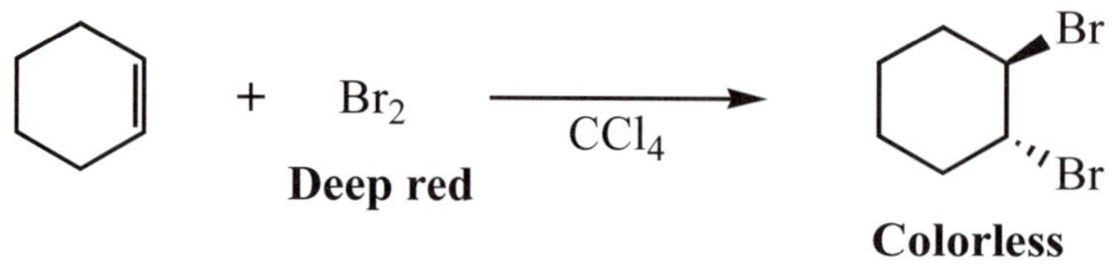

Deep red          $\xrightarrow{CCl_4}$          Colorless

**8.33**    *Write equations for the reaction of 1-butanol, a primary alcohol, with each reagent:*

(a) ~~~~~OH $\xrightarrow{\text{Na}}$ ~~~~~O⁻Na⁺  + H₂

(b) ~~~~~OH $\xrightarrow[\text{heat}]{\text{HBr}}$ ~~~~~Br  + H₂O

(c) ~~~~~OH $\xrightarrow[\text{H}_2\text{SO}_4;\ \text{heat}]{\text{K}_2\text{Cr}_2\text{O}_7}$ ~~~~~COOH  + Cr³⁺

(d) ~~~~~OH $\xrightarrow{\text{SOCl}_2}$ ~~~~~Cl  + SO₂ + Cl⁻

(e) ~~~~~OH $\xrightarrow{\text{PCC}}$ ~~~~~CHO  + Cr³⁺

**8.34**    *Write equations for the reaction of 2-butanol, a secondary alcohol, with each reagent:*

(a) [2-butanol] $\xrightarrow{\text{Na}}$ [sodium 2-butoxide]  + H₂

(b) [2-butanol] $\xrightarrow[\text{heat}]{\text{H}_2\text{SO}_4}$ [1-butene] + [trans-2-butene] + [cis-2-butene]

**Major product**

(c) [2-butanol] $\xrightarrow[\text{heat}]{\text{HBr}}$ [2-bromobutane]  + H₂O

(d) [2-butanol] $\xrightarrow[\text{H}_2\text{SO}_4;\ \text{heat}]{\text{K}_2\text{Cr}_2\text{O}_7}$ [butan-2-one]  + Cr³⁺

(e) [2-butanol] $\xrightarrow{\text{SOCl}_2}$ [2-chlorobutane]  + SO₂ + Cl⁻

(f) [2-butanol] $\xrightarrow{\text{PCC}}$ [butan-2-one]  + Cr³⁺

**8.35**  *When (R)-2-butanol is left standing in aqueous acid, it slowly loses its optical activity. When the organic material is recovered from the aqueous solution, only 2-butanol is found. Account for the observed loss of optical activity.*

**Secondary and tertiary alcohols, when treated with aqueous acid, undergo an $S_N1$ reaction with water. When (R)-2-butanol is protonated and loses water, an achiral planar carbocation results and the original stereochemistry is lost. Attack from either side results in a racemic mixture of the enantiomers.**

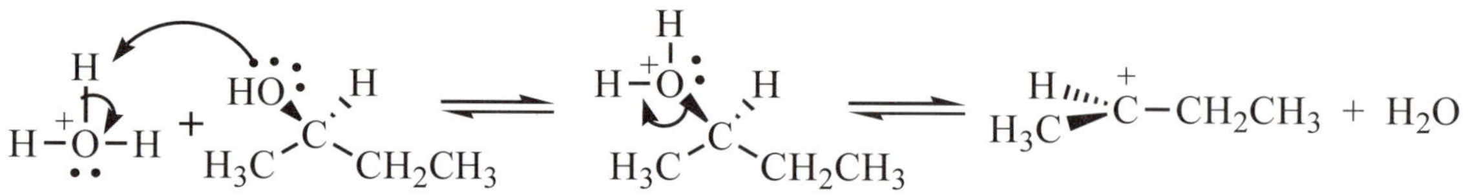

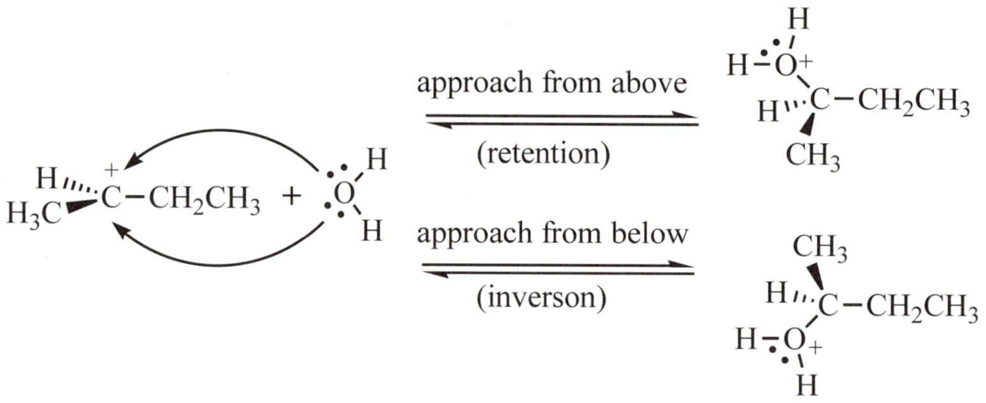

approach from above

(retention)

approach from below

(inverson)

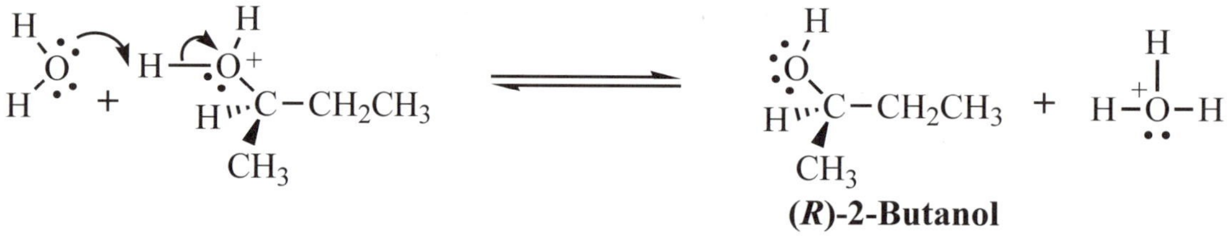

(R)-2-Butanol

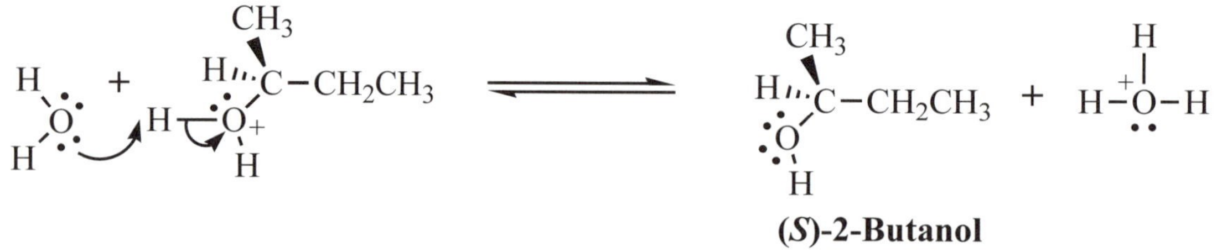

(S)-2-Butanol

**8.36**  *What is the most likely mechanism of the following reaction? Draw a structural formula for the reaction intermediate(s) formed during the reaction.*

**Step 1:  Protonation of the alcohol.**

**Step 2:  Loss of water from the protonated alcohol to form a 3° carbocation.**

**Step 3:  Attack of the chloride nucleophile on the carbocation to form product.**

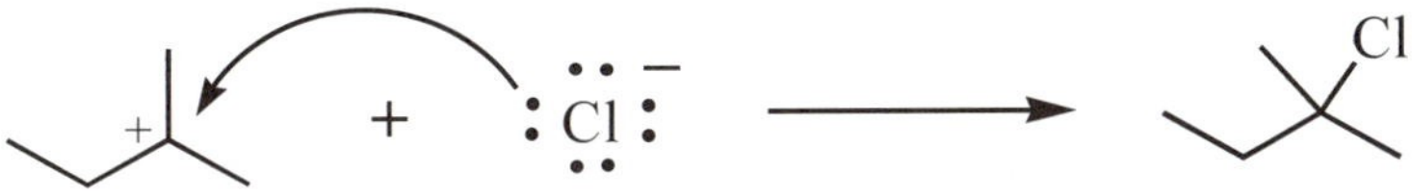

**8.37**    *Complete the equations for these reactions:*

(a)

(b)

(c)

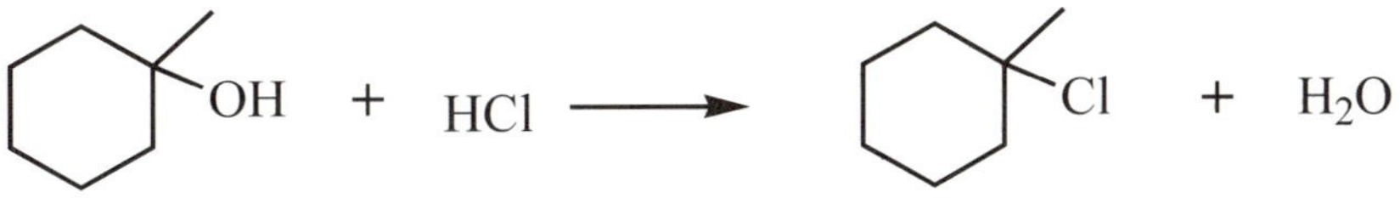

(d)

*(e)*

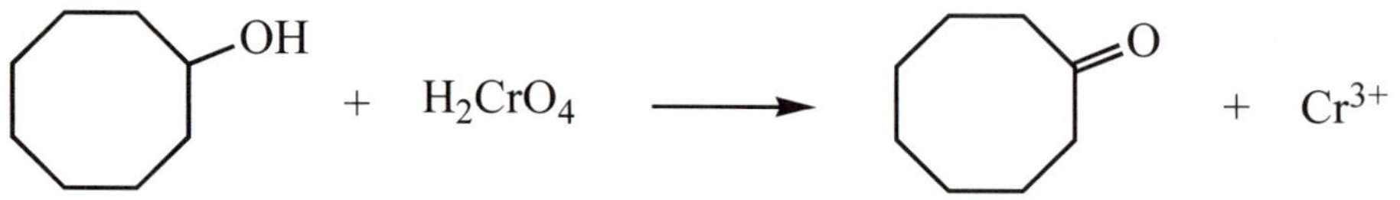
OH
+ H2CrO4
O
+ Cr3+

*(f)*

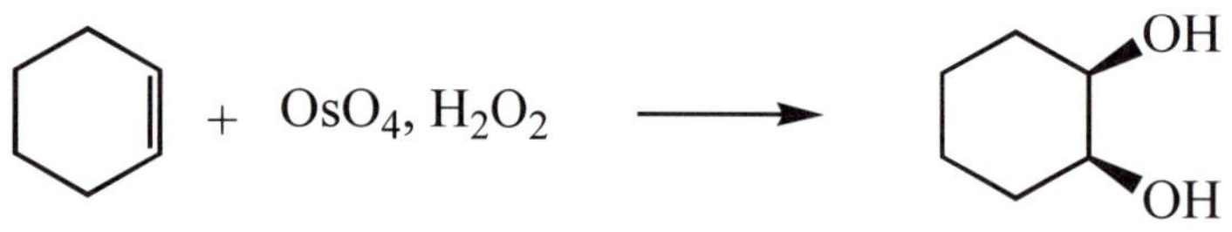
+ OsO4, H2O2
OH
OH

**8.38** *In the commercial synthesis of methyl tert-butyl ether (MTBE), once used as an antiknock, octane-improving gasoline additive, 2-methylpropene and methanol are passed over an acid catalyst to give the ether:*

$$CH_3\overset{\overset{\displaystyle CH_3}{|}}{C}{=}CH_2 \quad + \quad CH_3OH \xrightarrow[\textit{catalyst}]{\textit{acid}} CH_3\underset{\underset{\displaystyle CH_3}{|}}{\overset{\overset{\displaystyle CH_3}{|}}{C}}OCH_3$$

*2-Methylpropene (Isobutylene)*    *Methanol*    *2-Methoxy-2-methylpropane (Methyl tert-butyl ether, MTBE)*

*Propose a mechanism for this reaction.*

**Step 1:**

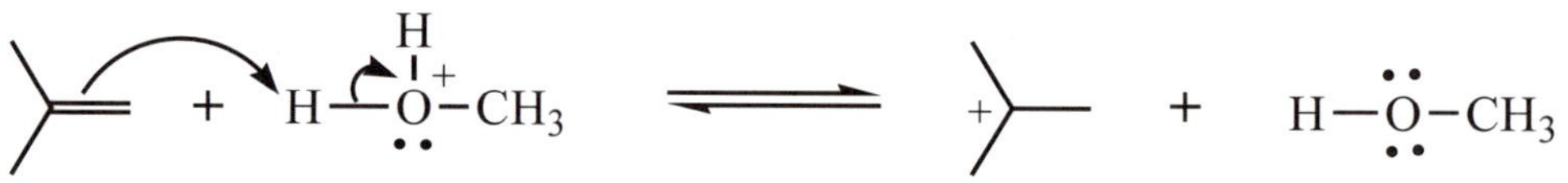

**Step 2:**

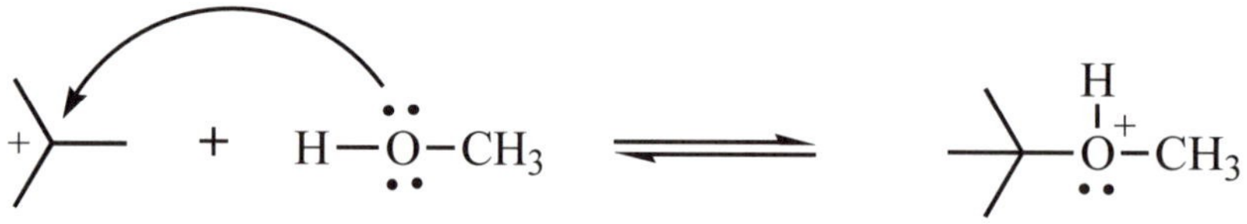

**Step 3:**

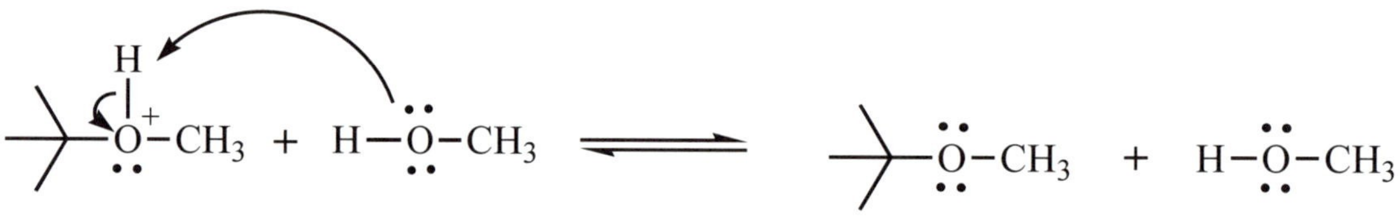

**8.39**  *Cyclic bromoalcohols, upon treatment with base, can sometimes undergo intramolecular S_N2 reactions to form bicyclic ethers. Determine whether each of the following compounds is capable of forming a bicyclic ether, and draw the product for those which can:*

(a)

Br,,. ⬡ ,OH   —base→   bicyclic ether   +   Br⁻

(b)

Br ⬡ ,OH   —base→   Does not undergo cyclization

(c)

,,,OH / Br cyclopentane   —base→   epoxide   +   Br⁻

**Syntheses**

**8.40**  *Show how to convert*

(a) *1-Propanol to 2-propanol in two steps.*

CH₃CH₂CH₂OH  $\xrightarrow[\text{heat}]{H_2SO_4}$  propene  $\xrightarrow[H_2O]{H_2SO_4}$  2-propanol (OH)

(b) *Cyclohexene to cyclohexanone in two steps.*

cyclohexene  $\xrightarrow[H_2O]{H_2SO_4}$  cyclohexanol (—OH)  $\xrightarrow{H_2CrO_4}$  cyclohexanone (=O)

(c) *Cyclohexanol to cis-1,2-cyclohexanediol in two steps.*

cyclohexanol (—OH)  $\xrightarrow[\text{heat}]{H_2SO_4}$  cyclohexene  $\xrightarrow[ROOH]{OsO_4}$  cis-1,2-cyclohexanediol (OH, OH)

(d) *Propene to propanone (acetone) in two steps.*

propene  $\xrightarrow[H_2O]{H_2SO_4}$  2-propanol (OH)  $\xrightarrow{H_2CrO_4}$  propanone (O)

**8.41**   *Show how to convert cyclohexanol to these compounds:*

(a) cyclohexanol $\xrightarrow[\text{heat}]{\textbf{H}_3\textbf{PO}_4}$ cyclohexene

(b) cyclohexanol $\xrightarrow[\text{heat}]{\textbf{H}_3\textbf{PO}_4}$ cyclohexene $\xrightarrow[\textbf{Pt}]{\textbf{H}_2}$ cyclohexane

(c) cyclohexanol $\xrightarrow{\textbf{H}_2\textbf{CrO}_4}$ cyclohexanone

**8.42**   *Show reagents and experimental conditions that can be used to synthesize these compounds from 1-propanol (any derivative of 1-propanol prepared in an earlier part of this problem may be used for a later synthesis):*

(a) 1-propanol $\xrightarrow{\text{PCC}}$ propanal

(b) 1-propanol $\xrightarrow{\text{H}_2\text{CrO}_4}$ propanoic acid

(c) 1-propanol $\xrightarrow[\text{heat}]{\text{H}_2\text{SO}_4}$ propene

(d) propene (from (c)) $\xrightarrow[\text{H}_2\text{O}]{\text{H}_2\text{SO}_4}$ 2-propanol

(e) 2-propanol (from (d)) $\xrightarrow{\text{HBr}}$ 2-bromopropane

(f) 1-propanol $\xrightarrow{\text{SOCl}_2}$ 1-chloropropane

(g) *from (d)*

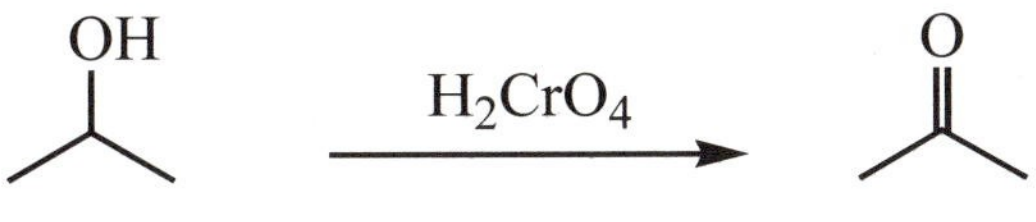

(h) *from (c)*

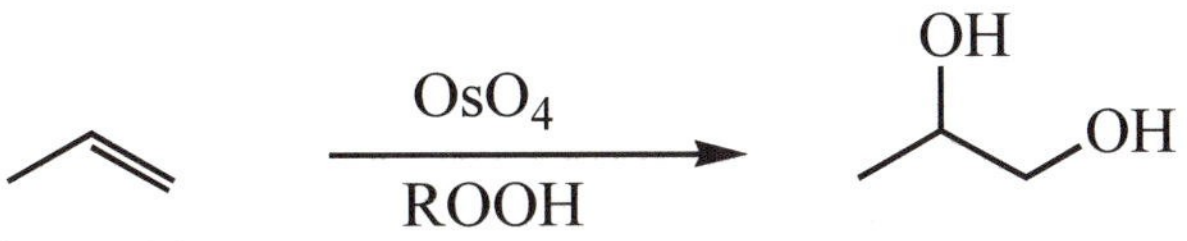

**8.43** *Show how to prepare each compound from 2-methyl-1-propanol (isobutyl alcohol):*

(a)

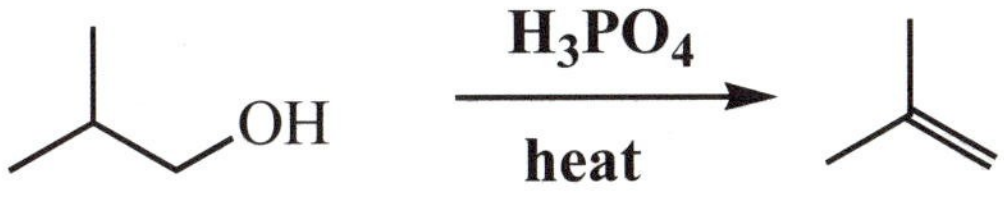

(b) 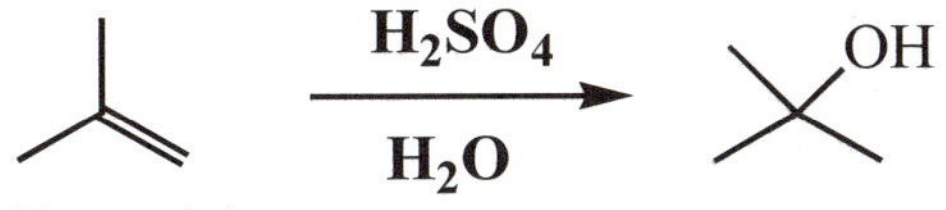 *From (a)*

(c) *From (a)*

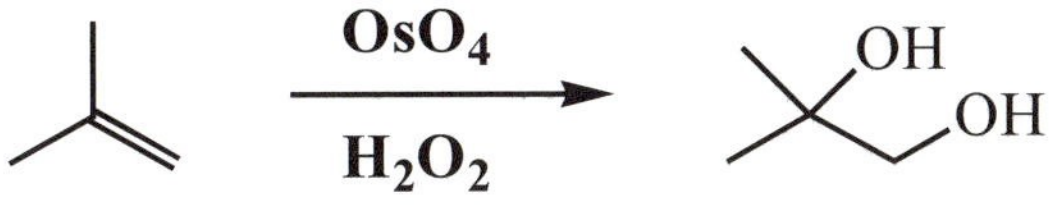

(d)

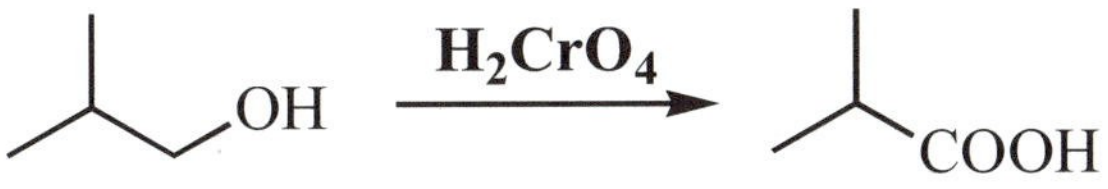

**8.44** *Show how to prepare each compound from 2-methylcyclohexanol. For any preparation involving more than one step, show each intermediate compound formed.*

(a)

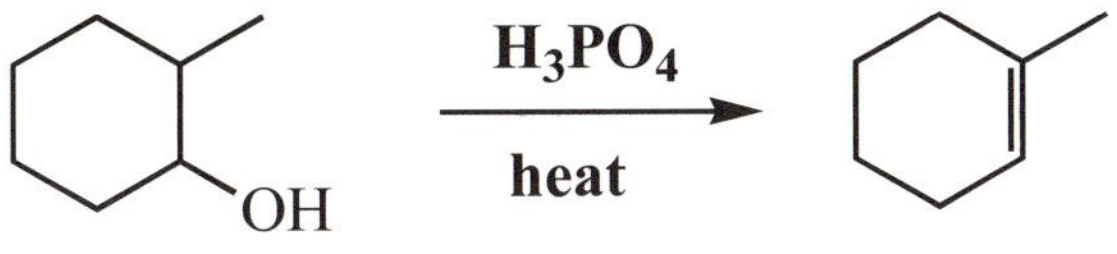

(b) 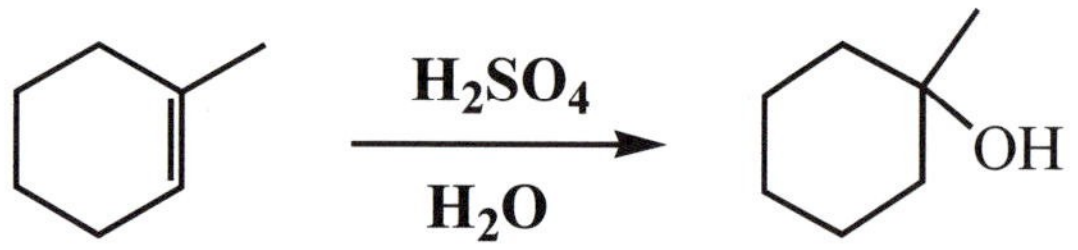   From (a)

(c)

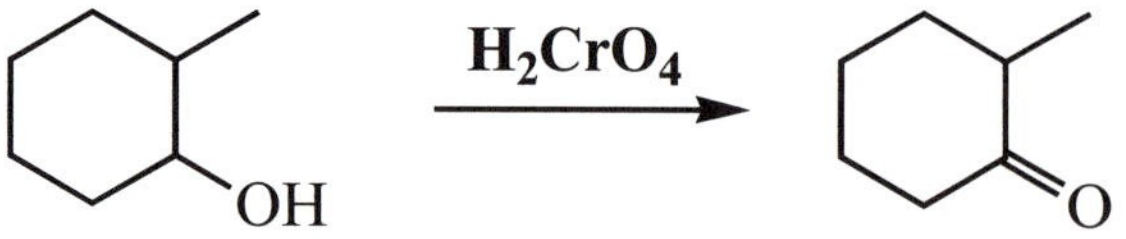

(d) 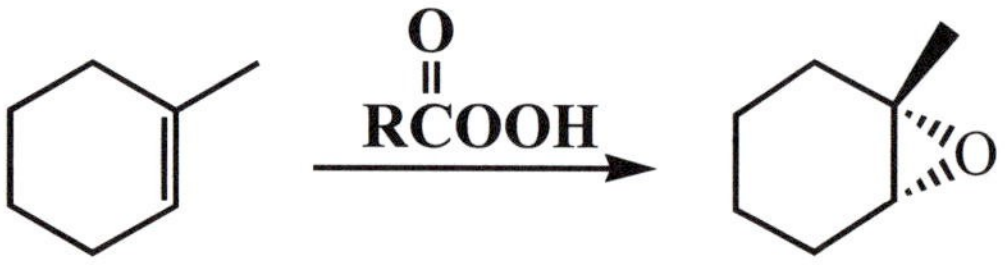   From (a)

(e) 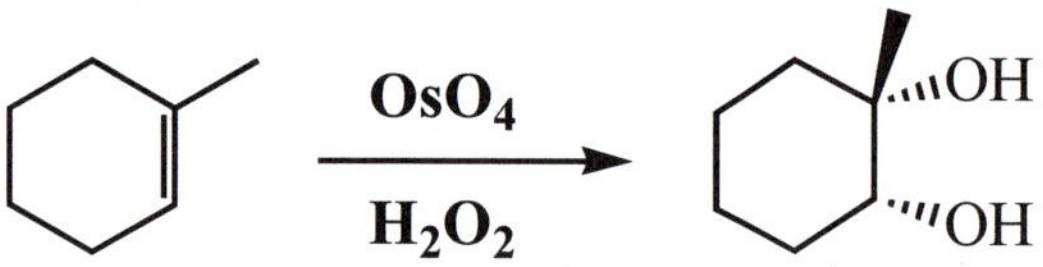   From (a)

(f) 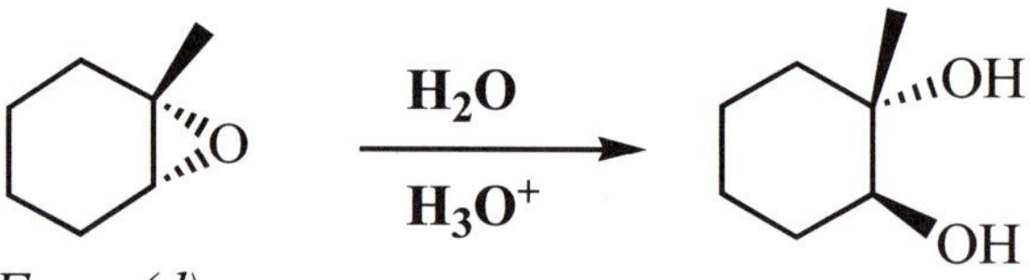   From (d)

**8.45**  *Show how to convert the alcohol on the left to compounds (a), (b), and (c).*

(a)

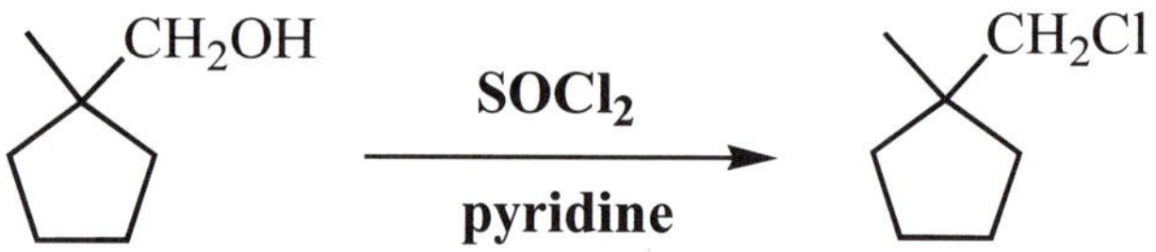

(b)

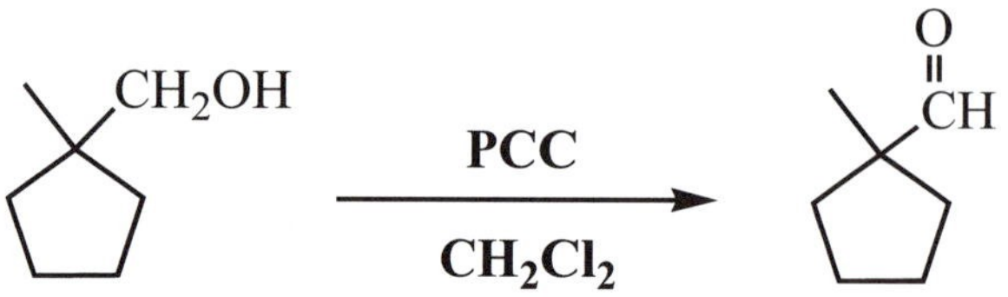

(c)

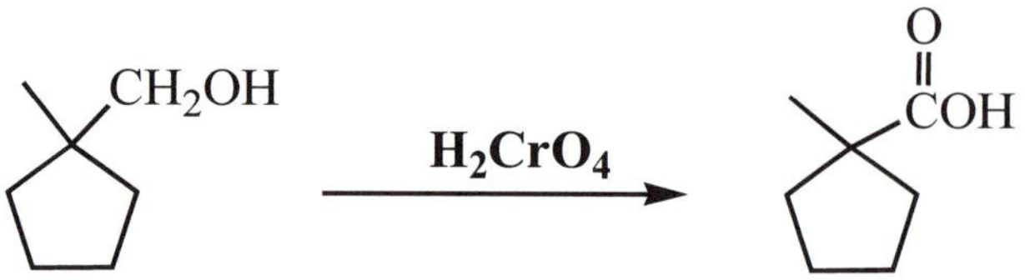

**8.46** *Disparlure, a sex attractant of the gypsy moth (Porthetria dispar), has been synthesized in the laboratory from the following (Z)-alkene:*

*(a) How might the (Z)-alkene be converted to disparlure?*

**Treating alkenes with peroxycarboxylic acids generates epoxides.  The stereochemistry about the double bond (cis) is preserved in the reaction, forming a cis disubstituted ring.**

(Z)-2-Methyl-7-octadecene                                    Disparlure

*(b) How many stereoisomers are possible for disparlure? How many are formed in the sequence you chose?*

**Disparlure has two stereocenters, therefore $2^2 = 4$ stereoisomers are possible. These include the (7R,8R), (7R,8S), (7S,8R), and (7S,8S) disparlure stereoisomers.  The starting material contained a cis double bond, therefore, only the *cis* epoxide is formed.  The peroxycarboxylic can approach the alkene from either side with equal probability, being that each face of the alkene is equivalent (use molecular models to convince yourself of this fact).  The result of this reaction is a racemic mixture of the (7R,8S) and (7S,8R) enantiomers.**

**8.47** *The chemical name for bombykol, the sex pheromone secreted by the female silkworm moth to attract male silkworm moths, is trans-10-cis-12-hexadecadien-1-ol.  (The compound has one hydroxyl group and two carbon-carbon double bonds in a 16-carbon chain.)*

*(a) Draw a structural formula for bombykol, showing the correct configuration about each carbon-carbon double bond.*

***trans*-10-*cis*-12-hexadecadien-1-ol**

*(b) How many cis-trans isomers are possible for the structural formula you drew in Part (a)?  All possible cis-trans isomers have been synthesized in the laboratory, but only*

*the one named bombykol is produced by the female silkworm moth, and only it attracts male silkworm moths.*

**There are four possibilities for cis-trans isomers: *trans*-10-*cis*-12-hexadecadien-1-ol, *trans*-10-*trans*-12-hexadecadien-1-ol, *cis*-10-*cis*-12-hexadecadien-1-ol, and *cis*-10-*trans*-12-hexadecadien-1-ol.**

**Looking Ahead**

**8.48**  *Compounds that contain an N-H group associate by hydrogen bonding.*

   (a) *Do you expect this association to be stronger or weaker than that between compounds containing an O-H group?*

   **Hydrogen bonding between O-H groups is stronger than between N-H  groups. The O-H bond is more polar than an N-H bond, thus stronger hydrogen-bonding attraction exists between hydroxyl groups.**

   (b) *Based on your answer to part (a), which would you predict to have the higher boiling point, 1-butanol or 1-butanamine?*

1-Butanol                1-Butanamine

   **1-Butanol will have a higher boiling point than 1-butanamine because the extent of intermolecular association from hydrogen bonding is greater in O-H containing compounds than in N-H containing compounds.**

**8.49**  *Write balanced equations for the reactions of phenol and cyclohexanol with NaOH:*

   (a) *Which compound is more acidic (Hint: see Table 2.2)?*

   **Phenol ($pK_a = 9.95$) is a stronger acid than cyclohexanol ($pK_a = 16$).**

*(b) Which conjugate base is more nucleophilic?*

**The conjugate base of cyclohexanol is a stronger nucleophile than phenoxide. When comparing the nucleophilic strength of similar atoms, nucleophilic strength increases with increased base strength.**

**8.50**   *Draw a resonance structure for each of the following compounds in which the heteroatom (O or S) is positively charged:*

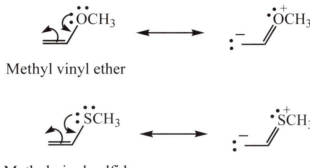

Methyl vinyl ether

Methyl vinyl sulfide

*(a) Compared with ethylene, how does each resonance structure influence the reactivity of the alkene towards an electrophile?*

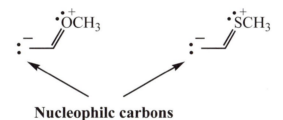

**Nucleophilc carbons**

**The nucleophilic carbons of the above relatively stable resonance structures that make up the hybrid are very reactive towards electrophiles at the end carbon because of their negative charges.**

*(b) Peracids are known to be electrophilic reagents.  Based on the resonance picture and your knowledge of periodic properties of the elements, would an epoxide be more likely to form with methyl vinyl ether or methyl vinyl sulfide?*

**Sulfur's lone pairs of electrons are not as tightly held compared to oxygen's electron lone pairs because sulfur is less electronegative, therefore, the methyl vinyl sulfide will be more reactive because it is more nucleophilic.**

(c)  *Would your answer to Part (b) be the same or different if only inductive effects were taken into consideration?*

**Oxygen is more electronegative than sulfur, and thus the carbon-oxygen bond is more polarized.  If inductive effects were only considered, the methyl vinyl ether would be predicted to be more reactive.**

**8.51**  *Rank the members within each set of reagents from most to least nucleophilic:*

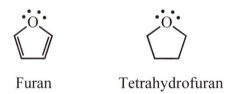

(b)   R−Ö:⁻          R−N̈H⁻          R−C̈H₂⁻

R−C̈H₂⁻   >   R−N̈H⁻   >   R−Ö:⁻

**8.52**  *Which of the following compounds is more basic?*

Furan          Tetrahydrofuran

**Tetrahydrofuran is more basic than furan.  The oxygen in furan holds one electron pair in an unhybridized *p* orbital and is involved in furan's aromaticity.  The other lone pair of electrons is held in an *sp²* hybridized orbital.  The oxygen tightly holds both of the electron pairs in furan.  Remember that as the *s* character of hybridized orbitals increases, the more tightly held those electrons are by the nucleus.  The oxygen lone pairs on tetrahydrofuran reside in *sp³* hybridized orbitals, which hold on to their electrons less tightly and are more available for bonding with protons.**

**8.53**  *If the reactivity of the following carbonyl compounds is directly proportional to the leaving group stability, rank the order of reactivity of each compound.*

**Most reactive** ⟶ **Least reactive**

## CHAPTER 9
### Solutions to Problems

**9.1**    *Write names for these compounds:*

(a)

**OH**

**2-Phenyl-2-propanol**

(b)

**(E)-3,4-Diphenyl-3-hexene**

(c)

COOH

CH₃

**3-Methylbenzoic acid
or *m*-Methylbenzoic acid**

**9.2**    *Predict the products resulting from vigorous oxidation of each compound by H₂CrO₄:*

(a)    $\xrightarrow{\text{H}_2\text{CrO}_4}$    

.COOH
.COOH

$+$   $Cr^{3+}$

(b)    $O_2N$ —    $\xrightarrow{\text{H}_2\text{CrO}_4}$    $O_2N$ — .COOH   $+$   $Cr^{3+}$

**9.3**    *Write a stepwise mechanism for the sulfonation of benzene. Use HSO₃⁺ as the electrophile.*

**Step 1:  Generation of the HSO₃⁺ electrophile.**

$$\text{HO}-\overset{\overset{O}{\|}}{\underset{\underset{O}{\|}}{S}}-\text{OH} \;+\; \text{H}-\overset{\overset{O}{\|}}{\underset{\underset{O}{\|}}{\overset{\cdot\cdot}{O}}}-S-\text{OH} \;\;\rightleftharpoons\;\; \text{HO}-\overset{\overset{O}{\|}}{\underset{\underset{O}{\|}}{S}}-\overset{\overset{H}{|}}{\overset{+}{O}}-H \;\;\rightleftharpoons\;\; \text{HO}-\overset{\overset{O}{\|}}{\underset{\underset{O}{\|}}{S^{+}}} \;+\; \text{SO}_3\text{H}^{-}$$

**Step 2:  Nucleophilic attack of benzene on HSO₃⁺ electrophile.**

$$\text{benzene} \;+\; \overset{\overset{O}{\|}}{\underset{\underset{O}{\|}}{S^{+}}}-\text{OH} \;\;\rightleftharpoons\;\; \left[ \;\overset{\text{SO}_3\text{H}}{\text{H}} \;\leftrightarrow\; \overset{\text{SO}_3\text{H}}{\text{H}} \;\leftrightarrow\; \overset{\text{SO}_3\text{H}}{\text{H}} \; \right]$$

**Resonance-stabilized intermediate**

**Step 3: Loss of proton to regenerate aromatic ring (benzenesulfonic acid) and sulfuric acid.**

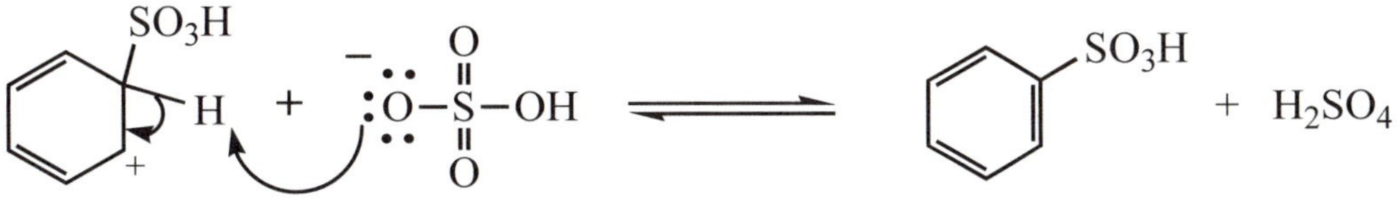

**9.4**   Write a structural formula for the product formed from Friedel-Crafts alkylation or acylation of benzene with:

*(a)*

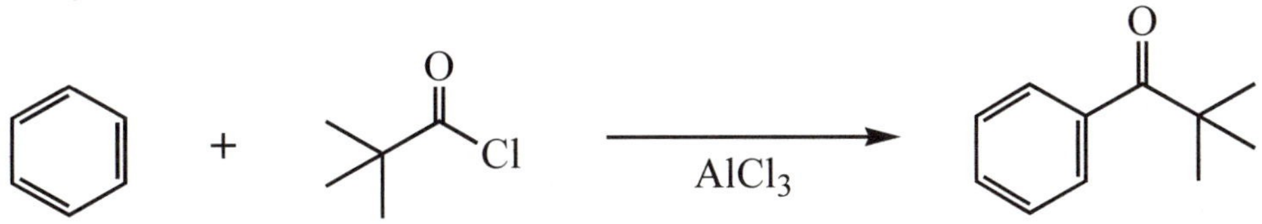

*(b)*

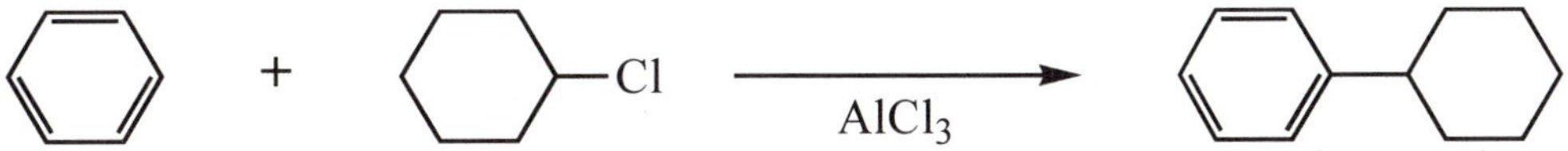

*(c)*

**9.5**   Write a mechanism for the formation of tert-butylbenzene from benzene and tert-butyl alcohol in the presence of phosphoric acid.

**Step 1:  Protonation of *tert*-butyl alcohol.**

**Step 2:  Generation of *tert*-butyl cation electrophile.**

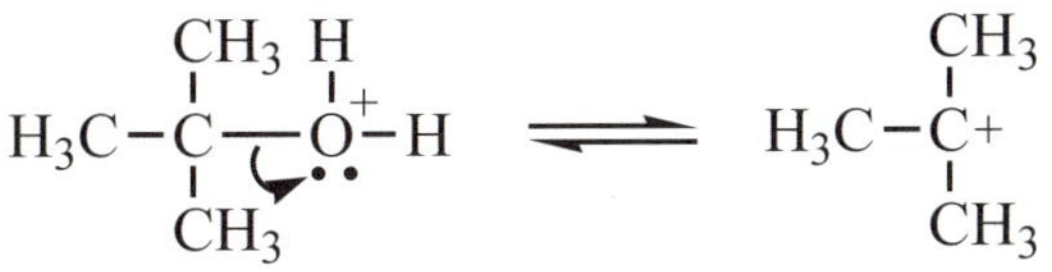

**Step 3:  Nucleophilic attack of the benzene in the *tert*-butyl cation electrophile.**

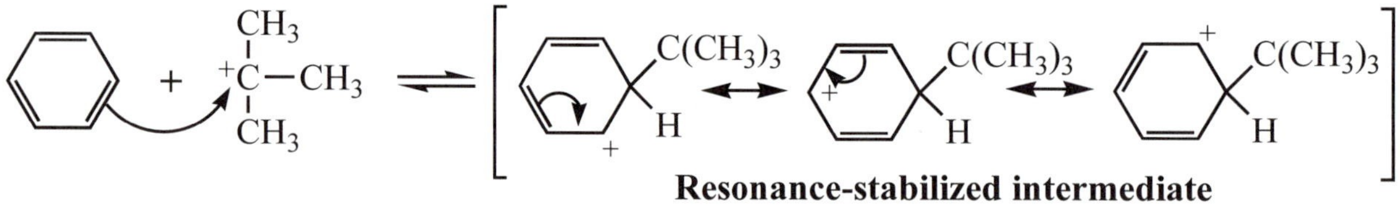

Resonance-stabilized intermediate

**Step 3:  Loss of proton to regenerate aromatic ring (*tert*-butyl benzene) and $H_3PO_4$.**

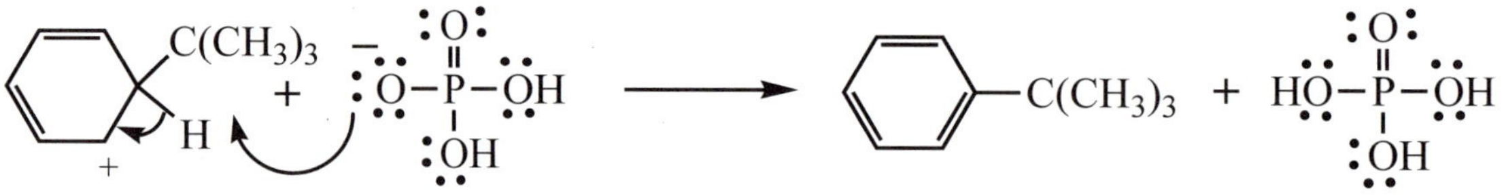

**9.6**   *Complete the following electrophilic aromatic substitution reactions.  Where you predict meta substitution, show only the meta product.  Where you predict ortho-para substitution, show both products:*

*(a)*

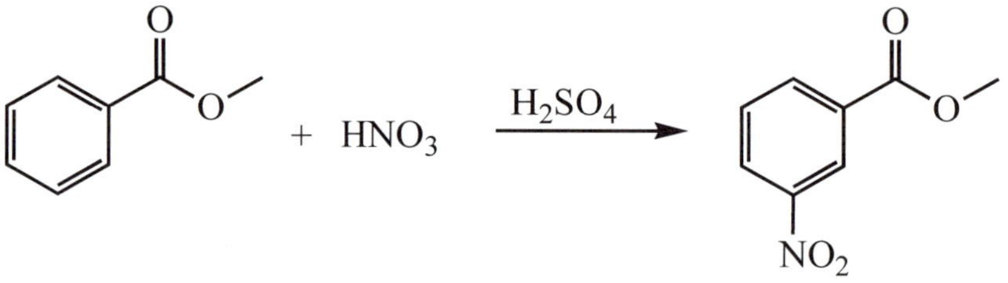

*(b)*

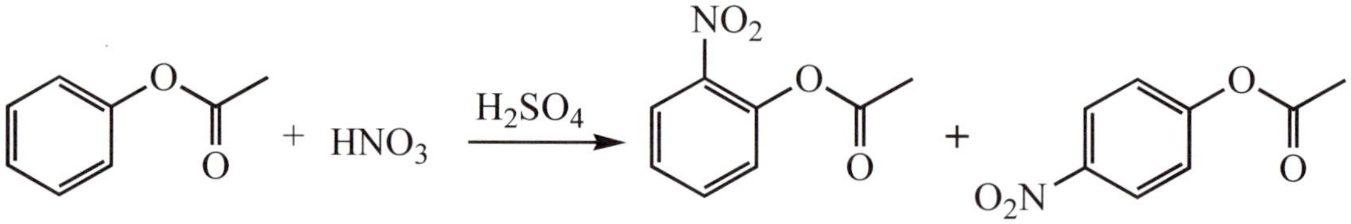

**9.7**   *Because the electronegativity of oxygen is greater than that of carbon, the carbon of a carbonyl group bears a partial positive charge, and its oxygen bears a partial negative charge.  Using this information, show that a carbonyl group is meta directing:*

**Ortho attack:**

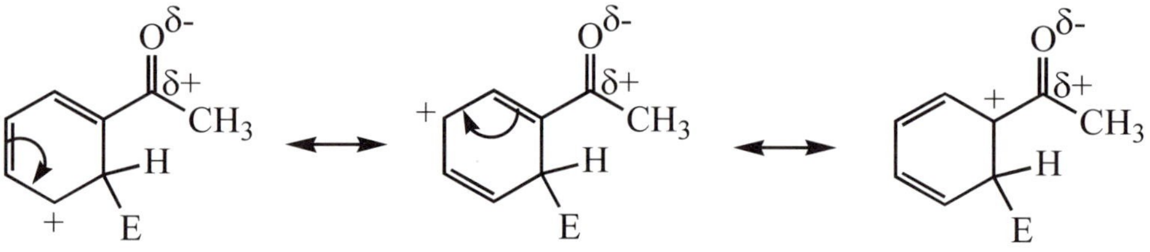

The third resonance structure places a positive charge adjacent to a partial positive charge on the carbonyl carbon atom. This is a destabilizing interaction that raises the overall energy of the intermediate and transition state, disfavoring the ortho attack.

**Para attack:**

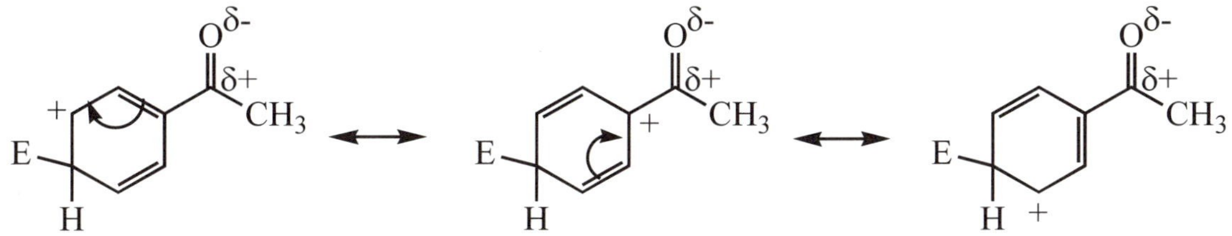

The second resonance structure places a positive charge adjacent to a partial positive charge on the carbonyl carbon atom. This destabilizing interaction raises the overall energy of the intermediate and transition state, disfavoring the para attack.

**Meta attack:**

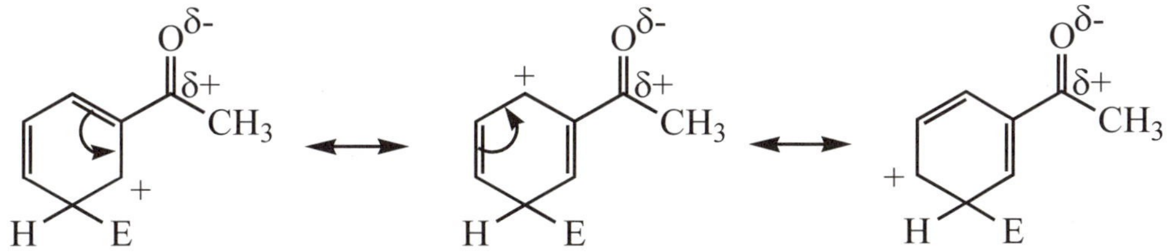

Close inspection of the resonance structures resulting from a meta attack reveals no destabilized resonance structures relative to the ortho and para attacks. Electrophilic aromatic substitutions favor meta attack when electron-withdrawing substituents (except for halogens) are bonded to the benzene ring.

**9.8**   *Predict the product of treating each compound with HNO₃/H₂SO₄.*

(a)

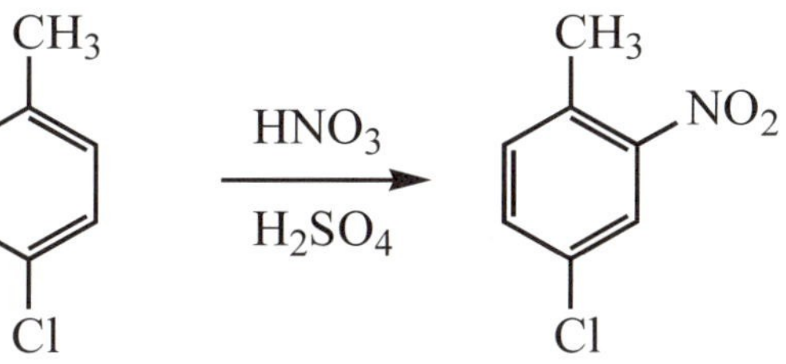

The methyl substituent is more strongly activating than Cl, therefore, the $NO_2^+$ electrophile will be directed ortho to the methyl group.

(b)

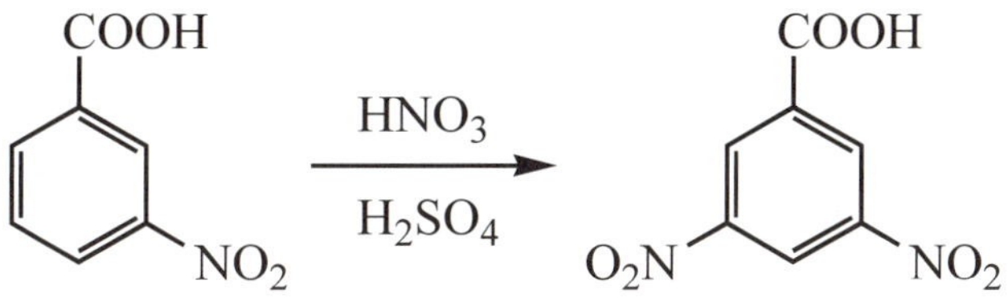

**9.9**     *Arrange these compounds in order of increasing acidity:  2,4-dichlorophenol, phenol, cyclohexanol.*

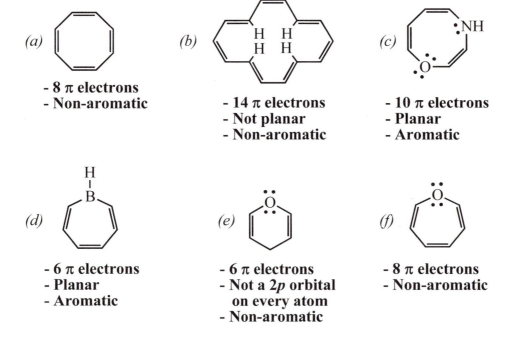

**Aromaticity**

**9.10**     *Which of the following compounds are aromatic?*

*(a)*

- 8 π electrons
- Non-aromatic

*(b)*

- 14 π electrons
- Not planar
- Non-aromatic

*(c)*

- 10 π electrons
- Planar
- Aromatic

*(d)*

- 6 π electrons
- Planar
- Aromatic

*(e)*

- 6 π electrons
- Not a 2p orbital
  on every atom
- Non-aromatic

*(f)*

- 8 π electrons
- Non-aromatic

Although compound (b) contains 14 π electrons (a Hückel number), a closer inspection of the molecule reveals that the protons inside the ring are held too close together and force the ring to be non-planar.  Structure (b) subjected to molecular modeling yields the following structure and confirms its non-planarity (only the carbon atoms are shown).  Because (b) is non-planar, it cannot be aromatic.

In compound (c), only one of the electron pairs on oxygen is held in a 2p orbital, the other electron pair is in an $sp^2$ hybridized orbital, perpendicular to the π bonds. The nitrogen lone pair in compound (c) is also held in a 2p orbital for a total of 10 π electrons in a planar ring making it aromatic.  In compound (d), the boron atom contributes an empty **2p** orbital, giving an aromatic compound with six π electrons held in a seven-membered ring with seven **2p** orbitals.  Compound **(f)**, is much like compound **(c)** where the oxygen holds one electron pair in a **2p** orbital and the other in an **sp²** hybridized orbital for a total of eight π electrons, giving a non-aromatic compound.

**9.11**  *Explain why cyclopentadiene (pK_a 16) is many orders of magnitude more acidic than cyclopentane (pK_a > 50). (Hint: Draw the structural formula for the anion formed by removing one of the protons on the -CH₂- group, and then apply the Hückel criteria for aromaticity.)*

cyclopentadiene                    cyclopentane

**Cyclopentadiene:**

resonance stabilized cyclopentadienyl anion

**Cyclopentane:**

The conjugate base of cyclopentadiene is a highly stable six π electron aromatic anion with five resonance structures delocalizing the negative charge onto five different carbons.  The cyclopentyl anion, a conjugate base of cyclopentane, is forced to bear the negative charge on one carbon, making it a much stronger conjugate base than the resonance stabilized aromatic cyclopentadienyl anion.  As a conjugate base's negative charge becomes more stable though delocalization and resonance, its base strength becomes weaker.  Acid strengths are inversely proportional to their conjugate base strengths.

**Nomenclature and Structural Formulas**

**9.12**    *Name these compounds:*

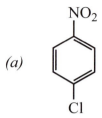

*(a)*

**4-Nitrochlorobenzene**
**(*p*-Chloronitrobenzene)**

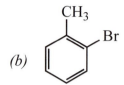

*(b)*

**2-Bromotoluene**
**(*o*-Bromotoluene)**

*(c)*

**3-Phenyl-1-propanol**

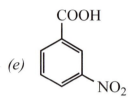

*(d)*

**2-Phenyl-3-buten-2-ol**

*(e)*

**3-Nitrobenzoic acid**
**(*m*-Nitrobenzoic acid)**

*(f)*

**1-Phenylcyclohexanol**

*(g)*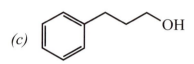

**(*E*)-1,2-Diphenylethene**
**(*trans*-1,2-Diphenylethene)**

*(h)*

**2,4-Dichlorotoluene**

**9.13**    *Draw structural formulas for these compounds:*

*(a) 1-Bromo-2-chloro-4-ethylbenzene*

*(b) 4-Iodo-1,2-dimethylbenzene*

*(c) 2,4,6-Trinitrotoluene*

*(d) 4-Phenyl-2-pentanol*

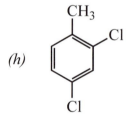

*(e) p-Cresol*

*(f) 2,4-Dichlorophenol*    *(g) 1-Phenylcyclopropanol*    *(h) Styrene (phenylethylene)*

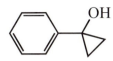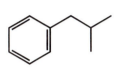

*(i) m-Bromophenol*    *(j) 2,4-Dibromoaniline*    *(k) Isobutylbenzene*    *(l) m-Xylene*

**9.14**   *Show that pyridine can be represented as a hybrid of two equivalent contributing structures.*

**9.15**   *Show that naphthalene can be represented as a hybrid of three contributing structures. Show also, by the use of curved arrows, how one contributing structure is converted to the next.*

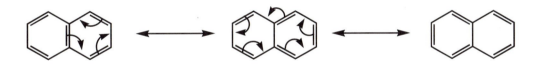

**9.16**   *Draw four contributing structures for anthracene.*

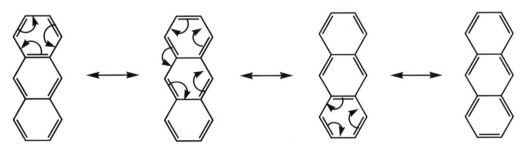

**Electrophilic Aromatic Substitution: Monosubstitution**

**9.17**    *Draw a structural formula for the compound formed by treating benzene with each of the following combinations of reagents.*

**9.18**    *Show three different combinations of reagents you might use to convert benzene to isopropylbenzene.*

**Isopropylbenzene**

In syntheses (2) and (3), H₃PO₄ can be used in place of H₂SO₄.

**9.19**    *How many monochlorination products are possible when naphthalene is treated with Cl₂/AlCl₃?*

**Two monochloronaphthalenes are possible when naphthalene is treated with Cl₂/AlCl₃.**

**9.20**   *Write a stepwise mechanism for the following reaction, using curved arrows to show the flow of electrons in each step.*

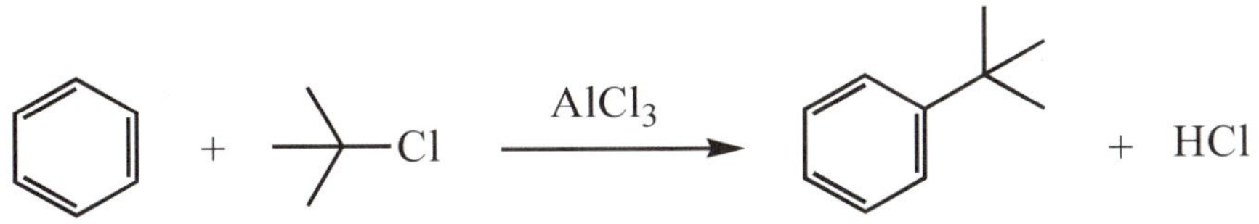

**Step 1:  Formation of Lewis acid-Lewis base complex.**

**Step 2:  Formation of the *tert*-butyl cation.**

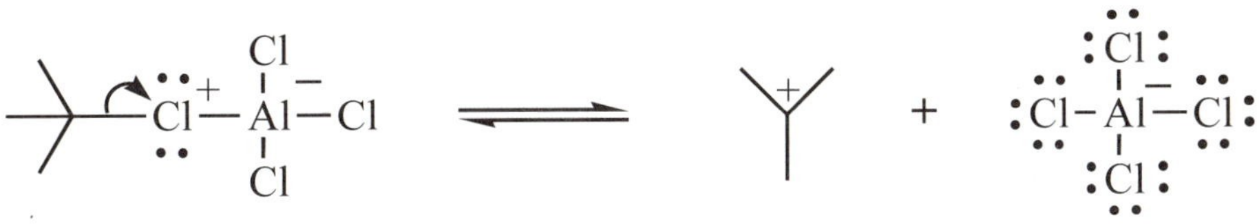

**Step 3:  Electrophilic attack on the benzene ring to form a resonance-stabilized carbocation intermediate with three resonance structures (try to draw the other two yourself).**

**Step 4:  Proton transfer to regenerate the aromatic ring.**

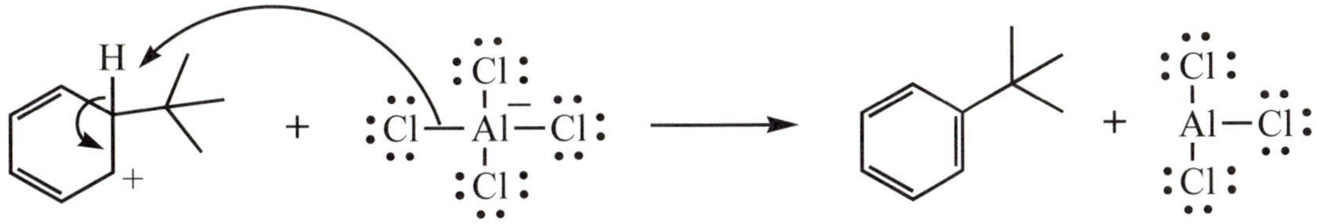

**9.21**  *Write a stepwise mechanism for the preparation of diphenylmethane by treating benzene with dichloromethane in the presence of an aluminum chloride catalyst.*

**Step 1:  Formation of Lewis acid-Lewis base complex.**

**Step 2:  Nucleophilic attack of benzene on the electrophilic Lewis acid-Lewis base complex and the formation of a resonance-stabilized carbocation.  There are two more resonance structures (try to draw the others yourself).**

**Step 3:  Deprotonation of the carbocation to give benzyl chloride, HCl, and AlCl₃.**

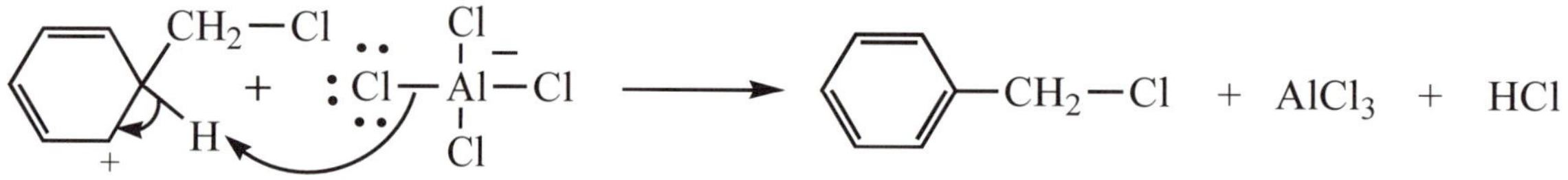

**Step 4:  Formation of Lewis acid-Lewis base complex between benzyl chloride and AlCl₃.**

**Step 5:  Dissociation of the complex to give a resonance-stabilized benzyl cation (only one out of five contributing structures shown) and AlCl₄⁻.**

**Step 6: Nucleophilic attack of the second molecule of benzene on the benzylic cation to form another resonance-stabilized carbocation (again, only one contributor is shown).**

**Step 7: Deprotonation of the carbocation intermediate to regenerate the aromatic ring (diphenylmethane), HCl, and AlCl₃.**

## Electrophilic Aromatic Substitution: Disubstitution

**9.22**    *When treated with Cl₂/AlCl₃, 1,2-dimethylbenzene (o-xylene) gives a mixture of two products. Draw structural formulas for these products.*

For this problem, assume monochlorination conditions. Because of symmetry considerations, there are two different products that can be produced.

**9.23**    *How many monosubstitution products are possible when 1,4-dimethylbenzene (p-xylene) is treated with Cl₂/AlCl₃? When m-xylene is treated with Cl₂/AlCl₃?*

**2-Chloro-1,4-dimethyl-
benzene**

**2-Chloro-1,3-dimethyl-benzene**          **1-Chloro-2,4-dimethyl-benzene**

Monochlorination of 1,4-dimethybenzene with $Cl_2/AlCl_3$ yields only one product, 2-chloro-1,4-dimethylbenzene.  Under the same conditions, 1,3-dimethylbenzene yields two products: 2-chloro-1,3-dimethylbenzene and 1-chloro-2,4-dimethylbenzene.

**9.24**  *Draw the structural formula for the major product formed on treating each compound with $Cl_2/AlCl_3$:*

(f)

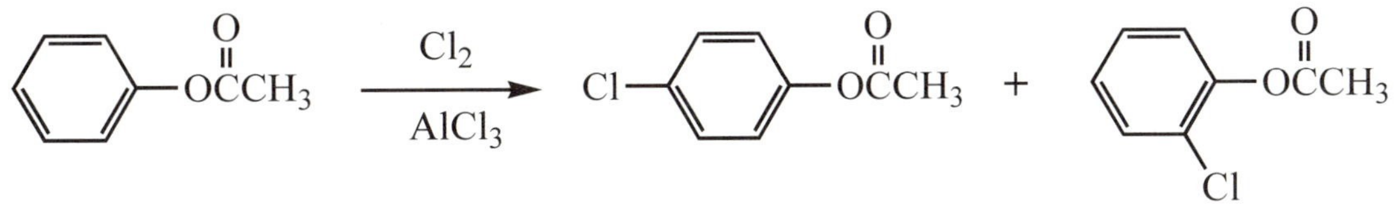

(g)

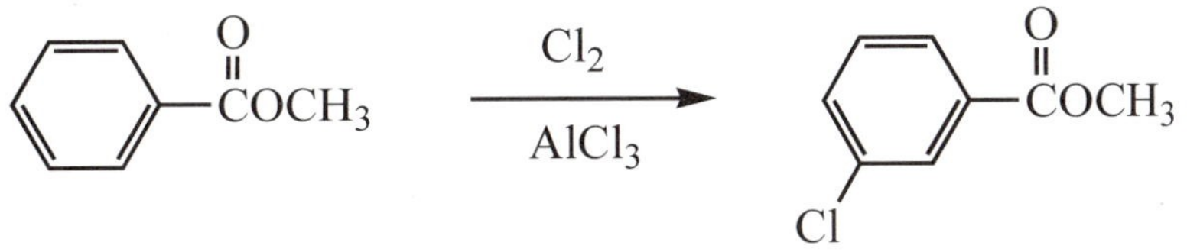

**9.25** *Which compound, chlorobenzene or toluene, undergoes electrophilic aromatic substitution more rapidly when treated with Cl₂/AlCl₃? Explain and draw structural formulas for the major product(s) from each reaction.*

**Toluene undergoes electrophilic aromatic substitution faster than chlorobenzene. Even though the chlorine substituent is an ortho-para director through the resonance contribution of a lone pair of electrons, its high electronegativity makes it an electron-withdrawing group and deactivates the benzene ring towards electrophilic attack. The methyl substituent is also an ortho-para director and an electron-releasing group; therefore it activates the benzene ring to electrophiles and reacts faster relative to chlorobenzene.**

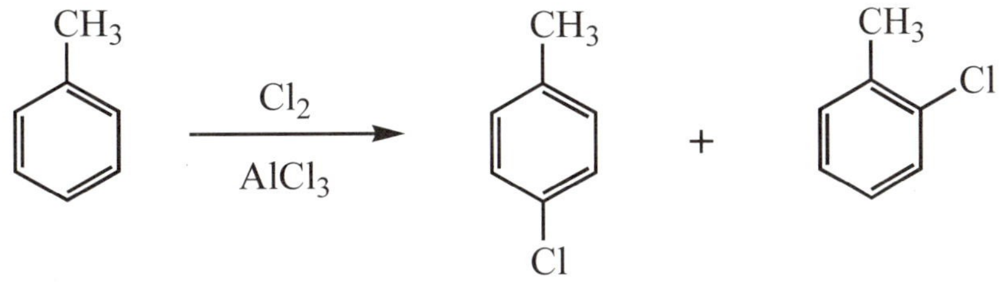

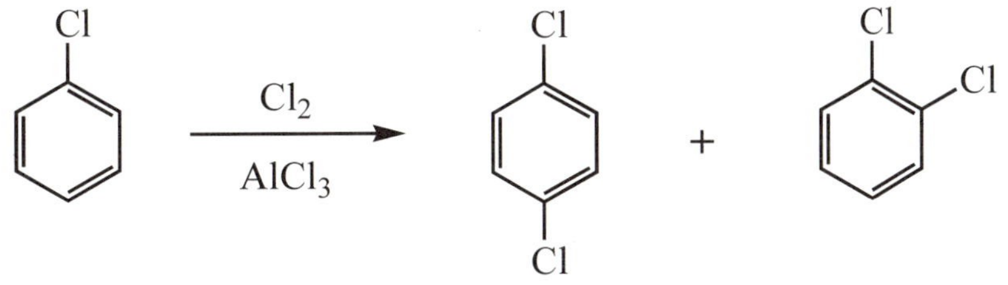

**9.26** *Arrange the compounds in each set in order of decreasing reactivity (fastest to slowest) toward electrophilic aromatic substitution:*

(a)

(A) 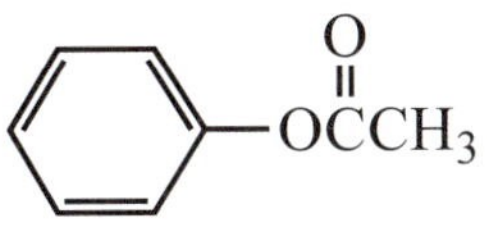            (B) 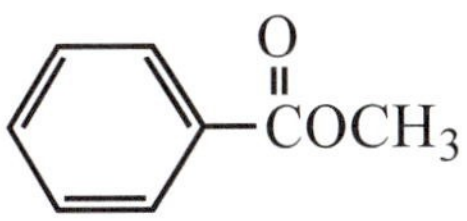            (C)

**Reactivity towards electrophilic aromatic substitution:  (B) > (A) > (C)**

*(b)*    –NO$_2$     (A)      –COOH   (B)     (C)

**Reactivity towards electrophilic aromatic substitution:  (C) > (B) > (A)**

*(c)*   –NH$_2$ (A)      –NHCCH$_3$ (B)      –CNHCH$_3$ (C)

**Reactivity towards electrophilic aromatic substitution:  (A) > (B) > (C)**

*(d)*   (A)      –CH$_3$ (B)      –OCH$_3$ (C)

**Reactivity towards electrophilic aromatic substitution:  (C) > (B) > (A)**

**9.27**   *Account for the observation that the trifluoromethyl group is meta directing, as shown in the following example:*

CF$_3$ + HNO$_3$ $\xrightarrow{H_2SO_4}$ CF$_3$ , NO$_2$ + H$_2$O

**The trifluoromethyl group is a strong electron-withdrawing group because the highly electronegative fluorine atoms pull electron density away from the attached carbon atom, making it electron deficient.**

This electron-withdrawing inductive effect also creates an electron deficient benzene π-system, making the benzene a poorer nucleophile and deactivating it to electrophilic attack.

**Intermediate from ortho attack:**

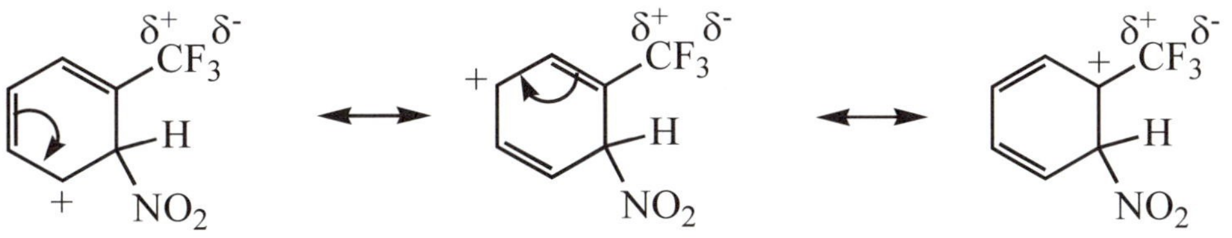

The third resonance structure places a positive charge adjacent to a partial positive charge on the –CF$_3$ carbon directly attached to the ring.  This is a destabilizing interaction that raises the overall energy of the intermediate and transition state, disfavoring the intermediate cation formation.

**Intermediate from meta attack:**

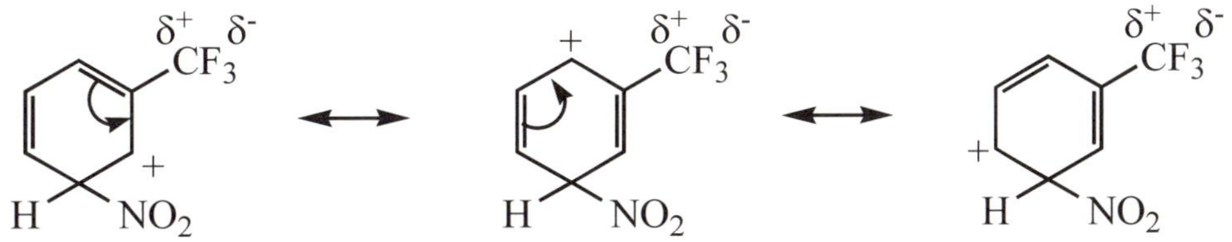

The resonance structures resulting from a meta attack reveal no destabilizing resonance structures relative to the ortho and para attacks.  The activation energy to the formation of the carbocation intermediate is much lower in the meta attack relative to ortho and para attacks.

**Intermediate from para attack:**

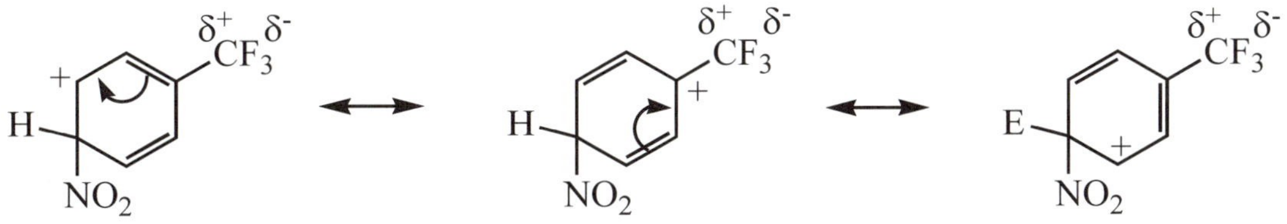

The second resonance structure places a positive charge adjacent to a partial positive charge on the –CF$_3$ carbon directly attached to the ring.  This destabilizing interaction raises the overall energy of the intermediate and transition state, disfavoring the para attack.

**9.28**    *Show how to convert toluene to these carboxylic acids:*

*(a)* **The synthesis of 4-chlorobenzoic acid requires the use of the methyl group of toluene to direct the chloro group para to the methyl.  Then, the methyl group can be oxidized to the carboxyl group with $K_2Cr_2O_7$ in sulfuric acid.**

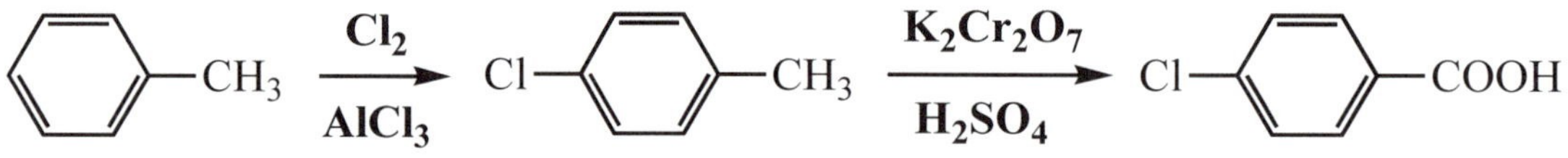

*(b)* **The synthesis of 3-chlorobenzoic acid is similar to that outlined in (a), except the order of the reactions is reversed.  Placement of the chloro group meta requires that the carboxyl group be in place before chlorination.  To accomplish this, the methyl group of toluene must be oxidized first to form the meta director, then followed with chlorination.**

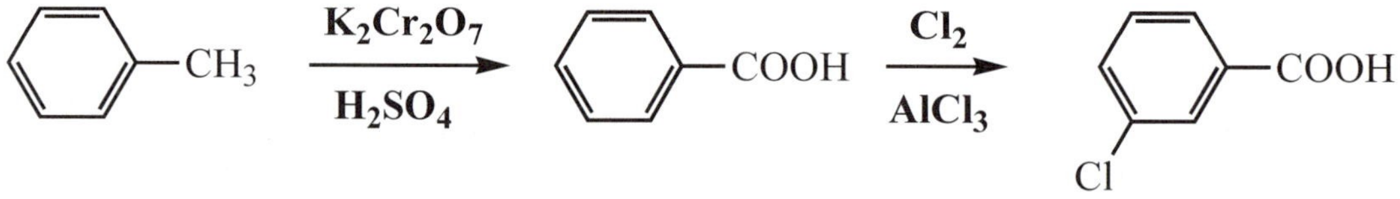

**9.29**    *Show reagents and conditions that can be used to bring about these conversions:*

*(a)*

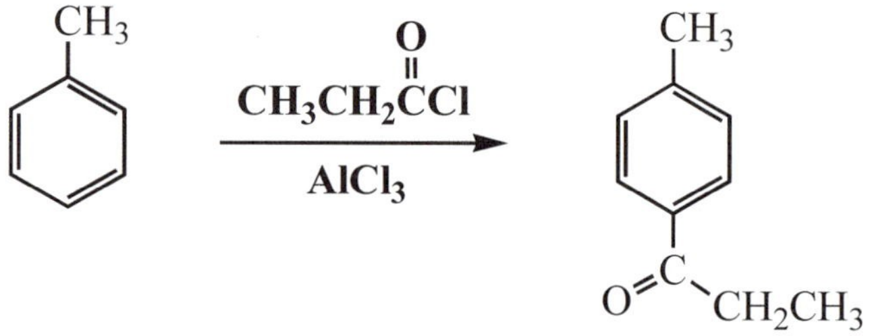

*(b)*

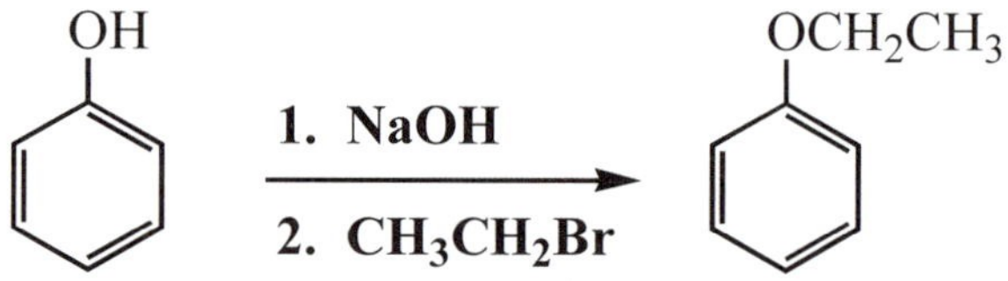

*(c)*

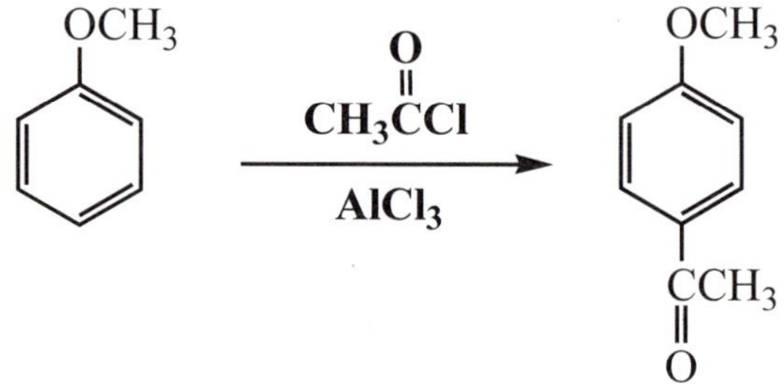

**9.30**   *Propose a synthesis of triphenylmethane from benzene as the only source of aromatic rings. Use any other necessary reagents.*

**The reaction of three moles of benzene with one mole of trichloromethane (chloroform) in the presence of aluminum chloride will give triphenylmethane.**

**9.31**   *Reaction of phenol with acetone in the presence of an acid catalyst gives bisphenol A, a compound used in the production of polycarbonate and epoxy resins (Sections 17.5C and 17.5F):*

Acetone                              Bisphenol A

*Propose a mechanism for the formation of bisphenol A.*

**Step 1:  The reaction begins with a protonation of acetone to form its conjugate acid, which may be written as a hybrid of two resonance contributing structures.**

**Step 2: The conjugate acid of acetone is an electrophile and reacts with phenol at the para position to give a resonance-stabilized cation (only one contributing structure is shown here: you should try to draw the others).**

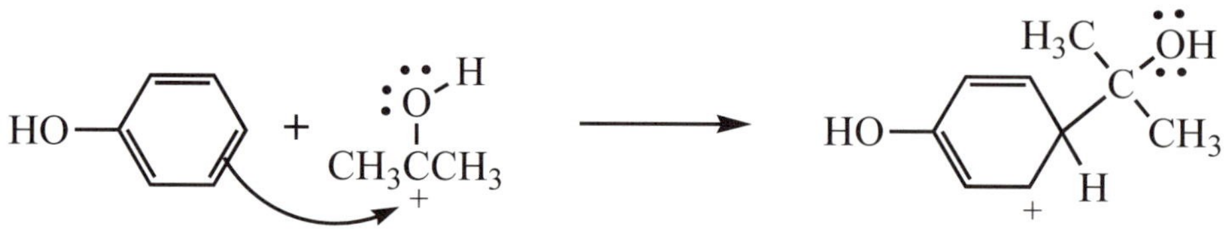

**Step 3: Deprotonation of the cation gives the intermediate 2-(4-hydroxyphenyl)-2-propanol.**

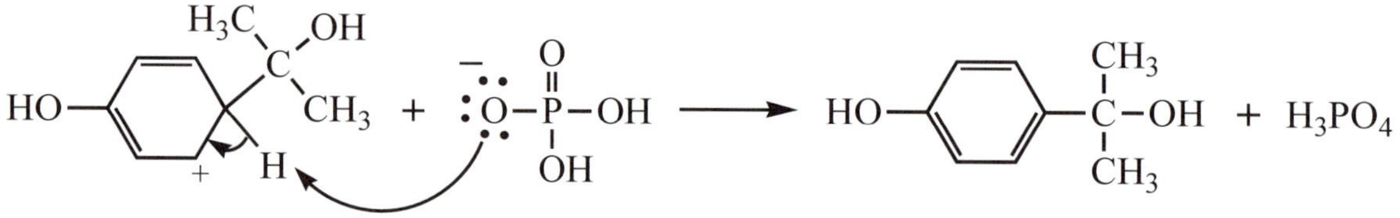

**Step 4: The tertiary alcohol is protonated.**

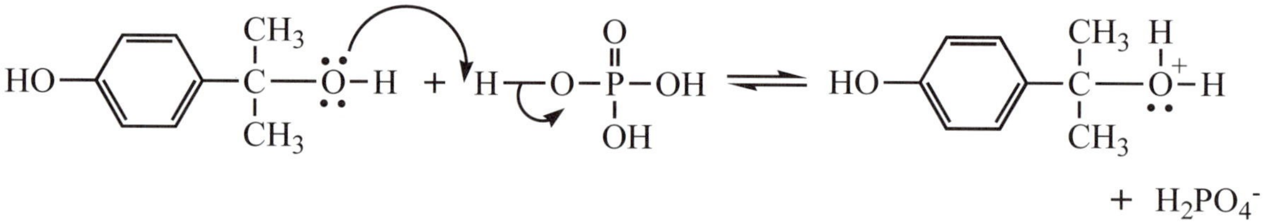

**Step 5: The protonated alcohol loses water to give a resonance-stabilized cation intermediate (only one contributor is shown; five others are possible).**

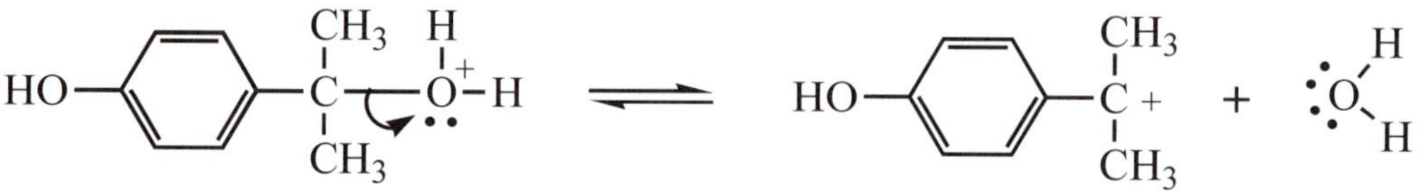

**Step 6: Attack of the phenol upon the carbocation gives a new resonance stabilized cation intermediate (only one contributor is shown, three others are possible).**

**Step 7: Deprotonation of the cation intermediate yields the product bisphenol A.**

+ $H_3PO_4$

**9.32** *2,6-Di-tert-butyl-4-methylphenol, more commonly known as butylated hydroxytoluene or BHT, is used as an antioxidant in foods to "retard spoilage." BHT is synthesized industrially from 4-methylphenol (p-cresol) by reaction with 2-methylpropene in the presence of phosphoric acid. Propose a mechanism for this reaction:*

4-Methylphenol     2-Methylpropene     2,6-Di-*tert*-butyl-4-methylphenol
(Butylated hydroxytoluene, BHT)

**Step 1: Formation of the *tert*-butyl cation.**

**Step 2: The *tert*-butyl cation reacts with the aromatic ring ortho to the strongly activating hydroxyl substituent, yielding a resonance-stabilized cation (only the most stable resonance contributing structure is shown. You should try to draw the other three).**

**Step 3:  Deprotonation of the resonance-stabilized cation yields 2-*tert*-butyl-4-methylphenol.**

$$+ \quad H_3PO_4$$

**Step 4:  2-*tert*-Butyl-4-methylphenol undergoes one more electrophilic aromatic substitution reaction with a *tert*-butyl cation yielding a resonance-stabilized cation (only the most stable resonance contributing structure is shown.  You should try to draw the other three).**

**Step 5: Deprotonation of the resonance-stabilized cation yields BHT.**

$$+ \quad H_3PO_4$$

**9.33**  *The first herbicide widely used for controlling weeds was 2,4-dichlorophenoxyacetic acid (2,4-D). Show how this compound might be synthesized from 2,4-dichlorophenol and chloroacetic acid, ClCH₂COOH.*

2,4-Dichlorophenol                2,4-Dichlorophenoxyacetic acid
                                          (2,4-D)

## Acidity of Phenols

**9.34**  *Use the resonance theory to account for the fact that phenol (pK_a 9.95) is a stronger acid than cyclohexanol (pK_a approximately 18).*

**Phenoxide ion, the conjugate base of phenol, is stabilized by resonance, with the lone pair and negative charge delocalized over four atoms.  The delocalized lone pair is less able to bond with protons.  As the conjugate base strength decreases, its corresponding acid becomes stronger.**

**Cyclohexoxide, the conjugate base of cyclohexanol, is not stabilized by resonance, thus the electron pair can easily bond with protons as a strong base, and thus its corresponding acid becomes weaker.**

**9.35**  *Arrange the compounds in each set in order of increasing acidity (from least acidic to most acidic):*

*(a)* (cyclohexyl)-OH < (phenyl)-OH < $CH_3COOH$

*(b)*  $H_2O$  <  $NaHCO_3$  <  (phenyl)-OH

*(c)* (phenyl)-$CH_2OH$  <  (phenyl)-OH  <  $O_2N$-(phenyl)-OH

**9.36**  *From each pair, select the stronger base.*

**As the p$K_a$ of the conjugate acid increases, the basicity of its corresponding base also increases.**

*(a)* (phenyl)-$O^-$  or  $\boxed{OH^-}$ 

**The resonance-stabilized phenoxide anion is a weaker base relative to an unstabilized hydroxide anion.**

10                  16     p$K_a$ (conjugate acid)

*(b)* (phenyl)-$O^-$  or  $\boxed{\text{(cyclohexyl)-}O^-}$ 

**The resonance-stabilized phenoxide anion is a weaker base relative to unstabilized cyclohexoxide anion.**

10                  16     p$K_a$ (conjugate acid)

*(c)*  $\boxed{\text{(phenyl)-}O^-}$  or  $HCO_3^-$ 

**The $HCO_3^-$ anion has greater resonance stabilization by delocalizing the negative charge over two electronegative oxygens. The phenoxide anion delocalizes the negative charge over one oxygen and three less electronegative carbons.**

10                  6.4     p$K_a$ (conjugate acid)

*(d)*  $\boxed{\text{(phenyl)-}O^-}$  or  $CH_3COO^-$ 

**The acetate anion has greater resonance stabilization by delocalizing the negative charge over two electronegative oxygens. The phenoxide anion delocalizes the negative charge over one oxygen and three less electronegative carbons.**

10                  4.7     p$K_a$ (conjugate acid)

**9.37**    *Account for the fact that water-insoluble carboxylic acids (pK$_a$ 4-5) dissolve in 10% sodium bicarbonate with the evolution of a gas, but water-insoluble phenols (pK$_a$ 9.5-10.5) do not show this chemical behavior.*

**When carbonic acid is formed, it decomposes to carbon dioxide and water according to the following equation:**

$$H_2CO_3 \longrightarrow CO_2 \ + \ H_2O$$

**Acid-base equilibria favor the side with the weaker acid and weaker base (the side containing the acid with the largest pK$_a$). Carboxylic acids react with bicarbonate to form a carboxylate anion (the conjugate base) and a weaker carbonic acid (the conjugate acid), which decomposes to carbon dioxide and water. Phenols are weaker acids than carbonic acid, so their equilibria favor the left side and will not dissolve into phenoxide anions along with the evolution of carbon dioxide.**

**9.38**    *Describe a procedure for separating a mixture of 1-hexanol and 2-methylphenol (o-cresol) and recovering each in pure form. Each is insoluble in water but soluble in diethyl ether.*

**This separation scheme uses the acidity of a phenolic hydroxyl (pK$_a$ ~ 10). Phenols can be completely deprotonated by aqueous sodium hydroxide and the resulting sodium phenoxide is soluble in water, as all sodium salts are. 1-Hexanol is not acidic enough to be completely deprotonated by aqueous sodium hydroxide and therefore remains unchanged. Addition of diethyl ether will partition the mixture into aqueous and organic phases. The 1-hexanol will migrate to the ether layer and the sodium phenoxide to the aqueous layer, where the phases are separated. Evaporation of the ether from the organic phase yields pure 1-hexanol. Acidification of the aqueous layer converts sodium phenoxide to phenol. The phenol is extracted from the aqueous mixture using ether. After separating the organic phase and evaporating the ether, pure 2-methylphenol is isolated.**

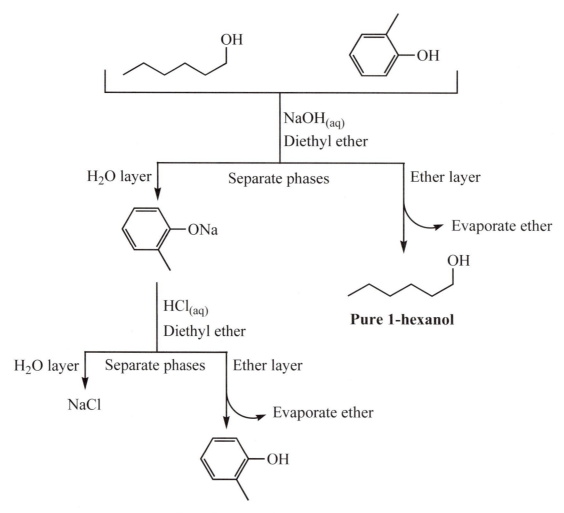

**Pure 1-hexanol**

**Pure 2-methylphenol**

## Syntheses

**9.39**   *Using styrene, C₆H₅CH=CH₂, as the only aromatic starting material, show how to synthesize these compounds.  In addition to styrene, use any other necessary organic or inorganic chemicals. Any compound synthesized in one part of this problem may be used to make any other compound in the problem.*

(d) [structure] Ph-CH(OH)CH₃  →(K₂Cr₂O₇ / H₂SO₄)→  Ph-COOH

(e) [structure] Ph-CH=CH₂  →(H₂ / Ni)→  Ph-CH₂CH₃

(f) [structure] Ph-CH=CH₂  →(OsO₄ / H₂O₂)→  Ph-CH(OH)CH₂OH  ←(H₂O/H₃O⁺)—

or  [structure] Ph-CH=CH₂  →(RCOOH)→  Ph-CH(-O-)CH₂ (epoxide)  —(H₂O/H₃O⁺)→

**9.40**  *Show how to synthesize these compounds starting with benzene, toluene, or phenol as the only sources of aromatic rings. Assume that in all syntheses you can separate mixtures of ortho-para products to give the desired isomer in pure form:*

(a) *m-Bromonitrobenzene*               (b) *1-Bromo-4-nitrobenzene*
(c) *2,4,6-Trinitrotoluene (TNT)*       (d) *m-Bromobenzoic acid*
(e) *p-Bromobenzoic acid*               (f) *p-Dichlorobenzene*
(g) *m-Nitrobenzenesulfonic acid*       (h) *1-Chloro-3-nitrobenzene*

(a) [benzene]  →(HNO₃ / H₂SO₄)→  [Ph-NO₂]  →(Br₂ / AlCl₃)→  [m-bromonitrobenzene: NO₂ and Br in meta]

(b) [benzene]  →(Br₂ / AlCl₃)→  [Ph-Br]  →(HNO₃ / H₂SO₄)→  [O₂N—C₆H₄—Br para]

(c) [toluene, CH₃]  →(HNO₃ / H₂SO₄)→  [2,4,6-trinitrotoluene: CH₃ with O₂N, NO₂ ortho and NO₂ para]

(d) [toluene, Ph-CH₃]  →(K₂Cr₂O₇ / H₂SO₄)→  [Ph-COOH]  →(Br₂ / AlCl₃)→  [m-bromobenzoic acid: COOH with Br meta]

*(e)* benzene-CH$_3$  $\xrightarrow[\text{AlCl}_3]{\text{Br}_2}$  Br-benzene-CH$_3$  $\xrightarrow[\text{H}_2\text{SO}_4]{\text{K}_2\text{Cr}_2\text{O}_7}$  Br-benzene-COOH

*(f)* benzene  $\xrightarrow[\text{AlCl}_3]{\text{Cl}_2}$  benzene-Cl  $\xrightarrow[\text{AlCl}_3]{\text{Cl}_2}$  Cl-benzene-Cl

*(g)* benzene  $\xrightarrow[\text{H}_2\text{SO}_4]{\text{HNO}_3}$  benzene-NO$_2$  $\xrightarrow[\text{heat}]{\text{H}_2\text{SO}_4}$  benzene with SO$_3$H and NO$_2$

**or** benzene  $\xrightarrow[\text{heat}]{\text{H}_2\text{SO}_4}$  benzene-SO$_3$H  $\xrightarrow[\text{H}_2\text{SO}_4]{\text{HNO}_3}$

*(h)* benzene  $\xrightarrow[\text{H}_2\text{SO}_4]{\text{HNO}_3}$  benzene-NO$_2$  $\xrightarrow[\text{AlCl}_3]{\text{Cl}_2}$  benzene with NO$_2$ and Cl

**9.41**  *Show how to synthesize these aromatic ketones starting with benzene or toluene as the only sources of aromatic rings. Assume in all syntheses that mixtures of ortho-para products can be separated to give the desired isomer in pure form.*

*(a)* benzene-CH$_3$  $\xrightarrow[\text{AlCl}_3]{\overset{\displaystyle O}{\text{CH}_3\overset{\|}{\text{C}}\text{Cl}}}$  H$_3$C-benzene-$\overset{\displaystyle O}{\overset{\|}{\text{C}}}$CH$_3$

*(b)* benzene  $\xrightarrow[\text{FeBr}_3]{\text{Br}_2}$  benzene-Br  $\xrightarrow[\text{AlCl}_3]{\overset{\displaystyle O}{\text{CH}_3\overset{\|}{\text{C}}\text{Cl}}}$  Br-benzene-$\overset{\displaystyle O}{\overset{\|}{\text{C}}}$CH$_3$

*(c)* benzene  $\xrightarrow[\text{AlCl}_3]{\overset{\displaystyle O}{\text{CH}_3\overset{\|}{\text{C}}\text{Cl}}}$  benzene-$\overset{\displaystyle O}{\overset{\|}{\text{C}}}$CH$_3$  $\xrightarrow[\text{FeBr}_3]{\text{Br}_2}$  Br-benzene-$\overset{\displaystyle O}{\overset{\|}{\text{C}}}$CH$_3$

**9.42**   *The following ketone, isolated from the roots of several members of the iris family, has an odor like that of violets and is used as a fragrance in perfumes. Describe the synthesis of this ketone from benzene.*

4-Isopropylacetophenone

**9.43**   *The bombardier beetle generates p-quinone, an irritating chemical, by the enzyme-catalyzed oxidation of hydroquinone using hydrogen peroxide as the oxidizing agent. Heat generated in this oxidation produces superheated steam, which is ejected, along with p-quinone, with explosive force.*

Hydroquinone                                *p*-Quinone

*(a) Balance the above equation.*

*(b) Show that this reaction of hydroquinone is an oxidation.*

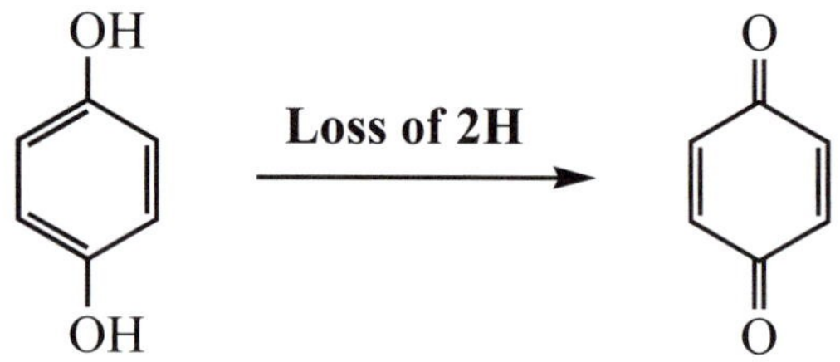

**9.44**   *Following is a structural formula for musk ambrette, a synthetic musk used in perfumes to enhance and retain fragrance.  Propose a synthesis for musk ambrette from m-cresol.*

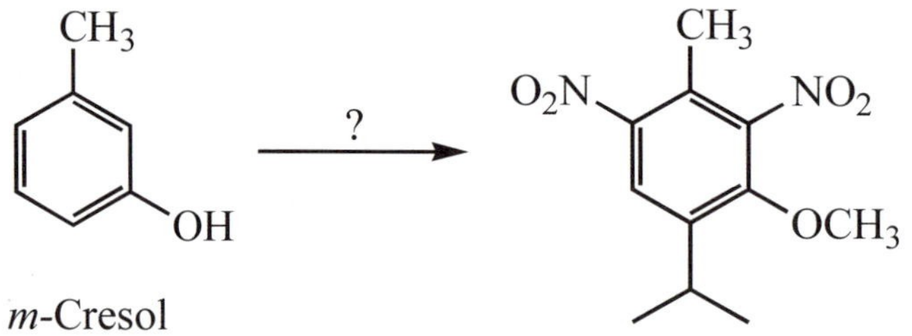

*m*-Cresol

Musk ambrette

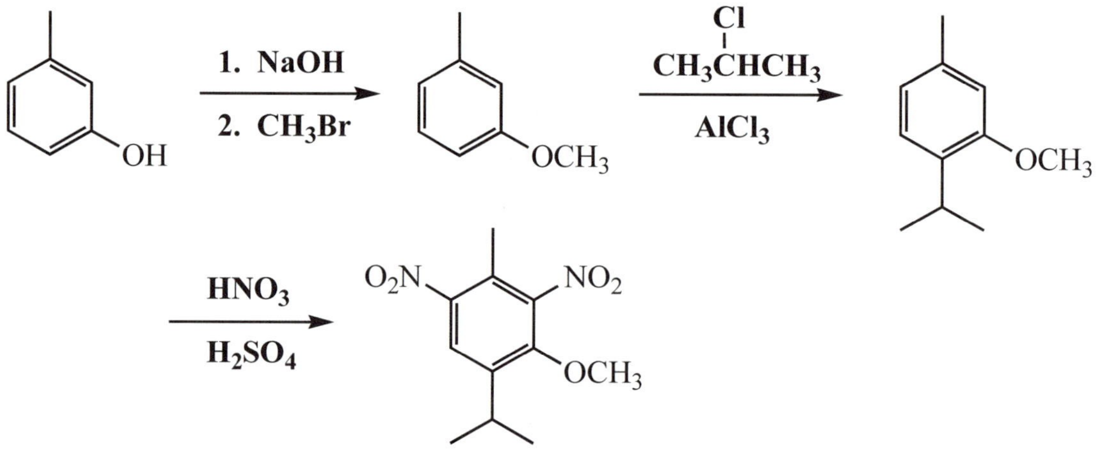

**9.45**   *1-(3-Chlorophenyl)propanone is a building block in the synthesis of bupropion, the hydrochloride salt of which is the antidepressant Wellbutrin. During clinical trials, researchers discovered that smokers reported a lessening in their craving for tobacco after one to two weeks on the drug. Further clinical trials confirmed this finding, and the drug is also marketed under the trade name Zyban as an aid in smoking cessation. Propose a synthesis for this building block from benzene. (We will see in Section 13.9 how to complete the synthesis of bupropion.)*

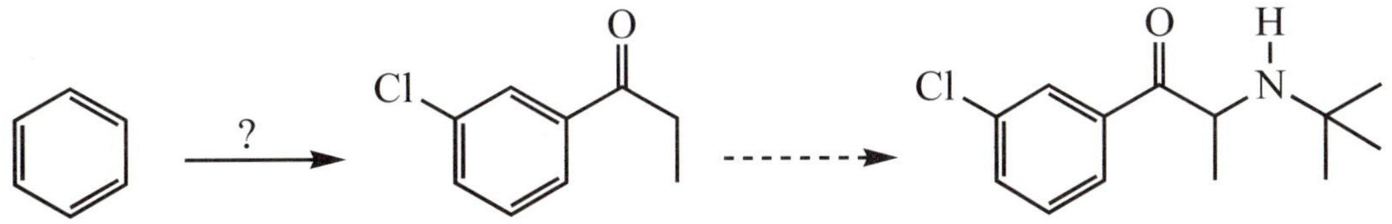

Benzene                1-(3-Chlorophenyl)-1-
propanone

Bupropion
(Wellbutrin, Zyban)

**Proposed synthesis of 1-(3-chlorophenyl)-1-propane:**

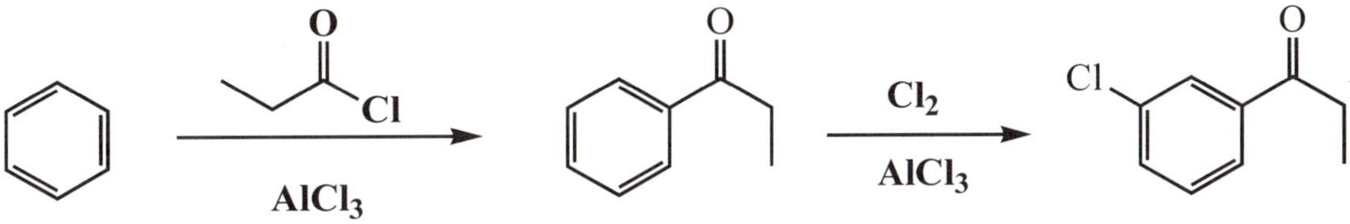

**Looking Ahead**

**9.46**   *Which of the following compounds can be made directly by using an electrophilic aromatic substitution reaction?*

**None of the above compounds can be synthesized directly using electrophilic aromatic substitution.**

(a) 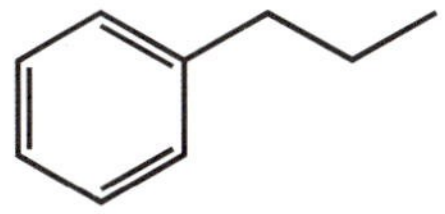   **The 1û carbocation electrophile ($CH_3CH_2\overset{+}{C}H_2$) rearranges to the more stable 2û carbocation cation ($CH_3\overset{+}{C}HCH_3$).**

(b) 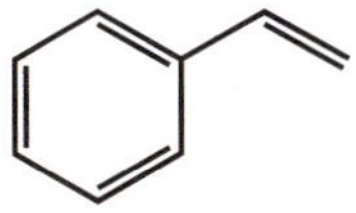   **The vinyl cation ($H_2\overset{+}{C}=CH$) is too unstable to form.**

(c) 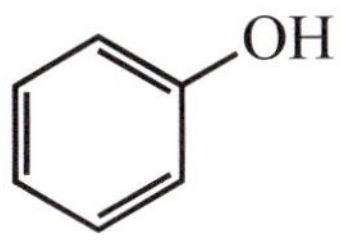   **The electrophile $OH^+$ is too unstable to form.**

(d) [structure: benzene ring with NH₂]          **The electrophile NH₂⁺ is too unstable to form.**

**9.47**   *Which compound is a better nucleophile?*

[structure: aniline with NH₂]     or     [structure: cyclohexanamine with NH₂]

Aniline                        Cyclohexanamine

**Cyclohexanamine is a better nucleophile and a stronger base than aniline.  Aniline's nitrogen electron lone pair is delocalized through resonance by the benzene ring, thus leaving it less available for electron pair donation as a base or nucleophile.  Cyclohexanamine's nitrogen lone pair is localized and thus more available for donation as a base or nucleophile.**

**9.48**   *Suggest a reason that the following arenes do not undergo electrophilic aromatic substitution when AlCl₃ is used in the reaction.*

(a) [structure: benzene with CH(OH)CH₂CH₃]     (b) [structure: benzene with CH₂SH]     (c) [structure: benzene with NH₂]

**AlCl₃ reacts violently with protic acids to evolve HCl.  In all three compounds, –OH, –SH, and –NH₂ groups are strong Lewis bases and form strong bonds with the AlCl₃ catalyst, diminishing the reactivity of the arene or the Lewis acid-Lewis base complex.**

**9.49**   *Predict the product of the following acid-base reaction:*

[structure: imidazole with N2 labeled] + H₃O⁺ ⟶ [structure: protonated imidazole with +N2] + H₂O

Imidazole

**Imidazole is an aromatic amine with a six π electron system.  Nitrogen (N1) donates its electron lone pair to the six π electron aromatic system while nitrogen (N2) has its lone pair perpendicular to the aromatic p-orbitals and does not participate in the**

**aromaticity of imidazole, leaving the lone pair available for use as a base or nucleophile without sacrificing the aromaticity. If the electron lone pair on N1 *was* protonated by an acid, the lone pair would be lost in bonding with the proton and the imidazole ring would no longer be aromatic and stable. Aromaticity is a very powerful influence on reaction mechanisms, molecular structure, and stability.**

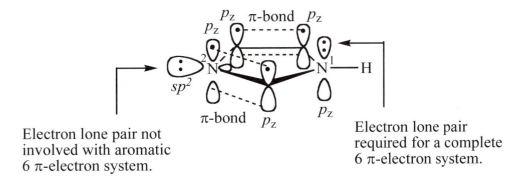

Electron lone pair not involved with aromatic 6 π-electron system.

Electron lone pair required for a complete 6 π-electron system.

**9.50**  *Which haloalkane reacts faster in an S$_N$1 reaction?*

Cl

or

Cl

**The formation of the more stable carbocation intermediate in an S$_N$1 reaction will occur faster. The resonance-stabilized carbocation is the most stable intermediate with charge delocalization over four carbon atoms through five resonance contributing structures (try drawing the other four resonance contributing structures).**

:Cl:

rate-determining
step

:Cl:⁻  +

**Resonance stabilized
carbocation intermediate**

:Cl:

rate-determining
step

:Cl:⁻  +

**2ûCarbocation intermediate**

## CHAPTER 10
### Solutions to Problems

**10.1**    *Identify all carbon stereocenters in coniine, nicotine, and cocaine.*

(a)

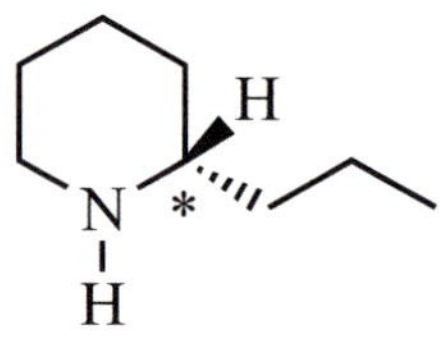

(*S*)-Coniine

(b)

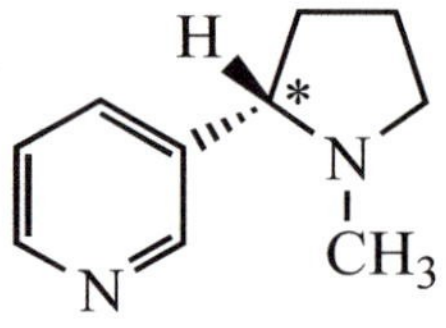

(*S*)-Nicotine

(c)

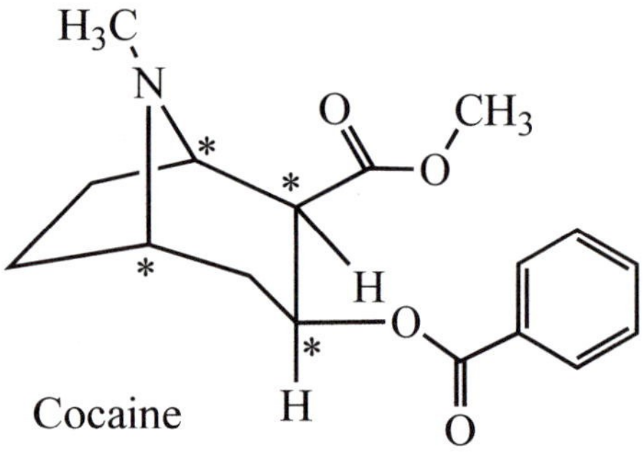

Cocaine

**10.2**    *Write a structural formula for each amine:*

(a) 2-Methyl-1-propanamine        (b) Cyclohexanamine 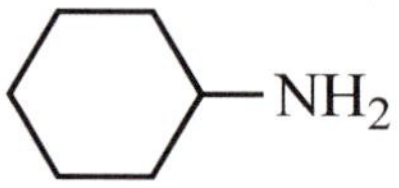        (c) (R)-2-Butanamine

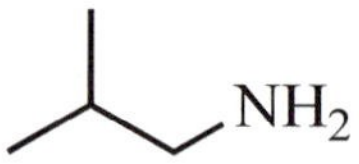

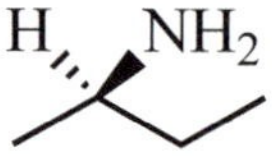

**10.3**    *Write a structural formula for each amine:*

(a) Isobutylamine        (b) Cyclohexylmethylamine 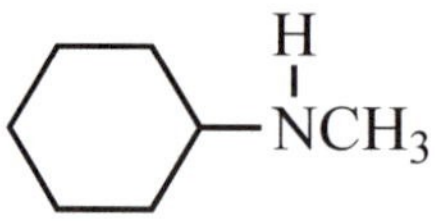        (c)  Benzylamine

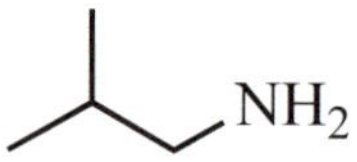

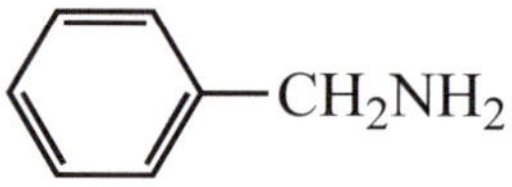

**10.4**    *Predict the position of equilibrium for this acid-base reaction:*

$$CH_3NH_3^+ \; + \; H_2O \; \rightleftharpoons \; CH_3NH_2 \; + \; H_3O^+$$

| $pK_a$ 10.64 | Weaker | Stronger | $pK_a$ -1.74 |
|---|---|---|---|
| Weaker acid | base | base | Stronger acid |

**Equilibrium favors formation of the weaker acid, so the equilibrium lies to the left.**

**10.5**   *Select the stronger acid from each pair of ions:*

*(a)* **(A) is the stronger acid.  4-Nitronaniline (p$K_b$ 13.0) is a weaker base than 4-methylaniline (p$K_b$ 8.92).   The decreased basicity of 4-nitroaniline is due to the electron-withdrawing and resonance effects of the para nitro group.  Because 4-nitroaniline is the weaker base, its conjugate acid (A) is the stronger acid.**

$$O_2N-\!\!\!\bigcirc\!\!\!-NH_3^+ \quad or \quad H_3C-\!\!\!\bigcirc\!\!\!-NH_3^+$$

(A)                              (B)

*(b)* **(C) is the stronger acid.  Aromatic heterocycles, such as pyridine (the conjugate base of C), are much weaker bases (p$K_b$ 8.75),  so the conjugate acids are stronger than conjugate acids of aliphatic amines, such as compound D, cyclohexanamine (p$K_b$ 3.34).  The lone pair of electrons in the $sp^2$ orbital on the nitrogen atom of pyridine has more s character; therefore more tightly held by the nucleus and are less available for bonding with a proton.**

$$\bigcirc\!\!\overset{+}{NH} \quad or \quad \bigcirc\!\!-NH_3^+$$

(C)                              (D)

**10.6**   *Complete each acid-base reaction and name the salt formed:*

*(a)*  $(CH_3CH_2)_3N \; + \; HCl \; \longrightarrow \; (CH_3CH_2)_3\overset{+}{N}H \; + \; Cl^-$

*(b)*  $\bigcirc\!\!NH \; + \; CH_3COOH \; \longrightarrow \; \bigcirc\!\!\overset{+}{N}H_2 \; + \; CH_3COO^-$

**10.7**   *As shown in Example 10.7, alanine is better represented as an internal salt. Suppose that the internal salt is dissolved in water.*

$$\underset{\underset{NH_2}{|}}{CH_3CHCOH} \overset{O}{\overset{\|}{\phantom{x}}} \quad \rightleftharpoons \quad \underset{\underset{+NH_3}{|}}{CH_3CHCO^-} \overset{O}{\overset{\|}{\phantom{x}}}$$

*(a) In what way would you expect the structure of alanine in aqueous solution to change if concentrated HCl were added to adjust the pH of the solution to 2.0?*

**At a pH of 2, all of the basic atoms will be protonated and the amino acid will have a positive charge overall.**

$$CH_3CHCOH$$
with a carbonyl O above (O double bond) and $+NH_3$ below

*(b) In what way would you expect the structure of alanine in aqueous solution to change if concentrated NaOH were added to bring the pH of the solution to 12.0?*

**At a pH of 12, all of the acidic protons will be deprotonated and the amino acid will have a negative charge overall.**

$$CH_3CHCO^-$$
with a carbonyl O above (O double bond) and $NH_2$ below

**10.8**  *Show how you can use the same set of steps in Example 10.8, but in a different order, to convert toluene to 3-hydroxybenzoic acid.*

Toluene                                                                    3-Hydroxy-
benzoic acid

**Step 1:  Oxidation of the benzylic carbon using chromic acid (Section 9.5).**

**Step 2:  Nitration of benzoic acid using nitric acid/sulfuric acid (Section 9.7B).**

**Step 3:  Reduction of the nitro group using either $H_2$ in the presence of a transition metal catalyst, or using Fe, Sn, or Zn in the presence of aqueous HCl (Section 10.8).**

**Step 4:  Treatment of the aromatic amine with $NaNO_2$/HCl forms the diazonium salt, which after warming the aqueous solution, yields the final product.**

The biggest difference between the synthesis of 3-hydroxybenzoic acid and the synthesis of 4-hydroxybenzoic acid as outlined in Example 10.8 is in the first two steps. For the synthesis of 3-hydroxybenzoic acid, the meta director –COOH must be established early in the synthesis to direct the nitration, subsequent reduction, and conversion of the diazonium salt to the meta substituted hydroxyl in the final product.

## Structure and Nomenclature

**10.9** *Draw a structural formula for each amine:*

*(a) (R)-2-Butanamine*

*(b) 1-Octanamine*

CH₃(CH₂)₆CH₂NH₂

*(c) 2,2-Dimethyl-1-propanamine*

*(d) 1,5-Pentanediamine*

*(e) 2-Bromoaniline*

*(f) Tributylamine*

(CH₃CH₂CH₂CH₂)₃N

*(g) N,N-Dimethylaniline*

*(h) Benzylamine*

*(i) tert-Butylamine*

*(j) N-Ethylcyclohexanamine*

*(k) Diphenylamine*

*(l) Isobutylamine*

**10.10** *Draw a structural formula for each amine:*

*(a) 4-Aminobutanoic acid*

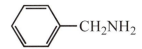

*(b) 2-Aminoethanol (ethanolamine)*

*(c) 2-Aminobenzoic acid*

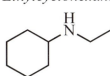

*(d) (S)-2-Aminopropanoic acid (alanine)*

*(e) 4-Aminobutanal*

*(f) 4-Amino-2-butanone*

**10.11** *Draw examples of 1°, 2°, and 3° amines that contain at least four sp³ hybridized carbon atoms. Using the same criterion, provide examples of 1°, 2°, and 3° alcohols. How does the classification system differ between the two functional groups?*

**The classification of 1°, 2°, and 3° amines is based on how many hydrogen atoms of ammonia are replaced by alkyl or aryl groups. Classification of alcohols as 1°, 2°, and 3° is based on how many substituents are bonded to the carbon bearing the alcohol hydroxyl.**

**Butylamine**
**a primary amine**

**Diethylamine**
**a secondary amine**

**Ethyldimethylamine**
**a tertiary amine**

**1-Butanol**
**a primary alcohol**

**2-Butanol**
**a secondary alcohol**

**2-Methyl-2-propanol**
**a tertiary alcohol**

**10.12** Classify each amino group as primary, secondary, or tertiary and as aliphatic or aromatic:

*(a)*

**Primary aromatic amine**

*Benzocaine*
*(a topical anesthetic)*

*(b)*

**Tertiary aliphatic amine**

**Secondary aromatic amine**

**Tertiary heterocyclic aromatic amine**

*Chloroquine*
*(a drug for the treatment of malaria)*

(c)

Serotonin
(a neurotransmitter)

**10.13** *Epinephrine is a hormone secreted by the adrenal medulla. Among epinephrine's actions, it is a bronchodilator. Albuterol, sold under several trade names, including Proventil and Salbumol, is one of the most effective and widely prescribed antiasthma drugs. The R enantiomer of albuterol is 68 times more effective in the treatment of asthma than the S enantiomer.*

(*R*)-Epinephrine
(Adrenaline)

(*R*)-Albuterol

*(a) Classify each amino group as primary, secondary, or tertiary.*

**Both amino groups are secondary aliphatic amines.**

*(b) List the similarities and differences between the structural formulas of these compounds.*

**The structural similarities are indicated in bold on the structures. Both are secondary aliphatic amines and they both are chiral with the (R) configuration. The differences include that adrenaline has ortho hydroxyls on the benzene ring whereas (*R*)-albuterol has a hydroxymethyl group ortho to the hydroxyl. Also, adrenaline has a methyl group on the amine while (*R*)-albuterol has a *tert*-butyl group on the amine.**

**10.14**  *There are eight constitutional isomers with molecular formula C₄H₁₁N. Name and draw structural formulas for each. Classify each amine as primary, secondary, or tertiary.*

**Primary amines:**

**Butylamine**          ***sec*-Butylamine**          **Isobutylamine**          ***tert*-Butylamine**

**Secondary amines:**

**Diethylamine**          **Methylpropylamine**          **Isopropylmethylamine**

**Tertiary amine:**

**Ethyldimethylamine**

**10.15**  *Draw a structural formula for each compound with the given molecular formula:*
   *(a) A 2⁰ arylamine, C₇H₉N*                    *(b) A 3⁰ arylamine, C₈H₁₁N*

   —NHCH₃                              —N(CH₃)₂

   *(c) A 1⁰ aliphatic amine, C₇H₉N*                    *(d) A chiral 1⁰ amine, C₄H₁₁N*

   —CH₂NH₂                     and

*(e) A 3¡ heterocyclic amine, C₅H₁₁N*

N—CH₃

*(f) A trisubstituted 1¡ arylamine, C₉H₁₃N*

—NH₂

(other isomers are possible)

*(g) A chiral quaternary ammonium salt, C₉H₂₂NCl*

Cl⁻   CH₂CH₃
         |+
H₃C—N—CH₂CH₂CH₃
       *|
       CH(CH₃)₂

CH₃  CH₂CH₃
 |      |+
CH₃CH₂CH—N—CH₃   Cl⁻
        *      |
              CH₂CH₃

**Although we have focused on carbon stereocenters, a quaternary ammonium salt that has a nitrogen atom with four different substituents is also a stereocenter.**

## Physical Properties

**10.16** *Propylamine, ethylmethylamine, and trimethylamine are constitutional isomers of molecular formula C₃H₉N:*

CH₃CH₂CH₂NH₂          CH₃CH₂NHCH₃          (CH₃)₃N

Propylamine                Ethylmethylamine          Trimethylamine
bp 48¡C                     bp 37¡C                    bp 3¡C

*Account for the fact that trimethylamine has the lowest boiling point of the three, and propylamine has the highest.*

**Trimethylamine lacks the N-H bonds that are required for hydrogen bonding, therefore cannot form intermolecular hydrogen bonds. The other two amines do have N-H bonds and participate in intermolecular hydrogen bonding. Propylamine has the highest boiling point because it can act as a hydrogen bond donor to two other molecules where ethylmethylamine can act as a hydrogen bond donor to only one other molecule. The boiling point in an isomeric series will increase with increasing intermolecular interactions.**

**10.17** *Account for the fact that 1-butanamine has a lower boiling point than 1-butanol:*

~~NH₂          ~~OH

1-Butanamine              1-Butanol
bp 78¡C                    bp 117¡C

The hydrogen bonds between N-H---N are not as strong as those for O-H---O because N-H bonds are less polar than O-H bonds.  Stronger intermolecular hydrogen bonds between molecules of 1-butanol take greater energy to separate for vaporization, thus the higher boiling point of the alcohol.

**10.18**  *Account for the fact that putrescine, a foul smelling compound produced by rotting flesh, ceases to smell upon treatment with two equivalents of HCl.*

**1,4-Butanediamine**
**(Putrescine)**

The relatively volatile 1,4-butanediamine is protonated twice when treated with two equivalents of HCl, forming an ammonium salt.  Ammonium salts are ionic and not volatile because of the strong intermolecular ionic interactions.

**Basicity of Amines**
**10.19**  *Account for the fact that amines are more basic than alcohols.*

Nitrogen is less electronegative than oxygen, so the lone pair of electrons on nitrogen is held less tightly and more available to bond with protons than either lone pair on oxygen; therefore, amines are more basic than alcohols.

**10.20**  *From each pair of compounds, select the stronger base:*

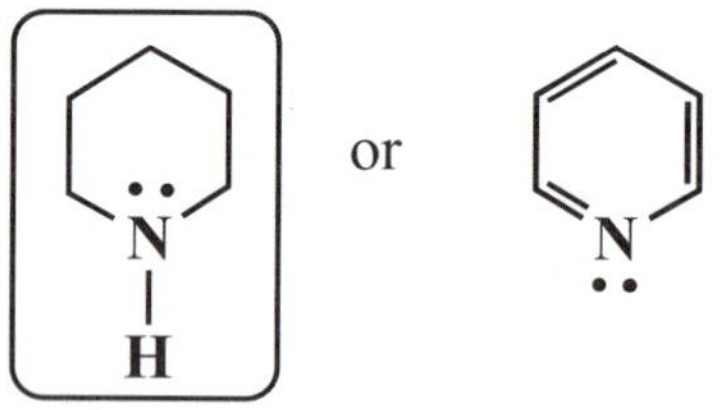

Piperidine is a stronger base than pyridine because the basic electron pair on piperidine is held less tightly by the $sp^3$ hybridized nitrogen than pyridine's more electronegative $sp^2$ hybridized nitrogen.

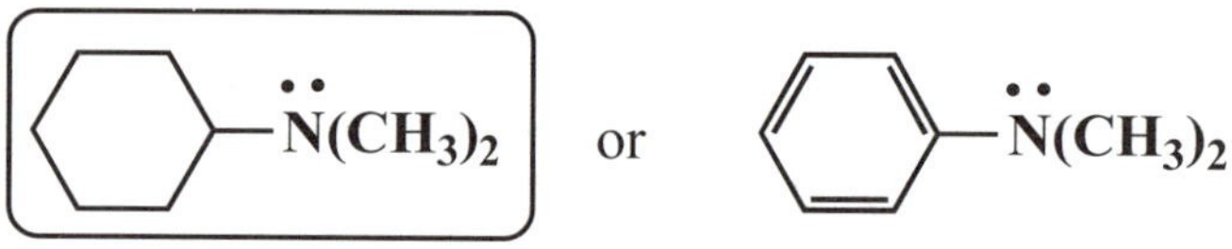

The nitrogen electron pair on *N,N*-dimethylaniline is delocalized by resonance into the aromatic ring, making it less available for bonding with protons. The nitrogen electron pair on cyclohexyldimethylamine is not delocalized by resonance thus making it more basic.

*(c)*

The nitrogen electron pair on 3-methylaniline is delocalized by resonance into the aromatic ring, making it less available for bonding with protons. The nitrogen electron pair on benzylamine is not delocalized by resonance thus making it more basic.

*(d)*

Although both 4-nitroaniline and 4-methylaniline nitrogen lone pairs are delocalized by resonance, the electron-withdrawing nitro group further delocalizes the electron pair, making it less available for bonding with protons. The methyl group on 4-methylaniline is an electron-releasing group and actually increases the basicity of the electron pair relative to 4-nitroaniline.

**10.21** *Account for the fact that substitution of a nitro group makes an aromatic amine a weaker base, but makes a phenol a stronger acid. For example, 4-nitroaniline is a weaker base than aniline, but 4-nitrophenol is a stronger acid than phenol.*

The nitro group withdraws electron density from the aromatic ring. For 4-nitroaniline, delocalization of the lone pair of electrons on the nitrogen through resonance makes it less able to bond with protons, making it a weaker base relative to the unsubstituted aniline.

The same effect is seen in the conjugate base of phenols. The negative charge on the conjugate base of 4-nitrophenol is delocalized by resonance into the benzene ring and the nitro group. By making the negative charge on the conjugate base more stable, the conjugate base becomes weaker and its acid becomes stronger.

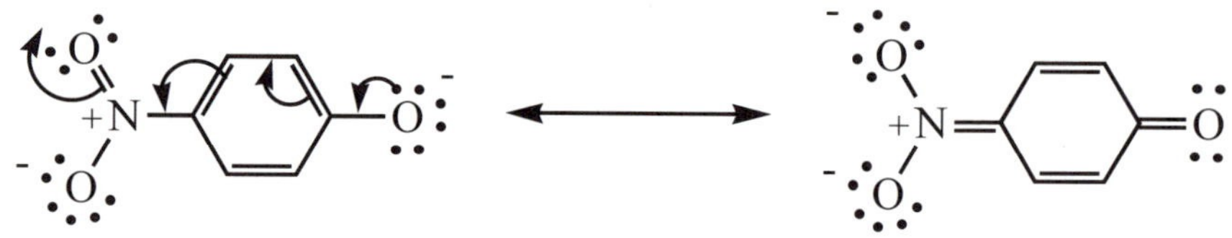

**10.22** *Select the stronger base in this pair of compounds:*

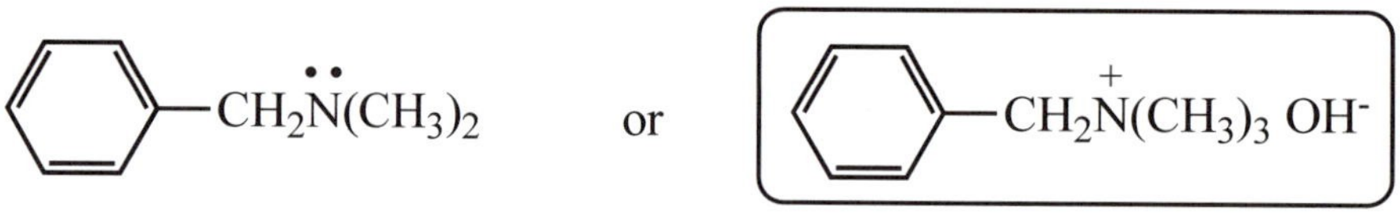

**Benzyltrimethylammonium hydroxide is a stronger base than benzyldimethylamine because the basic site is the hydroxide anion. The quaternary ammonium cation is simply the counter ion, much like the sodium cation is a counter ion for the hydroxide anion in sodium hydroxide. The basic site on benzyldimethylamine is the nitrogen lone pair of electrons. Hydroxide is a stronger base ($pK_b$ -1.7) than tertiary amines ($pK_b$ between 3-4).**

**10.23** *Complete the following acid-base reactions and predict the position of equilibrium for each. Justify your prediction by citing values of $pK_a$ for the stronger and weaker acid in each equilibrium. For values of acid ionization constants, consult Table 2.2 ($pK_a$'s of some inorganic and organic acids), Table 8.3 ($pK_a$'s of alcohols), Section 9.9B (acidity of phenols), and Table 10.2 (base strengths of amines). Where no ionization constants are given, make the best estimate from the aforementioned tables and section.*

**In acid-base reactions, the equilibrium favors the side with the weaker acid and weaker base. Remember, $pK_a + pK_b = 14$.**

*(a)* **The equilibrium lies to the right.**

CH₃COOH   +   (pyridine)   ⇌   CH₃COO⁻   +   (pyridinium N⁺–H)
                                            **(Weaker base)**

Acetic acid        Pyridine
**p$K_a$ 4.76**    **(Stronger base)**                          **p$K_a$ 5.25**
**(Stronger acid)**                                              **(Weaker acid)**

*(b)* **The equilibrium lies to the right.**

Phenol (OH)   +   (CH₃CH₂)₃N   ⇌   Phenol (O⁻)   +   (CH₃CH₂)₃NH⁺

Phenol            Triethylamine      **(Weaker base)**        **p$K_a$ 10.75**
**p$K_a$ 9.95**   **(Stronger base)**                          **(Weaker acid)**
**(Stronger acid)**

*(c)* **The equilibrium lies to the right.**

Ph–CH(NH₂)CH₃   +   lactic acid   ⇌   Ph–CH(NH₃⁺)CH₃   +   lactate⁻

1-Phenyl-2-       2-Hydroxypropanoic      **p$K_a$ ~11**        **(Weaker base)**
propanamine       acid                     **(Weaker acid)**
(Amphetamine)     (Lactic acid)
**(Stronger base)**   **p$K_a$ 3.08**
                  **(Stronger acid)**

*(d)* **The equilibrium lies to the right.**

Ph–CH(NHCH₃)CH₃   +   acetic acid   ⇌   Ph–CH(NH₂⁺CH₃)CH₃   +   acetate⁻

Methamphetamine   Acetic acid                                    **(Weaker base)**
**(Stronger base)**   **p$K_a$ 4.76**       **p$K_a$ ~11**
                  **(Stronger acid)**      **(Weaker acid)**

**10.24**  *The pK$_a$ of the morpholinium ion is 8.33:*

Morpholinium ion                    Morpholine

$pK_a = 8.33$

*(a) Calculate the ratio of morpholine to morpholinium ion in aqueous solution at pH 7.0.*

$$pH = -\log[H^+]$$

$$[H^+] = 10^{-pH}$$

$$K_a = \frac{[\text{Morpholine}][H^+]}{[\text{Morpholinium ion}]} = 10^{-8.33}$$

$$\frac{[\text{Morpholine}]}{[\text{Morpholinium ion}]} = \frac{K_a}{[H^+]} = \frac{10^{-8.33}}{10^{-7.00}} = 0.047$$

**Morpholine is a base and exists predominately as its conjugate acid, the morpholinium ion, in an aqueous solution at pH 7.0.**

*(b) At what pH are the concentrations of morpholine and morpholinium ion equal?*

**The concentrations of morpholine and morpholinium ion will be equal when the pH and the pK$_a$ are equal, which will occur at pH 8.33.**

**10.25**  *The pK$_b$ of amphetamine (Example 10.2) is approximately 3.2. Calculate the ratio of amphetamine to its conjugate acid at pH 7.4, the pH of blood plasma.*

**We know that** $[OH^-][H^+] = 10^{-14}$          **At pH 7.4,** $[OH^-] = \frac{10^{-14}}{[H^+]} = \frac{10^{-14}}{10^{-7.4}} = 10^{-6.6}$

$$K_b = \frac{[\text{Conjugate Acid}][OH^-]}{[\text{Amphetamine}]} = 10^{-3.2}$$

$$\frac{[\text{Conjugate Acid}]}{[\text{Amphetamine}]} = \frac{10^{-3.2}}{[OH^-]} = \frac{10^{-3.2}}{10^{-6.6}} \; 10^{3.4}$$

**Therefore, the ratio of [Amphetamine] : [Conjugate Acid] at pH 7.4 is $10^{3.4}$ (1:2,500). Although amphetamine is a base, it exists in blood plasma predominately as the conjugate acid form, even though the pH of blood plasma is slightly basic.**

**10.26** *Calculate the ratio of amphetamine to its conjugate acid at pH 1.0, such as might be present in stomach acid.*

$$pH + pOH = 14 \qquad pOH = 14 - pH \qquad pOH = 14.0 - 1.0 = 13.0$$

$$[OH] = 10^{-pOH} = 10^{-13}\,M$$

$$K_b = \frac{[\text{Conjugate Acid}][OH^-]}{[\text{Amphetamine}]} = 10^{-3.2}$$

$$\frac{[\text{Conjugate Acid}]}{[\text{Amphetamine}]} = \frac{10^{-3.2}}{[OH^-]} = \frac{10^{-3.2}}{10^{-13}} = 6.3 \times 10^{9}$$

**10.27** *Following is a structural formula of pyridoxamine, one form of vitamin B₆:*

Pyridoxamine
(Vitamin B₆)

*(a) Which nitrogen atom of pyridoxamine is the stronger base?*

**The primary amine indicated on the structure is more basic than the pyridine nitrogen atom because aliphatic amines are stronger bases than heterocyclic aromatic amines.**

*(b) Draw the structural formula of the hydrochloride salt formed when pyridoxamine is treated with one mole of HCl.*

**10.28** *Epibatidine, a colorless oil isolated from the skin of the Ecuadorian poison frog Epipedobates tricolor, has several times the analgesic potency of morphine. It is the first chlorine-containing, non-opioid (nonmorphine-like in structure) analgesic ever isolated from a natural source.*

Epibatidine

*(a) Which of the two nitrogen atoms of epibatidine is the more basic?*

**The aliphatic secondary amine has the more basic nitrogen atom because aliphatic amines are stronger bases than aromatic and heterocyclic aromatic amines.**

*(b) Mark all stereocenters in this molecule.*

**Epibatidine has three stereocenters. Each stereocenter is marked with an asterisk.**

**10.29** *Procaine was one of the first local anesthetics for infiltration and regional anesthesia:*

Procaine

*Its hydrochloride salt is marketed as Novocaine.*

*(a) Which nitrogen atom of procaine is the stronger base?*

**The tertiary amine is more basic than the primary aryl amine. Aryl amines are less basic than alkyl amines because the lone pair of electrons on the nitrogen atom are delocalized into the aromatic ring through resonance, thus making them less available for bonding with protons and Lewis acids.**

*(b) Draw the formula of the salt formed by treating procaine with one mole of HCl.*

*(c) Is procaine chiral? Would a solution of Novocaine in water be optically active or optically inactive?*

**Procaine is not chiral and an aqueous solution of procaine will be optically inactive.**

**10.30** *Treatment of trimethylamine with 2-chloroethyl acetate gives the neurotransmitter acetylcholine as its chloride salt:*

$$(CH_3)_3N \ + \ CH_3\overset{\overset{O}{\|}}{C}OCH_2CH_2Cl \longrightarrow \ C_7H_{16}ClNO_2$$
Acetylcholine chloride

*Propose a structural formula for this quaternary ammonium salt and a mechanism for its formation.*

**Acetylcholine, a quaternary ammonium chloride salt, is formed by an S$_N$2 mechanism between trimethylamine and 2-chloroethyl acetate.**

**Acetylcholine chloride**

**10.31** *Aniline is prepared by catalytic reduction of nitrobenzene:*

*Devise a chemical procedure based on the basicity of aniline to separate it from any unreacted nitrobenzene.*

**These compounds can first be dissolved in an organic solvent such as diethyl ether, and then separated by extracting with an aqueous solution of HCl (pH of 4 or less). Aniline, an amine and basic, will be protonated by the acid and converted to a water-soluble amine salt. The neutral nitrobenzene is soluble in the organic solvent.**

**The nitrobenzene is isolated after solvent evaporation. The aniline can be recovered by making the aqueous solution basic, thereby deprotonating the aniline and extracted with diethyl ether. Removal of ether yields pure aniline.**

**10.32** *Suppose that you have a mixture of these three compounds:*

$H_3C$——NO$_2$          $H_3C$——NH$_2$          $H_3C$——OH

4-Nitrotoluene            4-Methylaniline          4-Methylphenol
(*p*-Nitrotoluene)         (*p*-Toluidine)           (*p*-Cresol)

*Devise a chemical procedure based on their relative acidity or basicity to separate and isolate each in pure form.*

**These molecules can be separated by chemically reactive extractions into different aqueous solutions. First, the mixture is dissolved in an organic solvent such as diethyl ether, in which all three compounds are soluble. Then the ether solution is extracted with dilute aqueous HCl. Under acidic conditions, 4-methylaniline (a weak base) is protonated to the ammonium salt and dissolves in the aqueous solution. The aqueous solution is separated and treated with dilute NaOH and the water insoluble 4-methylaniline separates, and is recovered. The ether solution containing the remaining two components is treated with dilute NaOH. Under these conditions, 4-methylphenol (a weak acid) is deprotonated to its phenoxide ion and dissolves in the aqueous solution. Acidification of this aqueous solution with dilute HCl forms water-insoluble 4-methylphenol, which is then isolated. Evaporation of the remaining ether solution gives the 4-nitrotoluene.**

**10.33** *Following is a structural formula for metformin, the hydrochloride salt of which is marketed as the antidiabetic Glucophage:*

Metformin

*Metformin was introduced into clinical medicine in the United States in 1995 for the treatment of type 2 diabetes. More than 25 million prescriptions for this drug were written in 2000, making it the most commonly prescribed brand-name diabetes medication in the nation.*

*(a) Draw the structural formula for Glucophage.*

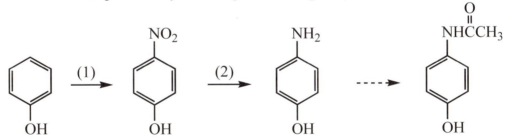

**Glucophage**

*(b) Would you predict Glucophage to be soluble or insoluble in water? Soluble or insoluble in blood plasma? Would you predict it to be soluble or insoluble in diethyl ether? In dichloromethane? Explain your reasoning.*

**Glucophage is an amine hydrochloride salt, which is water-soluble and being ionic, insoluble in non-polar organic solvents. Therefore, it is predicted that Glucophage will be soluble in water and blood plasma. Because Glucophage is a hydrochloride salt, it will be insoluble in non-polar solvents such as diethyl ether and dichloromethane.**

## Synthesis

**<u>10.34</u>** *4-Aminophenol is a building block in the synthesis of the analgesic acetaminophen. Show how this building block can be synthesized in two steps from phenol. (In Chapter 15, we will see how to complete the synthesis of acetaminophen.)*

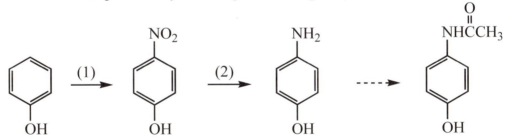

**(1) HNO$_3$/H$_2$SO$_4$     (2) H$_2$/Ni**

**10.35** *4-Aminobenzoic acid is a building block in the synthesis of the topical anesthetic benzocaine. Show how this building block can be synthesized in three steps from toluene (in Chapter 15, we will see how to complete the synthesis of benzocaine):*

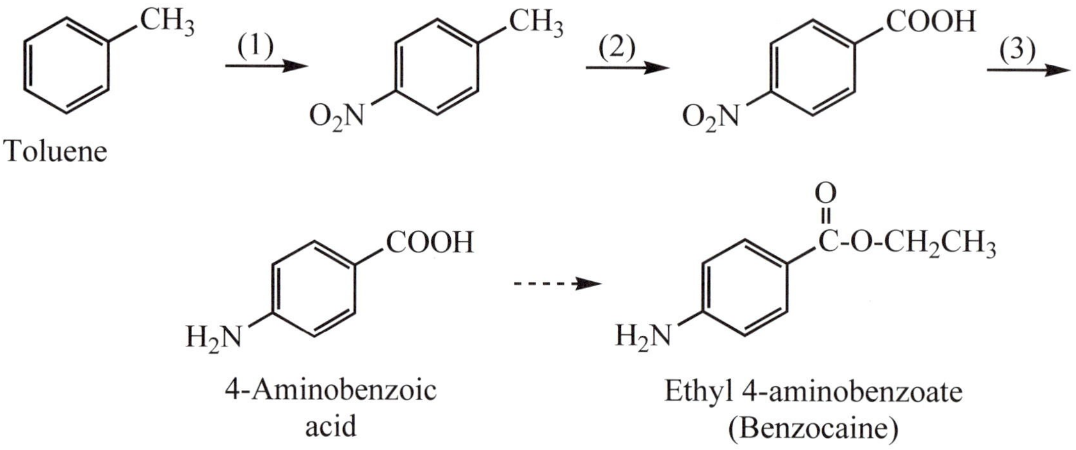

Toluene

**(1) HNO₃, H₂SO₄**        **(2) Na₂Cr₂O₇, H₂SO₄**        **(3) H₂, Ni**

4-Aminobenzoic
acid

Ethyl 4-aminobenzoate
(Benzocaine)

**10.36** *The compound 4-aminosalicylic acid is one of the building blocks needed for the synthesis of propoxycaine, one of the family of "caine" anesthetics. Some other members of this family of local anesthetics are procaine (Novocaine), lidocaine (Xylocaine), and mepivicaine (Carbocaine). 4-Aminosalicylic acid is synthesized from salicylic acid in five steps (in Chapter 15, we will see how to complete the synthesis of propoxycaine):*

Salicylic acid

4-Aminosalicylic acid

Propoxycaine

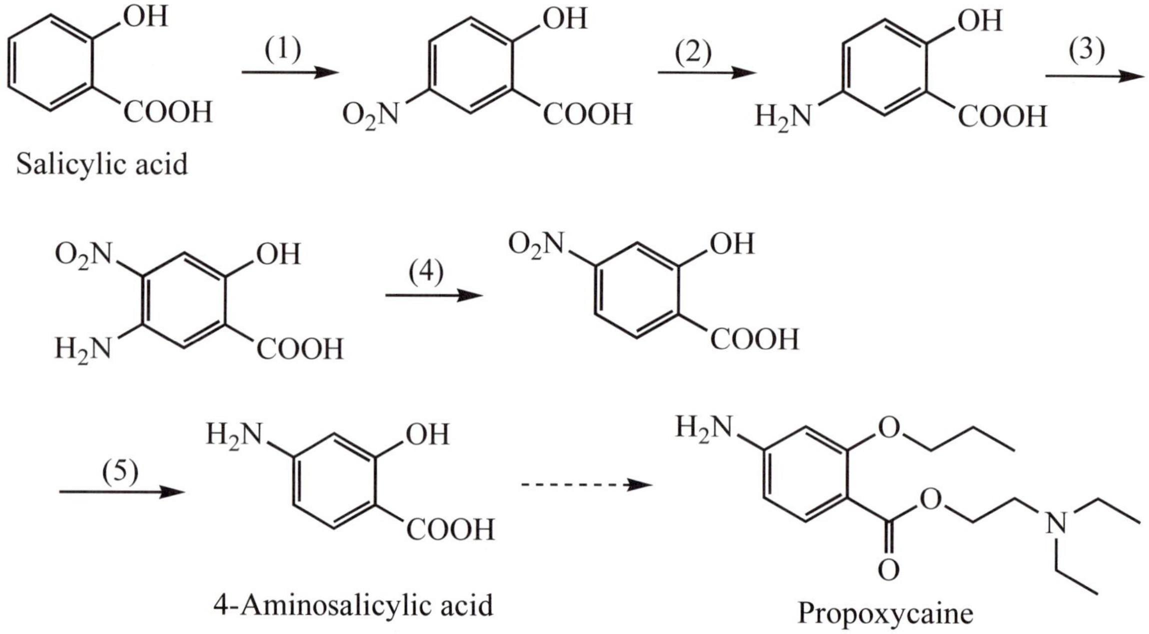

**(1) HNO₃, H₂SO₄    (2) H₂, Ni    (3) HNO₃, H₂SO₄**
**(4) 1. NaNO₂, HCl/0 °C    2. H₃PO₂    (5) H₂, Ni**

**10.37**  *A second building block for the synthesis of propoxycaine is 2-diethylaminoethanol:*

2-Diethylaminoethanol

*Show how this compound can be prepared from ethylene oxide and diethylamine.*

$$(CH_3CH_2)_2NH \ + \ \triangle \longrightarrow (CH_3CH_2)_2NCH_2CH_2OH$$

**10.38**  *Following is a two-step synthesis of the antihypertensive drug propranolol, a so-called β-blocker with vasodilating action:*

1-Naphthol              Epichlorohydrin

Propranolol
(Cardinol)

*Propranolol and other β-blockers have received enormous clinical attention because of their effectiveness in treating hypertension (high blood pressure), migraine headaches, glaucoma, ischemic heart disease, and certain cardiac arrhythmias.  The hydrochloride salt of propranolol has been marketed under at least 30 brand names, one of which is Cardinol (note the "card-" part of the name, after cardiac).*

*(a)  What is the function of potassium carbonate, K₂CO₃, in Step 1?  Propose a mechanism for the formation of the new oxygen-carbon bond in this step.*

**The K$_2$CO$_3$ is used to deprotonate the hydroxyl proton on naphthol (which has a pK$_a$ similar to that of phenol) to produce the nucleophilic 1-napthyloxide anion.**

**The formation of the new oxygen-carbon bond occurs via an S$_N$2 reaction:**

*(b) Name the amine used to bring about Step 2, and propose a mechanism for this step.*

**The amine used for the S$_N$2 reaction in Step 2 is isopropylamine.**

*(c) Is propranolol chiral? If so, how many stereoisomers are possible for it?*

**Propranolol has one stereocenter and is chiral. The stereocenter is identified in the introductory reaction scheme with an asterisk.**

**10.39** *The compound 4-ethoxyaniline, a building block of the over-the-counter analgesic phenacetin, is synthesized in three steps from phenol:*

4-Ethoxyaniline                                        Phenacetin

*Show reagents for each step in the synthesis of 4-ethoxyaniline. (In Chapter 15, we will see how to complete this synthesis).*

**(1) 1. NaOH  2. CH$_3$CH$_2$Br        (2) HNO$_3$/H$_2$SO$_4$        (3) H$_2$/Ni**

**10.40** *Radiopaque imaging agents are substances administered either orally or intravenously that absorb X rays more strongly than body material does. One of the best known of these agents is barium sulfate, the key ingredient in the "barium cocktail" used for imaging of the gastrointestinal tract. Among other X-ray imaging agents are the so-called triiodoaromatics. You can get some idea of the kinds of imaging for which they are used from the following selection of trade names: Angiografin, Gastrografin,*

*Cardiografin, Cholografin, Renografin, and Urografin. The most common of the triiodiaromatics are derivatives of these three triiodobenzenecarboxylic acids:*

3-Amino-2,4,6-
triiodobenzoic acid

3,5-Diamino-2,4,6-
triiodobenzoic acid

5-Amino-2,4,6-
triiodoisophthalic acid

*3-Amino-2,4,6-triiodobenzoic acid is synthesized from benzoic acid in three steps:*

3-Amino-
benzoic acid

3-Amino-2,4,6-
triiodobenzoic acid

*(a) Show reagents for Steps (1) and (2).*

**(1) HNO₃/H₂SO₄        (2) Ni/H₂**

*(b) Iodine monochloride, ICl, a black crystalline solid with a melting point of 27.2°C and a boiling point of 97°C, is prepared by mixing equimolar amounts of I₂ and Cl₂. Propose a mechanism for the iodination of 3-aminobenzoic acid by this reagent.*

**Step 1: Nucleophilic attack of the aromatic ring on ICl to form a cationic intermediate. There are three more resonance structures of the intermediate.**

**Step 2: Cation intermediate is deprotonated, re-establishing the aromatic ring.**

**Step 3: Nucleophilic attack of the aromatic ring on ICl to form a cation intermediate. There are other resonance structures.**

**Step 4: Cation intermediate is deprotonated, re-establishing the aromatic ring.**

**Step 5: Nucleophilic attack of the aromatic ring on ICl to form a cation intermediate. There are many other resonance structures.**

**Step 6: Cation intermediate is deprotonated, re-establishing the aromatic ring and forming the final product.**

*(c) Show how to prepare 3,5-diamino-2,4,6-triiodobenzoic acid from benzoic acid.*

**3,5-Diamino-2,4,6-triiodobenzoic acid**

*(d) Show how to prepare 5-amino-2,4,6-triiodoisophthalic acid from isophthalic acid (1,3-benzenedicarboxylic acid).*

**5-Amino-2,4,6-triiodoisophthalic acid**

**10.41**  *The intravenous anesthetic propofol is synthesized in four steps from phenol:*

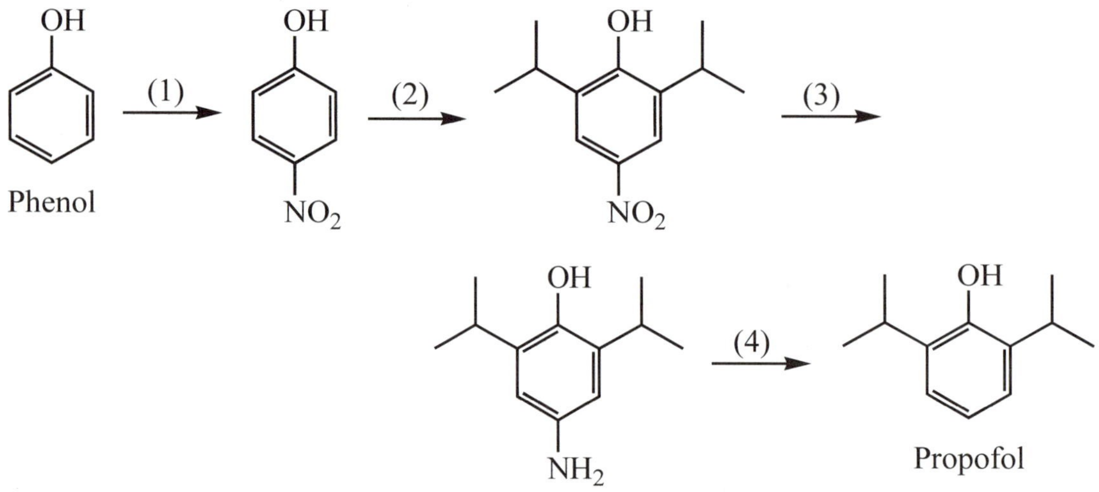

Phenol

*Show reagents to bring about each step.*

**(1)  HNO₃/H₂SO₄**           **(2) CH₃CH=CH₂/H₃PO₄**          **(3)  H₂/Ni**
**(4)  1.  NaNO₂, HCl/0 °C    2.  H₃PO₂**

**Looking Ahead**
**10.42**  *State the hybridization of the nitrogen atom in each of the following compounds:*

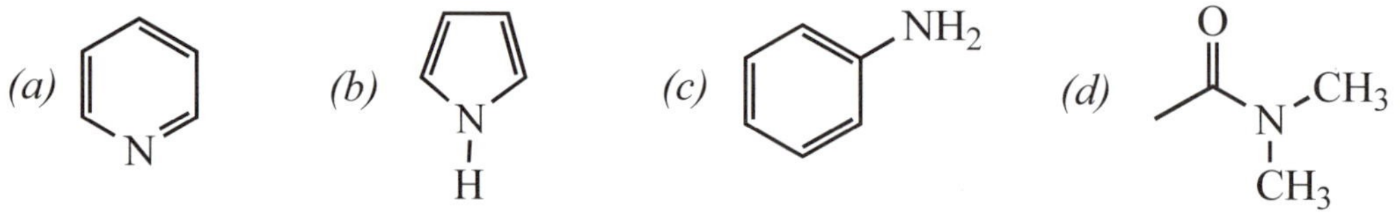

**All of the above nitrogens are *sp²* hybridized.  The unhybridized 2*p* orbital in (a) is
occupied by an electron pair that completes the aromaticity of the ring.  The
unhybridized 2*p* orbital in (b) is also occupied by an electron pair that completes the
aromaticity of the ring.  The unhybridized 2*p* orbital in (c) is occupied by two
electrons delocalized into the aromatic ring by resonance.**

**The unhybridized 2*p* orbital in (d) is occupied by two electrons that are delocalized
by the carbonyl through resonance.  In Chapter 19, we will see that this structural
feature is an important influence on the planarity of peptide bonds in proteins.**

**10.43** *Amines can act as nucleophiles. For each of the following molecules, circle the most likely atom that would be attacked by the nitrogen of an amine:*

**Nucleophiles will attack the carbons that have a partial positive or full positive charge.**

(a)   (b) 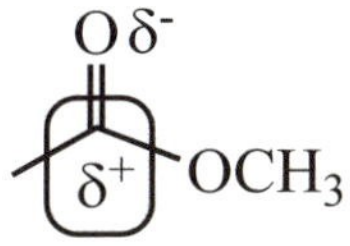   (c)

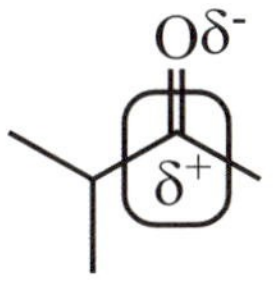

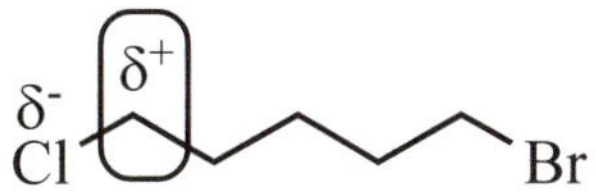

**10.44** *Draw a Lewis structure for a molecule with formula $C_3H_7N$ that does not contain a ring or an alkene (a carbon-carbon double bond).*

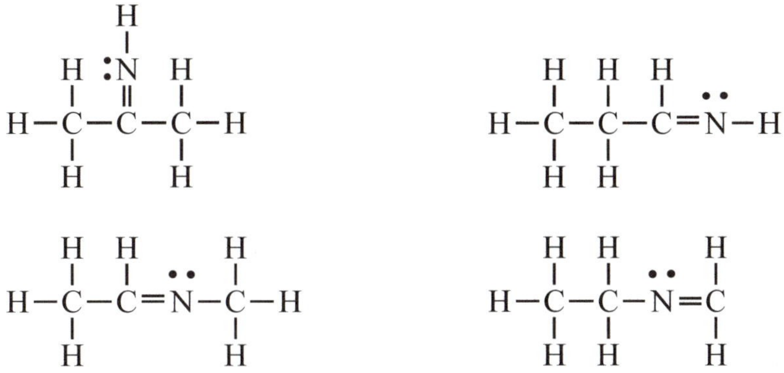

**10.45** *Rank the following leaving groups in order from best to worst:*

$$R-Cl \qquad R-O\overset{O}{\underset{}{\overset{\|}{C}}}-R \qquad R-OCH_3 \qquad R-N(CH_3)_2$$

**Leaving group ability increases as base strength decreases:**

$$Cl^- \; > \; {}^-O-\overset{O}{\overset{\|}{C}}-R \; > \; {}^-O-CH_3 \; > \; {}^-N(CH_3)_2$$

best leaving group ← worst leaving group

weakest base → strongest base

base strength

## CHAPTER 11
### *Solutions to Problems*

**11.1**   *Calculate the energy of red light (680 nm) in kilocalories per mole. Which form of radiation carries more energy, infrared radiation of wavelength 2.50 μm or red light of wavelength 680 nm?*

**The energy of red light at 680 nm = 42.1 kcal/mol which has a greater energy than infrared radiation at 2.50 μm (E = 11.4 kcal/mol as calculated in Example 1.1).**

**11.2**   *A compound shows strong, very broad IR absorption in the region 3200-3500 $cm^{-1}$ and strong absorption at 1715 $cm^{-1}$. What functional group accounts for both of these absorptions?*

**Carboxylic acids have two strong absorptions. The first absorption, a broad peak between 2400 and 3400 $cm^{-1}$ is due to the hydroxyl attached to the carbonyl carbon. The second absorbance is a strong peak around 1715 $cm^{-1}$ due to the C=O stretching vibration.**

**11.3**   *Propanoic acid and methyl ethanoate are constitutional isomers. Show how to distinguish between them by IR spectroscopy.*

$$\begin{array}{cc}
\quad O & \quad O \\
\quad \| & \quad \| \\
CH_3CH_2COH & CH_3COCH_3 \\
\text{Propanoic acid} & \text{Methyl ethanoate} \\
 & \text{(Methyl acetate)}
\end{array}$$

**Propanoic acid will have two major absorptions, a broad absorption between 2400-3400 $cm^{-1}$ due to the presence of a carboxylic acid -OH and a strong absorption around 1700-1725 $cm^{-1}$ from the carbonyl. Methyl acetate will not have the hydroxyl absorption.**

**11.4**   *What does the value of the wavenumber of the stretching frequency for a particular functional group indicate about the relative strength of the bond in that functional group?*

**As the wavenumber increases, so does the bond strength.**

**11.5**   *Calculate the index of hydrogen deficiency of cyclohexene, $C_6H_{10}$, and account for this deficiency by reference to its structural formula.*

**The index of hydrogen deficiency is two. The structural possibilities include two double bonds, a double bond and a ring, or two rings.**

**11.6**   *The index of hydrogen deficiency of niacin is 5. Account for this value by reference to the structural formula of niacin.*

Nicotinamide
(Niacin)

**Niacin has four double bonds (each worth one hydrogen deficiency) and a ring (worth one hydrogen deficiency) for a total index of hydrogen deficiency equal to five.**

**11.7**   *Complete the following table:*

| Class of Compound | Molecular Formula | Index of Hydrogen Deficiency | Reason for Hydrogen Deficiency |
|---|---|---|---|
| alkane | $C_nH_{2n+2}$ | 0 | (reference hydrocarbon) |
| alkene | $C_nH_{2n}$ | 1 | one pi bond |
| alkyne | $C_nH_{2n-2}$ | 2 | two pi bonds |
| alkadiene | $C_nH_{2n-2}$ | 2 | two pi bonds |
| cycloalkane | $C_nH_{2n}$ | 1 | one ring |
| cycloalkene | $C_nH_{2n-2}$ | 2 | one pi bond + one ring |

**11.8**   *Calculate the index of hydrogen deficiency of each compound.*

*(a) Aspirin, $C_9H_8O_4$*   **(6)**          *(b) Ascorbic acid (vitamin C), $C_6H_8O_6$*   **(3)**
*(c) Pyridine, $C_5H_5N$*   **(4)**          *(d) Urea, $CH_4N_2O$*   **(1)**
*(e) Cholesterol, $C_{27}H_{46}O$*   **(5)**          *(f) Trichloroacetic acid, $C_2HCl_3O$*   **(1)**

**11.9**   ***Compound A**, molecular formula, $C_6H_{10}$, reacts with $H_2/Ni$ to give **compound B**, $C_6H_{12}$. See also the IR spectrum of **compound A**. From this information about **compound A**, tell*

(a) *Its index of hydrogen deficiency.*

**The index of hydrogen deficiency for compound A is two.**

(b) *The number of its rings and/or pi bonds.*

**Compound A only added one molecule of hydrogen with $H_2$/Ni, therefore it can be concluded that compound A has one pi bond, leaving the other site of hydrogen deficiency a ring.**

(c) *What structural feature(s) would account for its index of hydrogen deficiency.*

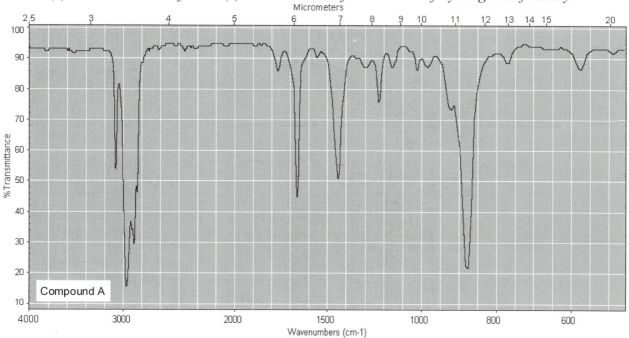

**From the hydrogenation data, it can be concluded that compound A contains only one C=C bond because it only added one molecule of $H_2$. Because rings do not hydrogenate under the conditions listed, the other hydrogen deficiency must be due to a ring. In addition, the C=C bond must be highly unsymmetrical, having a permanent dipole to explain the strong C=C stretching band around 1650 $cm^{-1}$. It would take additional chemical and spectroscopic data to unambiguously assign a structure to compound A. For example, [1]H-NMR and [13]C-NMR (Chapter 12) spectra would be extremely helpful. The combined data would allow you to determine that compound A is methylenecyclopentane.**

$$\text{(structure) } =CH_2$$

**Methylenecyclopentane**

**11.10**  *Compound C, molecular formula, $C_6H_{12}$, reacts with $H_2/Ni$ to give compound D, $C_6H_{14}$. See also the IR spectrum of compound C. From this information about compound C, tell (a) Its index of hydrogen deficiency.*

**The index of hydrogen deficiency for compound C is one.**

*(b) The number of its rings and/or pi bonds.*

**Compound may have one pi bond or one ring.**

*(c) What structural feature(s) would account for its index of hydrogen deficiency.*

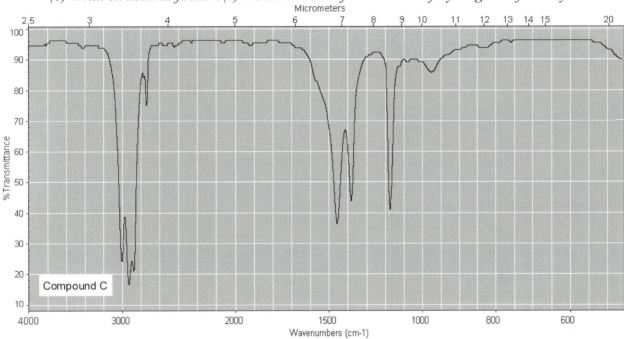

The infrared spectra of compound C indicates that the hydrogen deficiency is due to a C=C bond. The spectral evidence includes $C(sp^2)$-H stretching around 3020 cm$^{-1}$. The chemical data supports this conclusion because compound C added a hydrogen molecule upon hydrogenation with nickel catalyst. However, the absence of any C=C stretching bands near 1650 cm$^{-1}$ in the IR spectrum indicates that this must be an entirely symmetrical carbon-carbon double bond. It would take more information such as $^1$H and $^{13}$C NMR to unambiguously deduce the detailed structure of compound C. The combined data would eventually show that compound C was 2,3-dimethyl-2-butene.

**11.11**  *Following are infrared spectra of compounds E and F. One spectrum is of 1-hexanol and the other of nonane. Assign each compound its correct spectrum.*

**Compound E is nonane.  The only prominent absorptions observed in the IR spectrum are C($sp^3$)-H bends and stretches that occur between 2950 and 2850 cm$^{-1}$ and around 1450 cm$^{-1}$.**

**Compound F must be 1-hexanol.  In addition to the similar absorptions observed in nonane, Compound F has a very strong, broad absorption around 3340 cm$^{-1}$ that is associated with O-H stretching and an absorption around 1050 cm$^{-1}$ associated with C-O stretching.**

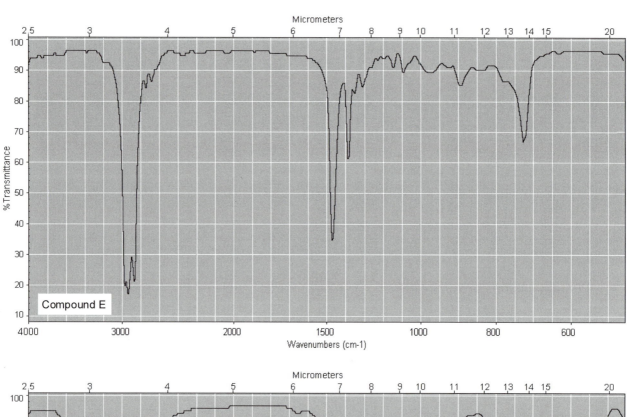

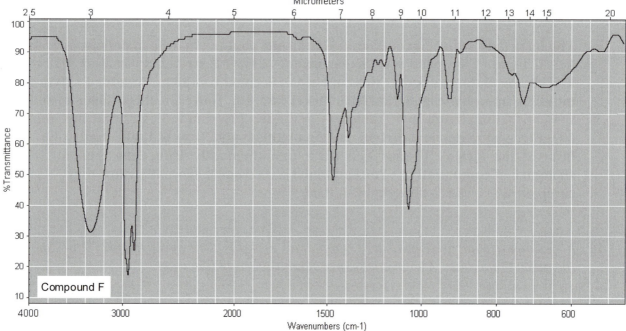

**11.12**     *2-Methyl-1-butanol and tert-butyl methyl ether are constitutional isomers of molecular formula C<sub>5</sub>H<sub>12</sub>O.  Assign each compound its correct infrared spectrum, **G** or **H**.*

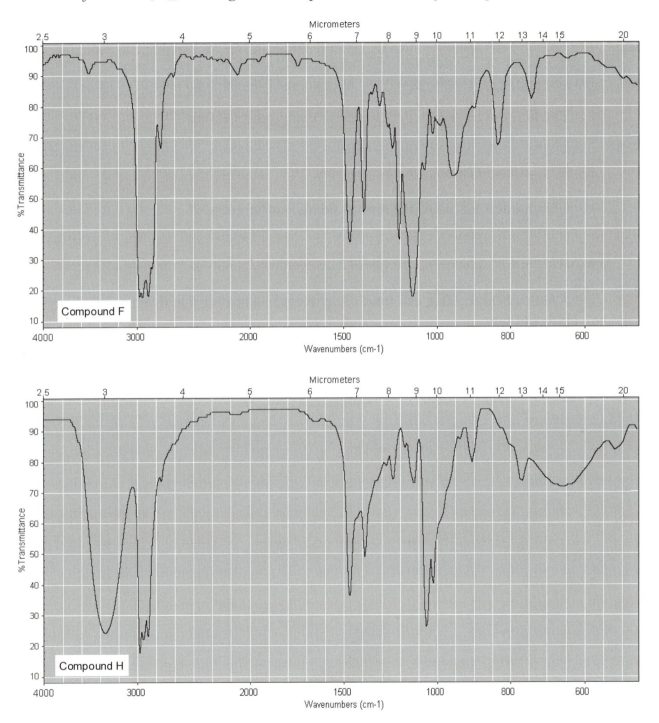

The molecules are similar except for the –OH group present in 2-methyl-1-butanol. The difference to notice in the IR spectrum will be the presence or absence of a broad peak between 3200 and 3400 cm$^{-1}$ associated with the –OH group.  The first spectrum for compound G corresponds to *tert*-butyl methyl ether because it lacks the broad peak in the IR spectrum between 3200 and 3400 cm$^{-1}$ and the second

**spectrum for compound H corresponds to 2-methyl-1-butanol because it has a broad peak at 3350 cm$^{-1}$, typical of alcohols.**

**11.13** *From examination of the molecular formula and IR spectrum of compound I, $C_9H_{12}O$, tell*

(a) *Its index of hydrogen deficiency.*

**The index of hydrogen deficiency for compound I is four.**

(b) *The number of its rings and/or pi bonds.*

**Compound I has a total of four pi bonds and/or rings.**

(c) *What one structural feature would account for this index of hydrogen deficiency.*

**A single benzene ring can (and often does) account for a hydrogen deficiency index of four.**

(d) *What oxygen-containing functional group it contains.*

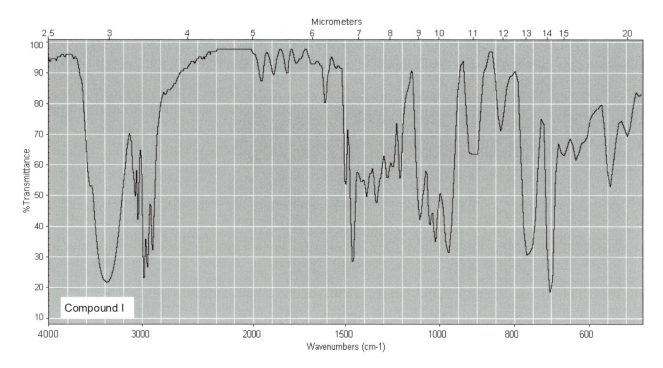

**The strong, broad absorption near 3400 cm$^{-1}$ indicates the presence of an –OH group, so this must be the oxygen containing functional group.  Again, other spectral methods would be required to assign a structure to compound I, but if**

you had $^1$H-and $^{13}$C-NMR data, it would confirm that compound I is 1-phenyl-1-propanol.

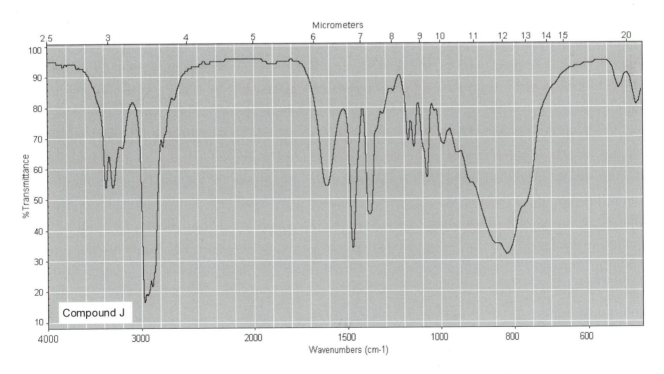

**1-Phenyl-1-propanol**

**11.14** *From examination of the molecular formula and IR spectrum of compound J, $C_5H_{13}N$, tell*

(a) *Its index of hydrogen deficiency.*

**The index of hydrogen deficiency for Compound J is zero.**

(b) *The number of its rings and/or pi bonds.*

**Compound J cannot have pi bonds or rings with a hydrogen deficiency of zero.**

(c) *The nitrogen-containing functional group(s) it might contain.*

**The index of hydrogen deficiency confirms that compound J must be an aliphatic amine. The presence of two broad N-H stretches at 3300 and 3400 cm$^{-1}$**

indicates that compound J is a primary aliphatic amine. Further investigations using $^1$H and $^{13}$C NMR will reveal that compound J is isopentylamine.

**3-Methyl-1-butanamine
(Isopentylamine)**

**11.15** *From examination of the molecular formula and IR spectrum of compound K, $C_6H_{12}O$, tell*

*(a) Its index of hydrogen deficiency.*

**The index of hydrogen deficiency for compound K is one.**

*(b) The number of its rings and/or pi bonds.*

**Based on the index of hydrogen deficiency, compound K can have only one ring or pi bond.**

*(c) What structural features would account for this index of hydrogen deficiency.*

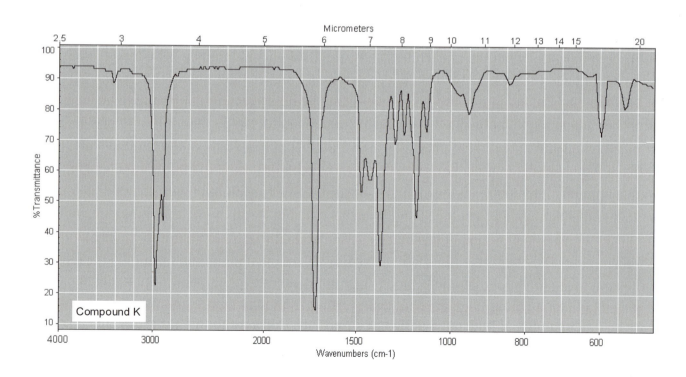

The very strong absorption band at about 1710 cm$^{-1}$ is characteristic of a C=O stretch, therefore the source of the hydrogen deficiency. With additional data from other spectroscopic methods, compound K is consistent with the structure of 4-methyl-2-pentanone.

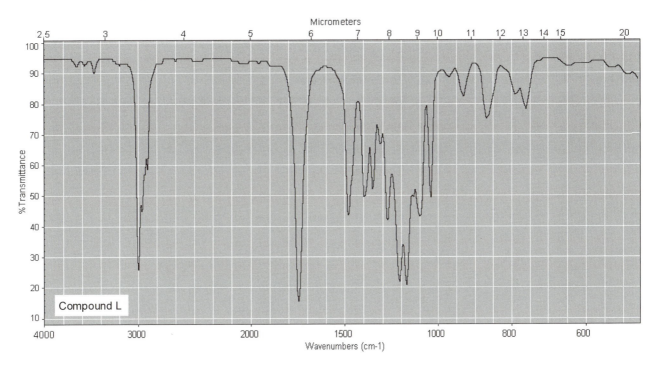

**4-Methyl-2-pentanone**

**11.16** *From examination of the molecular formula and IR spectrum of compound L, $C_6H_{12}O_2$, tell:*

(a) *Its index of hydrogen deficiency.*

**The index of hydrogen deficiency for compound L is one.**

(b) *The number of its rings and/or pi bonds.*

**Based on the index of hydrogen deficiency, compound L can have only one ring or pi bond.**

(c) *The oxygen-containing functional group(s) it might contain.*

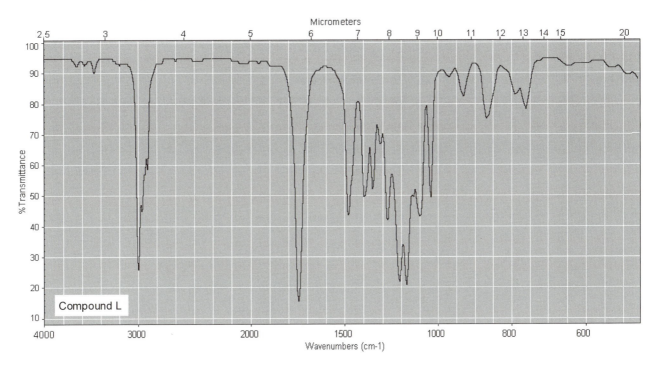

The key to solving this problem is to determine what functional group(s) contain oxygen in compound L. The strong absorption band at 1765 cm$^{-1}$ indicates the presence of a carbonyl group in compound L, possibly from a ketone, aldehyde, ester or carboxylic acid group. There is no strong –OH absorption band between 2400 and 3400 cm$^{-1}$, eliminating the possibility of a hydroxyl containing group such as a carboxylic acid or an alcohol. The strong absorptions between 1100 and 1200 cm$^{-1}$ indicate that compound L is an ester. Further analysis by $^1$H and $^{13}$C NMR will reveal that compound L is ethyl 2-methylpropanoate.

**Ethyl 2-methylpropanoate**

**11.17** *From examination of the molecular formula and IR spectrum of compound M, $C_3H_7NO$, tell:*

*(a) Its index of hydrogen deficiency.*

**The index of hydrogen deficiency for compound M is one.**

*(b) The number of its rings and/or pi bonds.*

**Based on the index of hydrogen deficiency, compound M must have a ring or a pi bond.**

*(c) The oxygen and nitrogen-containing functional group(s).*

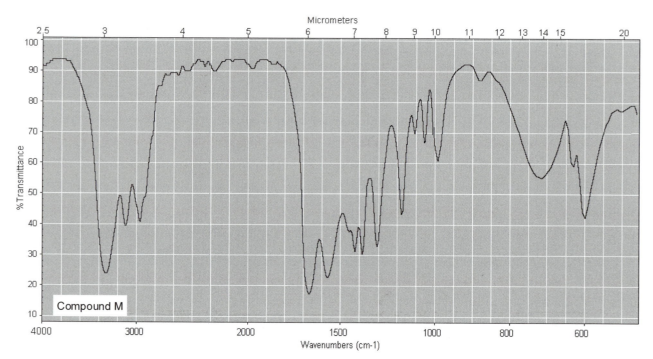

The oxygen and nitrogen-containing functional group is an amide. The absorption at about 1680 cm$^{-1}$ indicates an amide carbonyl while the single absorption between 3200 and 3400 cm$^{-1}$ indicates the N-H stretching of a secondary amide. Other spectroscopic and chemical techniques reveal that compound M is *N*-methylacetamide.

*N*-Methylacetamide

**11.18** *Show how IR spectroscopy can be used to distinguish between the compounds in each set.*

(a) *1-Butanol and diethyl ether*

1-Butanol will have a very strong, broad O-H absorption between 3200 and 3400 cm$^{-1}$, the diethyl ether will not show this absorption band.

(b) *Butanoic acid and 1-butanol*

Butanoic acid will have a strong C=O absorption between 1700 and 1725 cm$^{-1}$, 1-butanol will not show this absorption band.

(c) *Butanoic acid and 2-butanone*

Butanoic acid will have a very broad peak between 2400 and 3400 cm$^{-1}$ due to the O-H stretch of the carboxyl group and the 2-butanone will not show this absorption band.

(d) *Butanal and 1-butene*

Butanal has a strong absorbance between 1705 and 1740 cm$^{-1}$ due to the C=O stretching vibration, butene will not show this absorption band.

(e) *2-Butanone and 2-butanol*

2-Butanone will have a strong C=O absorption between 1705 and 1740 cm$^{-1}$, and the 2-butanol will have a broad, strong O-H absorption between 3200 and 3400 cm$^{-1}$ along with a medium C-O absorption between 1070 and 1150 cm$^{-1}$.

*(f)  Butane and 2-butene*

**Butane will show strong absorptions between 2850 and 3000 cm⁻¹ for C(sp³)-H stretching.  2-butene will show weak to medium absorptions between 3000 and 3100 cm⁻¹ for C(sp²)-H stretches and medium to weak absorptions between 1600-1680 cm⁻¹ due to C=C stretching.**

**11.19** *For each set of compounds, list one major feature that appears in the IR spectrum of one compound but not the other. In your answer, state what type of bond vibration is responsible for the spectral feature you list and its approximate position in the IR spectrum.*

*(a)*

**Benzoic acid will have a strong, broad O-H stretching absorption between 2400 and 3400 cm⁻¹, the benzaldehyde will not.**

*(b)*

**The amide on the left will have a strong absorption between 1630 and 1680 cm⁻¹ due to the C=O stretching, the amine on the right will not.**

*(c)*

**The carboxylic acid will have a strong, broad O-H stretching absorption between 2400 and 3400 cm⁻¹ from the acid OH and a strong, O-H stretch absorption between 3200 and 3400 cm⁻¹ from the alcohol -OH.  The cyclic ester will have neither of these absorptions.**

$$\text{(d)} \quad \overset{\displaystyle O}{\underset{\displaystyle \|}{C}}NH_2 \quad \text{and} \quad \overset{\displaystyle O}{\underset{\displaystyle \|}{C}}N(CH_3)_2$$

**The primary amide on the left will have two broad N-H stretch absorptions between 3200 and 3400 cm$^{-1}$, the tertiary amide on the right will not.**

**11.20** *Following is an infrared spectrum and a structural formula for methyl salicylate, the fragrant component of oil of wintergreen. On this spectrum, locate the absorption peak(s) due to:*

*(a) O-H stretching of the hydrogen-bonded -OH group(very broad and of medium intensity).*

**The O-H stretch is centered at 3200 cm$^{-1}$.**

*(b) C-H stretching of the aromatic ring (sharp and of weak intensity).*

**The aromatic C-H stretching is a hard to see weak signal at 3075 cm$^{-1}$ that is partially obscured by the broad –OH stretch at 3200 cm$^{-1}$. The sharp peak at 2980 cm$^{-1}$ is due to the C-H stretching of the –OCH$_3$ group.**

*(c) C=O stretching of the ester group (sharp and of strong intensity).*

**The carbonyl group C=O stretching band is centered at 1680 cm$^{-1}$. The carbonyl stretching bands for esters normally appear between 1735 and 1750 cm$^{-1}$. In the case of methyl salicylate, the fact that the carbonyl group is (1) conjugated with the pi bonds in the aromatic ring and (2) involved in internal hydrogen bonding combine to lower the stretching frequency to 1680 cm$^{-1}$.**

*(d) C=C stretching of the aromatic ring (sharp and of medium intensity).*

**IR absorbance bands for the aromatic C=C stretches are at 1441 cm$^{-1}$ and 1615 cm$^{-1}$.**

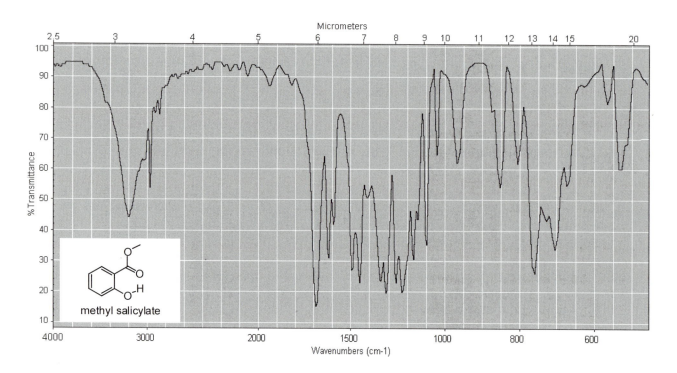

## Looking Ahead

**11.21**  *In the next chapter, transitions between energy levels corresponding to frequencies on the order of $3x10^8$ Hz are observed. Is this higher or lower energy than infrared radiation? Which region of the electromagnetic spectrum does this correspond to?*

**The frequency of $3x10^8$ Hz is lower in energy than the infrared region and exists in the near microwave/radio frequency region of the electromagnetic spectrum.**

**11.22**  *Predict the position of the C=O stretching absorption in acetate ion relative to that in acetic acid.*

|                    |                    |
|:------------------:|:------------------:|
|     Acetic acid    |     Acetate ion    |

**The C=O stretches for the acetate anion will occur at lower wave numbers (higher frequencies) relative to acetic acid due to greater resonance stabilization of the acetate anion. The resonance hybrid of the acetate anion has greater C-O single bond character in the carbonyl bond, thus increasing the frequency of stretching vibrations and lowering the wave number of absorption. The same effect is observed in the lowering of the C=O IR absorption stretching bands when a carbonyl C=O bond is conjugated to an adjacent pi bond.**

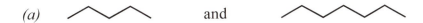

**11.23** *Determine whether IR spectroscopy can be used to distinguish between the following pairs of molecules. Assume that you do not have the reference spectrum of either molecule.*

*(a)* ⟨structure⟩          and          ⟨structure⟩

**IR spectroscopy cannot distinguish between pentane and heptane without a reference spectrum.**

*(b)* ⟨structure⟩          and          ⟨structure⟩

**The 1-methylphenol will have C-H bending absorptions between 735 and 770 cm⁻¹.  There are two C-H bending absorptions for 3-methylphenol between 680 and 730 cm⁻¹ and between 750 and 810 cm⁻¹.**

*(c)* ⟨structure⟩          and          ⟨structure⟩

**The symmetrical C=C stretch in *trans*-3-hexene does not result in change in molecular dipole, therefore, a C=C stretch between 1600 and 1680 cm⁻¹ will be absent.**

*(d)* ⟨structure⟩          and          ⟨structure⟩

**IR spectroscopy cannot distinguish between 2-pentanone and 3-pentanone without a reference spectrum.**

**11.24** *Following is the IR spectrum of L-tryptophan, a naturally occurring amino acid that is abundant in foods like turkey.  For many years, the L-tryptophan in turkey was believed to make people drowsy after Thanksgiving dinner.  Scientists now know that consumption of L-tryptophan only makes one drowsy if taken on an empty stomach.  Therefore it is unlikely that one's Thanksgiving Day turkey is the cause of drowsiness.  Notice that L-*

*tryptophan contains one stereocenter. It's enantiomer, D-tryptophan, does not occur in nature but can be synthesized in the laboratory. What would the IR spectrum of D-tryptophan look like?*

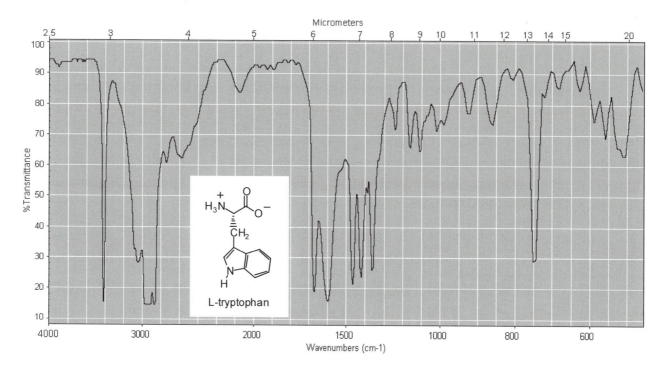

**The IR spectrum of *D*-tryptophan will be identical to that of *L*-tryptophan. The physical and chemical properties of enantiomers are the same in achiral environments, such as those that occur in a simple IR experiment.**

## CHAPTER 12
### *Solutions to Problems*

**12.1**　*Which of the following nuclei are capable of behaving like tiny bar magnets?*

(a)　$^{31}_{15}P$　　　　(b)　$^{195}_{78}Pt$

*(a)* **Phosphorous 31 has an odd atomic number and an odd atomic mass, therefore, would be expected to behave like a tiny bar magnet.**

*(b)* **Platinum 195 has an odd atomic mass, therefore would be expected to behave like a tiny bar magnet.**

**12.2**　*State the number of sets of equivalent hydrogens in each compound and the number of hydrogens in each set.*

**Letter superscripts are used to distinguish among nonequivalent hydrogen atoms. Use the "test atom" approach if you have difficulty understanding the answer.**

*(a)* **3-Methylpentane has four sets of equivalent hydrogen atoms ($H^a$ thru $H^d$). Two sets of equivalent hydrogen atoms are related by a mirror plane of symmetry (the methyl hydrogen atoms $H^a = H^{a'}$ for a total of 6H's and the methylene hydrogen atoms $H^b = H^{b'}$ for a total of four hydrogen atoms). There are three $H^d$ hydrogen atoms and only one $H^c$ hydrogen atom.**

$$H_3^aC-CH_2^{b'}-\overset{\overset{\displaystyle H^c}{|}}{\underset{\underset{\displaystyle CH_3^d}{|}}{C}}-CH_2^b-CH_3^a \qquad \text{3-Methylpentane}$$

mirror plane

*(b)* **2,2,4-Trimethylpentane has four sets of equivalent hydrogens. Three equivalent methyl groups give nine $H^a$ atoms, two methylene $H^b$ hydrogen atoms, one methine $H^c$ hydrogen, and two equivalent methyl groups give six $H^d$ atoms.**

$$H_3^aC-\overset{\overset{\displaystyle CH_3^a}{|}}{\underset{\underset{\displaystyle CH_3^a}{|}}{C}}-CH_2^b-\overset{\overset{\displaystyle CH_3^d}{|}}{\underset{\underset{\displaystyle CH_3^d}{|}}{CH^c}} \qquad \text{2,2,4-Trimethylpentane}$$

**12.3**   *Each compound gives only one signal in its $^1$H-NMR spectrum.  Propose a structural formula for each.*

(a) $C_3H_6O$

Index of hydrogen deficiency = 1

$$\underset{H_3C}{\,}\overset{\displaystyle O}{\underset{\displaystyle \phantom{x}}{\overset{\|}{C}}}\underset{CH_3}{\,}$$

(b) $C_5H_{10}$

Index of hydrogen deficiency = 1

(c) $C_5H_{12}$

Index of hydrogen deficiency = 0

$$H_3C-\underset{\underset{\displaystyle CH_3}{|}}{\overset{\overset{\displaystyle CH_3}{|}}{C}}-CH_3$$

(d) $C_4H_6Cl_4$

Index of hydrogen deficiency = 0

$$H_3C-CCl_2-CCl_2-CH_3$$

**12.4**   *The line of integration of the two signals in the $^1$H-NMR spectrum of a ketone of molecular formula $C_7H_{14}O$ shows a vertical rise of 62 and 10 chart divisions respectively.  Calculate the number of hydrogens giving rise to each signal, and propose a structural formula for this ketone.*

**The number of hydrogens associated with each signal is proportional to the number of chart divisions.  The ratio of signals is approximately 6:1, which corresponds to a 12:2 ratio for a total of 14 hydrogen atoms.  With all the hydrogen atoms accounted for, the larger signal represents 12 hydrogen atoms and the smaller signal represents two hydrogen atoms.  The structure consistent with the data is 2,4-dimethyl-3-pentanone.**

12 H signal

2 H signal  ⟶  $H-\underset{\underset{\displaystyle CH_3}{|}}{\overset{\overset{\displaystyle CH_3}{|}}{C}}-\overset{\overset{\displaystyle O}{\|}}{C}-\underset{\underset{\displaystyle CH_3}{|}}{\overset{\overset{\displaystyle CH_3}{|}}{C}}-H$  ⟵  2 H signal

12 H signal

**12.5**    *Following are two constitutional isomers of molecular formula $C_4H_8O_2$:*

$$
\underset{(1)}{CH_3CH_2O\overset{\overset{\textstyle O}{\|}}{C}CH_3}
\qquad\qquad
\underset{(2)}{CH_3CH_2\overset{\overset{\textstyle O}{\|}}{C}OCH_3}
$$

*(a) Predict the number of signals in the $^1$H-NMR spectrum of each isomer.*

**Each compound will exhibit three signals:  one signal for each of the different methyl groups and a third signal for the methylene group.**

*(b) Predict the ratio of areas of the signals in each spectrum.*

**The ratio of signals will be 3:3:2.**

*(c) Show how to distinguish between these isomers on the basis of chemical shift.*

**In both compounds 1 and 2, there will be methyl hydrogens far upfield with a signal between δ 0.8-1.0.  The difference occurs with the chemical shift of the downfield methyl and methylene groups, depending on which of the two is attached to the oxygen atom.  In compound 1, the methylene (bonded to oxygen) and downfield methyl will exhibit signals between δ 4.1 to 4.7 and between δ 2.1 to 2.3, respectively.   In compound 2, the methylene and downfield methyl (bonded to oxygen) will exhibit signals between δ 2.2 to 2.6 and between δ 3.7 to 3.9, respectively.**

**12.6**    *Following are pairs of constitutional isomers.  Predict the number of signals and the splitting pattern of each signal in the $^1$H-NMR spectrum of each isomer.*

*(a)* **Both compounds will display three signals:**

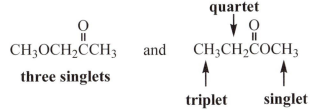

*(b)* **The compound on the left has only one signal.  The compound on the right has two signals, a triplet and a quintet.**

**quintet**
↓

Cl
|
CH₃CCH₃      and      ClCH₂CH₂CH₂Cl
|                            ↖    ↗
Cl
**one singlet**                **triplet**

**12.7**   *Explain how to distinguish between the members of each pair of constitutional isomers based on the number of signals in the* $^{13}$*C-NMR spectrum of each member.*

(a) **These molecules can be distinguished by comparing the number of** $^{13}$**C signals. The molecule on the left has only five nonequivalent carbons (two sets of equivalent carbons due to a mirror plane of symmetry) displaying five** $^{13}$**C signals.  The molecule on the right has seven nonequivalent carbons giving seven** $^{13}$**C signals.**

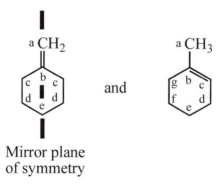

Mirror plane
of symmetry

(b) **The molecule on the left has lower symmetry and will have six different** $^{13}$**C signals, while the molecule on the right has greater symmetry (a perpendicular 180° rotation axis) and will have three different** $^{13}$**C signals.**

## Equivalency of Hydrogens
**12.8**   *Determine the number of signals you would expect to see in the* $^{1}$*H-NMR spectrum of each of the following compounds.*

**Note:  When molecule (b) is subjected to the test atom approach, H$^{d}$ and H$^{e}$ result in diastereomers.  Hydrogens of this type are referred to as diastereotopic hydrogen atoms and can give rise to separate signals.**

*(a)* $\overset{b}{\phantom{.}}\overset{b}{\phantom{.}}$ c $\underset{a}{\phantom{.}}\underset{d}{\phantom{.}}\underset{a}{\phantom{.}}$

**Four $^1$H signals**

*(b)* $\overset{b}{\phantom{.}}$ $\underset{a}{\phantom{.}}$ $\underset{c}{\phantom{.}}$ H$^d$ H$^e$

**Five H$^1$ signals**

*(c)* a $\overset{O}{\phantom{.}}$ $\underset{b}{\phantom{.}}$ b b b

**Two $^1$H signals**

*(d)* $\overset{a}{H}$ $\overset{O}{\phantom{.}}$ $\overset{b}{\phantom{.}}$ $\underset{b}{\phantom{.}}$ $\overset{O}{\phantom{.}}$ H$\overset{a}{\phantom{.}}$

**Two $^1$H signals**

*(e)* HO $\overset{b}{\phantom{.}}$ $\underset{c}{\phantom{.}}$ $\overset{d}{\phantom{.}}$ e e

**Five $^1$H signals**

*(f)* $\overset{a}{\phantom{.}}$ $\overset{b}{\phantom{.}}$ $\overset{H^d}{\underset{N}{\phantom{.}}}$ $\overset{a}{\phantom{.}}$ $\underset{a}{\phantom{.}}$ $\underset{c}{\phantom{.}}$ $\underset{c}{\phantom{.}}$ $\underset{b}{\phantom{.}}$ $\underset{a}{\phantom{.}}$

**Four $^1$H signals**

*(g)* $\overset{Cl}{\phantom{.}}$ $\overset{Cl}{\phantom{.}}$ $\underset{a}{\phantom{.}}$ $\underset{b}{\phantom{.}}$ c $\underset{b}{\phantom{.}}$ $\underset{a}{\phantom{.}}$ Cl

**Three $^1$H signals**

*(h)* HO $\overset{a}{\phantom{.}}$ $\overset{O}{\phantom{.}}$ $\overset{b}{\phantom{.}}$ c c $\underset{d}{\phantom{.}}$ d e

**Five $^1$H signals**

**12.9**   *Determine the number of signals you would expect to see in the $^{13}$C-NMR spectrum of each of the compounds in Problem 12.8.*

*(a)* $\overset{b}{\phantom{.}}$ $\overset{b}{\phantom{.}}$ c $\underset{a}{\phantom{.}}$ $\underset{d}{\phantom{.}}$ $\underset{a}{\phantom{.}}$

**Four $^{13}$C signals**

*(b)* $\overset{b}{\phantom{.}}$ $\overset{H}{\phantom{.}}$ e $\underset{a}{\phantom{.}}$ $\underset{c}{\phantom{.}}$ H $\underset{d}{\phantom{.}}$

**Five $^{13}$C signals**

*(c)* a $\overset{O}{\phantom{.}}$ $\underset{b}{\phantom{.}}$ c $\overset{d}{\phantom{.}}$ $\underset{d}{\phantom{.}}$ d

**Four $^{13}$C signals**

*(d)* H $\overset{a}{\phantom{.}}$ $\overset{O}{\phantom{.}}$ $\overset{b}{\phantom{.}}$ $\underset{b}{\phantom{.}}$ $\overset{a}{\phantom{.}}$ H $\underset{O}{\phantom{.}}$

**Two $^{13}$C signals**

*(e)* HO $\overset{a}{\phantom{.}}$ $\underset{b}{\phantom{.}}$ $\overset{d}{\phantom{.}}$ c $\underset{d}{\phantom{.}}$

**Four $^{13}$C signals**

*(f)* $\overset{a}{\phantom{.}}$ $\overset{b}{\phantom{.}}$ $\overset{H}{\underset{N}{\phantom{.}}}$ $\overset{a}{\phantom{.}}$ $\underset{a}{\phantom{.}}$ $\underset{c}{\phantom{.}}$ $\underset{c}{\phantom{.}}$ $\underset{b}{\phantom{.}}$ $\underset{a}{\phantom{.}}$

**Three $^{13}$C signals**

*(g)* $\overset{Cl}{\phantom{.}}$ $\overset{Cl}{\phantom{.}}$ $\underset{a}{\phantom{.}}$ $\underset{b}{\phantom{.}}$ c $\underset{b}{\phantom{.}}$ $\underset{a}{\phantom{.}}$ Cl

**Three $^{13}$C signals**

*(h)* HO $\overset{a}{\phantom{.}}$ $\overset{O}{\phantom{.}}$ $\overset{b}{\phantom{.}}$ c c $\underset{d}{\phantom{.}}$ d e

**Five $^{13}$C signals**

## Interpreting ¹H-NMR and ¹³C-NMR Spectra

**12.10** *Following are structural formulas for the constitutional isomers of xylene and three sets of ¹³C-NMR spectral. Assign each constitutional isomer its correct spectrum.*

**(a) Ortho-xylene will have four ¹³C signals, which corresponds to spectrum 2.**

(75 MHz, CDCl₃)

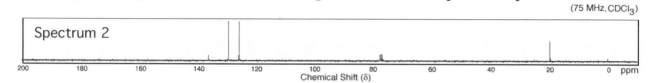

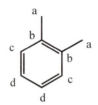

**(b) Meta-xylene will have five ¹³C signals, which corresponds to spectrum 3.**

(75 MHz, CDCl₃)

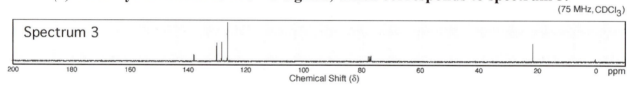

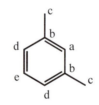

**(c) Para-xylene will have three ¹³C signals, which corresponds to spectrum 1.**

(75 MHz, CDCl₃)

**12.11** *Following is a $^1$H-NMR spectrum for* **compound A**, *with molecular formula $C_7H_{14}$.* **Compound A** *decolorizes a solution of bromine in carbon tetrachloride. Propose a structural formula of* **compound A**.

**For the spectral interpretations in this problem and the rest of the chapter, the chemical shift (δ) is given, followed by the relative integration, the multiplicity of the signal (singlet, doublet, triplet, etc.), and finally the identity of the hydrogen atoms giving rise to the signal (shown in bold).**

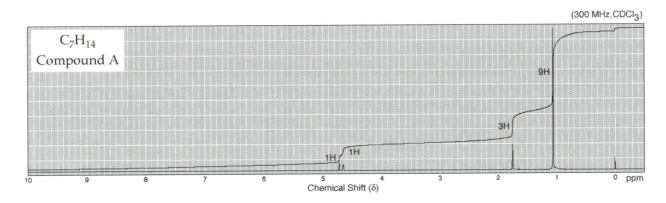

- Index of hydrogen deficiency = 1
- Chemical test suggests a double bond (decolorizes bromine solution).
- $^1$H-NMR shows four different signals (four different types of hydrogen atoms).
- The 3H and 9H singlet signals are most likely due to methyl and *tert*-butyl groups respectively, and not split by other hydrogen atoms two or three bonds away.
- The two signals around δ 4.7 are terminal vinylic hydrogen atoms, evident by the upfield chemical shift (δ 4.6-4.7 relative to δ 5.0-5.7 for hydrogen atoms bonded to substituted vinylic carbons).

2,3,3-Trimethyl-1-butene best fits the structure for compound A that is consistent with the data.

$$H \quad\quad C(CH_3)_3$$
$$H \quad\quad CH_3$$

**Compound A**

**12.12** *Following is a $^1$H-NMR spectrum of* **compound B**, *with the molecular formula $C_8H_{16}$.* **Compound B** *decolorizes a solution of $Br_2$ in $CCl_4$. Propose a structural formula for* **compound B**.

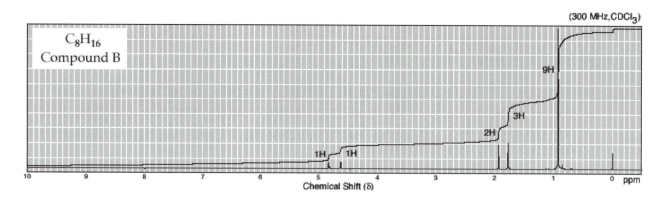

- **Index of hydrogen deficiency = 1**
- **Chemical test suggests a double bond (decolorizes bromine solution).**
- **$^1$H-NMR shows five different signals (five different types of hydrogen atoms).**
- **The 3H and 9H singlet signals are most likely due to hydrogen atoms on methyl and *tert*-butyl groups respectively, and not split by other hydrogen atoms two or three bonds away.**
- **The 2H singlet is an isolated –CH$_2$– (without adjacent hydrogen atoms)**
- **The two downfield signals are terminal vinylic hydrogen atoms, evident by the upfield chemical shift ($\delta$ 4.6-4.7 relative to $\delta$ 5.0-5.7 for hydrogens attached to substituted vinylic carbons).**

**2,4,4-Trimethyl-1-pentene best fits the structure for compound B that is consistent with the data.**

$$\begin{array}{c}H \quad\quad CH_2C(CH_3)_3\\ \diagdown \qu~ / \\ C=C \\ / \qu\quad \diagdown \\ H \qu\quad CH_3\end{array}$$

**Compound B**

**12.13**  *Following are $^1$H-NMR spectra for **compounds C** and **D**, each of molecular formula C$_4$H$_7$Cl. Each decolorizes a solution of Br$_2$ in CCl$_4$. Propose structural formulas for **compounds C** and **D**.*

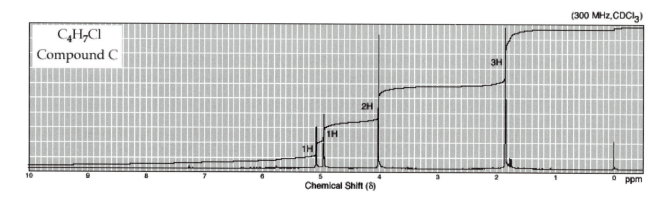

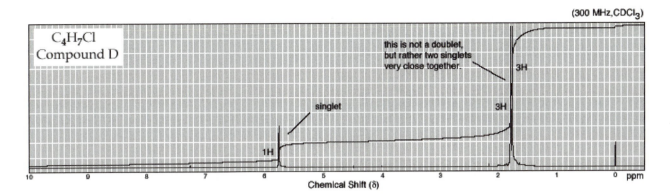

The index of hydrogen deficiency for both compounds C and D is one. Both compounds C and D decolorize bromine solution, indicating that they have a pi bond; and with an index of hydrogen deficiency of one, both C and D must have only one pi bond each.

The $^1$H-NMR spectrum for compound C shows a 3H singlet at $\delta$ 1.85 (most likely a –CH$_3$ group with no adjacent protons), a 2H singlet at $\delta$ 4.0 (most likely a –CH$_2$ with no adjacent hydrogen atoms) and two $^1$H signals at $\delta$ 4.93 and $\delta$ 5.09 (most likely due to nonequivalent terminal vinylic R$_2$C=CH$_2$ hydrogen atoms). The 2H singlet has a downfield shift, and therefore must have an electronegative chlorine atom bonded to it. Put the fragments together as outlined above by building the structure around the H$_2$C=C– fragment. Compound C is 3-chloro-2-methyl-1-propene.

The $^1$H-NMR spectrum for compound D shows two 3H singlets at $\delta$ 1.75 and $\delta$ 1.79 (most likely two nonequivalent –CH$_3$ groups with no adjacent protons) and one $^1$H signal at $\delta$ 5.76 (most likely due to a R$_2$C=CRH proton). There is only one vinylic hydrogen atom, therefore place the methyl groups and chlorine atom on the R$_2$C=C– fragment. Compound D is 1-chloro-2-methyl-1-propene.

|  |  |
|---|---|
| Compound C | Compound D |

**12.14** *Following are structural formulas for three alcohols of molecular formula $C_7H_{16}O$ and three sets of $^{13}C$-NMR spectral data. Assign each constitutional isomer to its correct spectral data.*

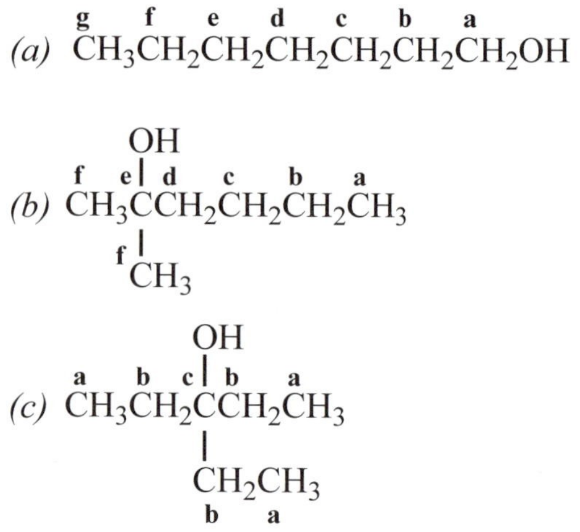

(a)   $\overset{g}{C}H_3\overset{f}{C}H_2\overset{e}{C}H_2\overset{d}{C}H_2\overset{c}{C}H_2\overset{b}{C}H_2\overset{a}{C}H_2OH$

| *Spectrum 1* | *Spectrum 2* | *Spectrum 3* |
|---|---|---|
| 74.66 | 70.97 | 62.93 |
| 30.54 | 43.74 | 32.79 |
| 7.73 | 29.21 | 31.86 |
|  | 26.60 | 29.14 |
|  | 23.27 | 25.75 |
|  | 14.09 | 22.63 |
|  |  | 14.08 |

(b)
$$\underset{f}{C}H_3\overset{OH}{\underset{e}{\overset{|}{C}}}\underset{d}{C}H_2\underset{c}{C}H_2\underset{b}{C}H_2\underset{a}{C}H_3$$
$$\underset{f}{\overset{|}{C}}H_3$$

(c)
$$\underset{a}{C}H_3\underset{b}{C}H_2\overset{OH}{\underset{c}{\overset{|}{C}}}\underset{b}{C}H_2\underset{a}{C}H_3$$
$$\underset{\underset{b\quad a}{CH_2CH_3}}{|}$$

**Comparing the sets of nonequivalent carbon atoms and thus the number of $^{13}C$ signals most easily identifies these constitutional isomers. Using the following analysis, it can be seen that compound (a) has seven sets of nonequivalent carbon atoms corresponding to spectrum 3. Compound (b) has six sets of nonequivalent carbon atoms corresponding to spectrum 2. Compound (c) has three sets of nonequivalent carbons corresponding to spectrum 1.**

**12.15** *Alcohol E, with molecular formula $C_6H_{14}O$, undergoes acid-catalyzed dehydration when warmed with phosphoric acid, giving **compound F**, with molecular formula $C_6H_{12}$, as the major product. A $^1H$-NMR spectrum of **compound E** shows peaks at $\delta\ 0.89$ (t, 6H), 1.12 (s, 3H), 1.38 (s, 1H), and 1.48 (q, 4H). The $^{13}C$-NMR spectrum of compound E shows peaks at $\delta\ 72.98, 33.72, 25.85,$ and $8.16$. Propose structural formulas for **compounds E** and F.*

**From the molecular formula, the calculated index of hydrogen deficiency is zero, so there are no rings or pi bonds. From the $^1H$-NMR, we learn that there are four different sets of nonequivalent hydrogen atoms and that the OH is not on a carbon with a C-H bond (there are no signals between $\delta$ 3.3 and $\delta$ 4.0). The easiest way to construct a molecule with a 6H triplet (two methyl groups) is to have two equal ethyl groups on the molecule. The $^{13}C$-NMR data reveals that there are four different sets of nonequivalent carbons in the molecule. A proposed structure for compound E is 3-methyl-3-pentanol and the chemical shifts associated with each set of hydrogen atoms are indicated on the structure below.**

$$OH^{1.38}$$

0.89   1.48   |   1.48   0.89
$$CH_3CH_2CCH_2CH_3$$
                 |
              $$CH_3$$ 1.12

$$\xrightarrow[\text{heat}]{H_3PO_4}$$

H₃C        CH₃
   \\        /
    C=C                  +   H₂O
   /        \\
  H        CH₂CH₃

**Compound E**                                            **Compound F**

**12.16** *Compound G, C₆H₁₄O, does not react with sodium metal and does not discharge the color of Br₂ in CCl₄.  Its ¹H-NMR spectrum of* **compound G** *consists of only two signals, a 12H doublet at δ 1.1 and a 2H septet at δ 3.6.  Propose a structural formula for* **compound G**.

**Compound G has an index of hydrogen deficiency of zero, indicating that there are no rings or pi bonds.  The fact that it does not discharge the color of bromine is no surprise (no pi bonds), but its lack of reactivity towards sodium metal indicates that compound G is not an alcohol.  The only other oxygen-containing functional group that involves a zero index of hydrogen deficiency is an acyclic alkyl ether.  A molecule of this size with only two sets of nonequivalent hydrogen atoms must be very symmetrical.  One set of hydrogen atoms is bonded to a carbon bonded to an oxygen (2H septet at δ 3.6).   The structure for compound G that best fits the above data is diisopropyl ether.**

CH₃ᵃ   CH₃ᵃ  ⟵——— δ 1.1
         |       |
δ 3.6 ——⟶ Hᵇ–C–O–C–Hᵇ
         |       |
        CH₃ᵃ   CH₃ᵃ

**Diisopropyl ether**

**12.17** *Propose a structural formula for each haloalkane.*
   *(a) C₂H₄Br₂    δ 2.5 (d, 3H) and 5.9 (q, 1H)*

2.5
H   Br
2.5  |    |   5.9
H–C–C–H
     |    |
     H   Br
     2.5

   *(b) C₄H₈Cl₂    δ 1.67 (d, 6H), 2.15 (q, 2H)*

2.15  2.15
H    H
1.67  |    |   1.67
CH₃–C–C–CH₃
     |    |
     Cl   Cl

*(c) C₅H₈Br₄   δ 3.6 (s, 8H)*

$$CH_2Br$$
$$BrCH_2-C-CH_2Br \quad \text{All of the hydrogens are equivalent.}$$
$$CH_2Br$$

*(d) C₄H₉Br   δ 1.1 (d, 6H), 1.9 (m, 1H), and 3.4 (d, 2H)*

$$\begin{array}{c} ^{1.1}\\ CH_3 \quad ^{3.4}\\ ^{1.1}\; CH_3-C-CH_2-Br\\ H\\ 1.9 \end{array}$$

*(e) C₅H₁₁Br   δ 1.1 (s, 9H) and 3.2 (s, 2H)*

All three methyl groups are equivalent: δ 1.1

$$CH_3-C-CH_2Br \quad (CH_3)_3, \; 3.2$$

*(f) C₇H₁₅Cl   δ 1.1 (s, 9H) and 1.6 (s, 6H)*

All three methyl groups are equivalent: δ 1.1

$$CH_3-C-C-Cl \quad 1.6$$

**12.18** *Following are structural formulas for esters (1), (2), and (3) and three ¹H-NMR spectra. Assign each compound its correct spectrum and assign all signals to their corresponding hydrogens.*

$$\underset{(1)}{CH_3COCH_2CH_3} \qquad \underset{(2)}{HCOCH_2CH_2CH_3} \qquad \underset{(3)}{CH_3OCCH_2CH_3}$$

**The ¹H-NMR spectrum for compound H corresponds to structure (2). Compound H has four sets of nonequivalent hydrogen atoms. The most obvious feature of the spectrum is the aldehyde hydrogen at δ 8.1, which is not present in the other structures or spectra.**

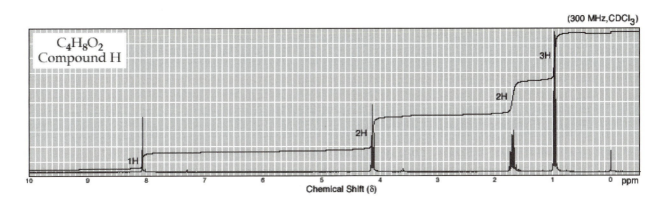

The $^1$H-NMR spectrum for compound I corresponds to structure (3). Compound I has three sets of nonequivalent hydrogen atoms. The most obvious feature of the spectrum is the chemical shift and splitting of the methyl groups. Structure (3) has a $CH_3O-$ group appearing as a 3H singlet at $\delta$ 3.7. This downfield methyl singlet signal would not occur for structures (1) and (2).

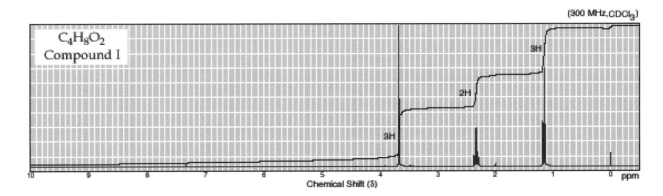

The $^1$H-NMR spectrum for compound J corresponds to structure (1). Compound J has three sets of nonequivalent hydrogen atoms. The most obvious feature of the spectrum is the chemical shift and splitting of the methylene groups bonded to the oxygen. Structure (1) has a $CH_3CH_2O-$ group appearing as a 2H quartet at $\delta$ 4.2. This downfield methylene quartet signal would not occur for structure (3).

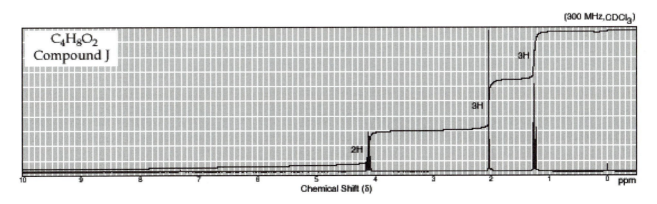

**12.19** *Compound K, $C_{10}H_{10}O_2$, is insoluble in water, 10% NaOH, and 10% HCl. A $^1H$-NMR spectrum of compound K shows signals at $\delta$ 2.55 (s, 6H) and 7.97 (s, 4H). A $^{13}C$-NMR spectrum of compound K shows four signals. From this information, propose a structural formula for K.*

**The chemical reactivity data rules out carboxylic acid and phenolic compounds. Compound K has an index of hydrogen deficiency of six, suggesting a benzene ring. A disubstituted benzene ring is supported by the 4H singlet $^1H$-NMR signal at $\delta$ 7.97. A compound with this many atoms and simple $^1H$- and $^{13}C$-NMR spectrum indicates a high degree of symmetry. After the benzene ring is considered, there are four carbon atoms, six hydrogen atoms, and two oxygen atoms with two sites of hydrogen deficiencies to account for. The structure of compound K is described below. Although the figure shows two mirror planes, the structure of compound K has other elements of symmetry such as a rotation axis. Use your model kits and try to observe them yourself and how they relate to hydrogen equivalency.**

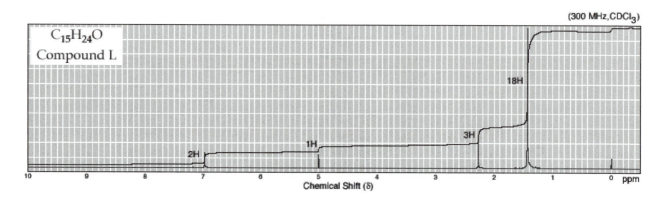

**12.20** *Compound L, $C_{15}H_{24}O$, is used as an antioxidant in many commercial food products, synthetic rubbers, and petroleum products. Propose a structural formula for compound L based on its $^1H$-NMR and $^{13}C$-NMR spectra.*

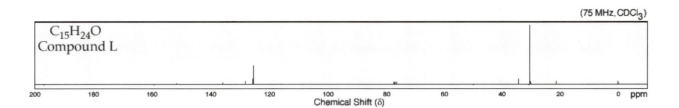

(75 MHz, CDCl₃)

C₁₅H₂₄O
Compound L

**Compound L has an index of hydrogen deficiency of four, suggesting a benzene ring. The two singlets at δ (1.3, 18H and 2.1, 3H) indicate *t*-butyl and methyl groups, respectively. The signal at δ 6.9, (s, 2H) are on a benzene ring and the signal at δ 5.0 (s, 1H) is a –OH group. Four ¹³C-NMR signals in the aromatic region arise from a symmetrically substituted benzene ring. Placing the identified groups on a benzene ring with the above information gives the structure for compound L.**

H₃C  5.0  CH₃
1.3 — H₃C OH CH₃ — 1.3
H₃C  CH₃
6.9  6.9
H  H
2.1
CH₃

Vertical mirror plane
of symmetry
(perpendicular to the paper) →

**12.21** *Propose a structural formula for these compounds, each of which contains an aromatic ring.*

*(a) C₉H₁₀O    δ 1.2 (t, 3H), 3.0 (q, 2H), and 7.4-8.0 (m, 5H)*

H  H
O  3.0  1.2
H— ‖ —CCH₂CH₃
7.4-8.0 —
H  H

*(b) $C_{10}H_{12}O_2$   $\delta$ 2.2 (s, 3H), 2.9 (t, 2H), 4.3 (t, 2H), and 7.3 (s, 5H)*

*(c) $C_{10}H_{14}$     $\delta$ 1.2 (d, 6H), 2.3 (s, 3H), 2.9 (septet, 1H), and 7.0 (s, 4H)*

**Note:  The benzene ring has two sets of nonequivalent hydrogen atoms.  Their chemical shifts are similar; therefore they appear as a singlet, although theoretically, they should appear as two sets of doublets.**

*(d) $C_8H_9Br$     $\delta$ 1.8 (d, 3H), 5.0 (q, 1H), and 7.3 (s, 5H)*

**12.22**  ***Compound M**, molecular formula $C_9H_{12}O$, readily undergoes acid-catalyzed dehydration to give **compound N**, with molecular formula $C_9H_{10}$.  A $^1H$-NMR spectrum of compound M shows signals at $\delta$ 0.91 (t, 3H), 1.78 (m, 2H), 2.26 (d, 1H), 4.55 (m, 1H), and 7.31 (m, 5H). From this information, propose structural formulas for **compounds M and N**.*

$$C_9H_{12}O \xrightarrow[\text{H}^+]{\text{heat}} C_9H_{10} + H_2O$$

| | Compound M | | Compound N |
|---|:---:|---|:---:|
| Index of hydrogen deficiency: | 4 | | 5 |

**Compound M dehydrates under acidic conditions, therefore, it is an alcohol. The product of this dehydration, compound N, is an alkene. Index of hydrogen deficiencies over four suggest the presence of a benzene ring (mono-substituted in this case), supported by a 5H signal at δ 7.31 for compound M. Out of the seven possible isomers of compound M, only 1-phenyl-1-propanol fit the above chemical and ¹H-NMR data.**

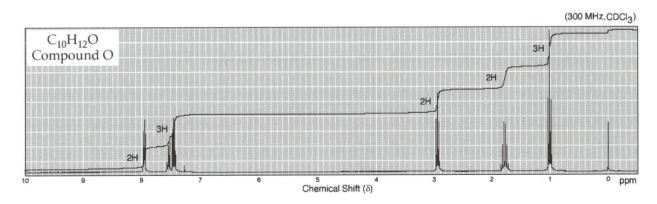

**Compound M**                          **Compound N**

**12.23** *Propose a structural formula for each ketone.*
   *(a) C₄H₈O      δ 1.0 (t, 3H), 2.1 (s 3H), and 2.4 (q, 2H)*

   **Index of hydrogen deficiency = 1**

$$\underset{\text{CH}_3\text{CH}_2\overset{\overset{\displaystyle O}{\|}}{\text{C}}\text{CH}_3}{\overset{1.0\quad 2.4 \qquad 2.1}{}}$$

   *(b) C₇H₁₄O      δ 0.9 (t, 6H), 1.6 (sextet, 4H), and 2.4 (t, 4H)*

   **Index of hydrogen deficiency = 1**

$$\underset{\text{CH}_3\text{CH}_2\text{CH}_2\overset{\overset{\displaystyle O}{\|}}{\text{C}}\text{CH}_2\text{CH}_2\text{CH}_3}{\overset{0.9\quad 1.6\quad 2.4 \qquad 2.4\quad 1.6\quad 0.9}{}}$$

**12.24** *Propose a structural formula for **compound O**, a ketone of molecular formula C₁₀H₁₂O.*

(300 MHz, CDCl₃)

C₁₀H₁₂O
Compound O

- **Index of hydrogen deficiency = 5**
- **A total integration of 5 for the signals between δ 7-8 suggest the presence of a monosubstituted benzene ring**
- **Signal at δ 1.0 (t, 3H) suggests a methyl group adjacent to a methylene group.**

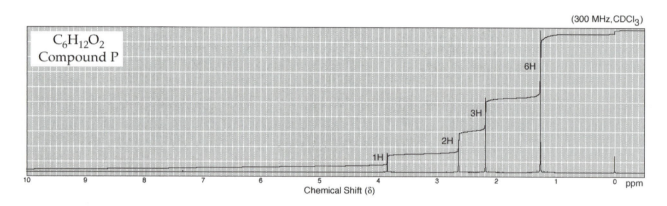

**Compound O**

**12.25** *Following is a $^1$H-NMR spectrum for **compound P**, with molecular formula $C_6H_{12}O_2$. **Compound P** undergoes acid-catalyzed dehydration to give **compound Q**, $C_6H_{10}O$. Propose structural formulas for **compounds P and Q**.*

(300 MHz, CDCl$_3$)

$C_6H_{12}O_2$
Compound P

6H
3H
2H
1H

10    9    8    7    6    5    4    3    2    1    0  ppm
Chemical Shift (δ)

$$C_6H_{12}O_2 \xrightarrow[\text{H}^+]{\text{heat}} C_6H_{10}O + H_2O$$

**Compound P**                **Compound Q**

Index of hydrogen
deficiency:                1                        2

**Compound P:**
- **Dehydrates under acidic conditions, which is consistent with an alcohol.**
- **Signal at δ 1.22 suggests two methyl groups with no adjacent hydrogen atoms.**
- **Signal at δ 2.18 indicates a methyl group with no adjacent hydrogen atoms.**
- **Signal at δ 2.62 is most likely due to a methylene group with no adjacent hydrogen atoms.**

**Compound P**                **Compound Q**

**12.26** *Propose a structural formula for **compound R**, with molecular formula $C_{12}H_{16}O$. Following is its $^1H$-NMR and $^{13}C$-NMR spectra.*

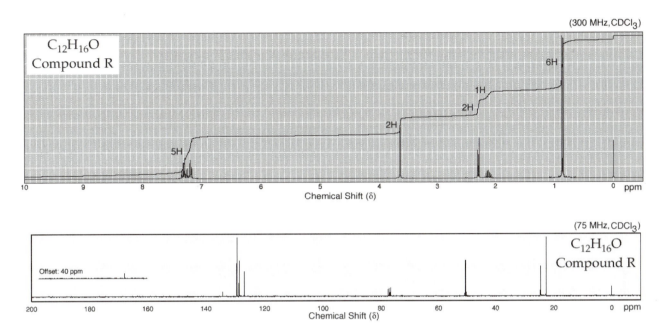

The signal in the $^{13}C$-NMR spectrum at δ 207 indicates the presence of a carbonyl group. With no other oxygen or nitrogen atoms and the absence of a $^1H$-NMR signal between δ 9-10 (aldehyde hydrogen) leads to the conclusion that the carbonyl is a ketone. The signal in the $^1H$-NMR spectrum at δ 0.84 (d, 6H) indicates two methyl groups adjacent to a –CH– group. There are also two –CH$_2$– groups, one that is not adjacent to two other hydrogen atoms (the singlet at δ 3.62) and the one next to a –CH– group (the doublet at δ 2.30). The multiplet at δ 2.12 must be this CH– group that is also adjacent to the two methyl groups. Five aromatic hydrogen atoms are found as a complex set of signals at δ 7.3. The only structure consistent with the data is 4-methyl-1-phenyl-2-pentanone.

**12.27**  *Propose a structural formula for each carboxylic acid.*

(a)  $C_5H_{10}O_2$

| $^1H$-NMR | $^{13}C$-NMR |
|-----------|--------------|
| 0.94 (t, 3H) | |
| 1.39 (m, 2H) | 33.89 |
| 1.62 (m, 2H) | 26.76 |
| 2.35 (t, 2H) | 22.21 |
| 12.0 (s, 1H) | 13.69 |

$$\underset{\text{0.94 1.39 1.62 2.35}}{CH_3CH_2CH_2CH_2}\overset{\overset{\displaystyle O}{\parallel}}{C}\underset{\uparrow \atop 180.7}{OH}\quad 12.0$$

(b)  $C_6H_{12}O_2$

| $^1H$-NMR | $^{13}C$-NMR |
|-----------|--------------|
| 1.08 (s, 9H) | 179.29 |
| 2.23 (s, 2H) | 46.82 |
| 12.1 (s, 1H) | 30.62 |
| | 29.57 |

$$\underset{\underset{1.08}{CH_3}}{\overset{\overset{1.08}{CH_3}}{CH_3-\underset{\displaystyle |}{\overset{\displaystyle |}{C}}-\underset{2.23}{CH_2}-\overset{\overset{\displaystyle O}{\parallel}}{C}\underset{\uparrow \atop 179.29}{OH}}}\quad 12.1$$

(c)  $C_5H_8O_4$

| $^1H$-NMR | $^{13}C$-NMR |
|-----------|--------------|
| 0.93 (t, 3H) | 170.94 |
| 1.80 (m, 2H) | 53.28 |
| 3.10 (t, 1H) | 21.90 |
| 12.7 (s, 2H) | 11.81 |

$$\underset{\underset{170.94}{\uparrow}}{\overset{\overset{\displaystyle O}{\parallel}}{HO-C}}-\underset{\underset{\underset{1.80 \quad 0.93}{CH_2CH_3}}{|}}{\overset{\overset{3.10}{\overset{\displaystyle H}{|}}}{C}}-\overset{\overset{\displaystyle O}{\parallel}}{C}OH$$
12.7          12.7

**12.28**  *Following are $^1H$-NMR and $^{13}C$-NMR spectra of **compound S**, with molecular formula $C_7H_{14}O_2$. Propose a structural formula for **compound S**.*

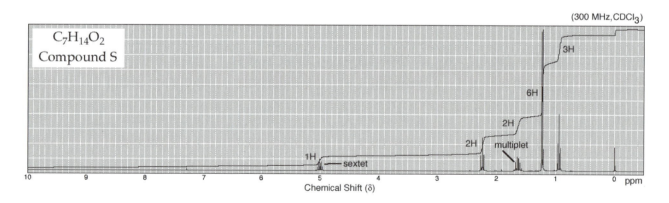

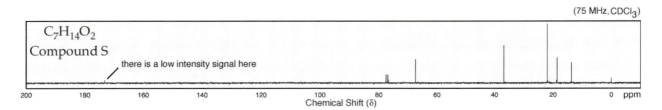

(75 MHz, CDCl₃)

$C_7H_{14}O_2$
Compound S
there is a low intensity signal here

**Compound S has an index of hydrogen deficiency of one. The $^{13}$C-NMR spectrum reveals six sets of nonequivalent carbon atoms, with a low intensity signal at δ 173 indicating a carboxyl group carbon atom and the signal at δ 67 is consistent with an $sp^3$ hybridized carbon atom bonded to an oxygen atom. The $^1$H-NMR spectrum shows no evidence of a carboxylic acid hydrogen signal, so the carboxyl group is an ester, which accounts for two of the oxygen atoms and the only site of hydrogen deficiency. The signal at δ 5.1 (m, 1H) indicates the carbon atom bound to the ester oxygen atom has a single hydrogen atom. The signal at δ 2.25 (t, 2H) indicates the –CH₂– group adjacent to the carbonyl of the ester is also next to another –CH₂– group, evident by its signal at δ 1.65 (m, 2H). The signal at δ 1.3 (d, 6H) is consistent with the two methyls on an isopropyl group, and correlates nicely with it being bonded to the –CH– bonded to the ester oxygen (the signal at δ 5.1). The signal at δ 0.95 (t, 3H) must be another methyl group, adjacent to the –CH₂– group (the signal at δ 1.65). The only structure that is consistent with all of the above analysis is isopropyl butyrate.**

$$
\begin{array}{c}
\overset{1.3}{CH_3} \\
\overset{0.95\ \ 1.65\ \ 2.25}{CH_3CH_2CH_2}\overset{O}{\underset{\parallel}{C}}-O-\overset{\mid}{\underset{\mid}{CH}}\,5.1 \\
CH_3 \\
1.3
\end{array}
$$

**12.29** *Propose a structural formula for each ester.*

(a) $C_6H_{12}O_2$

| $^1$H-NMR | $^{13}$C-NMR |
|---|---|
| 1.18 (d, 6H) | 177.16 |
| 1.26 (t, 3H) | 60.17 |
| 2.51 (m, 1H) | 34.04 |
| 4.13 (q, 2H) | 19.01 |
|  | 14.25 |

$$
\begin{array}{c}
\overset{2.51}{} \\
\overset{1.18}{CH_3}-\overset{H}{\underset{\mid}{C}}-\overset{O}{\underset{}{\overset{\parallel}{C}}}OCH_2CH_3\ \ ^{4.13\ \ 1.26} \\
\underset{1.18}{CH_3} \quad {}^{177.16} \quad {}^{60.17}
\end{array}
$$

*(b)* $C_7H_{12}O_4$

| $^1$H-NMR | $^{13}$C-NMR |
|-----------|--------------|
| 1.28 (t, 6H) | 166.52 |
| 3.36 (s, 2H) | 61.43 |
| 4.21 (q, 4H) | 41.69 |
|  | 14.07 |

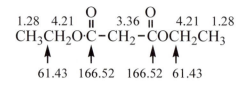

*(c)* $C_7H_{14}O_2$

| $^1$H-NMR | $^{13}$C-NMR |
|-----------|--------------|
| 0.92 (d, 6H) | 171.15 |
| 1.52 (m, 2H) | 63.12 |
| 1.70 (m 1H) | 37.31 |
| 2.09 (s, 3H) | 25.05 |
| 4.10 (t, 2H) | 22.45 |
|  | 21.06 |

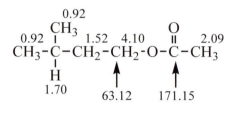

**12.30**  *Following are $^1$H-NMR and $^{13}$C-NMR spectra of **compound T**, with molecular formula $C_{10}H_{15}NO$. Propose a structural formula for this compound.*

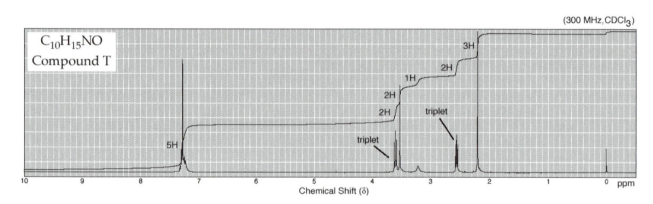

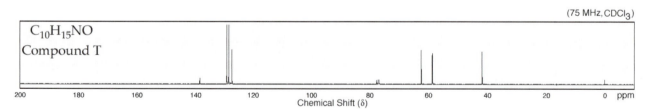

**Compound T has an index of hydrogen deficiency of four, which along with the signal at δ 7.3 indicates a monosubstituted benzene ring. The singlet at δ 2.2 (3H) in the $^1$H NMR spectrum represents a –CH$_3$ with no neighboring protons. The two triplets (each 2H) are consistent with –CH$_2$CH$_2$- in which the downfield signal is bonded to oxygen and the upfield signal is bonded to nitrogen. The broad signal at δ 3.2 is indicative of a –OH group. The other singlet at δ 3.5 represents a –CH$_2$- group with no neighboring protons. Placing each of these pieces together gives the**

following amino alcohol. The sp³-C region of the ¹³C NMR, at first glance, appears to only have 3 signals. But very careful inspection reveals that the signal at δ 58 is actually two closely spaced signals, illustrating that the interpretation of NMR spectroscopy is not always straightforward. The structure that fits the ¹H and ¹³C data is the following amine:

**Compound T**

**12.31** *Propose a structural formula for **amide U**, with molecular formula C₆H₁₃NO.*

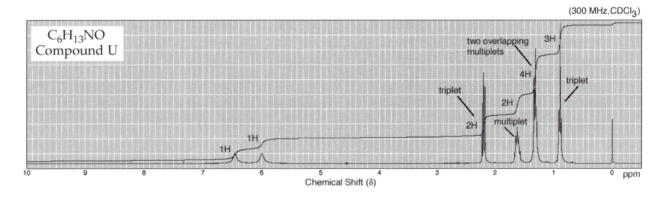

The signals at δ 0.9-2.2 are consistent with a CH₃CH₂CH₂CH₂CH₂– structure (no signals with an integration of 1H, which would be evidence of branching), with the last –CH₂– being adjacent to the carbonyl group of the amide, indicated by a chemical shift of δ 2.2. The two signals integrating to 1H each at δ 6.0 and δ 6.55 are from different amide hydrogens, indicating a primary amide. The only structure that fits the formula and ¹H-NMR data is hexanamide.

$$\underset{\textbf{Compound U}}{\overset{\displaystyle\underset{\text{0.9}}{\underbrace{\phantom{xxx}}}\overset{\text{1.4-1.8}}{\overbrace{\phantom{xxxxx}}}\underset{\text{2.2}}{\phantom{x}}\overset{\text{O}}{\underset{\|}{\phantom{C}}}\,\text{6.0, 6.55}}{\text{CH}_3\text{CH}_2\text{CH}_2\text{CH}_2\text{CH}_2\text{CNH}_2}}$$

**12.32**  *Propose a structural formula for the analgesic phenacetin, with molecular formula*
*C₁₀H₁₃NO₂, based on its ¹H-NMR spectrum.*

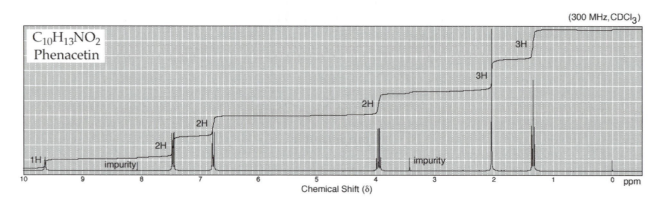

The 1H-NMR is what would be expected from the known structure of phenacetin.
The two signals δ 6.75 (d, 2H) and δ 7.50 (d, 2H) are characteristic of a 1,4-
disubstituted phenyl ring.  The signal at δ 9.65 (s, 1H) indicates a secondary amide
and the signal at δ 2.02 (s, 3H) comes from the acetyl methyl group on an acetamide
(an amide of acetic acid).  The typical ethyl splitting pattern for the signals at δ 1.32
and δ 3.96 suggests the presence of an ethyl group bonded to an oxygen atom,
evident by the downfield shift of the ethyl –CH₂– quartet at δ 3.96.

**Phenacetin**

**12.33**  *Propose a structural formula for **compound V**, an oily liquid of molecular formula*
*C₈H₉NO₂.  Compound V is insoluble in water and aqueous NaOH but dissolves in 10%*
*HCl.  When its solution in HCl is neutralized with NaOH, **compound V** is recovered*
*unchanged.  A ¹H-NMR spectrum of compound V shows signals at δ 3.84 (s, 3H), 4.18 (s,*
*2H), 7.60 (d, 2H), and 8.70 (d, 2H).*

Compound V has an index of hydrogen deficiency of five (a phenyl ring is likely).
The chemical data suggests an amine.  The two signals at δ 7.60 (d, 2H) and δ 8.7 (d,
2H) are characteristic of a 1,4-disubstituted phenyl ring.  The signal at δ 3.84 (s, 3H)
is a methyl group bonded to an oxygen atom, –OCH₃.  The signal at δ 4.18 (s, 2H) is
the primary amine hydrogen atoms.  This leaves only C and O left unaccounted for,
therefore with the –OCH₃ mentioned previously and a carbonyl group, it can be
concluded that the 1,4-disubstitution pattern involves an –NH₂ group and a methyl
ester.  The only structure consistent with the above data and analysis is methyl 4-
aminobenzoate

4.18       O    3.85
H₂N—⟨      ⟩—COCH₃
                ‖

**Compound V**

**12.34** *Following is a ¹H-NMR spectrum and a structural formula for anethole, C₁₀H₁₂O, a fragrant natural product obtained from anise. Using the line of integration, determine the number of protons giving rise to each signal. Show that this spectrum is consistent with the structure of anethole.*

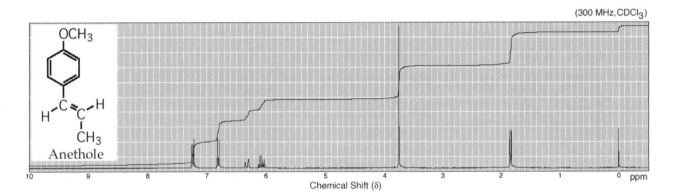

The chemical shift of the signals, the splitting patterns, and the number of hydrogen atoms corresponding to each signal are: δ 1.81 (d, 3H), 3.77 (s, 3H), 6.05 (m, 1H), 6.33 (d, 1H), 6.80 (d, 1H), 7.23 (d, 2H). The aromatic ring of anethole contains a propenyl group, which is para to a methoxy group. The 1,4-disubstitution pattern is responsible for the pair of doublets at δ 6.80 and δ 7.23. The vinylic proton nearest the aromatic ring will couple to only the hydrogen trans to it, giving the doublet at δ 6.33. The other vinylic hydrogen atom couples to both the trans hydrogen and to the hydrogen atoms of the methyl group, giving a complex signal at δ 6.05. The methyl hydrogen atoms on the propenyl group are split into a doublet by the adjacent vinylic hydrogen. Finally, the hydrogen atoms of the methoxy group are not coupled to anything and thus appear as a singlet at δ 3.77.

**12.35** *Propose a structural formula for* **compound W***, $C_4H_6O$, based on the following IR and* $^1$*H-NMR spectra.*

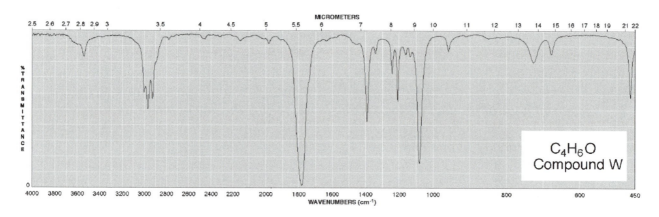

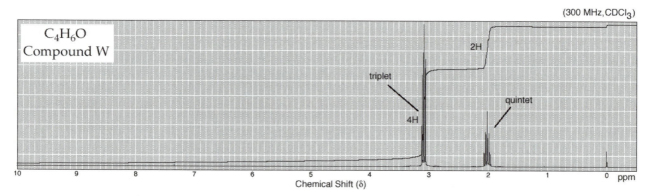

**Compound W has an index of hydrogen deficiency of two. The $^1$H-NMR and IR spectra show an absence of vinylic hydrogen atoms. The $^1$H-NMR also reveals the absence of methyl groups. The IR spectrum indicates the presence of a carbonyl, accounting for one site of hydrogen deficiency. Compound W may be a ring, which will account for the second site of hydrogen deficiency. The unusually high value for the carbonyl is due to the strained ring. Strained carbonyl C=O bonds will absorb IR at higher wave numbers (> 1750 cm$^{-1}$). The only structure consistent with the above IR and $^1$H-NMR data is cyclobutanone.**

**Compound W**

**12.36** *Propose a structural formula for compound X, $C_5H_{10}O_2$, based on the following IR and $^1H$-NMR spectra.*

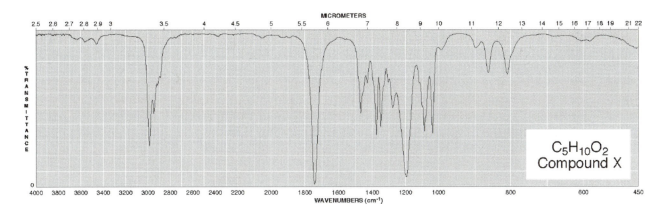

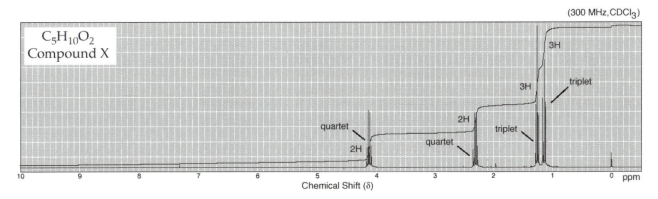

Compound X has an index of hydrogen deficiency of one. A notable feature of the IR spectrum is the strong peak at 1750 cm$^{-1}$, which indicates the presence of a carboxyl group and accounts for the only site of hydrogen deficiency. The absence of an –OH stretch between 2400 and 3400 cm$^{-1}$ rules out the possibility that the carboxyl group is a carboxylic acid. Two characteristic triplet-quartet pairs suggest two ethyl groups. The signals δ 4.13 (q, 2H) and δ 1.24 (t, 3H) come from the $CH_3CH_2O-$ group and δ 2.31 (q, 2H) and δ 1.19 (t, 3H) come from the $CH_3CH_2-$ group. The structure of ethyl propanoate is consistent with the above IR and $^1H$-NMR spectral data.

$$\underset{\text{1.19}}{CH_3}\underset{\text{2.31}}{CH_2}\overset{\overset{\displaystyle O}{\|}}{C}\underset{\text{4.13}}{OCH_2}\underset{\text{1.24}}{CH_3}$$

**Compound X**

**12.37** *Propose a structural formula for compound Y, $C_5H_9ClO_2$, based on the following IR and $^1$H-NMR spectra.*

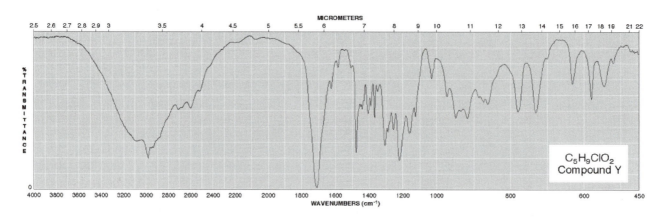

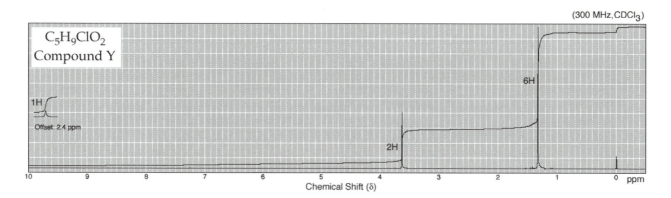

Compound Y has an index of hydrogen deficiency of one. A notable feature of the IR spectrum is a very broad peak at 3000 cm$^{-1}$ (O-H stretch) and a strong peak at 1710 cm$^{-1}$ (C=O stretch), which together indicate the presence of a carboxylic acid and accounts for the only site of hydrogen deficiency. The $^1$H-NMR confirms the presence of a carboxylic acid proton with a broad signal at $\delta$ 12.1 (s, 1H). The signal at $\delta$ 1.31 (s, 6H) indicates the presence of two methyl groups that have no adjacent hydrogen atoms. The signal at $\delta$ 3.61 (s, 2H) comes from a –CH$_2$– group with no adjacent hydrogen atoms. The only compound consistent with the above data is 3-chloro-3-methylbutanoic acid.

$$
\begin{array}{c}
1.31 \\
CH_3 \quad O \\
1.31 \mid \quad \parallel \quad 12.1 \\
CH_3CCH_2COH \\
\mid \quad 3.61 \\
Cl
\end{array}
$$

**Compound Y**

**12.38** *Propose a structural formula for compound Z, C₆H₁₄O, based on the following IR and ¹H-NMR spectra.*

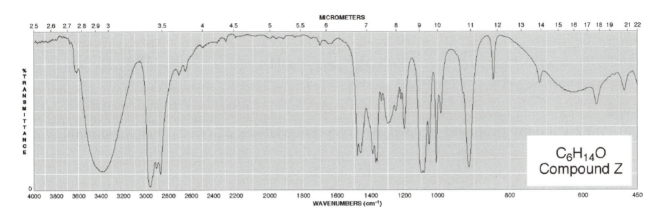

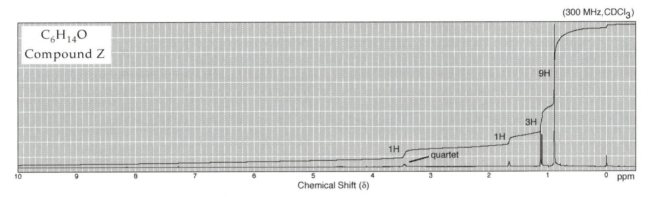

Compound Z has an index of hydrogen deficiency of zero. A notable feature of the IR spectrum is a very broad peak at 3400 cm⁻¹ (O-H stretch) indicating the presence of an alcohol hydroxyl group. The ¹H-NMR spectrum reveals a $(CH_3)_3C-$ group with a signal at δ 0.89 (s, 9H), which accounts for four of the six carbon atoms in Compound Z. The signal at δ 1.13 (d, 3H) indicates a methyl group coupled to an adjacent –CH– with a signal at δ 3.48 (q, 1H). The –CH– is also attached to the alcohol hydroxyl, evident by the downfield chemical shift. The signal at δ 1.65 (broad singlet, 1H) is due to the alcohol hydroxyl's proton. The data for compound Z is consistent with the structure for 3,3-dimethyl-2-butanol.

$$
\begin{array}{c}
\overset{0.89}{\text{CH}_3}\quad\overset{1.65}{\text{OH}} \\
\overset{0.89}{|}\qquad | \qquad {}^{1.13} \\
\text{CH}_3\text{C}\!\!-\!\!\!-\!\!\text{C}\!-\!\text{CH}_3 \\
|\qquad | \\
\text{CH}_3\quad\text{H} \\
0.89\quad 3.48
\end{array}
$$

**Compound Z**

**CHAPTER 13**
*Solutions to Problems*

**13.1**   *Write the IUPAC name for each compound.  Specify the configuration of (c).*

*(a)*

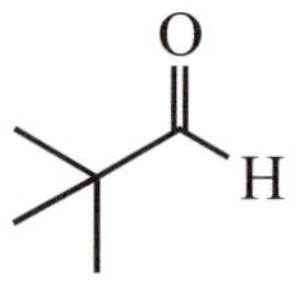

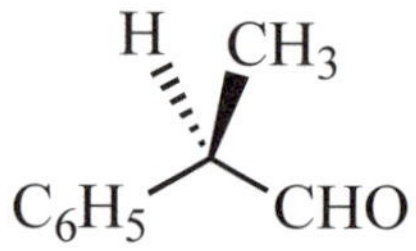

**2,2-Dimethylpropanal**

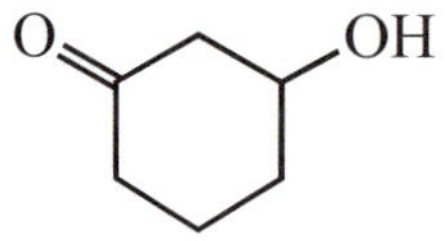

*(b)*

**3-Hydroxycyclohexanone**

*(c)*

**(R)-2-Phenylpropanal**

**13.2**   *Write structural formulas for all aldehydes with molecular formula C₆H₁₂O and give each its IUPAC name.  Which of these aldehydes are chiral?*

**There are eight aldehydes with this molecular formula.  The stereocenter is indicated with an asterisk and the chiral aldehydes are circled.**

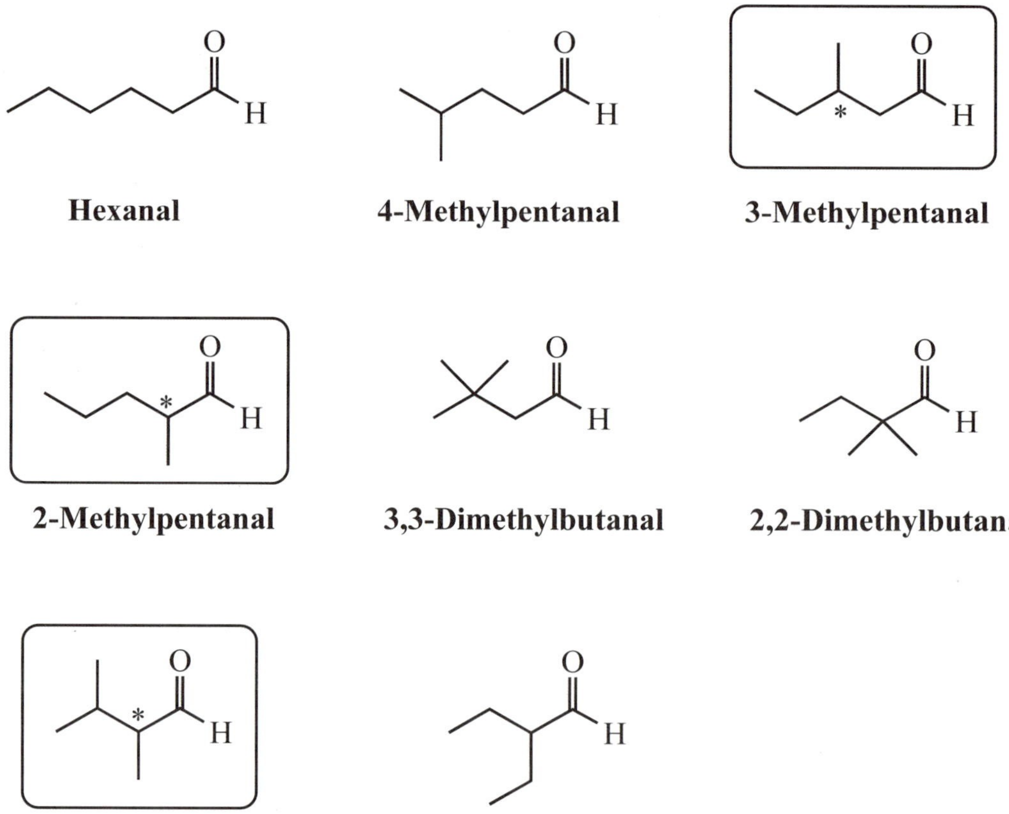

**Hexanal**

**4-Methylpentanal**

**3-Methylpentanal**

**2-Methylpentanal**

**3,3-Dimethylbutanal**

**2,2-Dimethylbutanal**

**2,3-Dimethylbutanal**

**2-Ethylbutanal**

**13.3**   *Write IUPAC names for these compounds, each of which is important in intermediary metabolism. Below each is the name by which it is more commonly known in the biological sciences.*

         OH                     O

*(a)*   CH$_3$CHCOOH      *(b)*  CH$_3$CCOOH     *(c)*  H$_2$NCH$_2$CH$_2$CH$_2$COOH

      Lactic acid               Pyruvic acid          γ-Aminobutyric acid

**2-Hydroxypropanoic acid    2-Oxopropanoic Acid     4-Aminobutanoic acid**

**13.4**   *Explain how these Grignard reagents react with molecules of their own kind to "self-destruct."*

   *(a)*  HO—⟨benzene ring⟩—MgBr      *(b)*

**These Grignard reagents do not exist because each has an acidic functional group that would be deprotonated by the highly basic Grignard reagent portion of the molecule. In (a), the phenol would be deprotonated to give a phenoxide, and in (b), the carboxylic acid would be deprotonated to give a carboxylate.**

**13.5**   *Show how these three compounds can be synthesized from the same Grignard reagent.*

*(a)*  ⟨cyclohexene⟩—MgBr   $\xrightarrow[\text{2. NH}_4\text{Cl/H}_2\text{O}]{\text{1. CH}_2\text{O}}$   ⟨cyclohexene⟩—CH$_2$OH

*(b)*  ⟨cyclohexene⟩—MgBr   $\xrightarrow[\text{2. NH}_4\text{Cl/H}_2\text{O}]{\text{1. CH}_3\text{CH}=\text{O}}$   ⟨cyclohexene⟩—CHCH$_3$ (with OH)

*(c)*  ⟨cyclohexene⟩—MgBr   $\xrightarrow[\text{2. NH}_4\text{Cl/H}_2\text{O}]{\text{1. cyclohexanone}}$   ⟨cyclohexene⟩—⟨cyclohexane⟩ (with HO)

**13.6**  *Hydrolysis of an acetal forms an aldehyde or ketone and two molecules of alcohol. Following are structural formulas for three acetals. Draw the structural formulas for the products of hydrolysis of each in aqueous acid.*

(a)

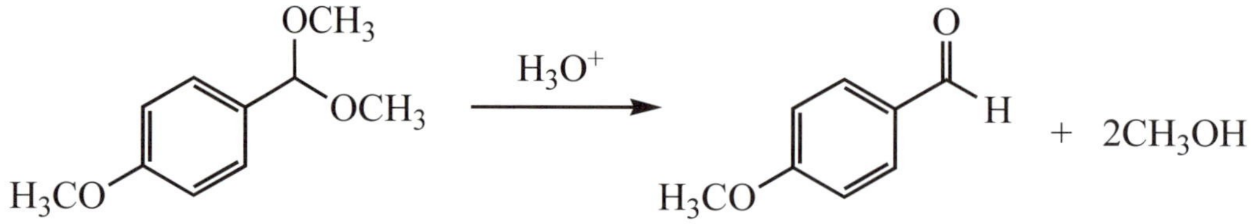

(b)

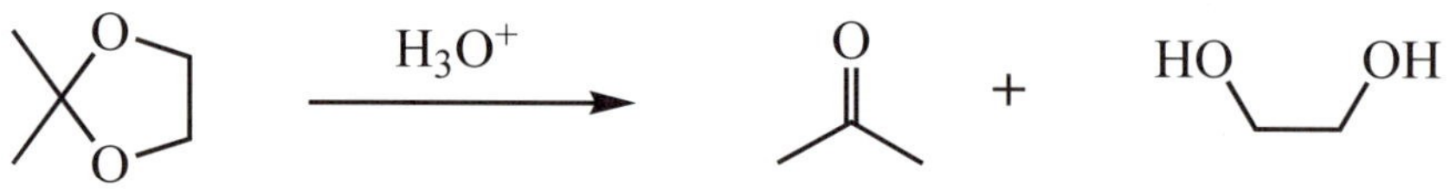

(c)

**13.7**  *Acid-catalyzed hydrolysis of an imine gives an amine and an aldehyde or ketone. When one equivalent of acid is used, the amine is converted to its ammonium salt. Write structural formulas for the products of hydrolysis of each imine using one equivalent of HCl.*

(a)

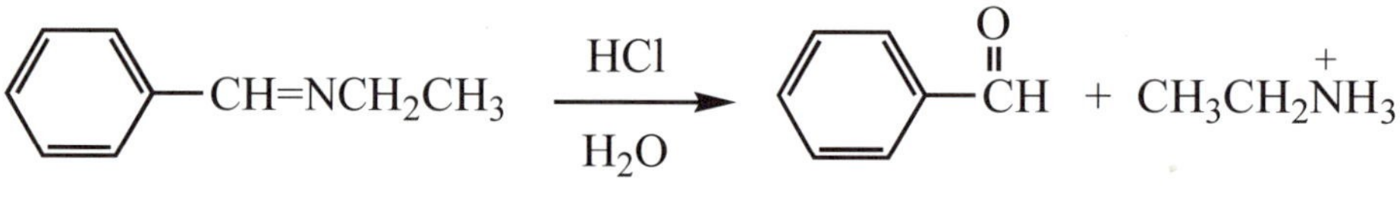

(b)

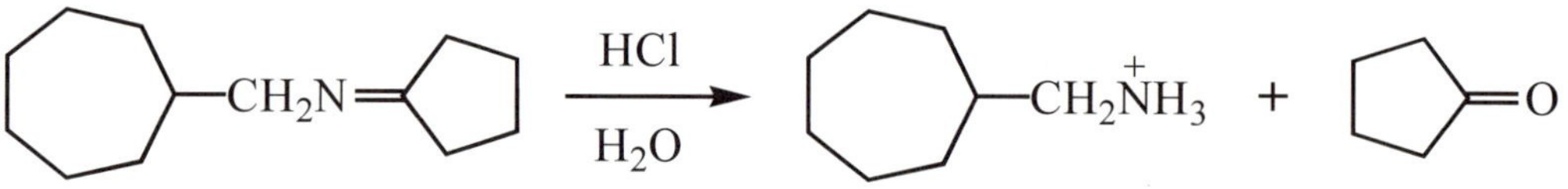

**13.8**  *Show how to prepare each amine by reductive amination of an appropriate aldehyde or ketone.*

(a)

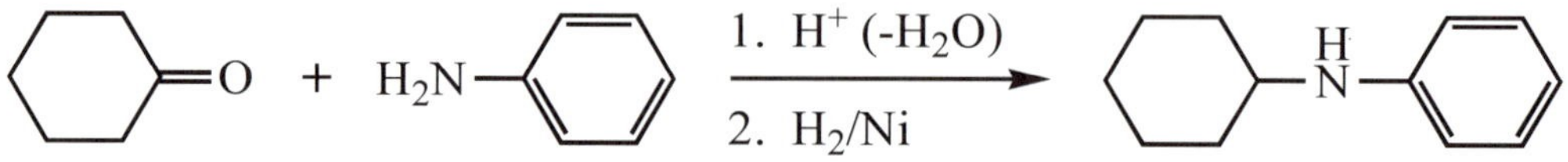

(b)

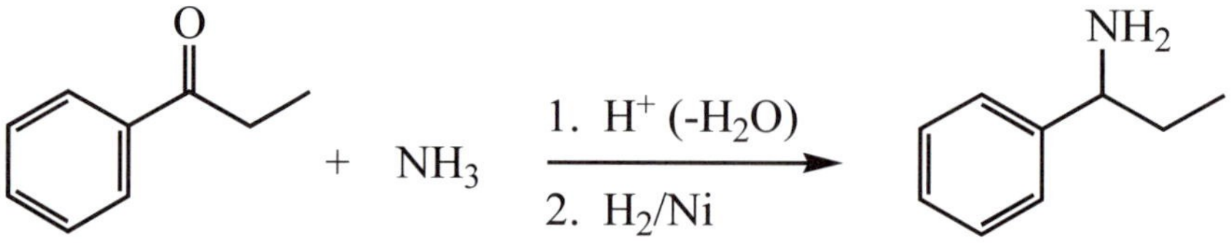

**13.9**   *Draw the structural formula for the keto form of each enol.*

(a)

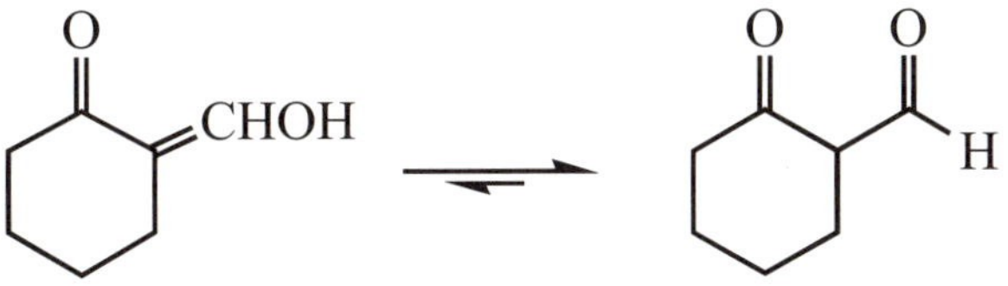

(b)

(c)

**13.10**  *Complete these oxidations.*

(a) *3-Oxobutanal*   +   $O_2$   $\longrightarrow$

+ $O_2$   $\longrightarrow$

(b) *3-Phenylpropanal*   +   *Tollens' reagent*   $\longrightarrow$

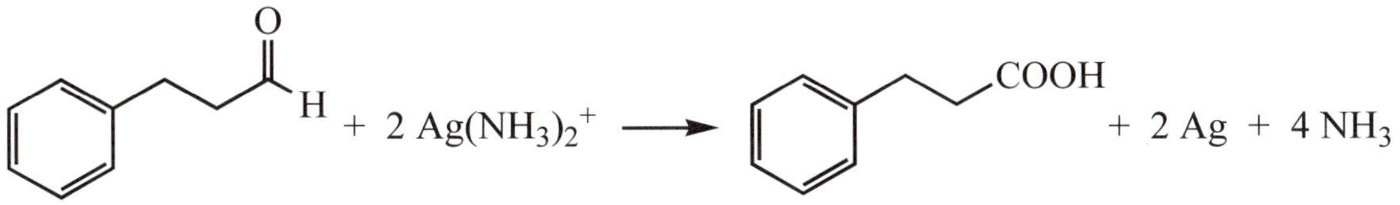

+ 2 $Ag(NH_3)_2^+$   $\longrightarrow$   + 2 Ag + 4 $NH_3$

**13.11**  *What aldehyde or ketone gives each alcohol on reduction by $NaBH_4$?*

(a)   $\xrightarrow{\text{NaBH}_4}$

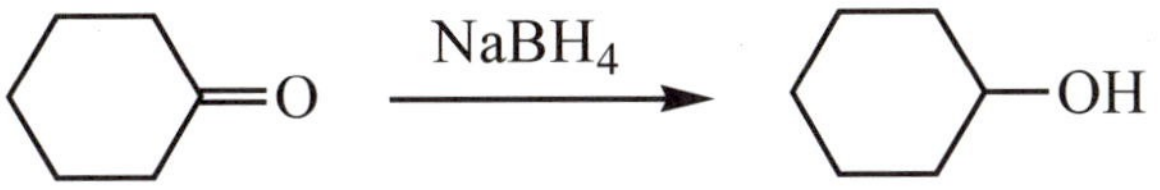

(b)   $-CH_2CH$   $\xrightarrow{\text{NaBH}_4}$   $-CH_2CH_2OH$

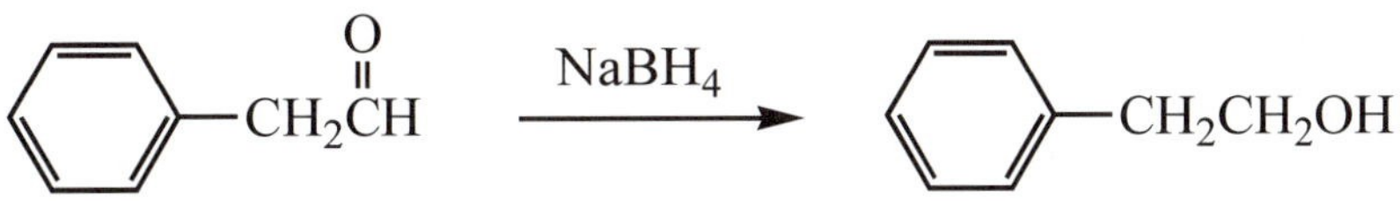

(c)   $\xrightarrow{\text{NaBH}_4}$

**Preparation of Aldehydes and Ketones (See Chapters 5, 8, and 9)**

**13.12** *Complete these reactions.*

(a) cyclooctanol $\xrightarrow[\text{H}_2\text{SO}_4]{\text{K}_2\text{Cr}_2\text{O}_7}$ cyclooctanone (—OH) $+$ $Cr^{3+}$

(b) cyclopentyl—$CH_2OH$ $\xrightarrow[\text{CH}_2\text{Cl}_2]{\text{PCC}}$ cyclopentyl—CHO $+$ $Cr^{3+}$

(c) cyclopentyl—$CH_2OH$ $\xrightarrow[\text{H}_2\text{SO}_4]{\text{K}_2\text{Cr}_2\text{O}_7}$ cyclopentyl—COOH $+$ $Cr^{3+}$

(d) benzene $+$ acid chloride $\xrightarrow{\text{AlCl}_3}$ ketone $+$ HCl

**13.13** *Show how you would bring about these conversions.*

(a) pentanol $\xrightarrow[\text{CH}_2\text{Cl}_2]{\text{PCC}}$ pentanal

(b) pentanol $\xrightarrow[\text{H}_2\text{SO}_4]{\text{K}_2\text{Cr}_2\text{O}_7}$ pentanoic acid

(c) 2-pentanol $\xrightarrow[\text{H}_2\text{SO}_4 \;\text{or}\; \text{PCC/CH}_2\text{Cl}_2]{\text{K}_2\text{Cr}_2\text{O}_7}$ 2-pentanone

(d) 1-butene $\xrightarrow[\text{H}_2\text{SO}_4]{\text{H}_2\text{O}}$ 2-butanol (OH) $\xrightarrow[\text{H}_2\text{SO}_4]{\text{K}_2\text{Cr}_2\text{O}_7}$ ketone

(e) benzene $+$ acetyl chloride $\xrightarrow{\text{AlCl}_3}$ acetophenone

*(f)*

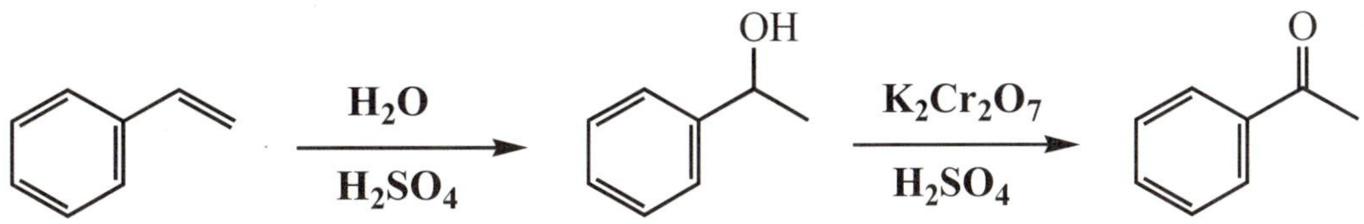
H₂O
H₂SO₄
OH
K₂Cr₂O₇
H₂SO₄
O

*(g)*

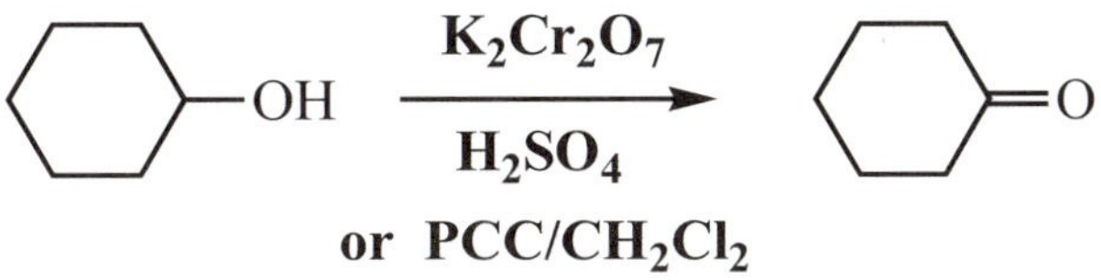
OH
K₂Cr₂O₇
H₂SO₄
or PCC/CH₂Cl₂
O

*(h)*

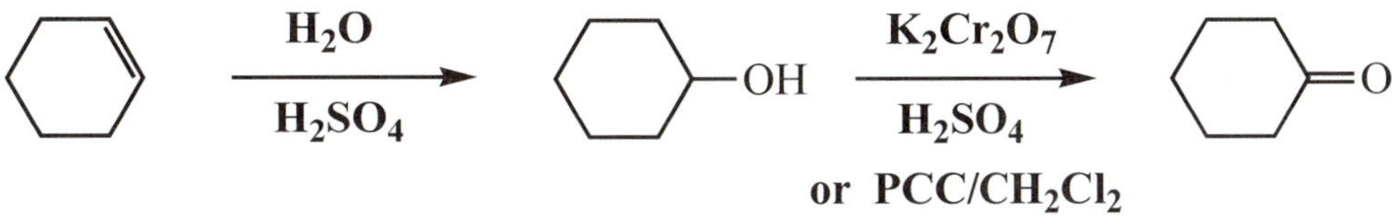
H₂O
H₂SO₄
OH
K₂Cr₂O₇
H₂SO₄
or PCC/CH₂Cl₂
O

**Structure and Nomenclature**

**13.14** *Draw a structural formula for the one ketone of molecular formula $C_4H_8O$ and for the two aldehydes of molecular formula $C_4H_8O$.*

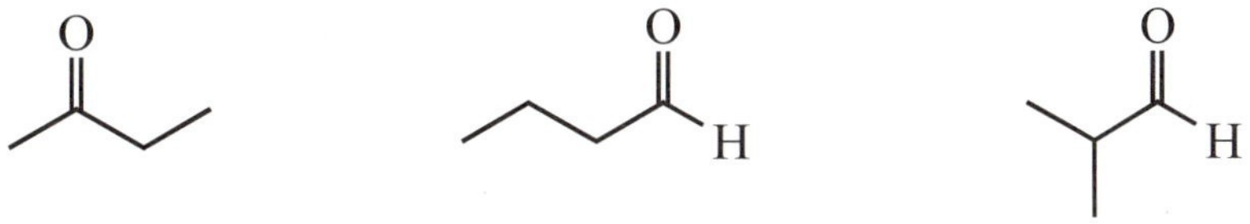
O
O
H
O
H

**13.15** *Draw structural formulas for the four aldehydes of molecular formula $C_5H_{10}O$. Which of these aldehydes are chiral?*

**The stereocenter is indicated with an asterisk and the chiral molecule is circled.**

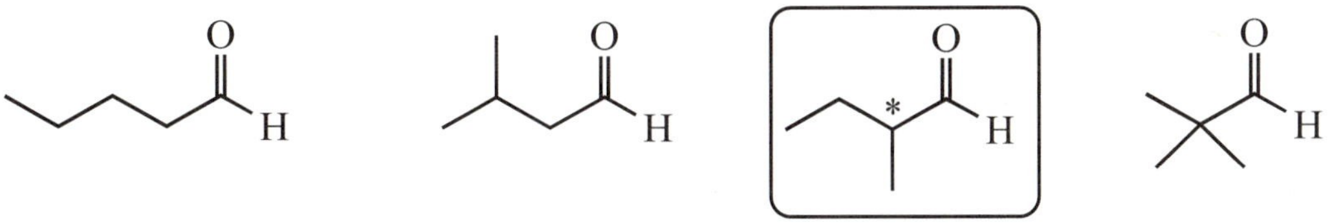
O
H
O
H
O
*
H
O
H

**13.16** *Name these compounds.*

*(a)*

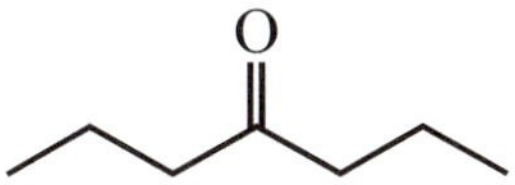
O

**4-Heptanone**

*(b)*

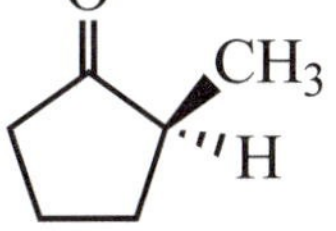
O
CH₃
H

**(*S*)-2-Methyl-cyclopentanone**

*(c)*

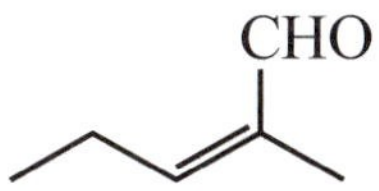
CHO

**(*Z*)-2-Methyl-2-pentenal (*cis*-2-Methyl-2-pentenal)**

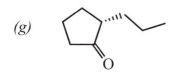

*(d)*

(*S*)-2-Hydroxypropanal          2-Methoxyacetophenone          2,2-Dimethyl-3-oxopropanoic acid

*(e)*

*(f)*

*(g)*

(*S*)-2-Propylcyclopentanone

**13.17** *Draw structural formulas for these compounds.*

*(a) 1-Chloro-2-propanone*

*(b) 3-Hydroxybutanal*

*(c) 4-Hydroxy-4-methyl-2-pentanone*

*(d) 3-Methyl-3-phenylbutanal*

*(e) (S)-3-bromocyclohexanone*

*(f) 3-Methyl-3-buten-2-one*

*(g)  5-Oxohexanal*

*(h) 2,2-Dimethylcyclohexanecarbaldehyde*

*(i) 3-Oxobutanoic acid*

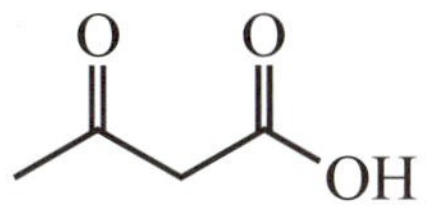

## Addition of Carbon Nucleophiles

**13.18** *Write an equation for the acid-base reaction between phenylmagnesium iodide and a carboxylic acid. Use curved arrows to show the flow of electrons in this reaction. In addition, show that this reaction is an example of a stronger acid and stronger base reacting to form a weaker acid and weaker base.*

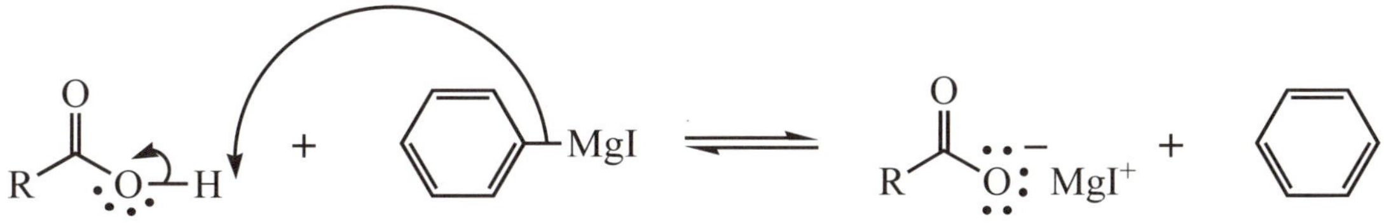

$pK_a$ 4-5          **Stronger base**          **Weaker base**          $pK_a$ 43
**Stronger acid**                                                      **Weaker acid**

**The right side of the equilibrium is favored because it has the weaker acid and base.**

**13.19** *Diethyl ether is prepared on an industrial scale by acid-catalyzed dehydration of ethanol. Explain why diethyl ether used in the preparation of Grignard reagents must be carefully purified to remove all traces of ethanol and water.*

$$2CH_3CH_2OH \xrightarrow[180°C]{H_2SO_4} CH_3CH_2OCH_2CH_3 + H_2O$$

**Grignard reagents are extremely basic and will deprotonate ethanol and water to form the hydrocarbon and hydroxide. As an example, consider the reaction in 13.18, replacing the carboxylic acid with water:**

**Stronger base**          $pK_a$ 15.7          **Weaker base**          $pK_a$ 43
                          **Stronger acid**                              **Weaker acid**

**The right side of the equilibrium is favored because it has the weaker acid and base.**

**13.20** *Draw structural formulas for the product formed by treatment of each compound with propylmagnesium bromide followed by hydrolysis in aqueous acid.*

(a) $CH_2O$

1. $\diagup\!\!\diagdown\!\!\diagup$ MgBr in ether
2. $H_3O^+$

→ product with OH

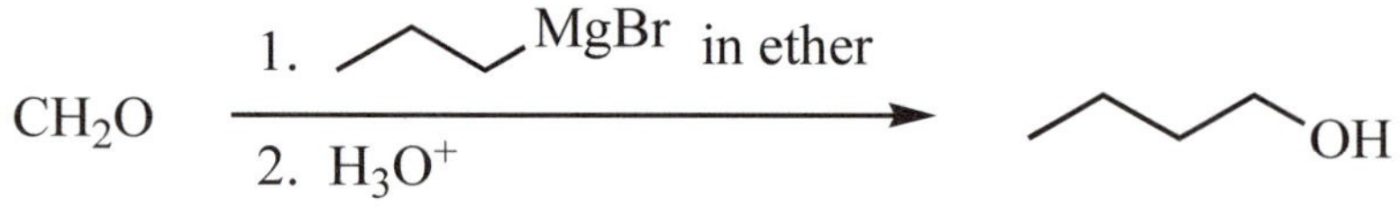

(b)

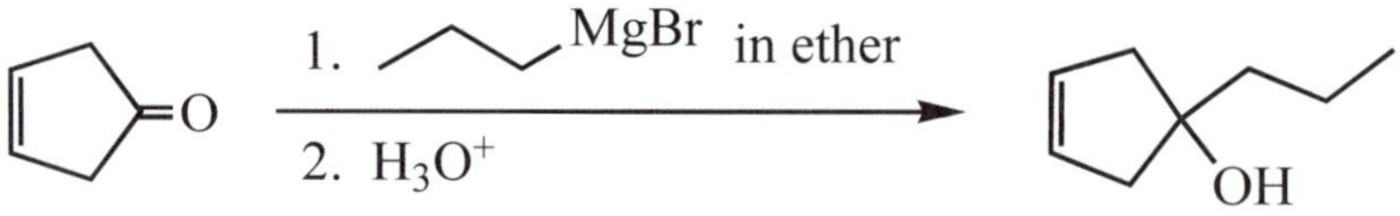

1. $\diagup\!\!\diagdown\!\!\diagup$ MgBr in ether
2. $H_3O^+$

(c)

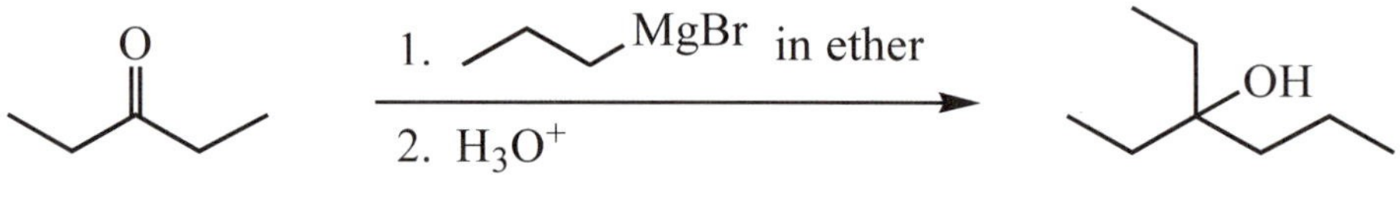

1. $\diagup\!\!\diagdown\!\!\diagup$ MgBr in ether
2. $H_3O^+$

(d) CHO

1. $\diagup\!\!\diagdown\!\!\diagup$ MgBr in ether
2. $H_3O^+$

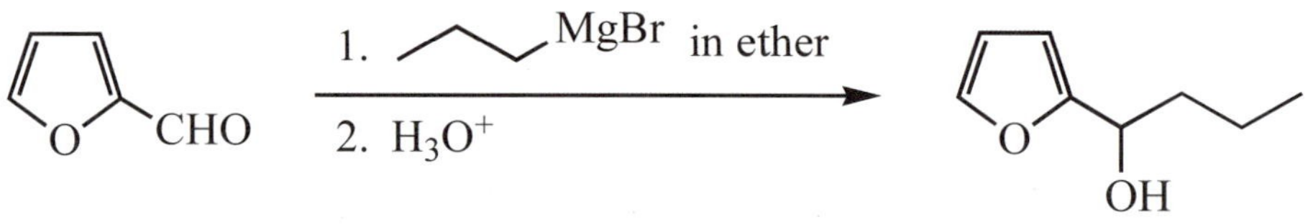

(e)

1. $\diagup\!\!\diagdown\!\!\diagup$ MgBr in ether
2. $H_3O^+$

$H_3CO$

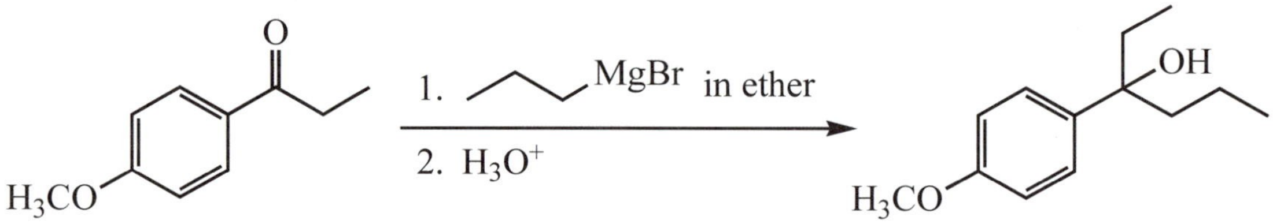

**13.21** *Suggest a synthesis for these alcohols starting from an aldehyde or ketone and an appropriate Grignard reagent. In parentheses below each target molecule is shown the number of combinations of Grignard reagent and aldehyde or ketone that might be used.*

(a)

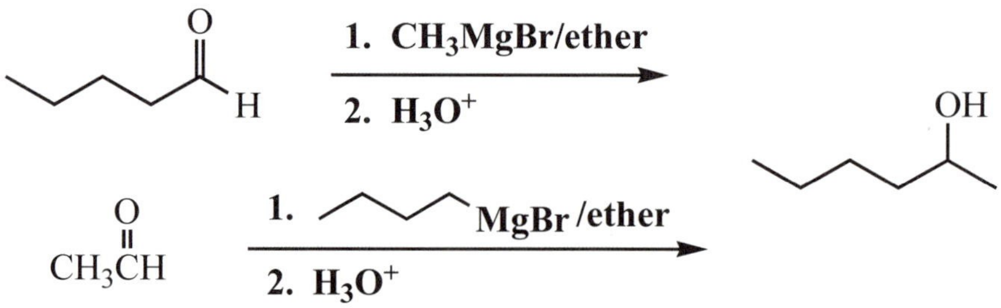

1. $CH_3MgBr$/ether
2. $H_3O^+$

→ OH

$CH_3CH$ (with O double bond)

1. $\diagup\!\!\diagdown\!\!\diagup$ MgBr /ether
2. $H_3O^+$

*(b)*

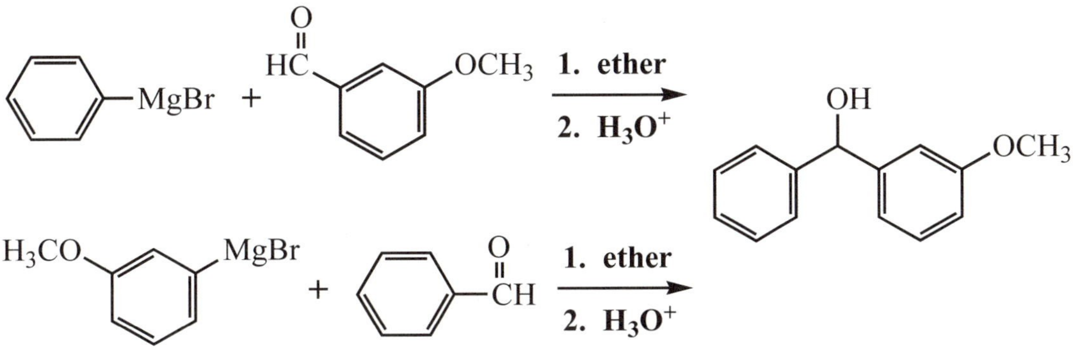

*(c)*

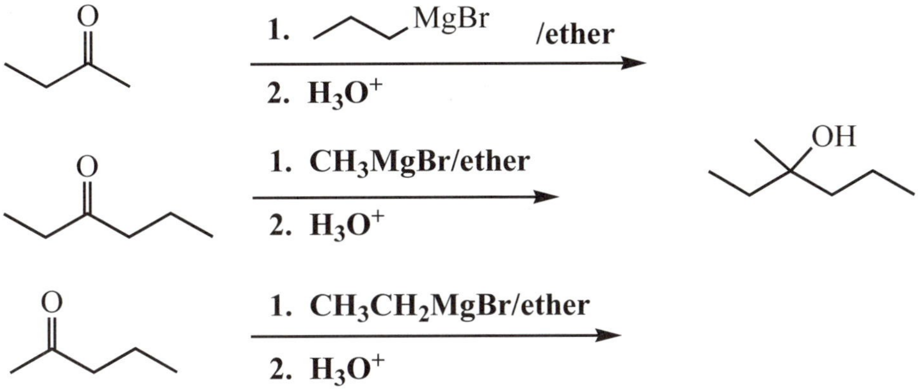

## Addition of Oxygen Nucleophiles

**13.22** *5-Hydroxyhexanal forms a six-membered cyclic hemiacetal, which predominates at equilibrium in aqueous solution.*

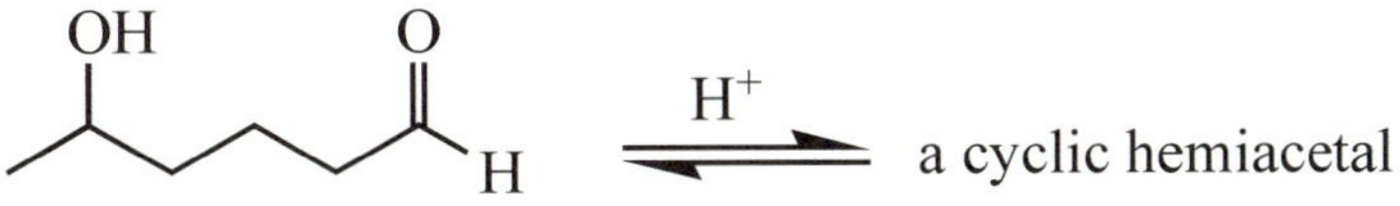

5-Hydroxyhexanal

*(a) Draw a structural formula for this cyclic hemiacetal.*

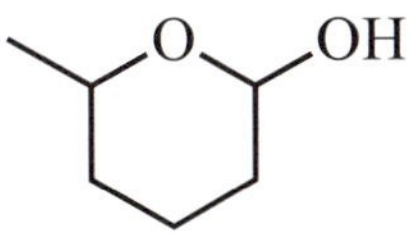

*(b) How many stereoisomers are possible for 5-hydroxyhexanal?*

**There are two enantiomers possible for 5-hydroxyhexanal.**

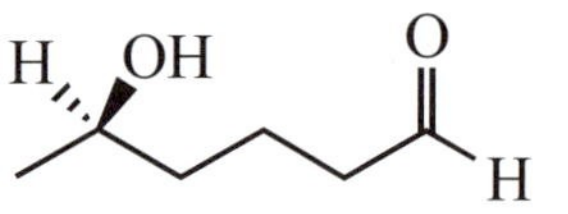

(R)-5-Hydroxyhexanal 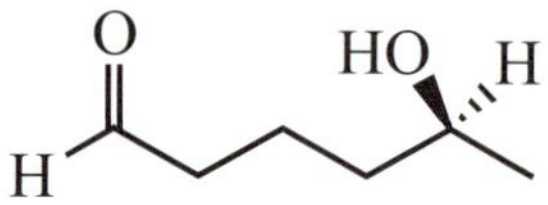                    (S)-5-Hydroxyhexanal

*(c) How many stereoisomers are possible for this cyclic hemiacetal?   (d) Draw
alternative chair conformations for each stereoisomer. (e) Which alternative chair
conformation for each stereoisomer is the more stable?*

**There are four stereoisomers possible for the cyclic hemiacetal. The criteria used
to determine the more stable conformation of each configuration is based on the
structure with the fewest 1,3-diaxial repulsions.**

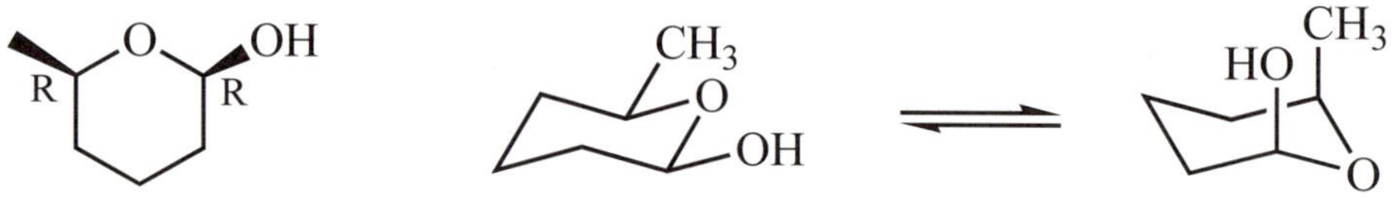

**More stable**

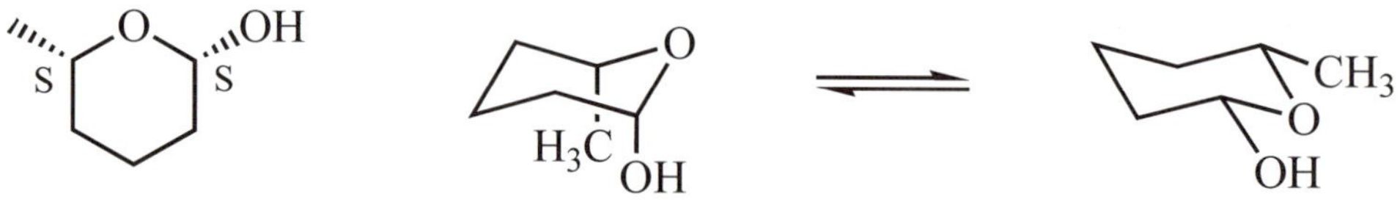

**More stable**

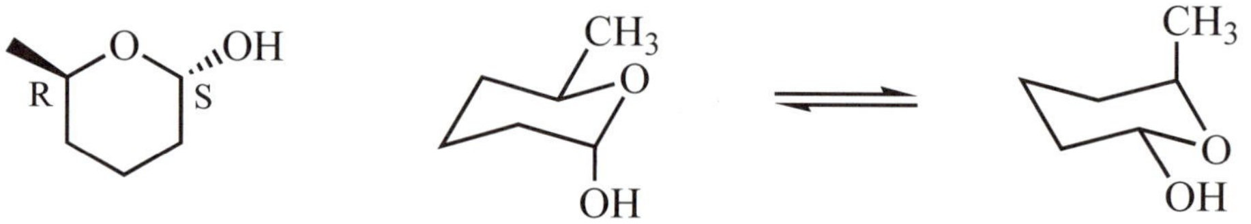

**Approximately the same stability**

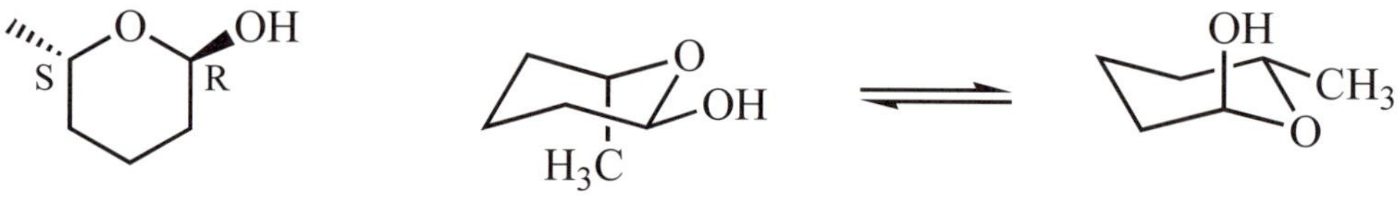

**Approximately the same stability**

**13.23** *Draw structural formulas for the hemiacetal and then the acetal formed from each pair of
reactants in the presence of an acid catalyst.*

*(a)* 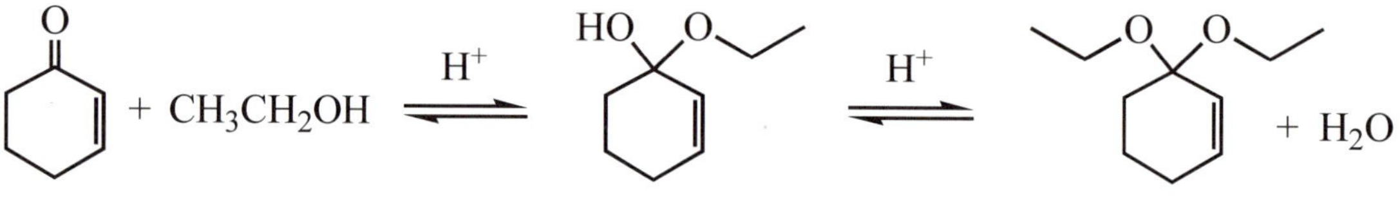 + $CH_3CH_2OH$  ⇌  ⇌  + $H_2O$

(b)

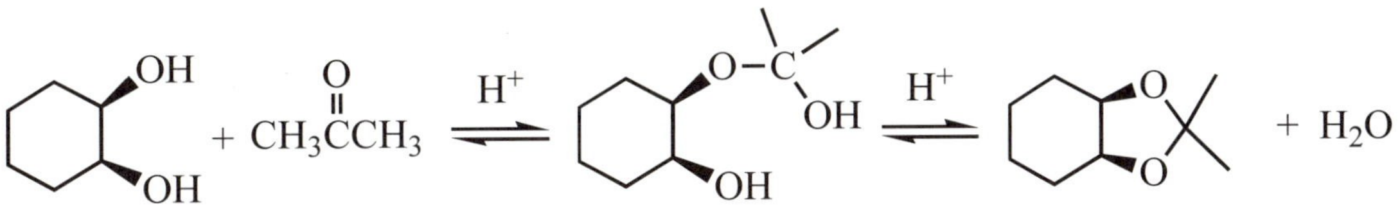

(c)

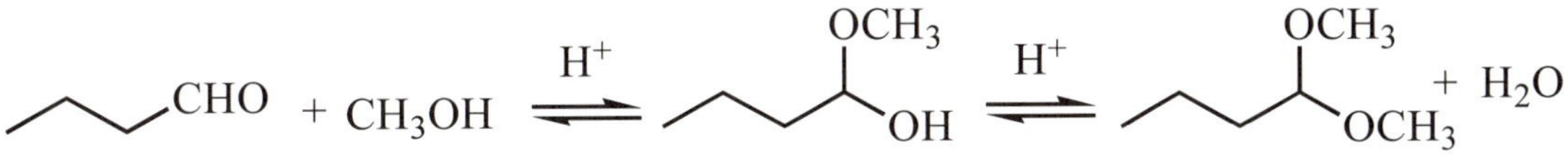

**13.24** *Draw structural formulas for the products of hydrolysis of each acetal in aqueous acid.*

(a)

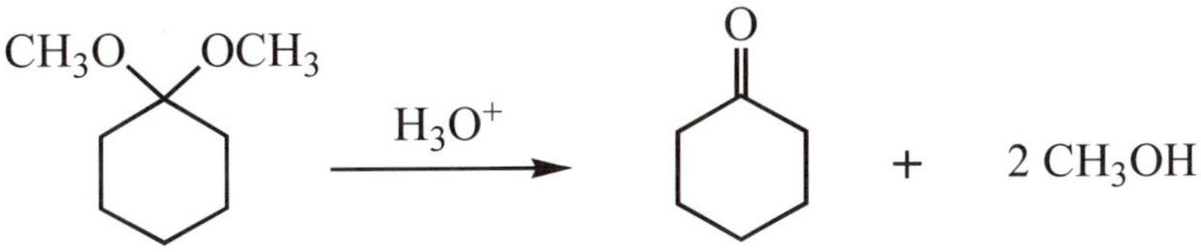

(b)

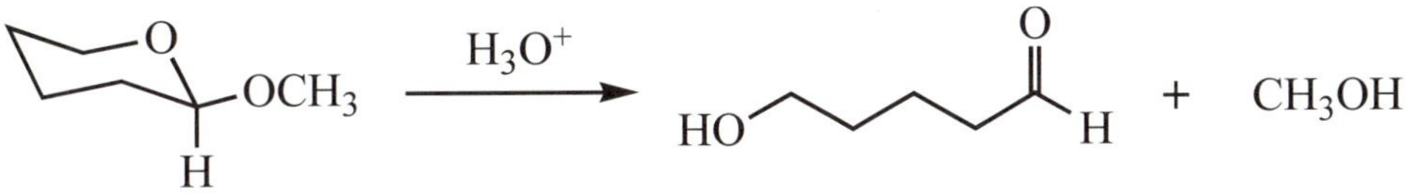

(c)

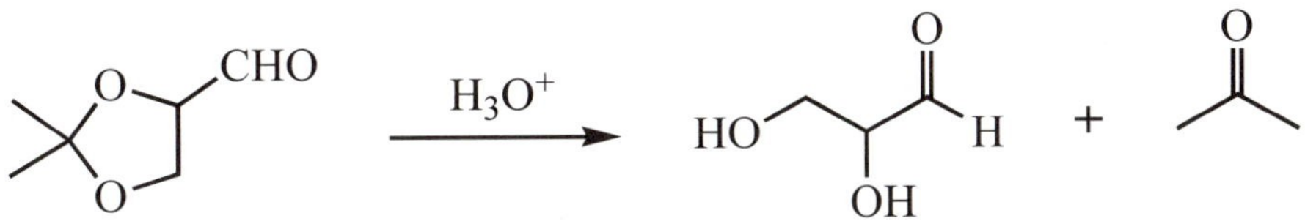

**13.25** *The following compound is a component of jasmine fragrance. From what carbonyl-containing compound and alcohol is it derived?*

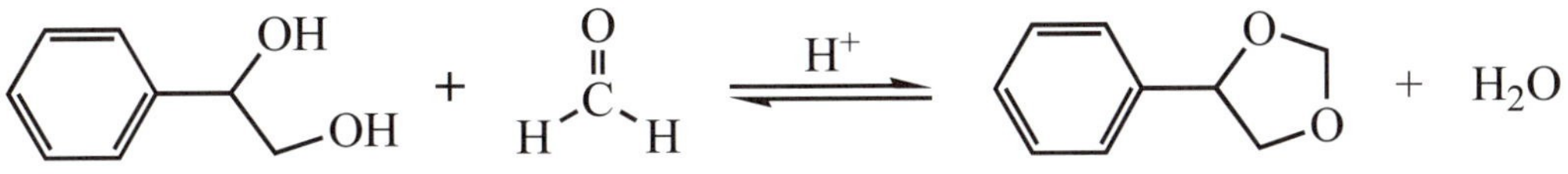

**Diol**          **Carbonyl**

**13.26** *Propose a mechanism for formation of the cyclic acetal from treatment of acetone with ethylene glycol in the presence of an acid catalyst. Your mechanism must be consistent with the fact that the oxygen atom of the water molecule is derived from the carbonyl oxygen of acetone.*

Acetone            Ethylene glycol

**The essential elements for mechanism of cyclic acetal formation are the same as that for an acyclic acetal formation.**

**Step 1: Protonation of the carbonyl oxygen forms a reactive electrophilic intermediate that is stabilized by resonance.**

**Step 2: One of the nucleophilic oxygen atoms of the ethylene glycol attacks the protonated carbonyl species to give a protonated hemiacetal.**

**Step 3:  A proton is lost to give the hemiacetal intermediate.**

**Step 4: The hemiacetal is protonated on the hydroxyl group.  Note that protonation of the ether oxygen could also occur, but that would simply be the reverse of the previous step and not lead to product so it is not shown.**

**Step 5: Water easily leaves to give another highly electrophilic intermediate that is stabilized by resonance.**

**Step 6: The other hydroxyl oxygen atom attacks the electrophilic carbon to give a protonated, cyclic intermediate.**

**Step 7:  Loss of a proton gives the final cyclic acetal product.**

**13.27** *Propose a mechanism for the formation of a cyclic acetal from 4-hydroxypentanal and one equivalent of methanol. If the carbonyl oxygen of 4-hydroxypentanal is enriched with oxygen-18, do you predict that the oxygen label appears in the cyclic acetal or in the water? Explain.*

**Step 1: Protonation of the carbonyl oxygen to give a resonance-stabilized cation.**

**Step 2: The hydroxyl group attacks the carbon atom of the protonated carbonyl group, which forms a protonated cyclic hemiacetal.**

**Step 3: Loss of proton from the protonated acetal leads to the hemiacetal.**

**Step 4: Protonation of the hydroxyl group converts it to a good leaving group.**

**Step 5: Loss of water yields a new resonance-stabilized carbocation. Note that the oxygen-18 label appears in the water.**

**Step 6: Nucleophilic attack of the methanol on the electrophilic carbon atom yields a protonated acetal.**

**Step 7: The final step involves the loss of proton from the protonated acetal to yield the acetal product.**

## Addition of Nitrogen Nucleophiles

**13.28** *Show how this secondary amine can be prepared by two successive reductive aminations:*

$$(1)\ NH_3,\ Ni/H_2 \qquad (2)\ PhCHO,\ Ni/H_2$$

**13.29** *Show how to convert cyclohexanone to each amine.*

(c)

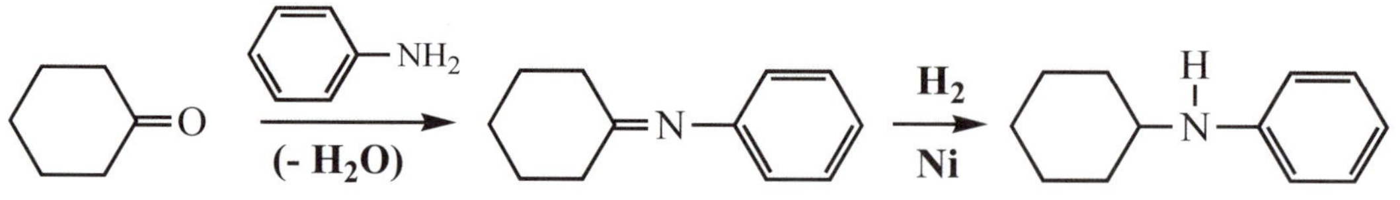

**13.30** *Following are structural formulas for amphetamine and methamphetamine. The major central nervous system effects of amphetamine and amphetamine-like drugs are locomotor stimulation, euphoria and excitement, stereotyped behavior, and anorexia. Show how each drug can be synthesized by reductive amination of an appropriate aldehyde or ketone.*

(a)

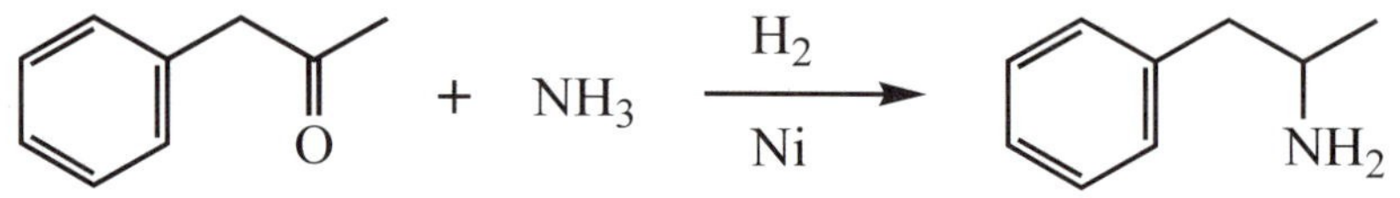

Amphetamine

(b)

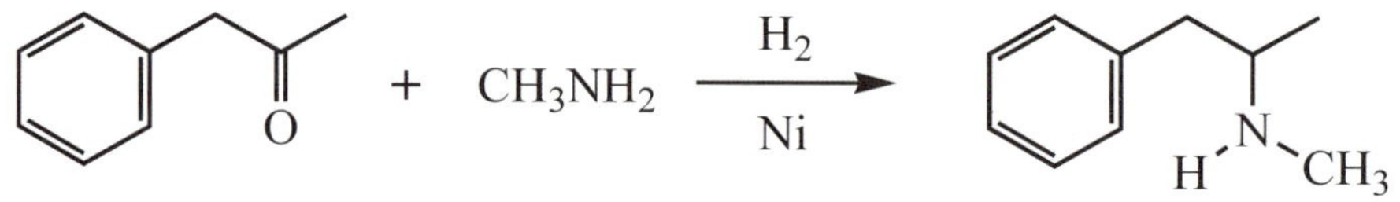

Methamphetamine

**13.31** *Rimantadine is effective in preventing infections caused by the influenza A virus and in treating established illness. It is thought to exert its antiviral effect by blocking a late stage in the assembly of the virus. Following is the final step in the synthesis of this compound.*

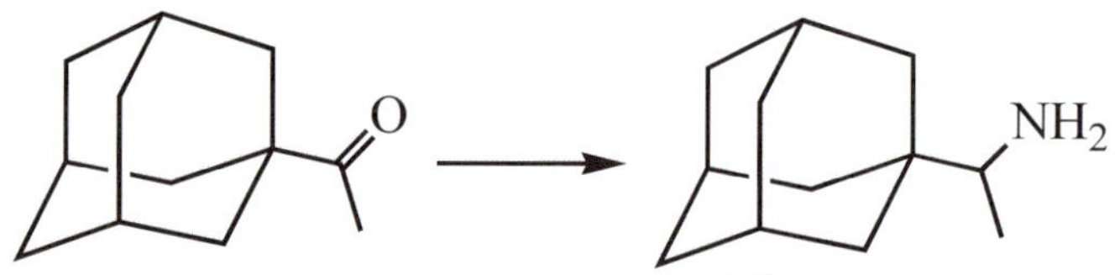

Rimantadine
(an antiviral agent)

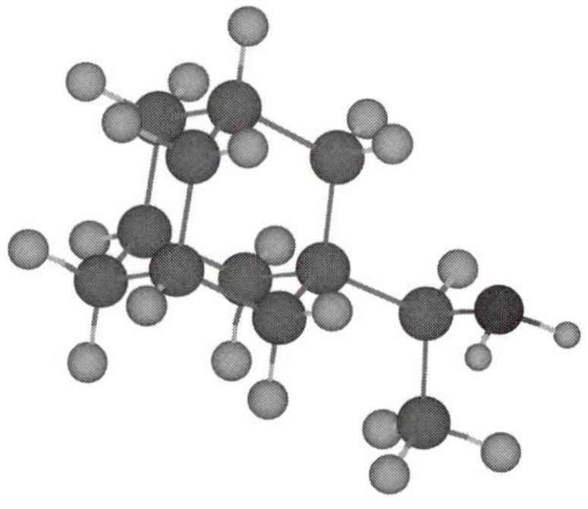

*(a) Describe experimental conditions to bring about this conversion*

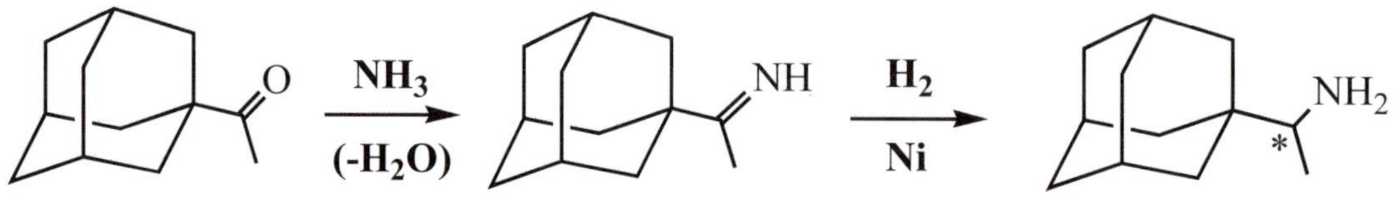

*(b) Is rimantadine chiral?*

**Yes, rimantadine is chiral.  The stereocenter is indicated with an asterisk.**

**13.32** *Methenamine, a product of the reaction of formaldehyde and ammonia, is a prodrug, a compound that is inactive by itself but is converted to an active drug in the body by a biochemical transformation.  The strategy behind use of methenamine as a prodrug is that nearly all bacteria are sensitive to formaldehyde at concentrations of 20 mg/mL or higher. Formaldehyde cannot be used directly in medicine, however, because an effective concentration in plasma cannot be achieved with safe doses.  Methenamine is stable at pH 7.4 (the pH of blood plasma) but undergoes acid-catalyzed hydrolysis to formaldehyde and ammonium ion under the acidic conditions of the kidneys and the urinary tract.  Thus, methenamine can be used as a site-specific drug to treat urinary infections.*

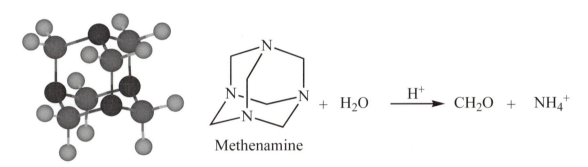

Methenamine

*(a) Balance the equation for the hydrolysis of methenamine to formaldehyde and ammonium ion.*

$$\text{(methenamine)} + 10\ H_2O \longrightarrow 6\ CH_2O + 4\ NH_4^+ + 4\ OH^-$$

*(b) Does the pH of an aqueous solution of methenamine increase, remain the same, or decrease as a result of its hydrolysis?  Explain.*

**When methenamine is hydrolyzed, ammonium hydroxide is formed. Ammonium hydroxide is a base so the pH will increase.**

*(c) Explain the meaning of the following statement:  The functional group in methenamine is the nitrogen analog of an acetal.*

**Acetals are 1,1-diethers, where a single carbon atom is bonded to the oxygen atoms of two ether functional groups.  In the case of methenamine, each carbon atom is bonded to two nitrogen atoms of two amine functional groups.**

*(d) Account for the observation that methenamine is stable in blood plasma but undergoes hydrolysis in the urinary tract.*

**Blood plasma is buffered to a slightly basic pH of 7.4. Methenamine is relatively stable to hydrolysis at an alkaline pH, thus stable under normal blood plasma conditions. Recall that acetals are also stable towards bases. On the other hand, both methenamine and acetals are readily hydrolyzed at acidic pH. The urinary tract is more acidic, so methenamine is hydrolyzed more rapidly there.**

## Keto-Enol Tautomerism

__13.33__ *The following molecule belongs to a class of compounds called enediols; each carbon of the double bond carries an -OH group. Draw structural formulas for the α-hydroxyketone and the α-hydroxyaldehyde with which this enediol is in equilibrium.*

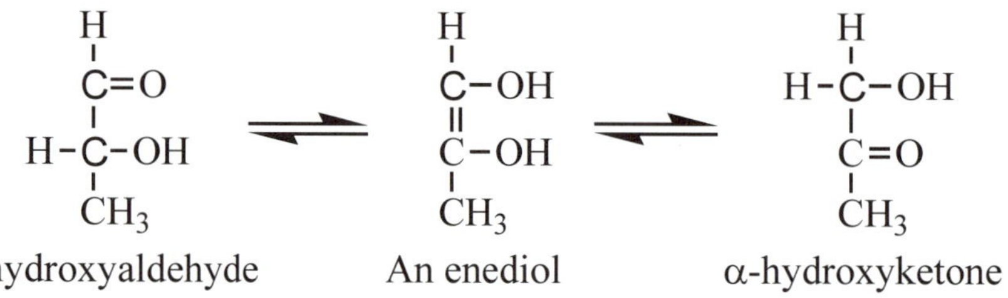

α-hydroxyaldehyde          An enediol          α-hydroxyketone

__13.34__ *In dilute aqueous acid, (R)-glyceraldehyde is converted into an equilibrium mixture of (R,S)-glyceraldehyde and dihydroxyacetone. Propose a mechanism for this isomerization.*

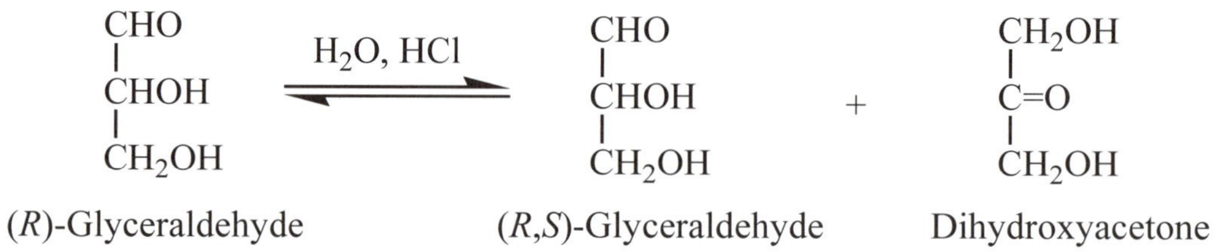

(R)-Glyceraldehyde          (R,S)-Glyceraldehyde          Dihydroxyacetone

**This equilibrium is an example of a keto-enol tautomerism:**

**Step 1:    The aldehyde carbonyl is protonated by hydronium ion, producing a resonance-stabilized cation intermediate.**

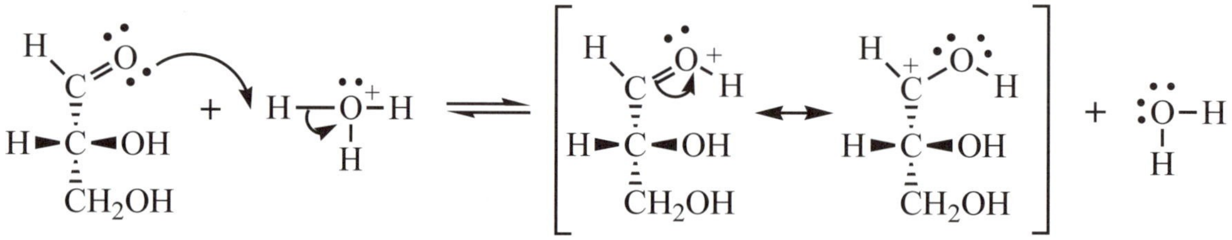

**Step 2:** Deprotonation of the resonance-stabilized intermediate leads to the enediol intermediate. The *Z* isomer is shown, but the *E* isomer can be formed as well.

**Step 3:** Protonation of the enediol intermediate at the original carbon can occur from either face of the alkene with equal probability, thereby, leading to a racemic mixture of glyceraldehydes after the loss of a proton.

(*R*)-Glyceraldehyde   (*S*)-Glyceraldehyde

**Step 4:** Formation of dihydroxyacetone occurs if the enediol intermediate in Step 3 is protonated at carbon 1 instead of carbon 2. The enediol intermediate is first protonated at carbon 1, forming the resonance-stabilized cation intermediate.

**Step 5: Loss of a proton from the resonance-stabilized intermediate yields dihydroxyacetone.**

$$CH_2OH$$
$$|$$
$$C=\overset{+}{\underset{\cdot\cdot}{O}}-H$$
$$|$$
$$CH_2OH$$

$\xrightarrow{-H^+}$

$$CH_2OH$$
$$|$$
$$C=\overset{\cdot\cdot}{\underset{\cdot\cdot}{O}}$$
$$|$$
$$CH_2OH$$

## Oxidation/Reduction of Aldehydes and Ketones

**13.35** *Draw a structural formula for the product formed by treatment of butanal with each set of reagents.*

(a) butanal $\xrightarrow[\text{2. } H_2O]{\text{1. } LiAlH_4}$ butanol (OH)

(b) butanal $\xrightarrow[\text{CH}_3\text{OH/H}_2\text{O}]{\text{NaBH}_4}$ butanol (OH)

(c) butanal $\xrightarrow[\text{Pt}]{\text{H}_2}$ butanol (OH)

(d) butanal $\xrightarrow[\text{NH}_3/\text{H}_2\text{O}]{\text{Ag(NH}_3)_2^+}$ butanoic acid (OH)

(e) butanal $\xrightarrow{\text{H}_2\text{CrO}_4}$ butanoic acid (OH)

(f) butanal $\xrightarrow[\text{H}_2/\text{Ni}]{\text{C}_6\text{H}_5\text{NH}_2}$ N-phenylbutylamine

**13.36** *Draw a structural formula for the product of the reaction of p-bromoacetophenone with each set of reagents in Problem 13.35.*

(a)

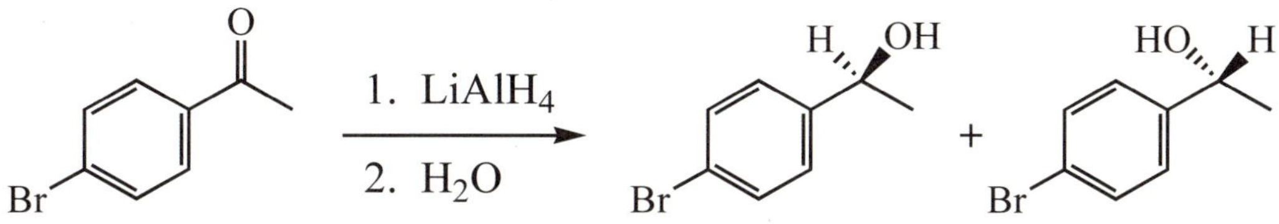

1. LiAlH$_4$
2. H$_2$O

(b)

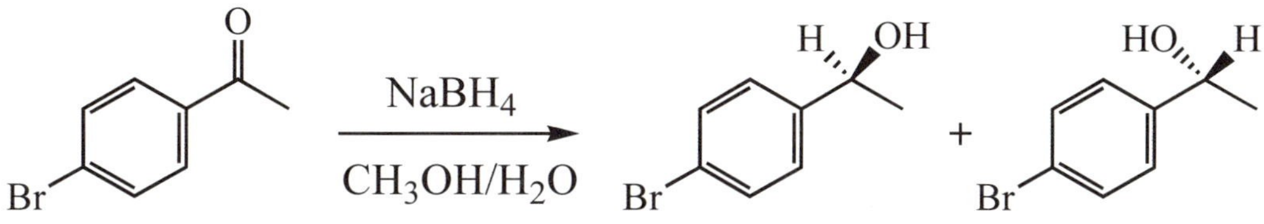

NaBH$_4$

CH$_3$OH/H$_2$O

(c)

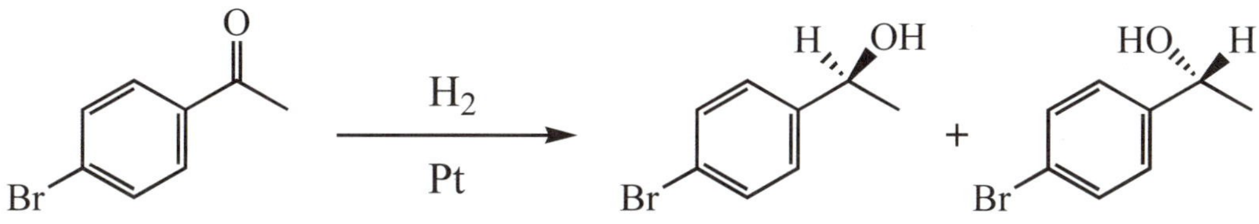

H$_2$

Pt

(d)

Ag(NH$_3$)$_2$$^+$

NH$_3$/H$_2$O

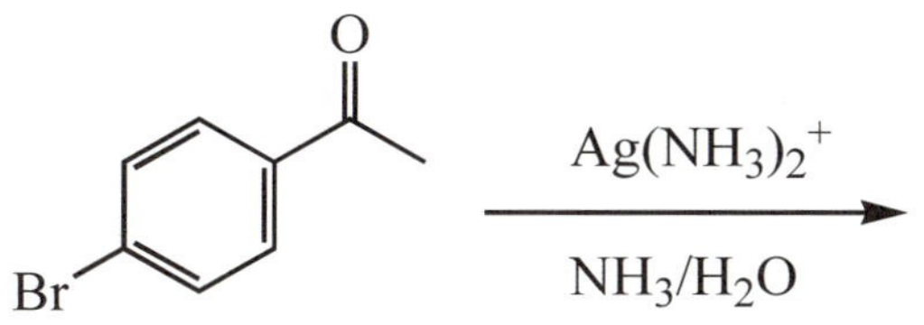

**No reaction:  Tollens' reagent only oxidizes aldehydes.**

(e)

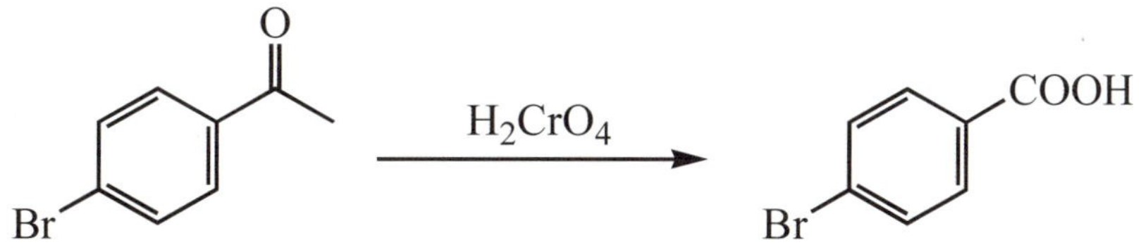

H$_2$CrO$_4$

(f)

C$_6$H$_5$NH$_2$

H$_2$/Ni

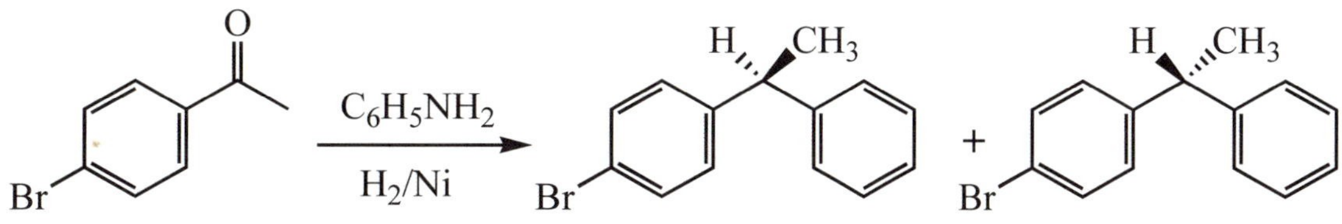

**Synthesis**

<u>**13.37**</u> *Show reagents and conditions to bring about the conversion of cyclohexanol to cyclohexanecarbaldehyde.*

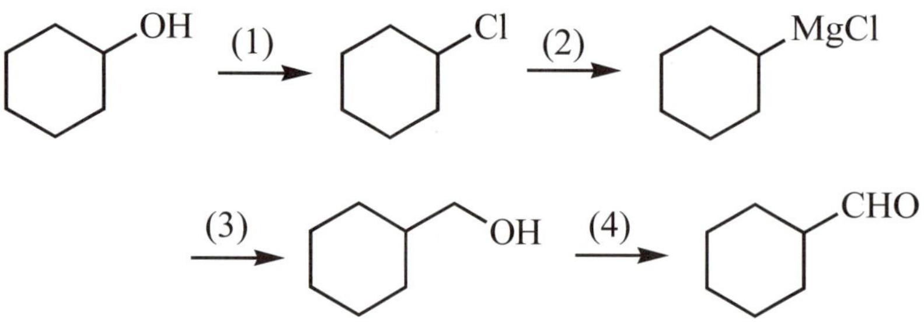
OH
(1)
Cl
(2)
MgCl
(3)
OH
(4)
CHO

**The synthetic steps are outlined below:**
**(1) $SOCl_2$ in pyridine (2) Mg in ether (3) 1.) $CH_2O$ 2.) $H_3O^+$ (4) PCC**

<u>**13.38**</u> *Starting with cyclohexanone, show how to prepare these compounds. In addition to the given starting material, use any other organic or inorganic reagents as necessary.*

*(a)*
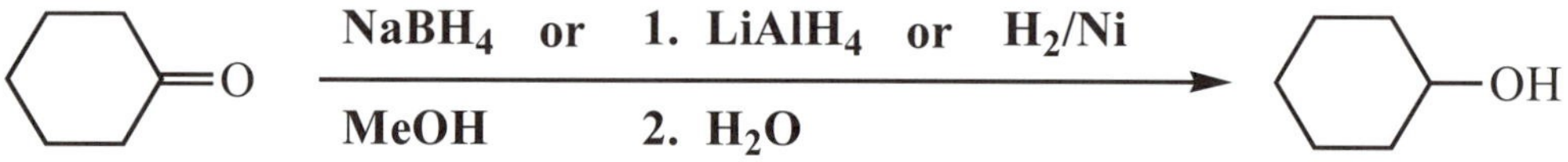
O
NaBH4 or 1. LiAlH4 or H2/Ni
MeOH 2. H2O
OH

*(b)*
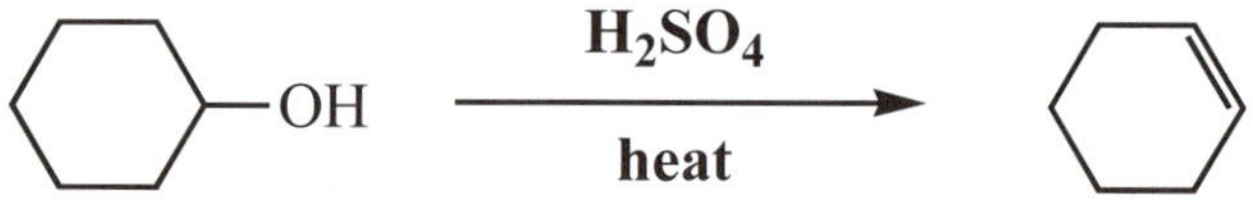
OH
H2SO4
heat

**From (a)**

*(c)*
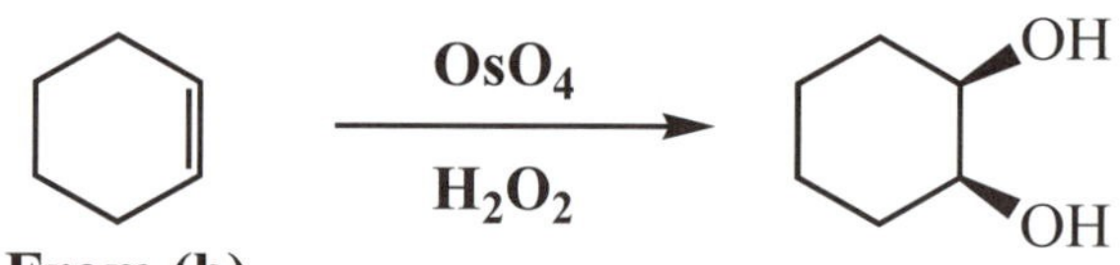
OsO4
H2O2
OH
OH

**From (b)**

*(d)*
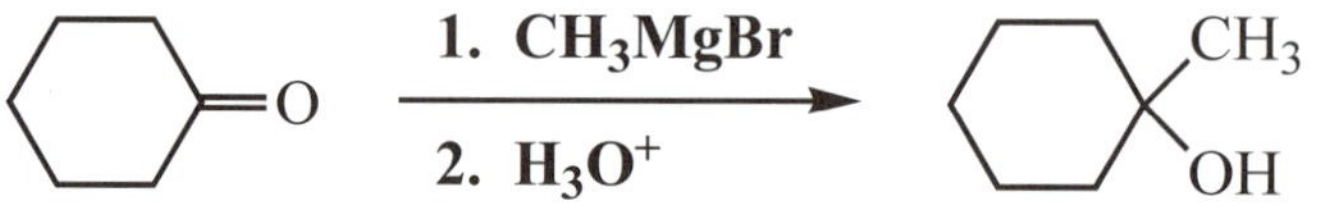
O
1. CH3MgBr
2. H3O+
CH3
OH

*(e)*
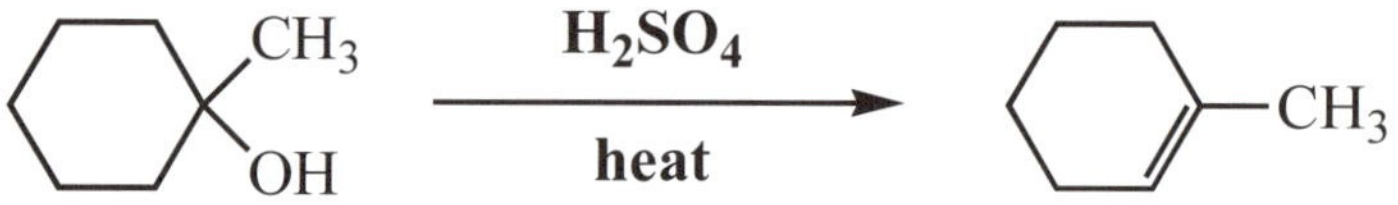
CH3
OH
H2SO4
heat
CH3

**From (d)**

*(f)* $\xrightarrow[\text{2. } H_3O^+]{\text{1. PhMgBr}}$

*(g)* **From (f)** $\xrightarrow[\text{heat}]{H_2SO_4}$

*(h)* **From (b)** $\xrightarrow{RCO_3H}$

*(i)* **From (h)** $\xrightarrow{H_3O^+}$

**13.39** *Show how to bring about these conversions.  In addition to the given starting material, use any other organic or inorganic reagents as necessary.*

*(a)* $C_6H_5\overset{\overset{\displaystyle O}{\|}}{C}CH_2CH_3$ $\xrightarrow[\text{MeOH} \quad \text{2. } H_2O]{\text{NaBH}_4 \text{ or } \text{1. LiAlH}_4 \text{ or } H_2/\text{Ni}}$ $C_6H_5\overset{\overset{\displaystyle OH}{|}}{C}HCH_2CH_3$

$\xrightarrow[\text{heat}]{H_2SO_4}$ $C_6H_5CH{=}CHCH_3$

*(b)* $\xrightarrow[\text{MeOH} \quad \text{2. } H_2O]{\text{NaBH}_4 \text{ or } \text{1. LiAlH}_4 \text{ or } H_2/\text{Ni}}$

$\xrightarrow[\text{pyridine}]{SOCl_2}$ $\xrightarrow[\substack{\text{2. } CH_2O \\ \text{3. } H_3O^+}]{\text{1. Mg/ether}}$

(c)

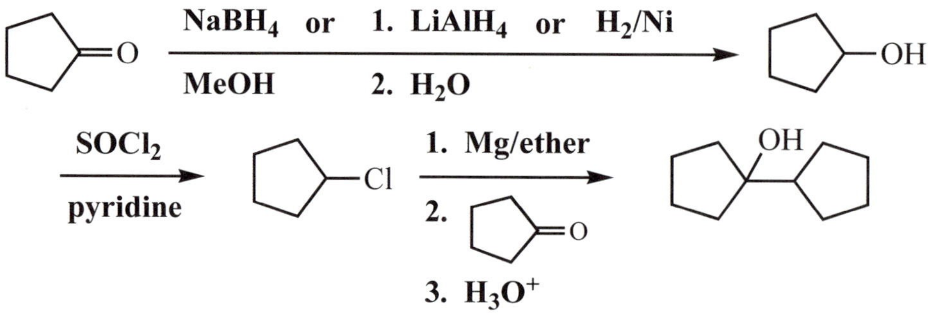

(d)

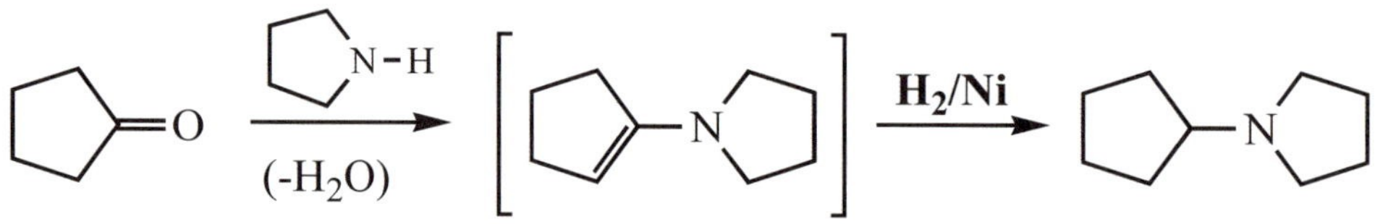

**13.40** *Many tumors of the breast are estrogen-dependent. Drugs that interfere with estrogen binding have antitumor activity and may even help prevent tumor occurrence. A widely used antiestrogen drug is tamoxifen.*

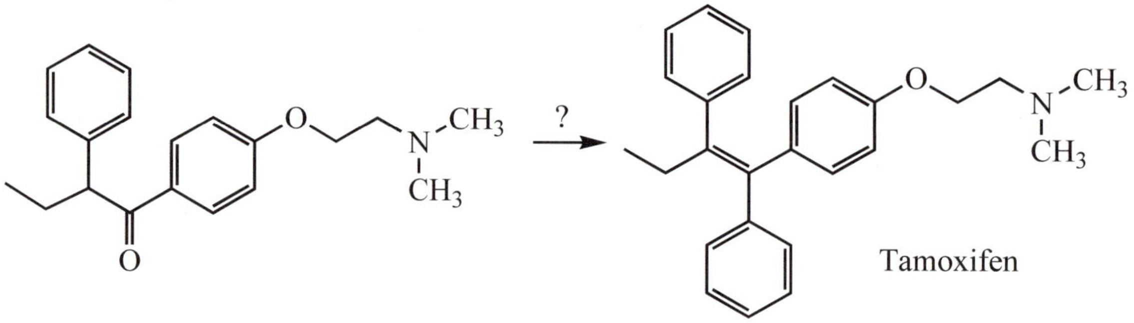

Tamoxifen

(a) *How many stereoisomers are possible for tamoxifen?*

**Cis-trans (*E*,*Z*) isomerism is possible about the carbon-carbon double bond in tamoxifen, therefore *E* and *Z* stereoisomers are possible.**

(b) *Specify the configuration of the stereoisomer shown here.*

**The alkene in tamoxifen has the *Z* configuration because the groups with the higher priority are on the same side of the double bond.**

(c) *Show how tamoxifen can be synthesized from the given ketone using a Grignard reaction followed by dehydration.*

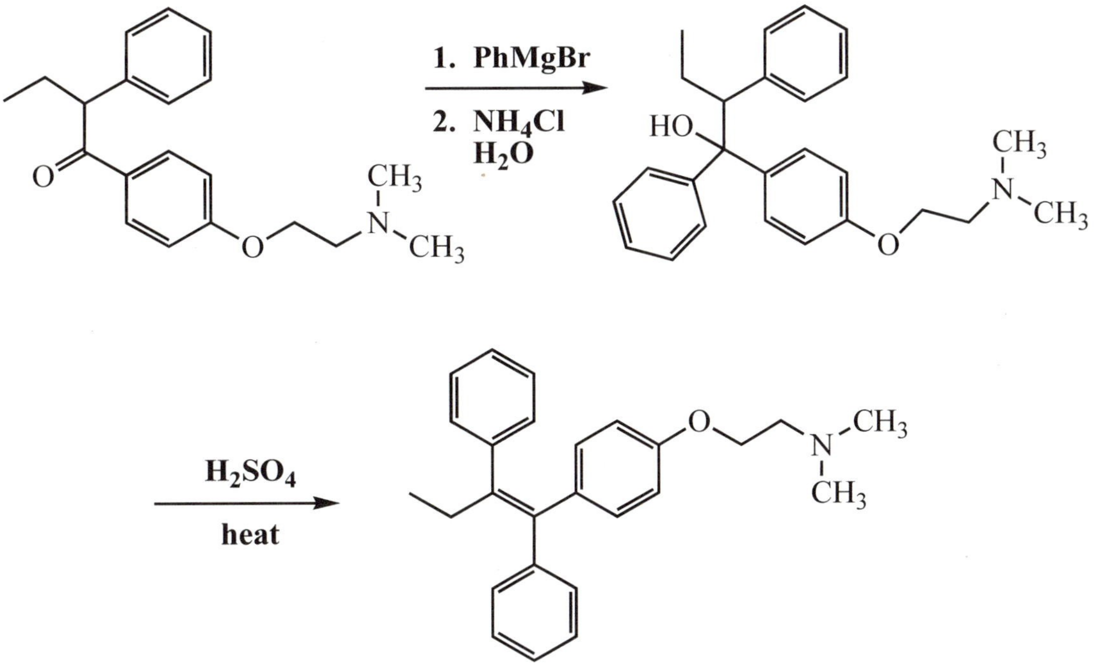

**1. PhMgBr**

**2. NH₄Cl H₂O**

$$\xrightarrow{\text{H}_2\text{SO}_4 \atop \text{heat}}$$

**13.41** *Following is a possible synthesis of the antidepressant bupropion (Wellbutrin). Show reagents to bring about each step in this synthesis.*

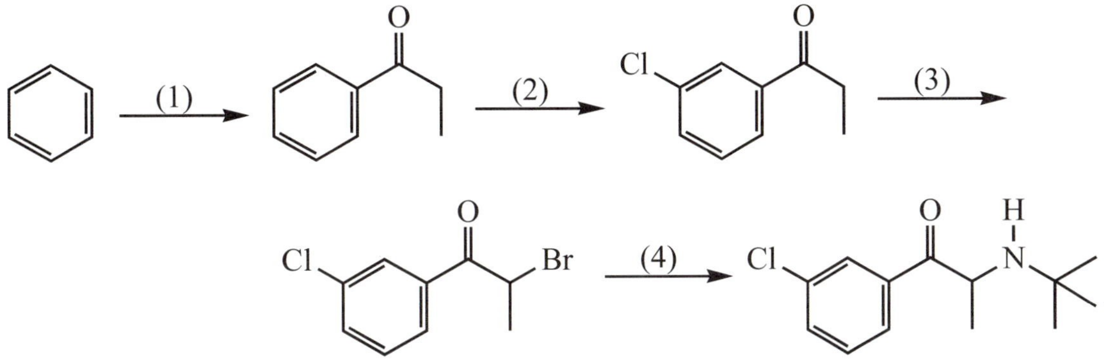

Bupropion
(Wellbutrin)

**The synthetic steps are outlined as follows:**

(1) CH₃CH₂$\overset{\overset{\text{O}}{\|}}{\text{C}}$Cl /AlCl₃     (2) Cl₂/AlCl₃     (3) Br₂/CH₃COOH     (4) 2 (CH₃)₃CNH₂

**13.42** *The synthesis of chlorpromazine in the 1950s and the discovery soon thereafter of its antipsychotic activity opened the modern era of biochemical investigations of the pharmacology of the central nervous system. One of the compounds prepared in the* *search for more effective antipsychotics was amitriptyline. Surprisingly, amitriptyline shows antidepressant activity rather than antipsychotic activity.*

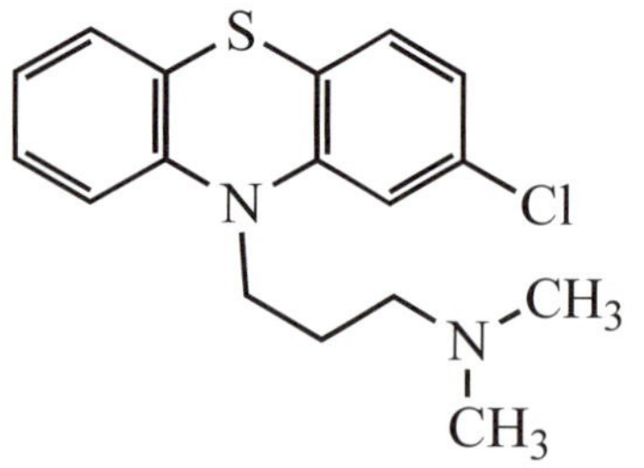

Chlorpromazine 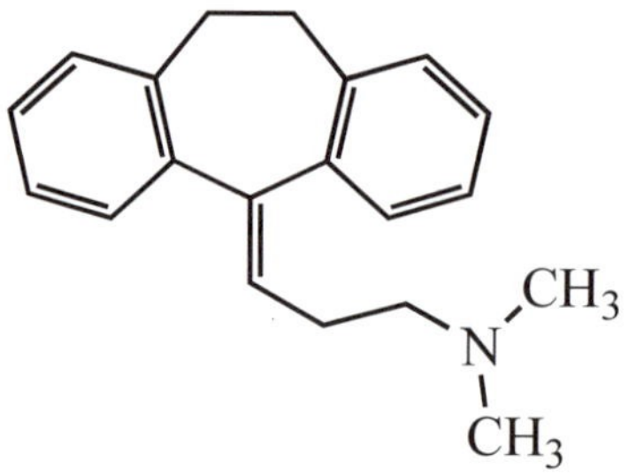                  Amitriptyline

*It is now known that amitriptyline inhibits the re-uptake of norepinephrine and serotonin from the synaptic cleft. Because the re-uptake of these neurotransmitters is inhibited, their effects are potentiated. They remain available to interact with serotonin and norepinephrine receptor sites longer and continue to cause excitation of serotonin and norepinephrine-mediated neural pathways. Following is a synthesis for amitriptyline.*

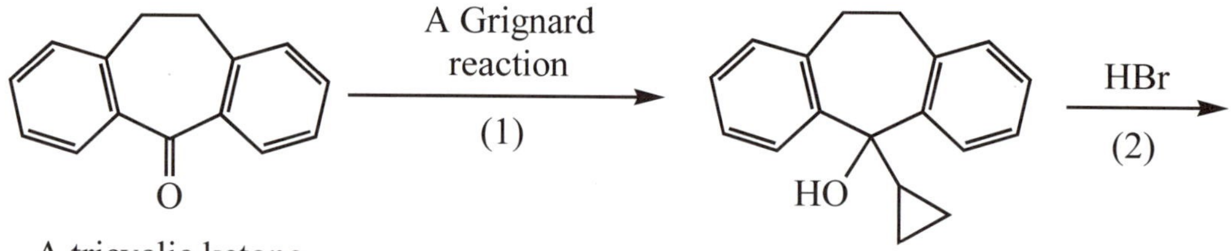

A tricyclic ketone

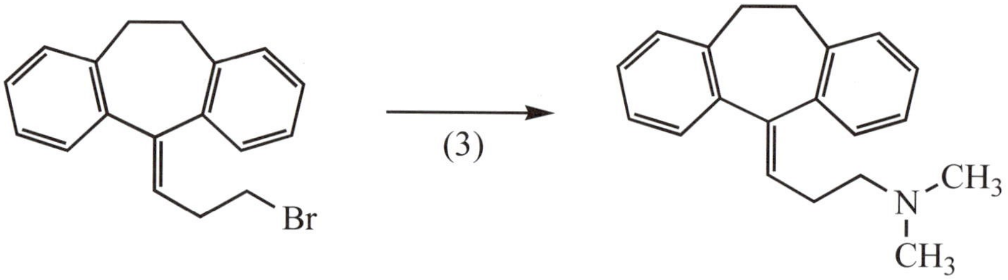

Amitriptyline

*(a) Propose a reagent for Step 1.*

1. 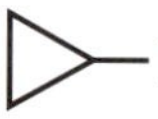  ▷—MgBr in ether   2. $H_3O^+$

*(b) Propose a mechanism for Step 2. Note: It is not acceptable to propose a primary carbocation as an intermediate.*

**Mechanism Step 1: Protonation of the hydroxyl by HBr to give a protonated alcohol intermediate**

**Mechanism Step 2: The protonated alcohol loses water to form a resonance-stabilized cation intermediate.**

**Mechanism Step 3: Attack of the bromide anion on the cyclopropyl ring of the cation intermediate leads to the product of step 2.**

*(c) Propose a reagent for Step (3).*

**Two equivalents of $(CH_3)NH$ will produce the final product amitriptyline.**

**13.43** *Following is a synthesis for diphenhydramine. The hydrochloride salt of this compound, best known by its trade name of Benadryl, is an antihistamine.*

*(a) Propose reagents for Steps 1 and 2.*

    **(1) (CH₃)₂NH**    **(2) SOCl₂/pyridine**

*(b) Propose reagents for Steps 3 and 4.*

*(c) Show that Step 5 is an example of nucleophilic aliphatic substitution.  What type of mechanism, $S_N1$ or $S_N2$ is more likely for this reaction?  Explain.*

**The substrate is a primary alkyl halide, which can only undergo an $S_N2$ reaction. To insure a facile $S_N2$ reaction, the diphenylmethanol needs to be deprotonated by a strong base to yield an alkoxide, which is a strong nucleophile.**

**13.44** *Following is a synthesis for the antidepressant venlafaxine.*

Venlafaxine

*(a) Propose a reagent for Step 1 and name the type of reaction that takes place*

**The Friedel-Crafts acylation reaction is used for Step 1 and involves:**

$$\underset{\text{CH}_3\overset{\displaystyle\text{O}}{\overset{\displaystyle\|}{\text{C}}}\text{Cl and AlCl}_3}{}$$

*(b) Propose reagents for Steps 2 and 3.*

**(2) $Cl_2$ in acetic acid      (3) 2 equiv. $(CH_3)_2NH$**

*(c) Propose reagents for Steps 4 and 5.*

**(4) $NaBH_4$ in methanol      (5) $SOCl_2$ in pyridine**

*(d) Propose a reagent for Step 6 and name the type of reaction that takes place.*

**The Grignard reaction is used to complete Step 6. The product from Step 5 is first treated with magnesium in ether. The resulting Grignard reagent is then treated with cyclohexanone. Once the carbonyl addition reaction is complete, aqueous acid is added to the alkoxide intermediate to form venlafaxine.**

**Spectroscopy**

**13.45** *Compound A, $C_5H_{10}O$, is used as a flavoring agent for many foods that possess a chocolate or peach flavor. Its common name is isovaleraldehyde and it gives $^{13}C$ NMR peaks at $\delta$ 202.7, 52.7, 23.6, and 22.6. Provide a structural formula for isovaleraldehyde and give its IUPAC name.*

22.6
23.6
22.6 52.7
O
202.7
H

**4-Methylbutanal
(Isovaleraldehyde)**

**13.46** *Compound C, $C_9H_{18}O$, is used in the automotive industry to retard the flow of solvent and thus improve the application of paints and coatings. It yields $^{13}C$ NMR peaks at $\delta$ 210.5, 52.4, 24.5, and 22.6. Provide a structure and IUPAC name for C.*

**Compound C has an index of hydrogen deficiency of one. The signal at $\delta$ 210.5 indicates an aldehyde or ketone, which accounts for its only site of hydrogen deficiency. There are only four sets of nonequivalent carbons. The $^{13}C$ data suggests that 2,6-Dimethyl-4-heptanone is the structure for compound C**

22.6          22.6
24.5          24.5
22.6  52.4      52.4  22.6
O
210.5

**2,6-Dimethyl-4-heptanone**

## Looking Ahead

**13.47** *Reaction of a Grignard reagent with carbon dioxide followed by treatment with aqueous HCl gives a carboxylic acid. Propose a structural formula for the bracketed intermediate formed by the reaction of phenylmagnesium bromide with $CO_2$ and propose a mechanism for its formation.*

MgBr
+ $CO_2$ →
[ Intermediate
(not isolated) ]
HCl, $H_2O$ →
O
C
OH

**Proposed mechanism and intermediate:**

MgBr
:O:
C
:O:
→
[
:O:
O:⁻ MgBr⁺
]

**Intermediate**

**13.48** *Rank the following carbonyls in order of increasing reactivity to nucleophilic attack. Explain your reasoning.*

**Each carbonyl compound can be described as a resonance hybrid composed of its component resonance contributing structures. The ketone (2-butanone) has an electron deficient carbon represented in its resonance structure.**

**The other two carbonyl compounds, an amide (*N*,*N*-dimethylacetamide) and an ester (methyl acetate) have additional resonance structures that involve the participation of electron lone pairs delocalized by the carbonyl oxygen, which stabilizes the compound and causes the carbonyl carbon to be less electrophilic, thus less reactive toward nucleophilic attack. As the participation of the lone pairs in resonance stabilization increases, the carbonyl carbon becomes less electrophilic. Nitrogen atoms are less electronegative than oxygen atoms, thus nitrogen lone pairs are more available for resonance stabilization and leave amides less reactive to nucleophilic attack than esters.**

**Based on the above reasoning, the order of increasing reactivity to nucleophilic attack is:**

**13.49** *Provide the enol form of this ketone and predict the direction of equilibrium.*

**The equilibrium favors the right (the enol phenol). The keto tautomer is usually favored in a keto-enol equilibrium, but the more stable phenol is favored in this equilibria because of its resonance and aromatic stabilization.**

**13.50** *Draw the cyclic hemiacetal formed by reaction of the highlighted -OH group with the aldehyde group.*

**13.51** *Propose a mechanism for the acid catalyzed reaction of the following hemiacetal with an amine acting as a nucleophile.*

**We will study these reactions further in Chapter 18.**

**Step 1:  Protonation of the hemiacetal hydroxyl yields a protonated hemiacetal.**

**Step 2:  Loss of H₂O forms the resonance-stabilized cation intermediate.**

**Step 3:  Nucleophilic attack by ethylamine on the cation intermediate.**

**Step 4:  Deprotonation of ammonium cation intermediate forms product.**

## CHAPTER 14
### *Solutions to Problems*

**14.1**  *Each of these compounds has a well-recognized common name.  A derivative of glyceric acid is an intermediate in glycolysis (Section 22.6).  Maleic acid is an intermediate in the tricarboxylic acid (TCA) cycle.  Mevalonic acid is an intermediate in the biosynthesis of steroids (Section 19.4).  Write the IUPAC name for each compound.  Be certain to show the configuration for each.*

(a)

Glyceric acid

**(*R*)-2,3-Dihydroxypropanoic acid**

(b)

Maleic acid

***cis*-2-Butenedioic acid**
**or (*Z*)-Butenedioic acid**

(c)

Mevalonic acid

**(*R*)-3,5-Dihydroxy-3-methylpentanoic acid**

**14.2**  *Match each compound with its appropriate $pK_a$ value.*

**2,2-Dimethylpropanoic acid has a $pK_a$ comparable to that of an unsubstituted aliphatic carboxylic acid.  Lactic acid is a stronger acid for two reasons:  the electron-withdrawing inductive effect of the hydroxyl group on an $sp^3$ carbon atom adjacent to the carboxyl group and the fact that the hydroxyl group can also stabilize the carboxylate anion of the deprotonated carboxylic acid via an intramolecular hydrogen bond.  Trifluoroacetic acid is an even stronger acid because of the combined inductive effects of the three fluorine atoms.**

|  | $CH_3$ | | OH |
|---|---|---|---|
|  | $CH_3CCOOH$ | $CF_3COOH$ | $CH_3CHCOOH$ |
|  | $CH_3$ | | |
|  | 2,2-Dimethyl-propanoic acid | Trifluoro-acetic acid | 2-Hydroxy-propanoic acid (Lactic acid) |
| $pK_a$ | 5.03 | 0.22 | 3.08 |

**14.3**   *Write an equation for the reaction of each acid in Example 14.3 with ammonia and name the salt formed.*

(a)   ![structure] $\sim\sim$COOH  +  NH$_3$  $\longrightarrow$  $\sim\sim$COO$^-$NH$_4{}^+$

Butanoic acid                         **Ammonium butanoate**

(b)   2-Hydroxypropanoic acid structure  + NH$_3$ $\longrightarrow$  Ammonium 2-hydroxypropanoate structure

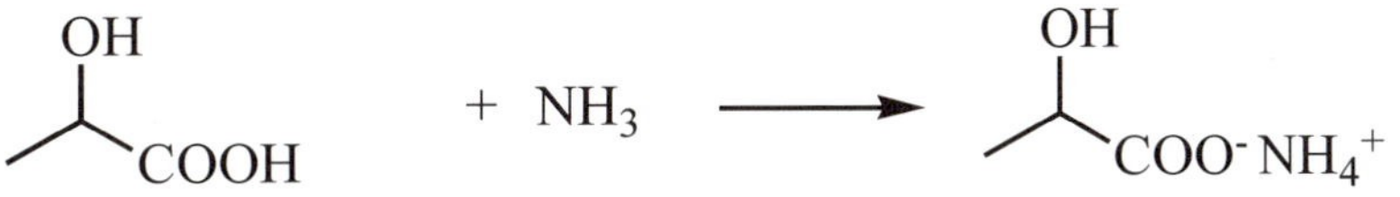

2-Hydroxypropanoic acid          **Ammonium 2-hydroxypropanoate**
(Lactic acid)                         **(Ammonium lactate)**

**14.4**   *Complete these Fischer esterification reactions:*

(a)   structure  +  HO—cyclohexyl  $\underset{}{\overset{H^+}{\rightleftharpoons}}$  ester structure  +  H$_2$O

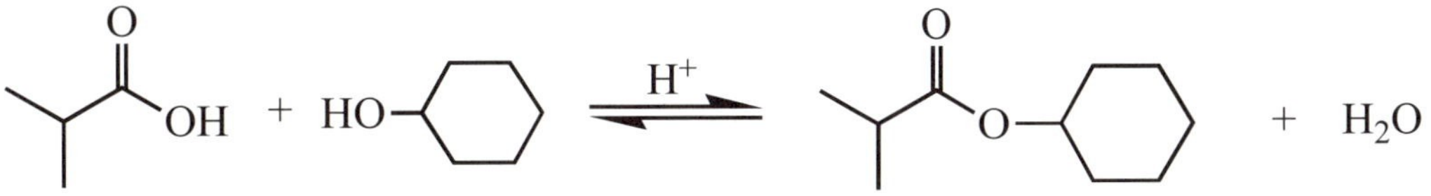

(b)   HO $\sim\sim$ OH  $\overset{H^+}{\rightleftharpoons}$  lactone structure  +  H$_2$O

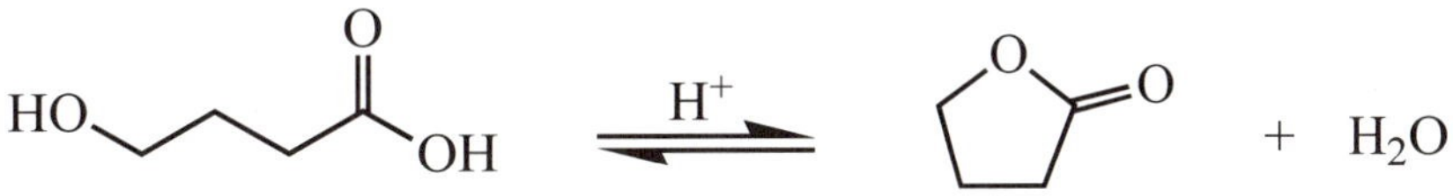

**14.5**   *Complete each equation:*

(a)   (benzene ring with COOH and OCH$_3$)  +  SOCl$_2$  $\longrightarrow$  (benzene ring with CCl=O and OCH$_3$)  +  SO$_2$  +  HCl

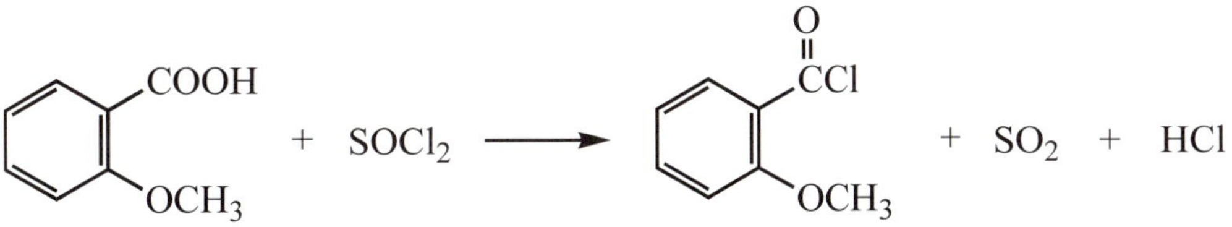

(b)   (cyclohexyl-OH)  +  SOCl$_2$  $\longrightarrow$  (cyclohexyl-Cl)  +  SO$_2$  +  HCl

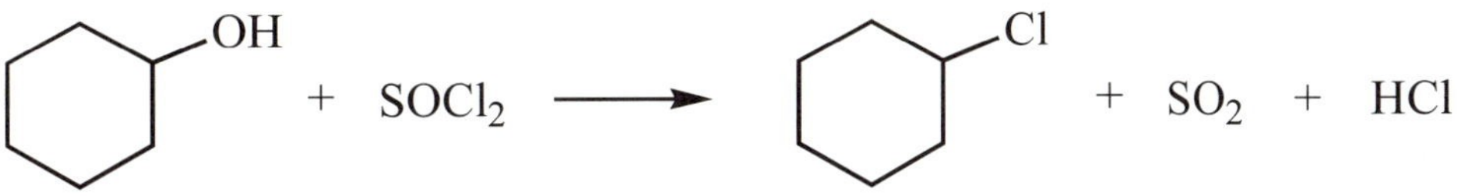

**14.6**  *Draw the structural formula for the indicated β-ketoacid.*

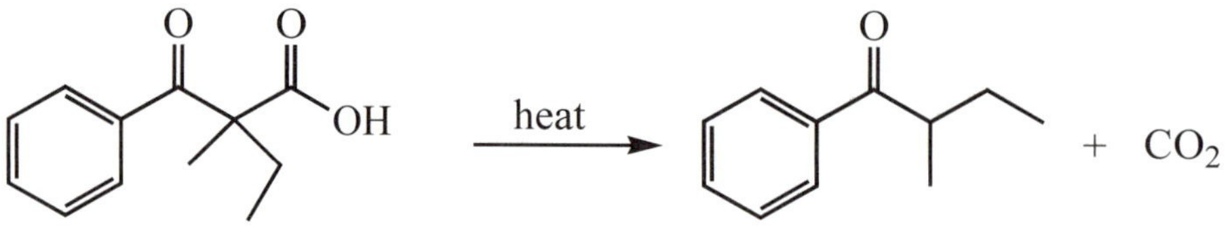

β-ketoacid

## Structure and Nomenclature

**14.7**  *Name and draw structural formulas for the four carboxylic acids with molecular formula C₅H₁₀O₂. Which of these carboxylic acids is chiral?*

$C_5H_{10}O_2$. *Which of these carboxylic acids is chiral?*

**There are four isomers of $C_5H_{10}O_2$ that are carboxylic acids, of which one is chiral. The stereocenter is indicated with an asterisk.**

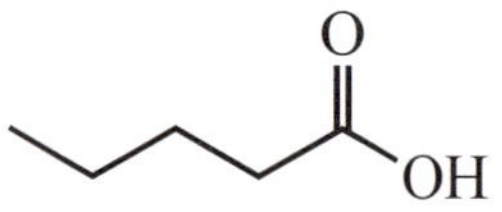

**Pentanoic acid** 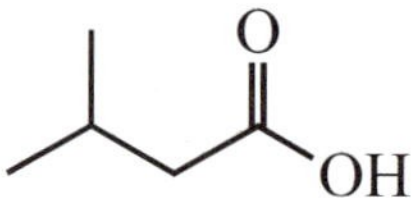                    **3-Methylbutanoic acid**

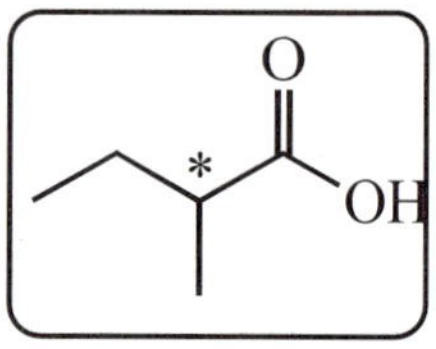

**2-Methylbutanoic acid** 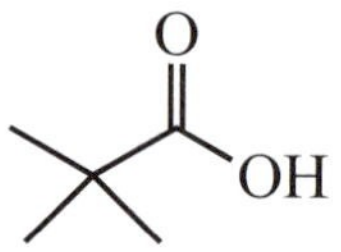            **2,2-Dimethylpropanoic acid**

**14.8**  *Write the IUPAC name for each compound.*

(a)  **1-Cyclohexenecarboxylic acid**

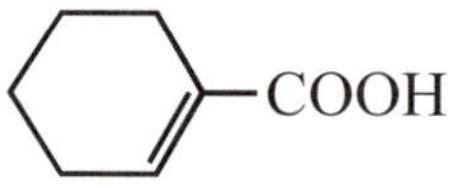

(b)  **4-Hydroxypentanoic acid**

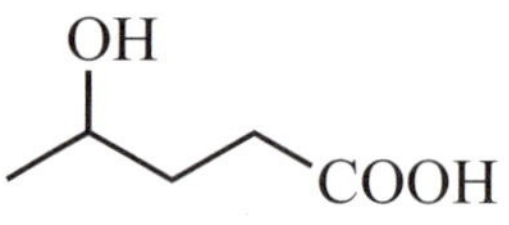

(c)  **(2E) 3,7--Dimethyl-2,6-octadienoic acid**

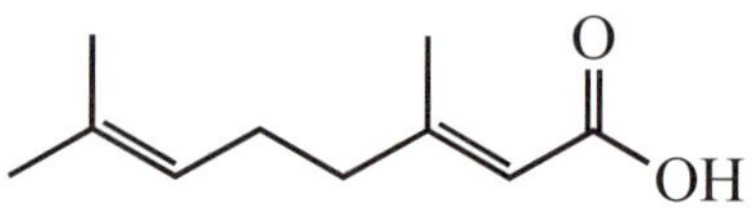

(d)  **1-Methylcyclopentane-carboxylic acid**

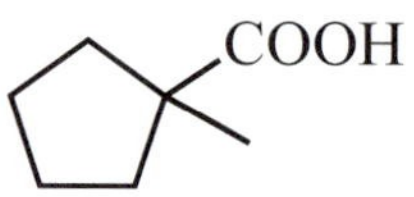

(e)  **Ammonium hexanoate**

(f)  **2-Hydroxybutanedioic acid**

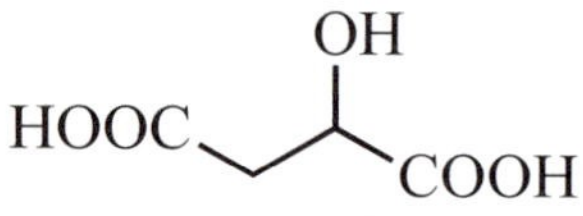

**14.9**  *Draw a structural formula for each carboxylic acid.*

*(a) 4-Nitrophenylacetic acid*

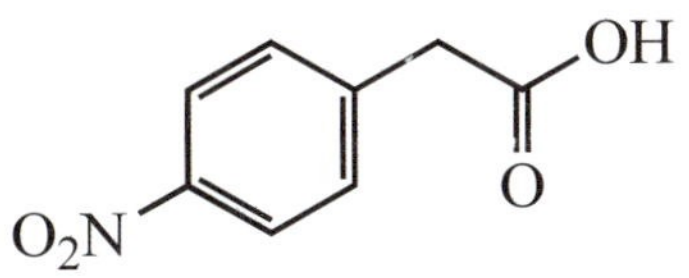

*(b) 4-Aminopentanoic acid*

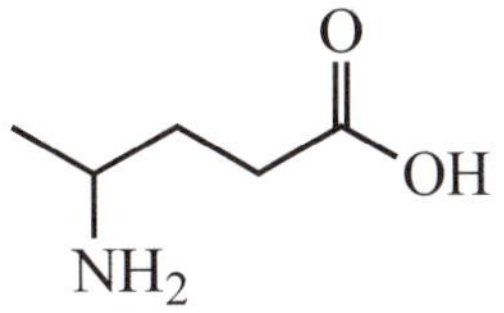

*(c) 3-Chloro-4-phenylbutanoic acid*

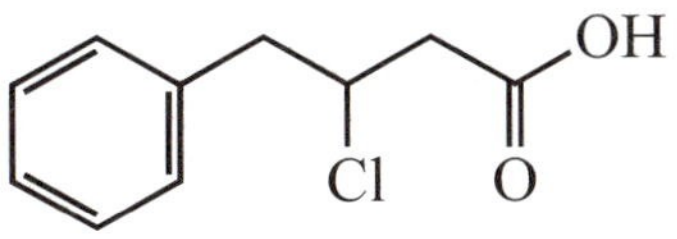

*(d) cis-3-Hexenedioic acid*

*(e) 2,3-Dihydroxypropanoic acid*

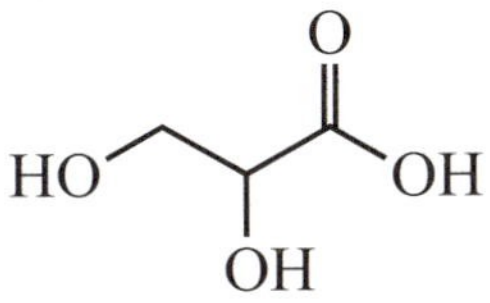

*(f) 3-Oxohexanoic acid*

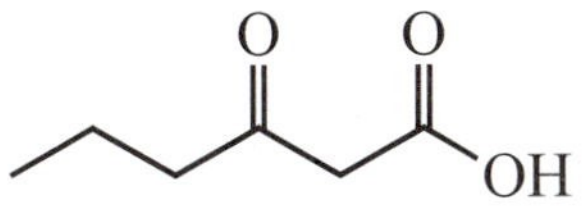

*(g) 2-Oxocyclohexanecarboxylic acid*

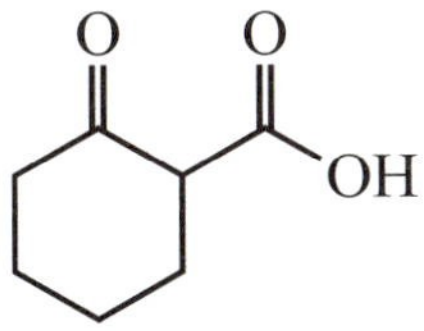

*(h) 2,2-Dimethylpropanoic acid*

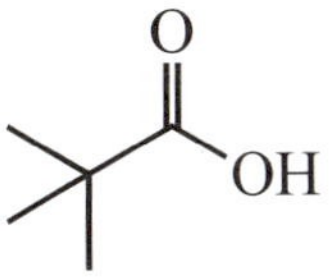

**14.10**  *Megatomoic acid, the sex attractant of the female black carpet beetle, has the structure*

$$CH_3(CH_2)_7CH=CHCH=CHCH_2COOH$$

Megatomoic acid

*(a) What is its IUPAC name?*

**Its IUPAC name is 3,5-Tetradecadienoic acid.**

*(b) State the number of stereoisomers possible for this compound.*

**Megatomoic acid has two carbon-carbon double bonds and each can have an *E* or *Z* (trans or cis) configuration; therefore, it has four possible stereoisomers.**

**14.11**  *The IUPAC name of ibuprofen is 2-(4-isobutylphenyl)propanoic acid.  Draw a structural formula of ibuprofen.*

**2-(4-Isobutylphenyl)propanoic acid**
**(Ibuprofen)**

**14.12**  *Draw structural formulas for these salts.*

*(a) Sodium benzoate*         *(b) Lithium acetate*         *(c) Ammonium  acetate*

$CH_3CO^-Li^+$         $CH_3CO^-NH_4^+$

*(d)       Disodium adipate*         *(e) Sodium salicylate*         *(f) Calcium butanoate*

**14.13**  *The monopotassium salt of oxalic acid is present in certain leafy vegetables, including rhubarb.  Both oxalic acid and its salts are poisonous in high concentrations.  Draw a structural formula of monopotassium oxalate.*

**Monopotassium oxalate**

**14.14** *Potassium sorbate is added as a preservative to certain foods to prevent bacteria and molds from causing food spoilage and to extend the foods' shelf life. The IUPAC name of potassium sorbate is potassium (E,E)-2,4-hexadienoate. Draw a structural formula of potassium sorbate.*

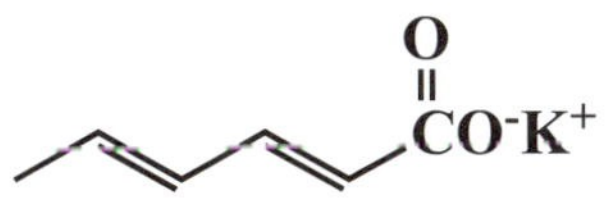

**Potassium (2E,4E)-2,4-hexadienoate**
**(Potassium sorbate)**

**14.15** *Zinc 10-undecenoate, the zinc salt of 10-undecenoic acid, is used to treat certain fungal infections, particularly tinea pedis (athlete's foot). Draw a structural formula of this zinc salt.*

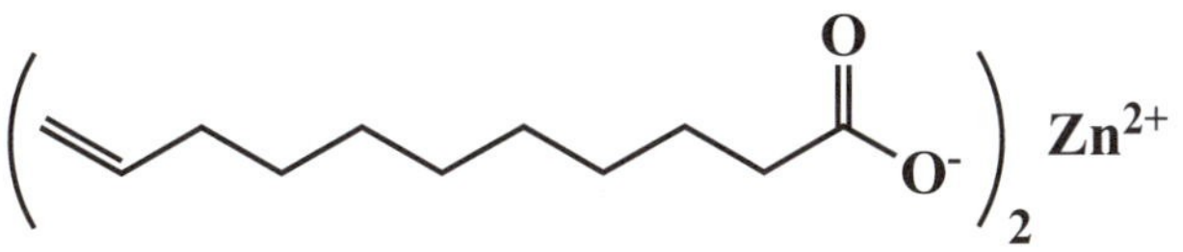

**Zinc 10-undecenoate**

## Physical Properties

**14.16** *Arrange the compounds in each set in order of increasing boiling point:*

**As the hydrogen bonding capability increases, so does the boiling point. Both carboxylic acids and alcohols act as hydrogen bond donors and acceptors, thus have higher capacities to hydrogen bond than ketones, aldehydes, and ethers. Carboxylic acids have a greater degree of hydrogen bonding due to the highly polar nature of the carboxyl group. Predicting and comparing boiling points is most useful with a series of compounds of similar molecular weight.**

(a) $CH_3(CH_2)_5COOH$       $CH_3(CH_2)_6CHO$      $CH_3(CH_2)_6CH_2OH$

    **$CH_3(CH_2)_6CHO$**  <  **$CH_3(CH_2)_6CH_2OH$**  <  **$CH_3(CH_2)_5COOH$**

     **bp 171 ûC**             **bp 195 ûC**            **bp 223 ûC**

(b) $CH_3CH_2COOH$      $CH_3CH_2CH_2CH_2OH$      $CH_3CH_2OCH_2CH_3$

    **$CH_3CH_2OCH_2CH_3$**  <  **$CH_3CH_2CH_2CH_2OH$**  <  **$CH_3CH_2COOH$**

     **bp 35 ûC**             **bp 117 ûC**            **bp 141 ûC**

## Preparation of Carboxylic Acids

**14.17** *Draw a structural formula for the product formed by treating each compound with warm chromic acid, $H_2CrO_4$.*

(a) 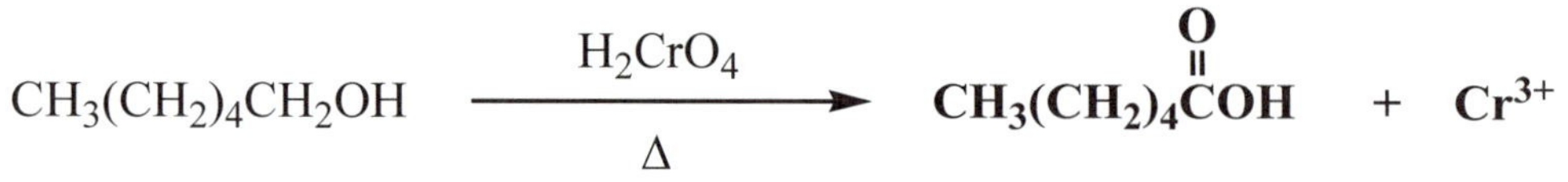 $CH_3(CH_2)_4CH_2OH \xrightarrow[\Delta]{H_2CrO_4} CH_3(CH_2)_4\overset{\overset{O}{\|}}{C}OH + Cr^{3+}$

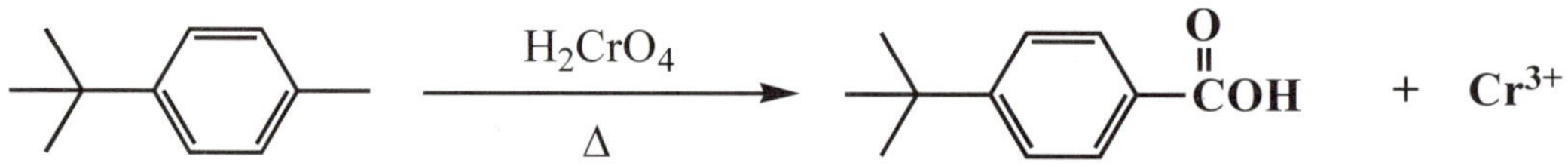

(b)

$\xrightarrow[\Delta]{H_2CrO_4}$

$+ Cr^{3+}$

(c) 

$\xrightarrow[\Delta]{H_2CrO_4}$

$+ Cr^{3+}$

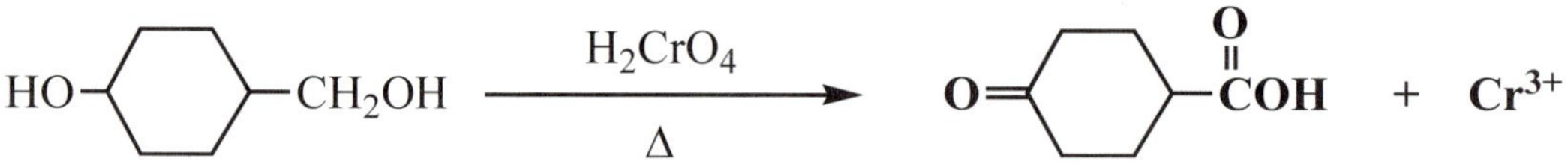

**14.18** *Draw a structural formula for a compound of the given molecular formula that, on oxidation by chromic acid, gives the carboxylic acid or dicarboxylic acid shown.*

(a) 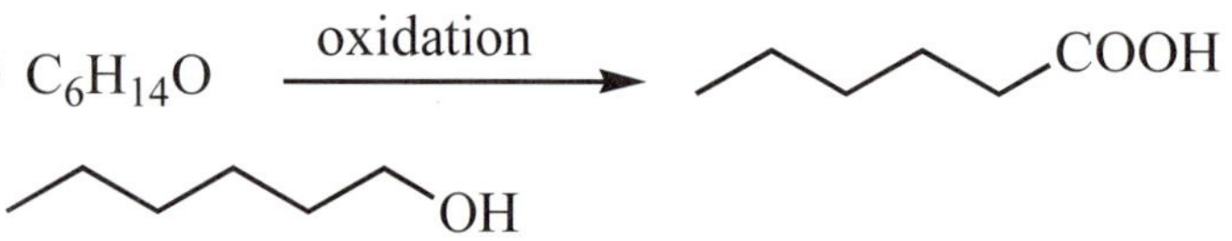 $C_6H_{14}O$ $\xrightarrow{\text{oxidation}}$

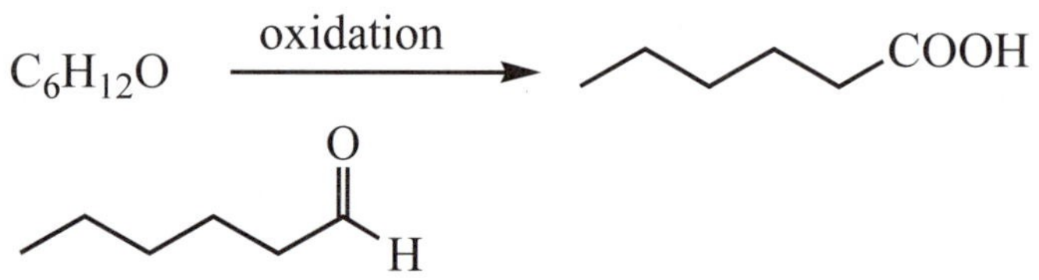

(b) $C_6H_{12}O$ $\xrightarrow{\text{oxidation}}$

(c) $C_6H_{14}O_2$ $\xrightarrow{\text{oxidation}}$

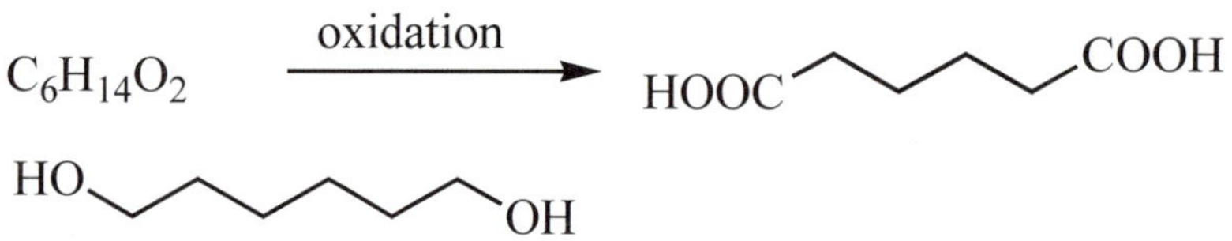

## Acidity of Carboxylic Acids

**14.19** *Which is the stronger acid in each pair?*

(a) *Phenol (p$K_a$ 9.95) or* **benzoic acid (p$K_a$ 4.17)**

**Recall that p$K_a$ is the -$\log_{10}$ of $K_a$.  As p$K_a$ decreases, acid strength increases; therefore benzoic acid is the stronger acid.**

*(b)* **Lactic acid ($K_a$ 8.4 x 10$^{-4}$)** *or ascorbic acid ($K_a$ 7.9 x 10$^{-5}$)*

As $K_a$ increases, acid strength also increases; therefore, lactic acid is the stronger acid.

**14.20** *Arrange these compounds in order of increasing acidity: benzoic acid, benzyl alcohol, and phenol.*

It is important to remember the ballpark $pK_a$ values for alcohols ($pK_a$ 16-18), phenols ($pK_a$ 10-11), and carboxylic acids ($pK_a$ 4-5). Using these values, an order of acidity can be can be established for the following compounds:

benzyl alcohol  <  phenol  <  benzoic acid

**14.21** *Assign the acid in each set its appropriate $pK_a$.*

*(a)* COOH and COOH—NO$_2$

$pK_a$  4.19        3.14

*(b)* COOH—NO$_2$ and COOH—NH$_2$

$pK_a$  3.14        4.92

*(c)* CH$_3$CCH$_2$COOH and CH$_3$CCOOH

$pK_a$  3.58        2.49

*(d)* CH$_3$CHCOOH (OH) and CH$_3$CH$_2$COOH

$pK_a$  3.08        4.78

**14.22** *Complete these acid-base reactions:*

*(a)* —CH$_2$COOH + NaOH ⟶ —CH$_2$COO$^-$Na$^+$ + H$_2$O

*(b)* CH$_3$CH=CHCH$_2$COOH + NaHCO$_3$ ⟶

CH$_3$CH=CHCH$_2$COO$^-$Na$^+$ + H$_2$O + CO$_2$

*(c)* [structure: benzene ring with COOH and OH groups] + NaHCO$_3$ $\longrightarrow$ [structure: benzene ring with COO$^-$Na$^+$ and OH groups] + H$_2$O + CO$_2$

*(d)* CH$_3$$\overset{\text{OH}}{\underset{|}{\text{CH}}}$COOH + H$_2$NCH$_2$CH$_2$OH $\longrightarrow$ CH$_3$$\overset{\text{OH}}{\underset{|}{\text{CH}}}$COO$^-$ + H$_3$$\overset{+}{\text{N}}$CH$_2$CH$_2$OH

*(e)* CH$_3$CH=CHCH$_2$COO$^-$Na$^+$ + HCl $\longrightarrow$ CH$_3$CH=CHCH$_2$COOH + Na$^+$Cl$^-$

**14.23** *The normal pH range for blood plasma is 7.35 - 7.45. Under these conditions, would you expect the carboxyl group of lactic acid (pK$_a$ 3.08) to exist primarily as a carboxyl group or as a carboxylate anion? Explain.*

**Recall that:**

$$K_a = \frac{[A^-][H^+]}{[HA]} \quad \text{so dividing both sides by } [H^+] \text{ gives} \quad \frac{[A^-]}{[HA]} = \frac{K_a}{[H^+]}$$

**Here, the [H$^+$] can be calculated from the pH, [HA] is the concentration of lactic acid, and [A$^-$] is the concentration of the carboxylate conjugate base. If the ratio of $K_a/H^+$ < 1, then [HA] will be the predominate form. Recall that pH = -log$_{10}$[H$^+$], so a pH of 7.4 corresponds to a [H$^+$] = 4.0 x 10$^{-8}$ M.**

$$pK_a = -\log_{10} K_a \quad \text{so for lactic acid,} \quad K_a = 10^{-(Ka)} = 8.4 \times 10^{-4}$$

$$\frac{[A^-]}{[HA]} = \frac{K_a}{[H^+]} = \frac{8.4 \times 10^{-4}}{4.0 \times 10^{-8}} = 2.1 \times 10^4$$

**Based on the calculations, lactic acid will exist primarily as the carboxylate anion in blood plasma.**

**14.24** *The pK$_a$ of ascorbic acid (Section 16.6) is 4.76. Would you expect ascorbic acid dissolved in blood plasma, pH 7.35 - 7.45, to exist primarily as ascorbic acid or as ascorbate anion? Explain.*

**Using the relationships and calculations outlined in Problem 14.23, at pH = 7.4, [H$^+$] = 4.0 x 10$^{-8}$:**

$$\frac{[A^-]}{[HA]} = \frac{K_a}{[H^+]} = \frac{1.7 \times 10^{-5}}{4.0 \times 10^{-8}} = 4.3 \times 10^2$$

**Based on the calculations, ascorbic acid will exist primarily as the carboxylate anion in blood plasma.**

**14.25** *Excess ascorbic acid is excreted in the urine, the pH of which is normally in the range 4.8 - 8.4. What form of ascorbic acid would you expect to be present in urine of pH 8.4, ascorbic acid or ascorbate anion?*

**Using the relationships and calculations outlined in Problem 14.23.  At pH = 8.4, $[H^+] = 4.0 \times 10^{-9}$:**

$$\frac{[A^-]}{[HA]} = \frac{K_a}{[H^+]} = \frac{7.9 \times 10^{-5}}{4.0 \times 10^{-9}} = 2.0 \times 10^4$$

**Ascorbic acid will exist primarily as the ascorbate anion in urine at pH of 8.4.**

**14.26** *The pH of human gastric juice is normally in the range 1.0 - 3.0.  What form of lactic acid (pK$_a$ 3.08), would you expect to be present in the stomach, lactic acid or its anion?*

**Using the relationships and calculations outlined in Problem 14.23, at pH = 3.0, $[H^+] = 1.0 \times 10^{-3}$:**

$$\frac{[A^-]}{[HA]} = \frac{K_a}{[H^+]} = \frac{8.4 \times 10^{-4}}{1.0 \times 10^{-3}} = 0.85$$

**Based on the calculations, lactic acid will exist approximately 50% as the protonated form and 50% as the lactate anion in gastric juices at pH < 3.0.  At lower pH values, the protonated form will predominate.**

**14.27** *Following are two structural formulas for the amino acid alanine (Section 20.1).  Is alanine better represented by structural formula A or B? Explain.*

$$CH_3-\underset{\underset{NH_2}{|}}{CH}-\overset{\overset{O}{\|}}{C}-OH \qquad\qquad CH_3-\underset{\underset{NH_3^+}{|}}{CH}-\overset{\overset{O}{\|}}{C}-O^-$$

(A)                                    (B)

**Amino acids have both acidic and basic components on the same molecule and exist as the internal salt called a zwitterion (structure B).**

**14.28** *In Chapter 19, we discuss a class of compounds called amino acids, so named because they contain both an amino group and a carboxyl group. Following is a structural formula for the amino acid alanine in the form of an internal salt.*

$$
\begin{array}{c}
O \\
\parallel \\
CH_3CHCO^- \\
\mid \\
NH_3^+
\end{array}
\quad \text{Alanine}
$$

*What would you expect to be the major form of alanine present in aqueous solution at:*

*(a) pH 2.0*

**Under a strongly acidic pH, all of the basic groups on an amino acid will be fully protonated.**

$$
\begin{array}{c}
O \\
\parallel \\
CH_3CHCOH \\
\mid \\
NH_3^+
\end{array}
$$

*(b)  pH 5-6*

**At a pH close to neutral, an amino acid with one carboxylic acid group and one amine group will exist as the internal salt.**

$$
\begin{array}{c}
O \\
\parallel \\
CH_3CHCO^- \\
\mid \\
NH_3^+
\end{array}
$$

*(c) pH 11.0.*

**Under a strongly alkaline pH, all of the acidic groups on an amino acid will be fully deprotonated.**

$$
\begin{array}{c}
O \\
\parallel \\
CH_3CHCO^- \\
\mid \\
NH_2
\end{array}
$$

**Reactions of Carboxylic Acids**

**14.29** *Give the expected organic products formed when phenylacetic acid, PhCH$_2$COOH, is treated with each reagent.*

(a) PhCH$_2$COH $\xrightarrow{\text{SOCl}_2}$ PhCH$_2$CCl + SO$_2$ + HCl

(b) PhCH$_2$COH $\xrightarrow{\text{NaHCO}_3,\ \text{H}_2\text{O}}$ PhCH$_2$CO$^-$ Na$^+$ + CO$_2$ + H$_2$O

(c) PhCH$_2$COH $\xrightarrow{\text{NaOH,\ H}_2\text{O}}$ PhCH$_2$CO$^-$ Na$^+$ + H$_2$O

(d) PhCH$_2$COH $\xrightarrow{\text{NH}_3,\ \text{H}_2\text{O}}$ PhCH$_2$CO$^-$ NH$_4^+$

(e) PhCH$_2$COH $\xrightarrow[\text{2. H}_2\text{O}]{\text{1. LiAlH}_4}$ PhCH$_2$CH$_2$OH

(f) PhCH$_2$COH $\xrightarrow[\text{2. H}_2\text{O}]{\text{1. NaBH}_4}$ **no reaction**

(g) PhCH$_2$COH $\xrightarrow[\text{H}_2\text{SO}_4\ (\text{catalyst})]{\text{CH}_3\text{OH}}$ PhCH$_2$COCH$_3$ + H$_2$O

(h) PhCH$_2$COH $\xrightarrow[\text{3 atm pressure}]{\text{H}_2/\text{Ni at 25°C}}$ **no reaction**

**14.30** *Show how to convert trans-3-phenyl-2-propenoic acid (cinnamic acid) to these compounds.*

(a) Ph—CH=CH—COOH $\xrightarrow[\text{2. H}_2\text{O}]{\text{1. LiAlH}_4}$ Ph—CH=CH—CH$_2$OH

(b) Ph—CH=CH—COOH $\xrightarrow[\text{Pt 25°C \ 2 atm}]{\text{H}_2}$ Ph—CH$_2$CH$_2$—COOH

*(c)*

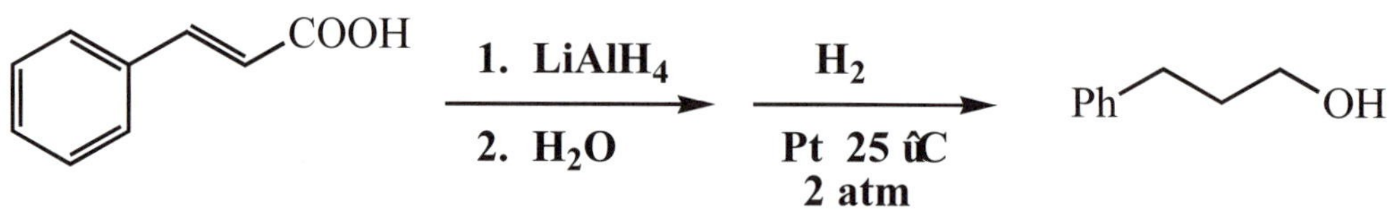


**14.31** *Show how to convert 3-oxobutanoic acid (acetoacetic acid) to these compounds.*

*(a)* $CH_3C(=O)CH_2COOH \xrightarrow[\textbf{MeOH}]{\textbf{NaBH}_4} CH_3CH(OH)CH_2COOH$

*(b)* $CH_3C(=O)CH_2COOH \xrightarrow[\textbf{2. H}_2\textbf{O}]{\textbf{1. LiAlH}_4} CH_3CH(OH)CH_2CH_2OH$

*(c)* $CH_3C(=O)CH_2COOH \xrightarrow[\textbf{MeOH}]{\textbf{NaBH}_4} CH_3CH(OH)CH_2COOH \xrightarrow[\textbf{heat}]{\textbf{H}_2\textbf{SO}_4} CH_3CH{=}CHCOOH$

**14.32** *Complete these examples of Fischer esterification. Assume an excess of the alcohol.*

*(a)*

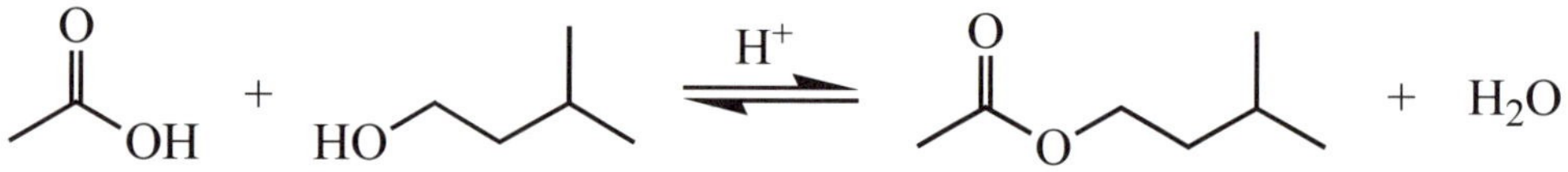


*(b)*

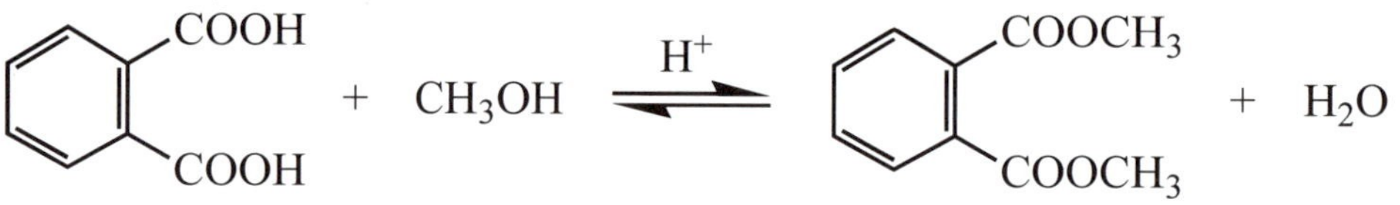


*(c)*

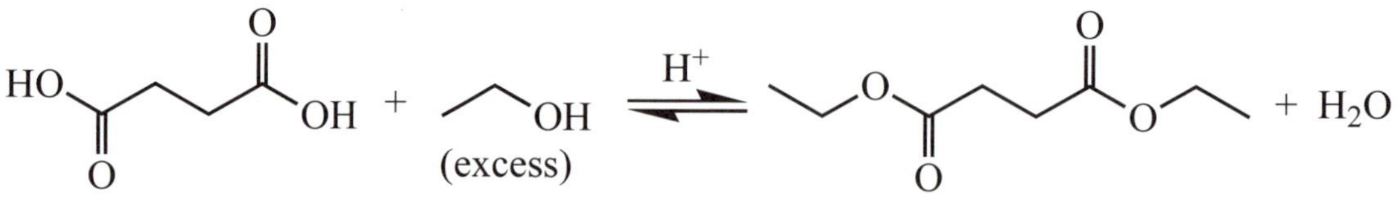


**14.33** *Formic acid is one of the components responsible for the sting of biting ants and is injected under the skin by bee and wasp stings. A way to relieve the pain is to rub the area of the sting with a paste of baking soda ($NaHCO_3$) and water, which neutralizes the acid. Write an equation for this reaction.*

**The sodium bicarbonate acts as a base and deprotonates the formic acid to form sodium formate, carbon dioxide, and water. This reaction is valid for all carboxylic acids.**

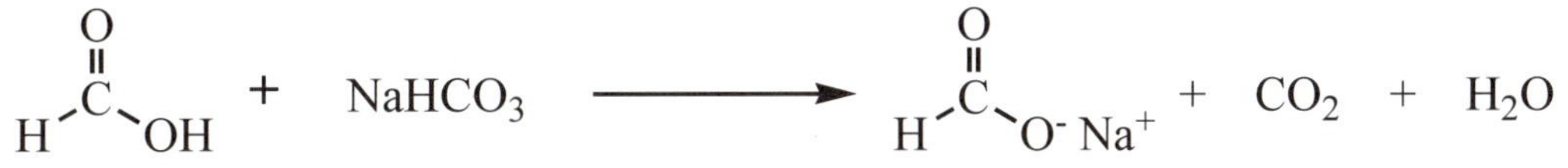

**14.34** *Methyl 2-hydroxybenzoate (methyl salicylate) has the odor of oil of wintergreen. This ester is prepared by Fischer esterification of 2-hydroxybenzoic acid (salicylic acid) with methanol. Draw a structural formula of methyl 2-hydroxybenzoate.*

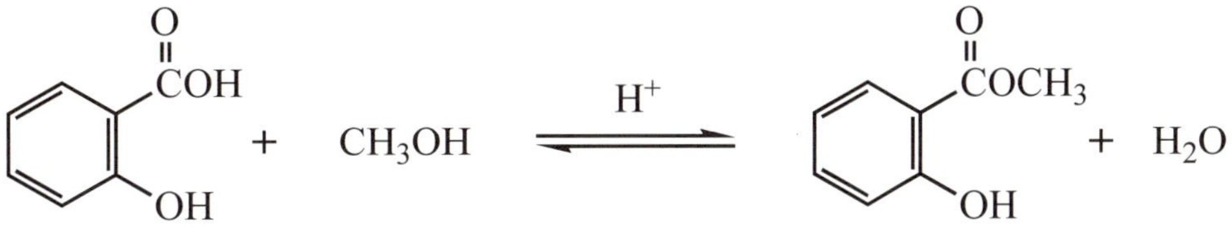

**Methyl salicylate**

**14.35** *Benzocaine, a topical anesthetic, is prepared by treatment of 4-aminobenzoic acid with ethanol in the presence of an acid catalyst followed by neutralization. Draw a structural formula of benzocaine.*

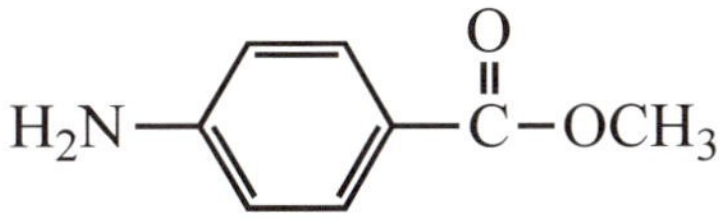

**Benzocaine**

**14.36** *Examine the structural formulas of pyrethrin and permethrin (See Chemical Connections 14C for the structures of pyrethrin and permethrin)*

*(a) Locate the ester groups in each compound.*

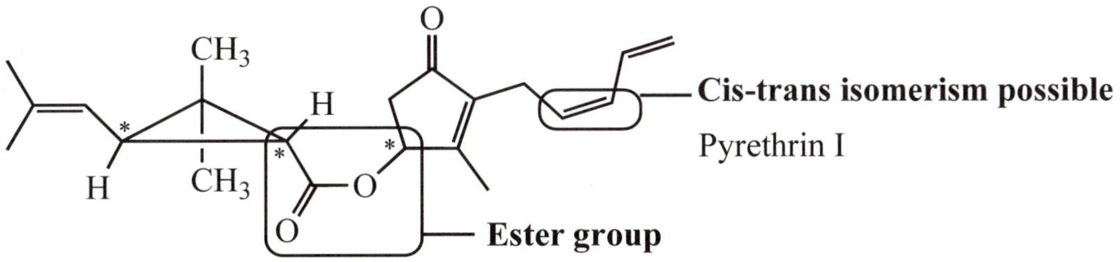

**Cis-trans isomerism possible**

Pyrethrin I

**Ester group**

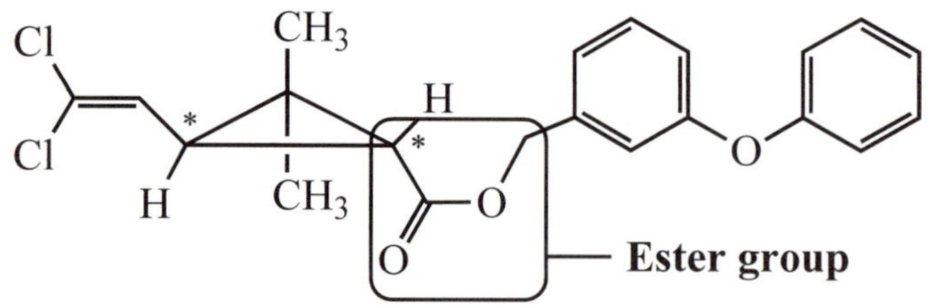

Permethrin

Ester group

*(b) Is pyrethrin chiral? How many stereoisomers are possible for it?*

**Pyrethrin is chiral and the three stereocenters are labeled on the above structure with asterisks. Cis-trans isomerism is possible from the circled carbon-carbon double bond to give a total of $2^4 = 16$ stereoisomers possible.**

*(c) Is permethrin chiral? How many stereoisomers are possible for it?*

**Permethrin is chiral and the stereocenters are labeled on the above structure with an asterisk. There are two stereocenters with a total of four possible stereoisomers.**

**14.37** *A commercial Clothing & Gear Insect Repellant gives the following information about permethrin, its active ingredient.*

*Cis/trans ratio: Minimum 35% (+/-) cis and maximum 65% (+/-) trans*

*(a) To what does the cis/trans ratio refer?*

**The cis/trans ratio refers to the percentage ratio of the cis enantiomer to the trans enantiomer in the commercial mixture.**

*(b) To what does the designation "(+/-)" refer?*

**The (+) and (-) notation identifies the direction a chiral stereoisomer rotates plane polarized light.**

**14.38** *From what carboxylic acid and alcohol is each ester derived?*

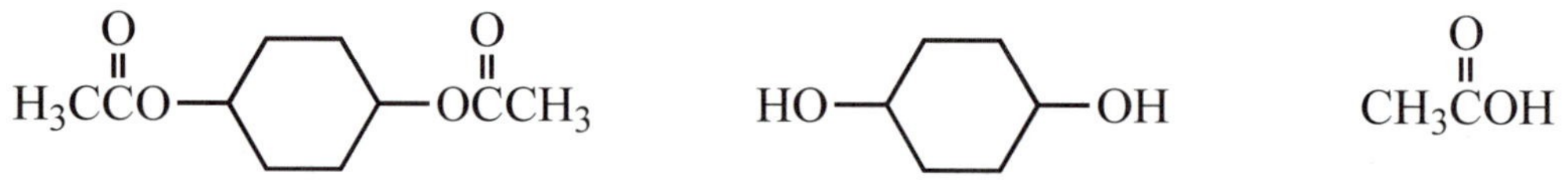

**Alcohol**

**Carboxylic acid**

(b)  $CH_3O\overset{O}{\overset{||}{C}}CH_2CH_2\overset{O}{\overset{||}{C}}OCH_3$          $CH_3OH$

**Alcohol**

$HO\overset{O}{\overset{||}{C}}CH_2CH_2\overset{O}{\overset{||}{C}}OH$

**Carboxylic
acid**

(c)  $\overset{O}{\overset{||}{C}}OCH_3$          $CH_3OH$

**Alcohol**

$\overset{O}{\overset{||}{C}}OH$

**Carboxylic
acid**

(d)  $CH_3CH_2CH{=}CH\overset{O}{\overset{||}{C}}OCH(CH_3)_2$

$H{-}\underset{\underset{CH_3}{|}}{\overset{\overset{CH_3}{|}}{C}}{-}CH_2OH$

**Alcohol**

$CH_3CH_2CH{=}CH\overset{O}{\overset{||}{C}}OH$

**Carboxylic
acid**

**14.39**  *When treated with an acid catalyst, 4-hydroxybutanoic acid forms a cyclic ester (a
lactone).  Draw the structural formula of this lactone.*

**A lactone
product**

**14.40**  *Draw a structural for the product formed on thermal decarboxylation of each compound.*

(a)  $C_6H_5\overset{O}{\overset{||}{C}}CH_2COOH$ $\xrightarrow{\text{heat}}$ $\overset{O}{\overset{||}{C}}CH_3$  +  $CO_2$

(b)  $C_6H_5CH_2\underset{\underset{COOH}{|}}{C}HCOOH$ $\xrightarrow{\text{heat}}$ $-CH_2CH_2COOH$  +  $CO_2$

*(c)*

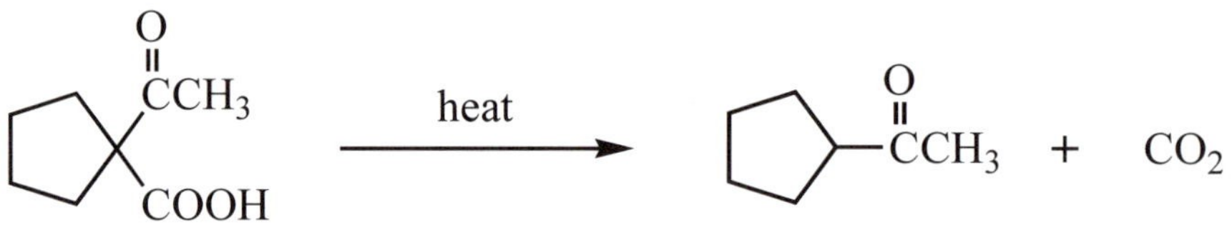

$$\text{heat}$$

## Synthesis

**14.41** *Methyl 2-aminobenzoate, a flavoring agent with the taste of grapes (see Chemistry In Action 14A), can be prepared from toluene by the following series of steps. Show how you might bring about each step in this synthesis.*

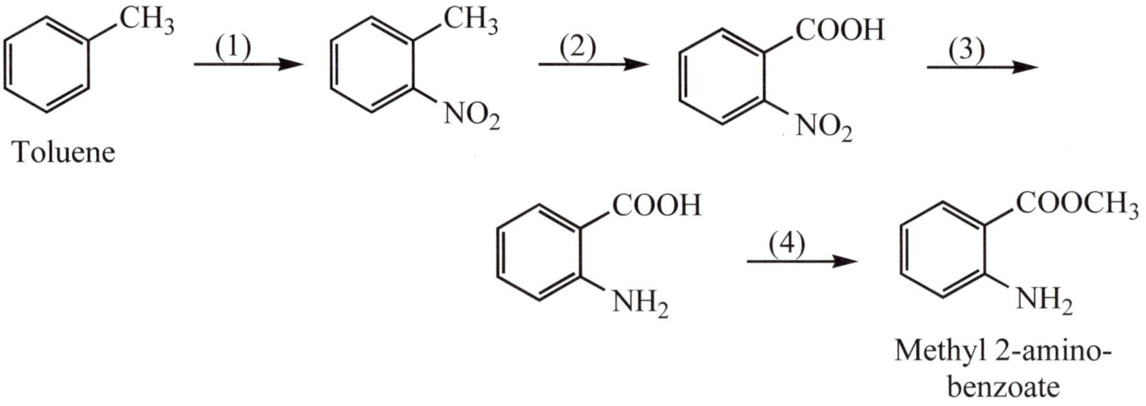

Toluene

Methyl 2-amino-
benzoate

**(1) HNO₃, H₂SO₄    (2) Na₂Cr₂O₇, H₂SO₄    (3) H₂/Ni    (4) CH₃OH, H⁺**

**14.42** *Methylparaben and propylparaben are used as preservatives in foods, beverages, and cosmetics. Show how the synthetic scheme in Problem 14.41 can be modified given each of these compounds.*

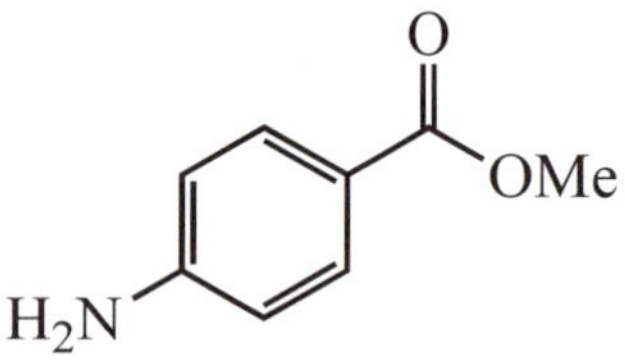

Methyl 4-aminobenzoate
(Methylparaben)

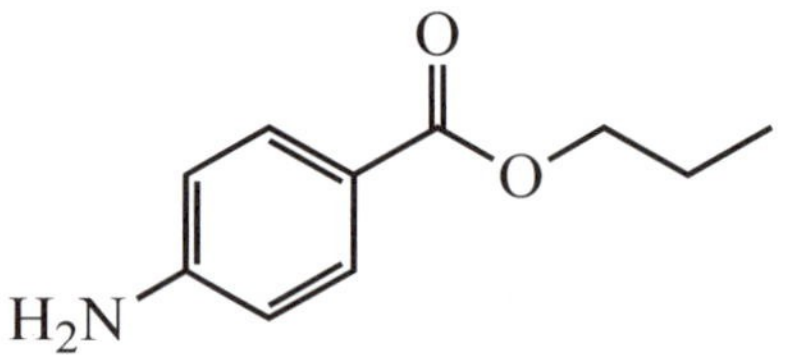

Propyl 4-aminobenzoate
(Propylparaben)

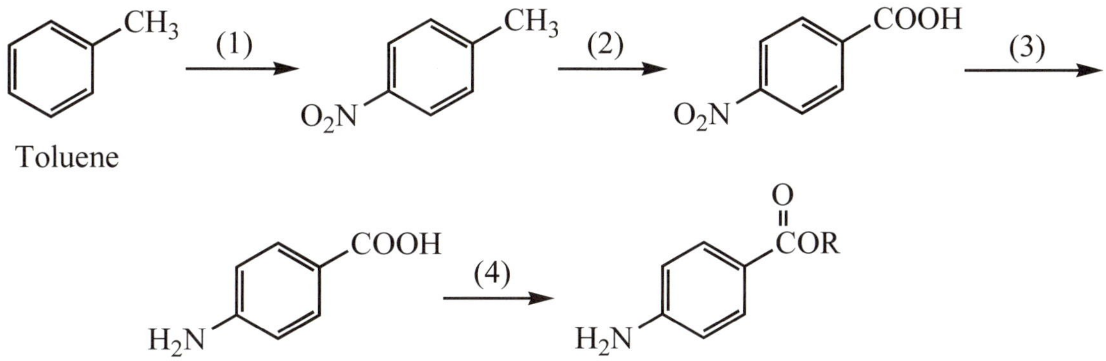

Toluene

R = OCH$_3$          **Methylparaben**
R = OCH$_2$CH$_2$CH$_3$   **Propylparaben**

**Steps 1 thru 3 are the same for both compounds and go as follows:**

**(1)  HNO$_3$, H$_2$SO$_4$     (2)  Na$_2$Cr$_2$O$_7$, H$_2$SO$_4$     (3)  H$_2$/Ni**

**For the synthesis of methylparaben, Step 4 involves a Fischer esterification with excess CH$_3$OH and an acid catalyst. The synthesis of propylparaben also involves a Fischer esterification in Step 4, but with excess 1-propanol and an acid catalyst.**

**14.43**  *Procaine (its hydrochloride is marketed as Novocain) was one of the first local anesthetics developed for infiltration and regional anesthesia. It is synthesized by the following Fischer esterification. Draw a structural formula for procaine.*

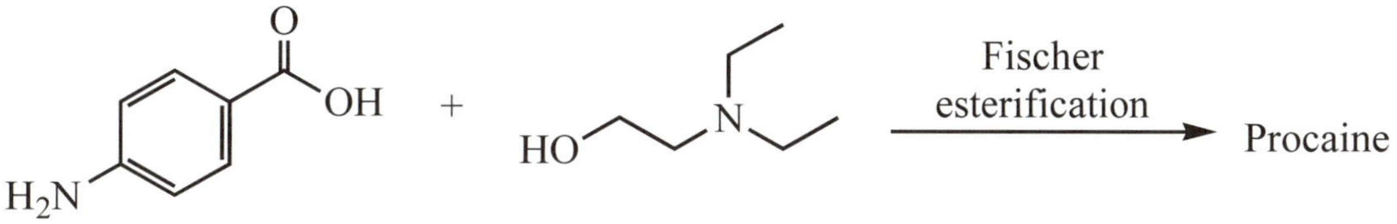

*p*-Aminobenzoic
acid

2-Diethylaminoethanol

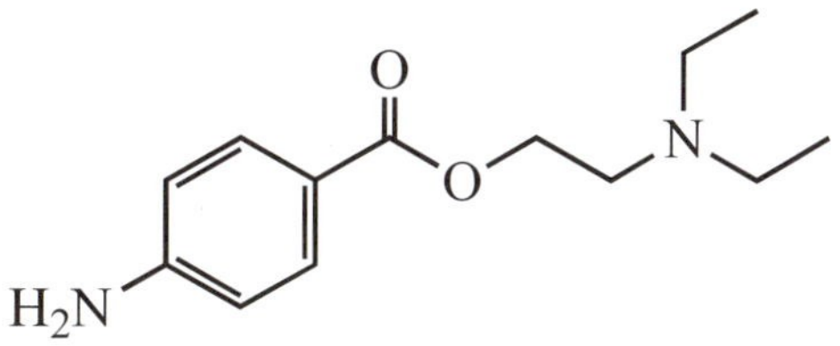

Procaine

**Procaine**

**14.44**  *Meclizine is an antiemetic - it helps prevent or at least lessen the vomiting associated with motion sickness, including seasickness. Among the names of its over-the-counter* *preparations are Bonine, Sea-Legs, Antivert, and Navicalm. Meclizine can be synthesized by the following series of steps.*

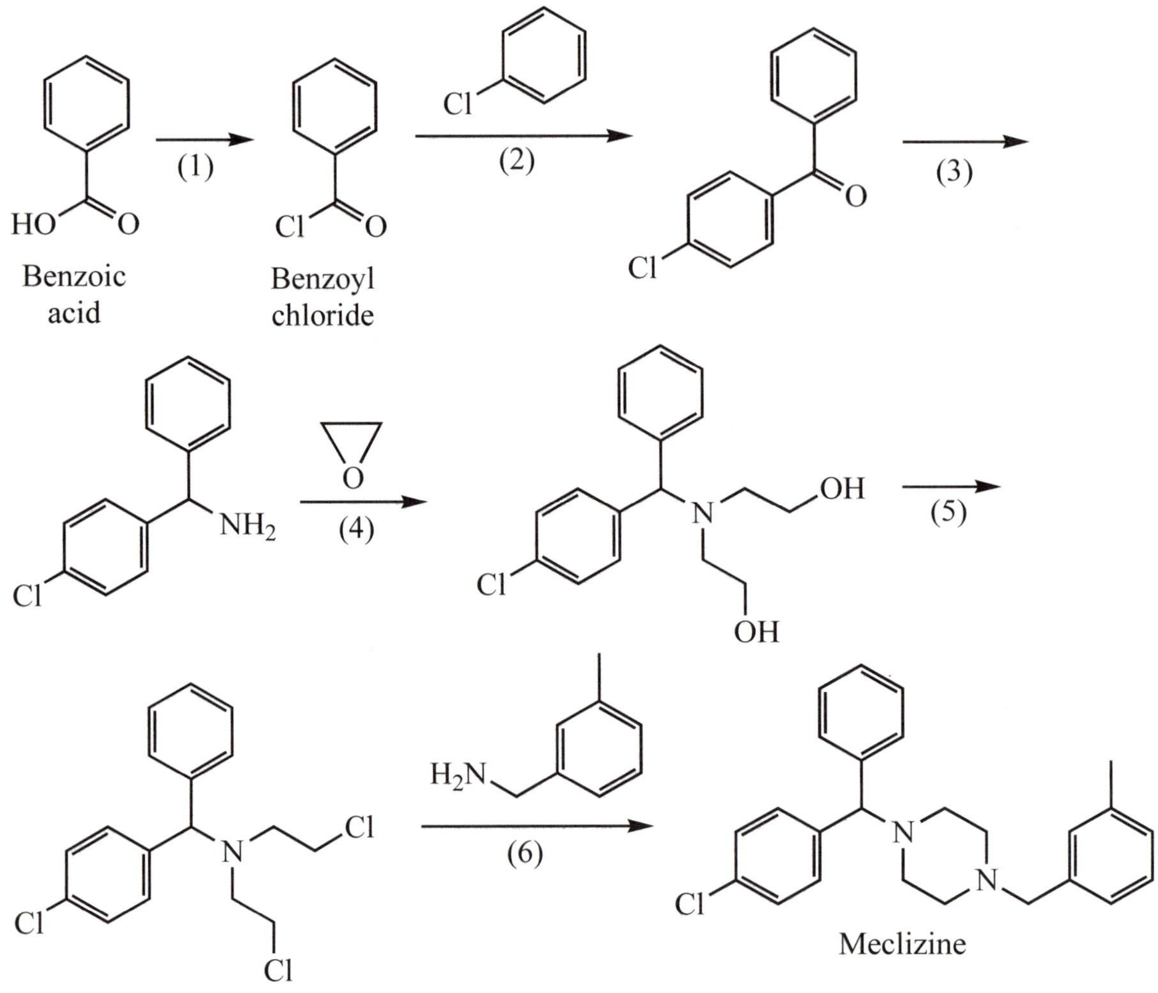

(a) *Propose a reagent for Step 1.*

**SOCl$_2$ in pyridine**

(b) *The catalyst for Step 2 is AlCl$_3$. Name the type of reaction that occurs in Step 2.*

**Friedel-Crafts acylation**

(c) *Propose reagents for Step 3.*

**NH$_3$, Ni/H$_2$**

*(d) Propose a mechanism for Step 4 and show that it is an example of nucleophilic aliphatic substitution.*

**Step 1:  Nucleophilic attack of the amine on the epoxide ring, with the relief of ring strain as the driving force for the reaction.**

**Step 2:  Proton transfer from an acidic ammonium cation to a basic alkoxide anion.**

**Step 3:  Nucleophilic attack of the amine on the epoxide ring, with the relief of ring strain as the driving force of the reaction.**

**Step 4: Proton transfer from an acidic ammonium cation to a basic alkoxide anion.**

*(e) Propose a reagent for Step 5.*

**SOCl₂ in pyridine**

*(f) Show that Step 6 is also an example of nucleophilic aliphatic substitution.*

**14.45** *Chemists have developed several syntheses for the anti-asthmatic drug albuterol (Proventil). One of these syntheses starts with salicylic acid, the same acid that is the starting material for the synthesis of aspirin.*

Salicylic acid

Albuterol

*(a) Propose a reagent and a catalyst for Step 1. What name is given to this type of reaction?*

**This reaction is a Friedel-Crafts acylation and it uses a reagent combination of:**

$$CH_3\overset{O}{\overset{\|}{C}}Cl \text{ and } AlCl_3 \text{ catalyst}$$

*(b) Propose a reagent for Step 2.*

**Bromine ($Br_2$) in acetic acid**

*(c) Name the amine used to bring about Step 3.*

**2-Methyl-2-propanamine (*tert*-butylamine)**

*(d) Step 4 is a reduction of two functional groups. Name the functional groups reduced and tell what reducing agent will accomplish this reduction.*

**In Step 4, a carboxylic acid and a ketone are reduced to primary and secondary alcohols, respectively. The reduction is accomplished by adding $LiAlH_4$ followed by the addition of water upon completion of the reduction by $LiAlH_4$.**

**Looking Ahead**

**14.46**  *Explain why α-amino acids, the building blocks of proteins (chapter 20), are nearly 1000 times more acidic than aliphatic carboxylic acids.*

| α-amino acid | aliphatic acid |
| :---: | :---: |
| p$K_a$ Å2 | p$K_a$ Å5 |

> **The protonated form of the amino acid exists with an ammonium ($-NH_3^+$) substituent adjacent to the carboxyl group.**

> **The electron-withdrawing inductive effect of the adjacent $-NH_3^+$ group helps stabilize the conjugate base of the amino acid relative to the conjugate base of the aliphatic carboxylic acid.  The more stable a conjugate base, the stronger the acid from which it came.**

**14.47**  *Which is more difficult to reduce with LiAlH$_4$, a carboxylic acid or a carboxylate ion?*

> **The carboxylate anion is more difficult to reduce than the carboxylic acid because the negative charge on the carboxylate makes the carbonyl carbon less electrophilic towards the hydride ion.**

**14.48**  *Show how an ester can react with H$^+$/H$_2$O to give a carboxylic acid and an alcohol. Hint: this is the reverse of the Fischer esterification.*

**The mechanism for the hydrolysis of an ester:**

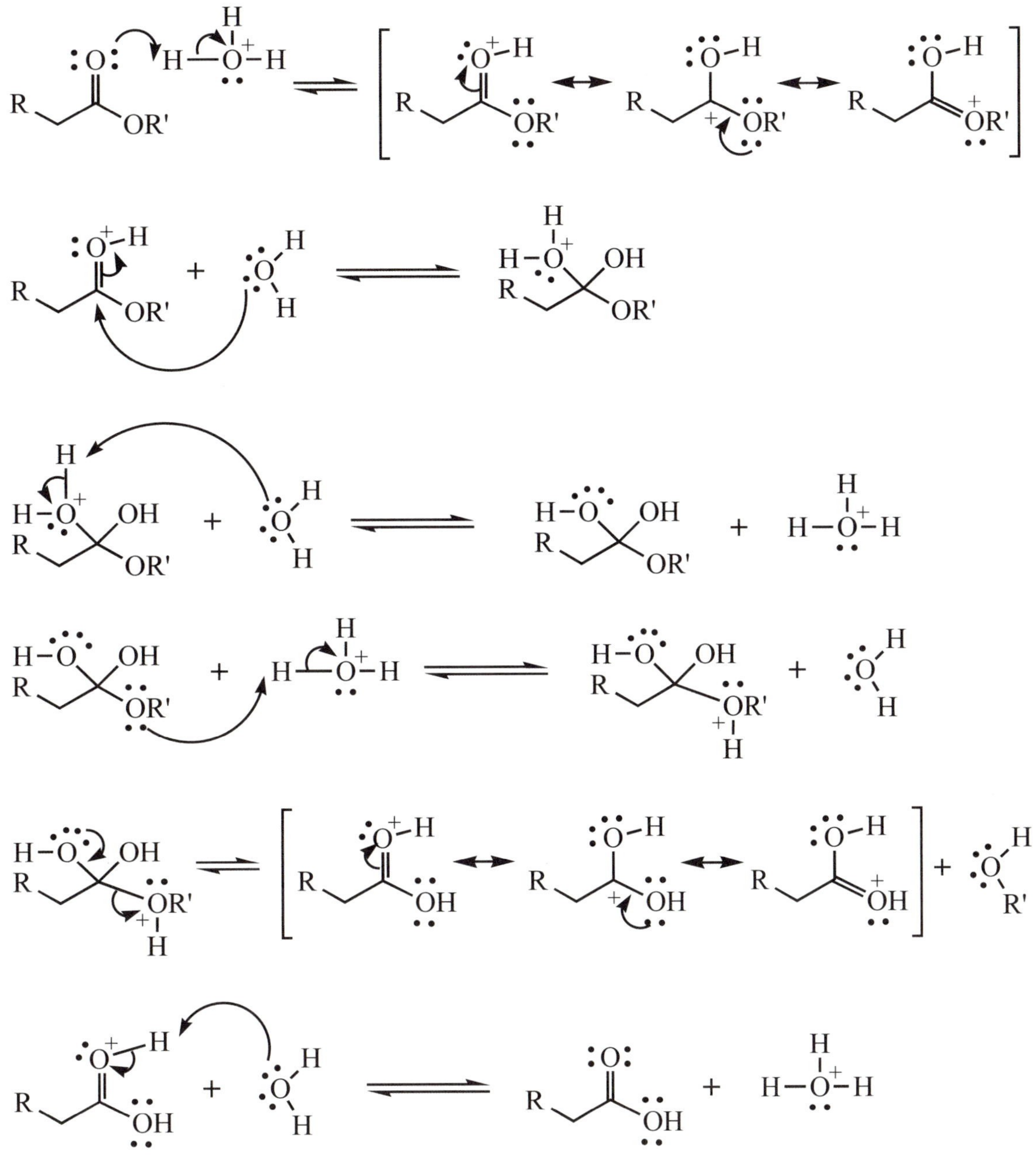

**14.49** *In chapter 13, we saw how Grignard reagents readily attack the carbonyl carbon of ketones and aldehydes. Should the same process occur with Grignards and carboxylic acids? With esters?*

**Grignard reagents do not attack the carbonyl carbon of carboxylic acids, but instead, act as strong bases and deprotonate carboxylic acid hydroxyl to form the carboxylate anion. Grignard reagents will react with esters of carboxylic acids and attack the carbonyl carbon twice to form tertiary alcohols (except in the case of**

formic acid esters, which will form secondary alcohols) upon subsequent hydrolysis of the alkoxide intermediate. We will discuss this reaction in Chapter 15.

**14.50** *In section 14.7, it was suggested that the mechanism for the Fischer esterification of carboxylic acids would be a model for many of the reactions of the functional derivatives of carboxylic acids. One such reaction, the reaction of an acid halide with water, is shown below. Suggest a mechanism for this reaction.*

**Step 1: Nucleophilic attack by water on the carbonyl carbon forms the tetrahedral intermediate.**

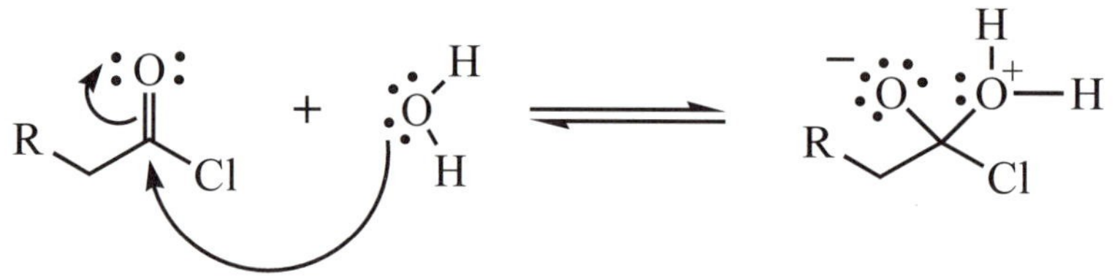

**Step 2: The tetrahedral intermediate is deprotonated by water.**

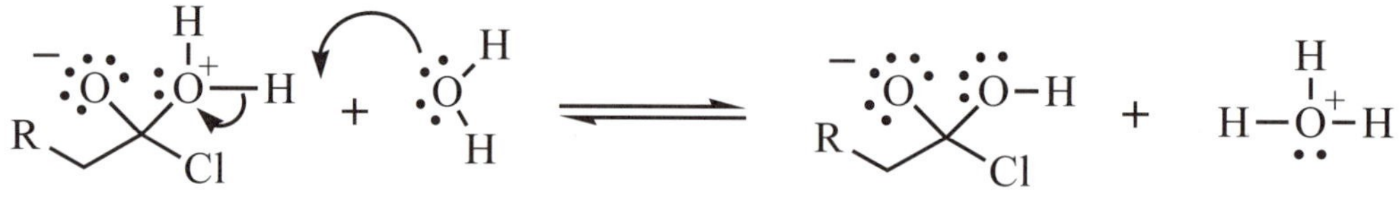

**Step 3: The tetrahedral intermediate collapses to form the carboxylic acid and a chloride anion. Hydronium ion ($H_3O^+$) and chloride anion represent HCl ionized in water.**

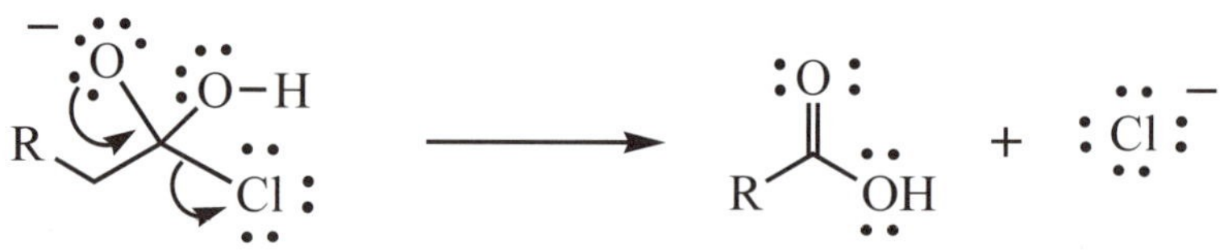

# CHAPTER 15
## Solutions to Problems

**15.1**    *Draw a structural formula for each compound.*

*(a) N-Cyclohexylacetamide*

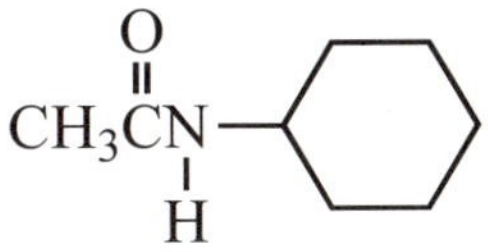

*(b) sec-Butyl acetate*

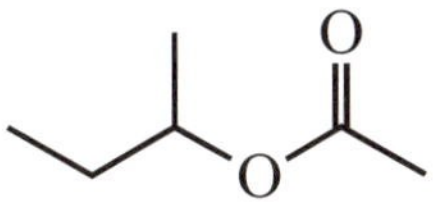

*(c) Cyclobutyl butanoate*

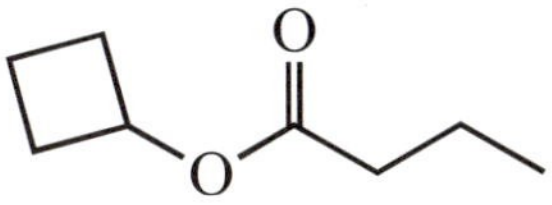

*(d)  N-(2-Octyl)succinimide*

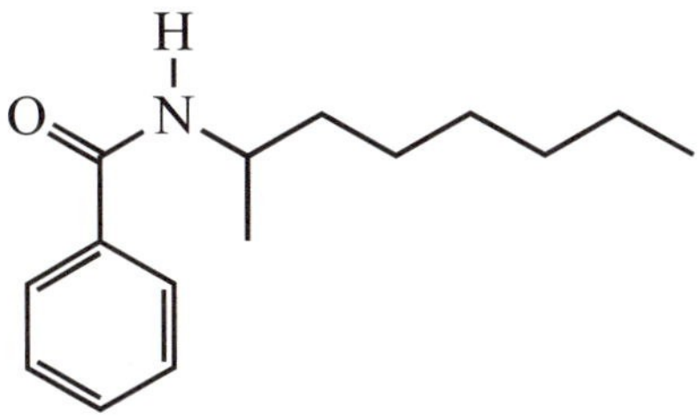

*(e)    Diethyl adipate*

$$C_2H_5OC(CH_2)_4COC_2H_5$$

*(f) Propanoic anhydride*

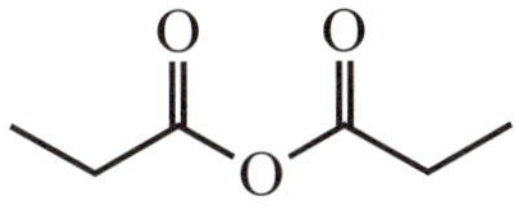

**15.2**    *Complete and balance equations for hydrolysis of each ester in aqueous solution. Show each product as it is ionized under the given experimental conditions.*

*(a)*

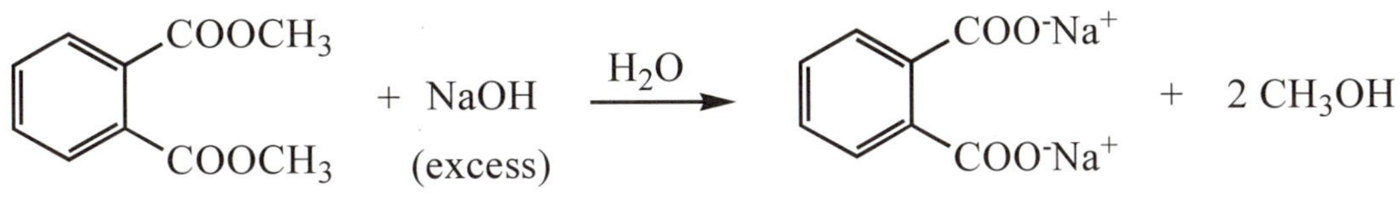

$+ \ NaOH \xrightarrow{\ H_2O\ }$ (excess)

$+ \quad 2 \ CH_3OH$

*(b)*

$+ \ H_2O \xrightarrow{\ HCl\ }$

$+ \ 2 \ CH_3CH_2OH$

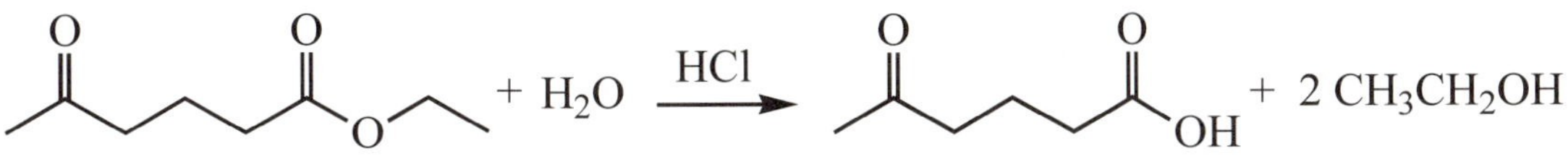

**15.3**    *Complete equations for the hydrolysis of the amides in Example 15.3 in concentrated aqueous NaOH. Show all products as they exist in aqueous NaOH, and show the number of moles of NaOH required for hydrolysis of each amide.*

*(a)* $CH_3CN(CH_3)_2 \xrightarrow[H_2O]{NaOH} CH_3CO^- Na^+ \ + \ (CH_3)_2NH$

*(b)*

$\xrightarrow[H_2O]{NaOH}$

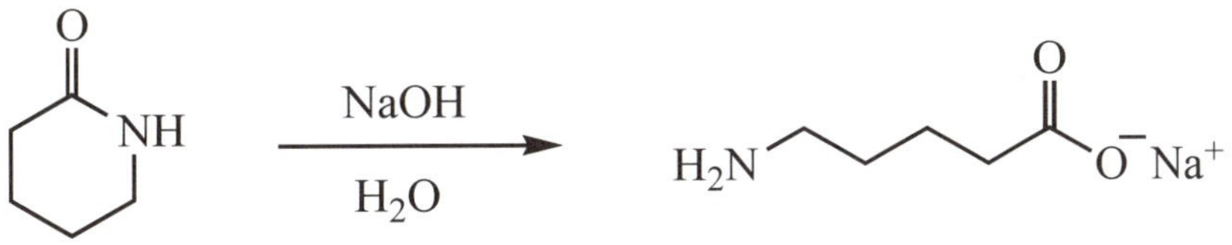

**15.4**   *Complete these equations.  The stoichiometry of each is given in the equation.*

(a) $CH_3\overset{O}{\overset{\|}{C}}O$—⬡—$O\overset{O}{\overset{\|}{C}}CH_3$ + $2NH_3$ ⟶ HO—⬡—OH + $CH_3\overset{O}{\overset{\|}{C}}NH_2$

(b) [lactone structure] + $NH_3$ ⟶ HO⌒⌒⌒$\overset{O}{\overset{\|}{C}}NH_2$

**15.5**   *Show how to prepare each alcohol by treatment of an ester with a Grignard reagent.*

(a) $H\overset{O}{\overset{\|}{C}}OCH_3$    $\xrightarrow[\text{2. } H_3O^+]{\text{1. 2 [cyclopentyl]–MgBr}}$    [dicyclopentyl carbinol with OH]

(b) $Ph\overset{O}{\overset{\|}{C}}OCH_3$    $\xrightarrow[\text{2. } H_3O^+]{\text{1. [allyl]MgBr}}$    [4-phenyl-1,6-heptadien-4-ol structure with OH and Ph]

**15.6**   *Show how to convert hexanoic acid to each amine in good yield.*

(a) ⌒⌒⌒$\overset{O}{\overset{\|}{C}}$OH    $\xrightarrow{SOCl_2}$    ⌒⌒⌒$\overset{O}{\overset{\|}{C}}$Cl

$\xrightarrow{\text{2 } (CH_3)_2NH}$   ⌒⌒⌒$\overset{O}{\overset{\|}{C}}$N(⌐)   $\xrightarrow[\text{2. } H_2O]{\text{1. } LiAlH_4}$   ⌒⌒⌒⌒N(⌐)

(b) ⌒⌒⌒$\overset{O}{\overset{\|}{C}}$OH    $\xrightarrow{SOCl_2}$    ⌒⌒⌒$\overset{O}{\overset{\|}{C}}$Cl

$\xrightarrow{\text{2 } \text{(CH}_3)_2CH-NH_2}$   ⌒⌒⌒$\overset{O}{\overset{\|}{C}}\overset{}{\underset{H}{N}}$CH(CH$_3$)$_2$   $\xrightarrow[\text{2. } H_2O]{\text{1. } LiAlH_4}$   ⌒⌒⌒⌒$\overset{}{\underset{H}{N}}$CH(CH$_3$)$_2$

**15.7**  *Show how to convert (R)-2-phenylpropanoic acid to these compounds:*

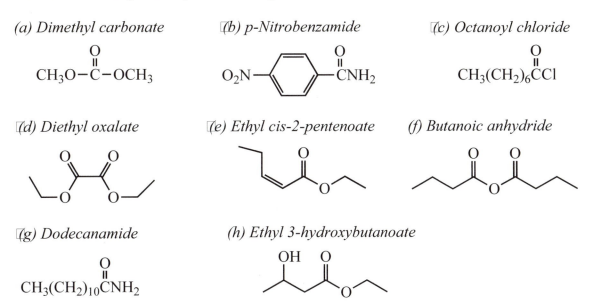

**15.8**  *Draw a structural formula for each compound.*

(a) Dimethyl carbonate

$$CH_3O-\overset{\overset{O}{\|}}{C}-OCH_3$$

(b) p-Nitrobenzamide

$$O_2N-\underset{}{\text{⬡}}-\overset{\overset{O}{\|}}{C}NH_2$$

(c) Octanoyl chloride

$$CH_3(CH_2)_6\overset{\overset{O}{\|}}{C}Cl$$

(d) Diethyl oxalate

(e) Ethyl cis-2-pentenoate

(f) Butanoic anhydride

(g) Dodecanamide

$$CH_3(CH_2)_{10}\overset{\overset{O}{\|}}{C}NH_2$$

(h) Ethyl 3-hydroxybutanoate

**15.9**  *Write the IUPAC name for each compound.*

(a)

$$\text{⬡}-\overset{\overset{O}{\|}}{C}O\overset{\overset{O}{\|}}{C}-\text{⬡}$$

**Benzoic anhydride**

(b) $CH_3(CH_2)_{14}\overset{\overset{O}{\|}}{C}OCH_3$

**Methyl hexadecanoate**

(c) $CH_3(CH_2)_4\overset{\overset{O}{\|}}{C}NHCH_3$

***N*-Methylhexanamide**

(d) H$_2$N—⟨benzene ring⟩—$\overset{\displaystyle O}{\overset{\|}{C}}$NH$_2$

**4-Aminobenzamide**

(e) CH$_2$(COOCH$_2$CH$_3$)$_2$

**Diethyl propanedioate
(Diethyl malonate)**

(f) PhCH$_2\overset{\displaystyle O}{\overset{\|}{C}}\underset{\underset{\displaystyle CH_3}{|}}{C}H\overset{\displaystyle O}{\overset{\|}{C}}OCH_3$

**Methyl 2-methyl-3-oxo-
4-phenylbutanoate**

**15.10** *When oil from the head of the sperm whale is cooled, spermaceti, a translucent wax with a white, pearly luster, crystallizes from the mixture. Spermaceti, which makes up 11% of whale oil, is composed mainly of hexadecyl hexadecanoate (cetyl palmitate). At one time, spermaceti was widely used in the making of cosmetics, fragrant soaps, and candles. Draw a structural formula of cetyl palmitate.*

$$CH_3(CH_2)_{14}CH_2O\overset{\displaystyle O}{\overset{\|}{C}}(CH_2)_{14}CH_3$$

**Hexadecyl hexadecanoate
(Cetyl palmitate)**

## Physical Properties

**15.11** *Acetic acid and methyl formate are constitutional isomers. Both are liquids at room temperature: one with a boiling point of 32°C, the other with a boiling point of 118°C. Which of the two has the higher boiling point?*

**Acetic acid has a higher boiling point (118 °C) than methyl formate (32 °C). Acetic acid has strong intermolecular hydrogen bonding which must be broken before boiling can occur. There is no intermolecular hydrogen bonding between methyl formate molecules, thus less energy is needed to break up intermolecular association leading to a lower boiling point relative to acetic acid.**

**15.12** *Acetic acid has a boiling point of 118°C, whereas its methyl ester has a boiling point of 57°C. Account for the fact that the boiling point of acetic acid is higher than that of its methyl ester, even though acetic acid has a lower molecular weight.*

**Acetic acid has a higher boiling point (118 °C) than its methyl ester, methyl acetate (57 °C) because of the strong intermolecular hydrogen bonding in carboxylic acids. Acetic acid has strong intermolecular hydrogen bonding which must be broken before boiling can occur. There is no intermolecular hydrogen bonding in methyl acetate, thus less energy is needed to break up intermolecular association leading to a lower boiling point relative to acetic acid.**

## Reactions

**15.13** *Arrange these compounds in order of increasing reactivity toward nucleophilic acyl substitution.*

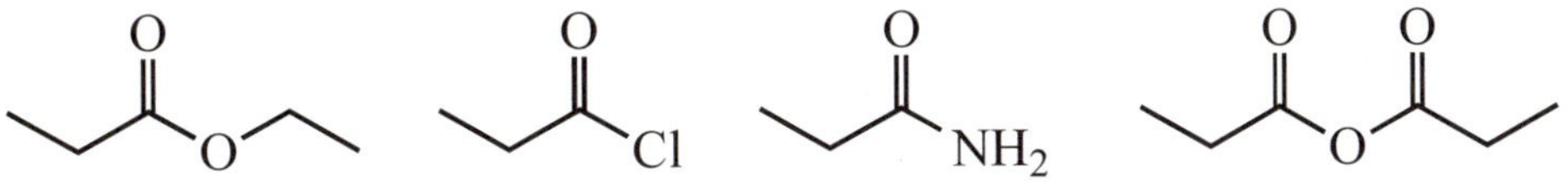

(1)          (2)          (3)          (4)

**In general, the order of reactivity toward nucleophilic acyl substitution is:**

**acid chlorides > acid anhydrides > esters  > amides**

**Therefore, the order for the molecules listed above is:**

**(ranked from least to most reactive)  3 < 1 < 4 < 2**

**15.14** *A carboxylic acid can be converted to an ester by Fischer esterification. Show how to synthesize each ester from a carboxylic acid and an alcohol by Fischer esterification.*

(a)

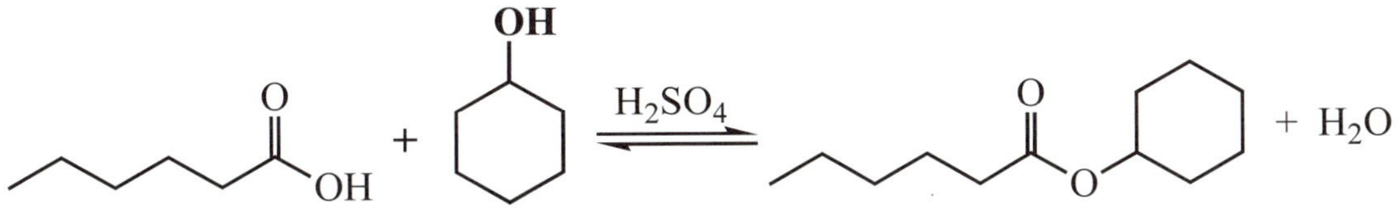

(b)

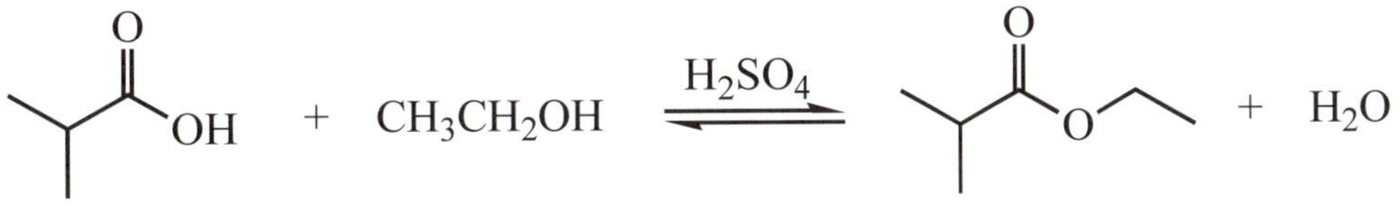

**15.15** *A carboxylic acid can also be converted to an ester in two reactions by first converting the carboxylic acid to its acid chloride and then treating the acid chloride with an alcohol. Show how to prepare each ester in Problem 15.14 from a carboxylic acid and an alcohol by this two-step scheme.*

(a)

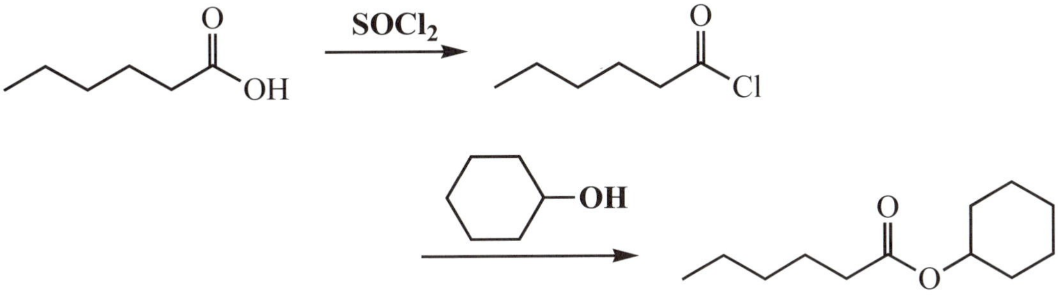

*(b)*

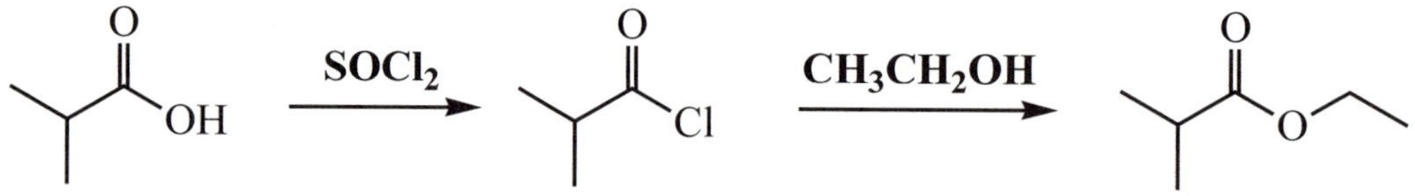

**15.16** *Show how to prepare these amides by reaction of an acid chloride with ammonia or an amine.*

*(a)*

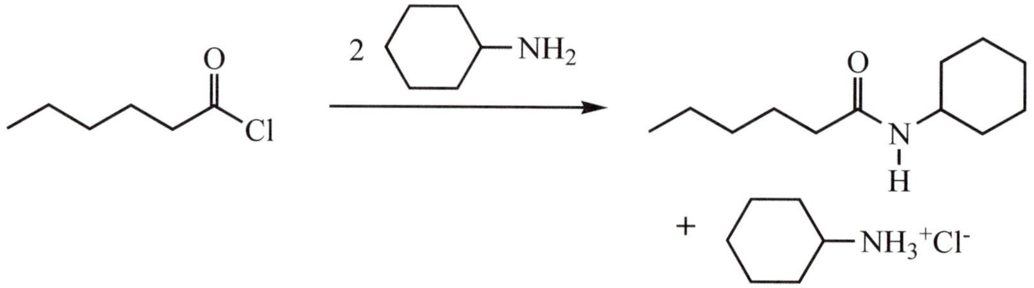

*(b)*

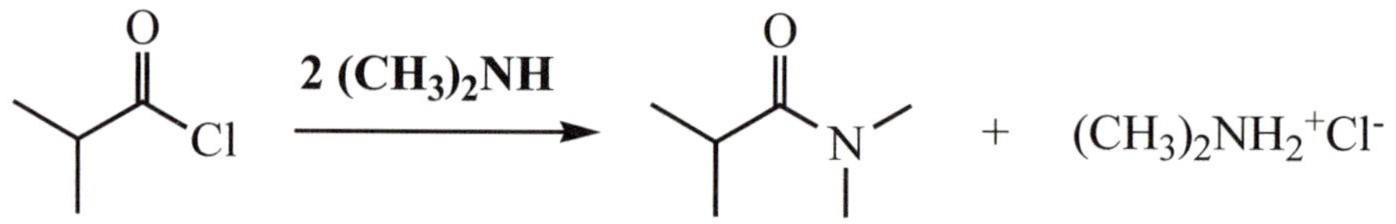

*(c)*

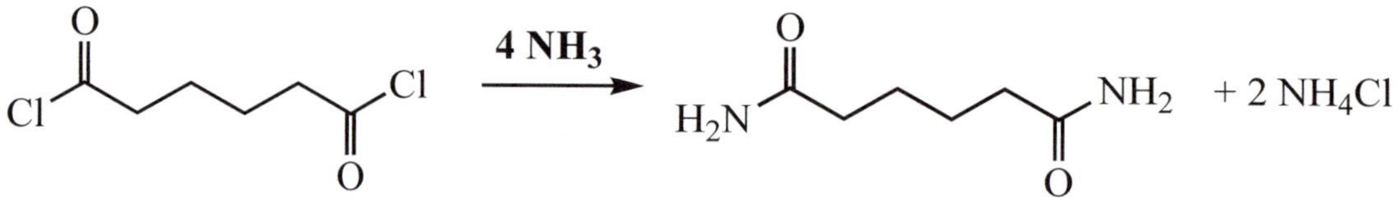

**15.17** *Write a mechanism for the reaction of butanoyl chloride and ammonia to give butanamide and ammonium chloride.*

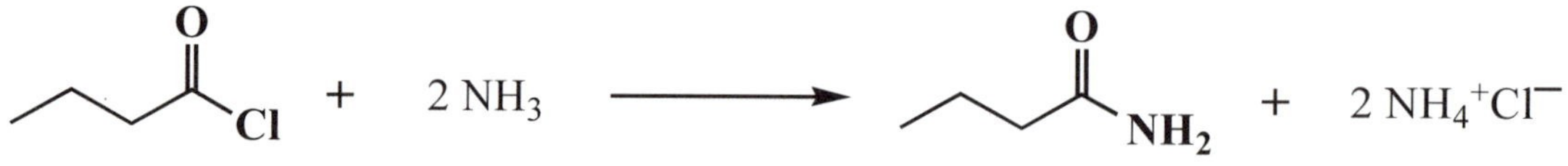

**Step 1:  Nucleophilic acyl attack by the first ammonia molecule on the carbonyl carbon to form an unstable tetrahedral intermediate.**

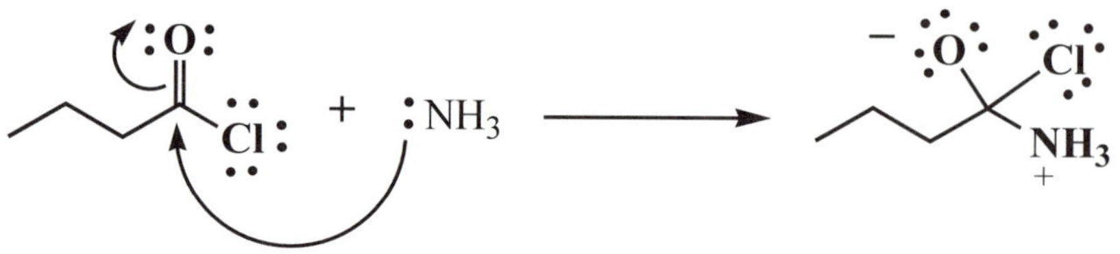

**Step 2:  The tetrahedral intermediate collapses with chloride leaving and the formation of a protonated amide.**

**Step 3:  The second molecule of ammonia removes the proton from the protonated amide producing the amide and an ammonium cation.**

**15.18**  *What product is formed when benzoyl chloride is treated with these reagents?*

(e) 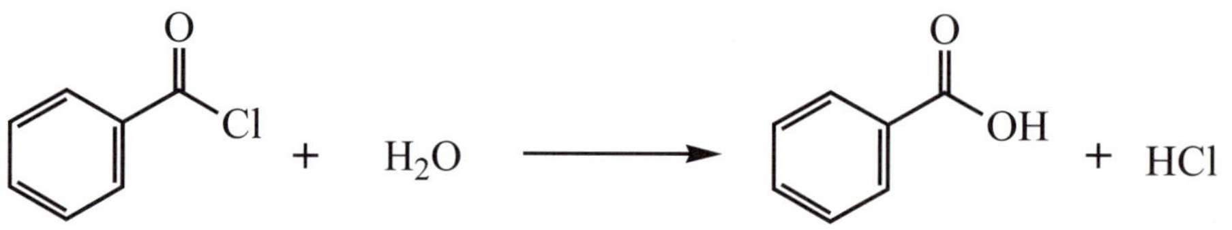 + H$_2$O ⟶ + HCl

(e) 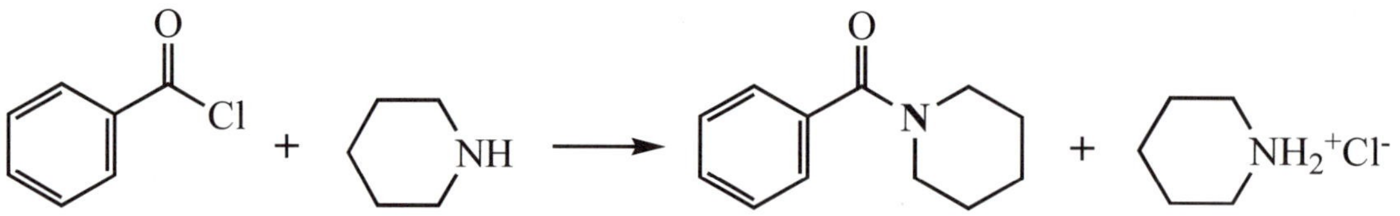 + ⟶ +

**15.19** *Write the product(s) of treatment of propanoic anhydride with each reagent.*

(a) 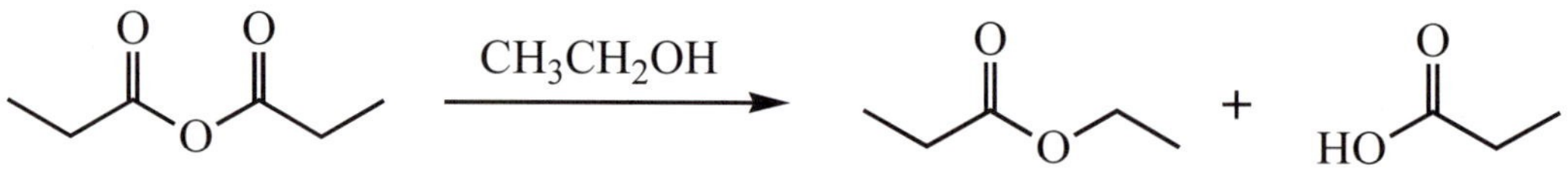 $\xrightarrow{\text{CH}_3\text{CH}_2\text{OH}}$ +

(b) 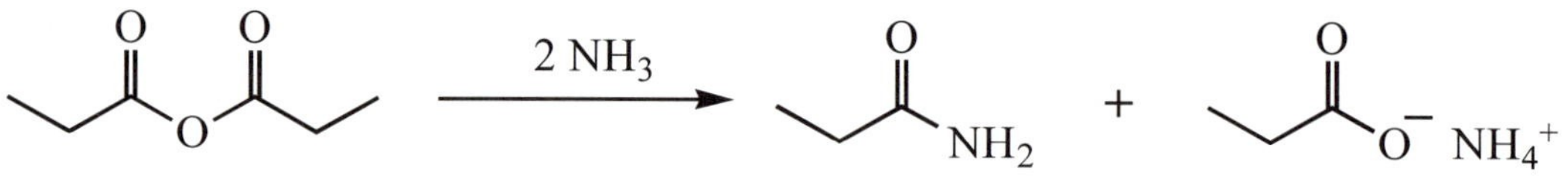 $\xrightarrow{2\text{ NH}_3}$ +

**15.20** *Write the product of treatment of benzoic anhydride with each reagent.*

(a) 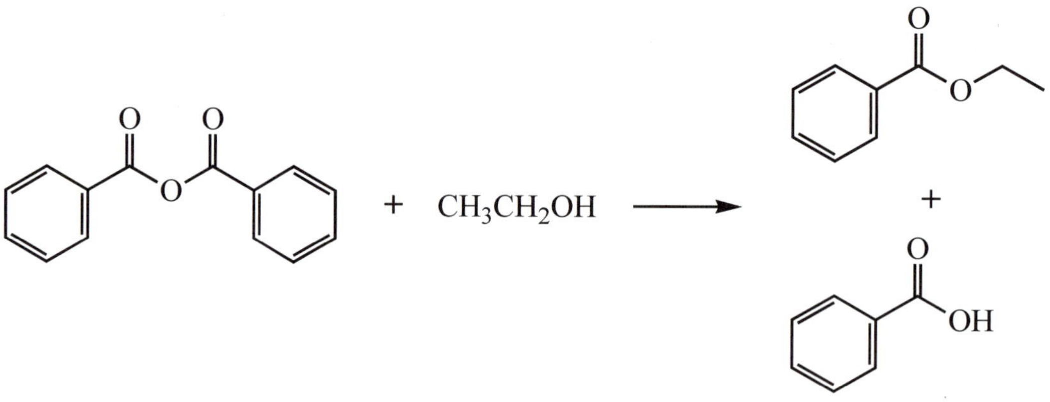 + CH$_3$CH$_2$OH ⟶

(b)

**15.21** *The analgesic phenacetin is synthesized by treating 4-ethoxyaniline with acetic anhydride. Write an equation for the formation of phenacetin.*

**15.22** *The analgesic acetaminophen is synthesized by treatment of 4-aminophenol with one equivalent of acetic anhydride. Write an equation for the formation of acetaminophen. (Hint: Remember from Section 7.6A that an -NH$_2$ group is a better nucleophile than an -OH group.)*

**15.23** *Nicotinic acid, more commonly named niacin, is one of the B vitamins. Show how nicotinic acid can be converted to ethyl nicotinate and then to nicotinamide.*

Nicotinic acid                                  Ethyl nicotinate                    Nicotinamide
(Niacin)

**15.24** *Complete these reactions.*

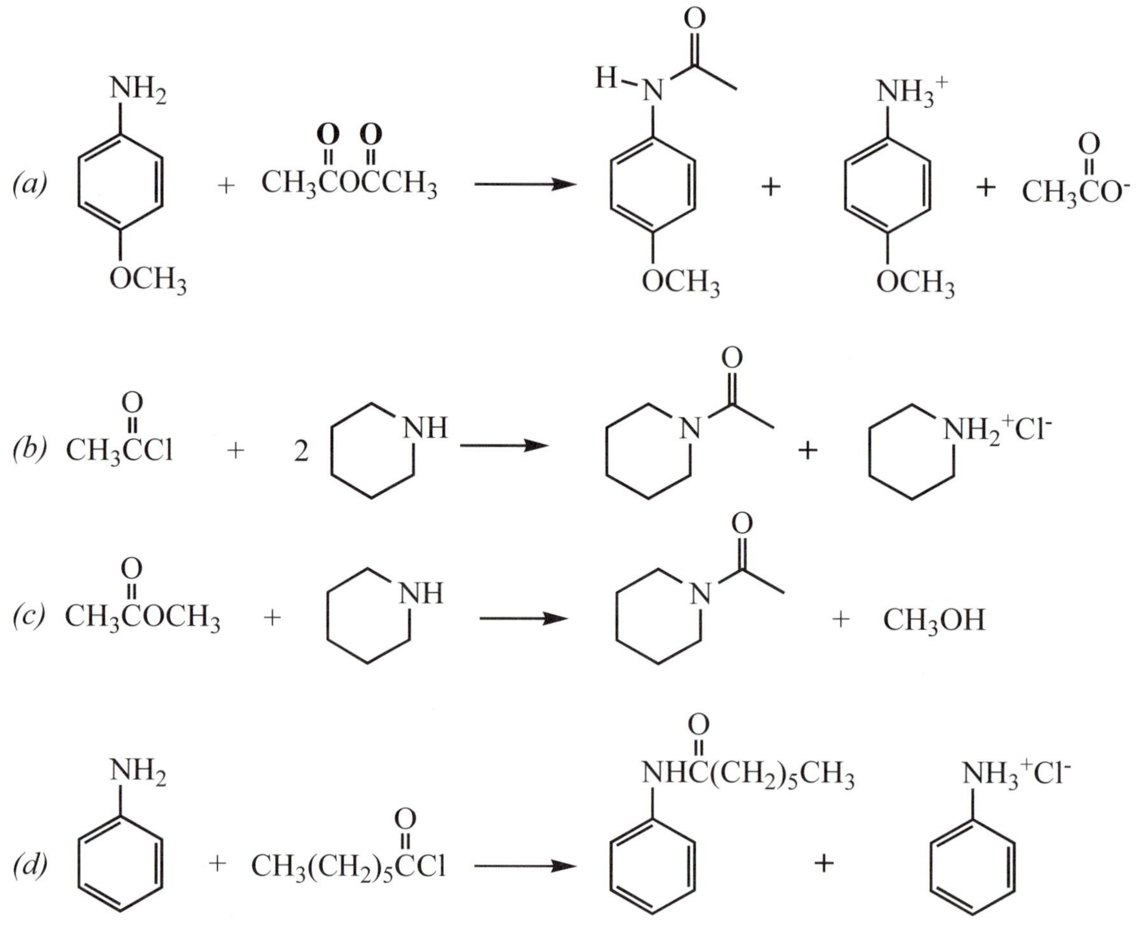

(a)

(b)

(c)

(d)

**15.25** *What product is formed when ethyl benzoate is treated with these reagents?*

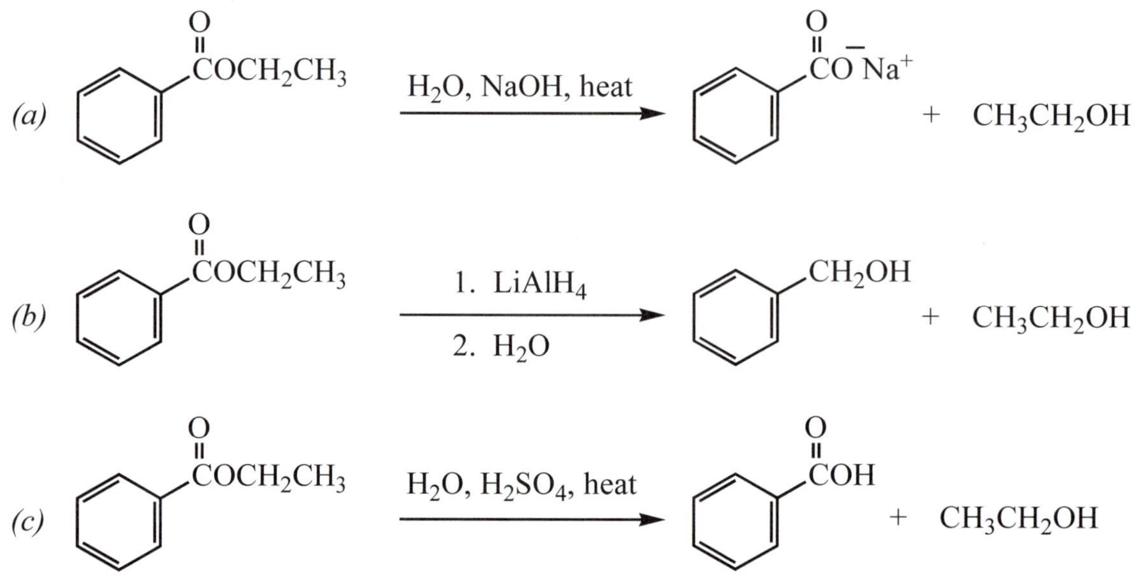

(a)

(b)

(c)

(d)

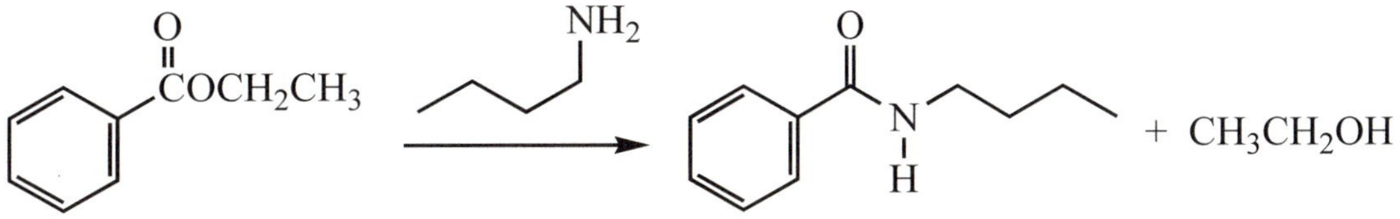

(e)

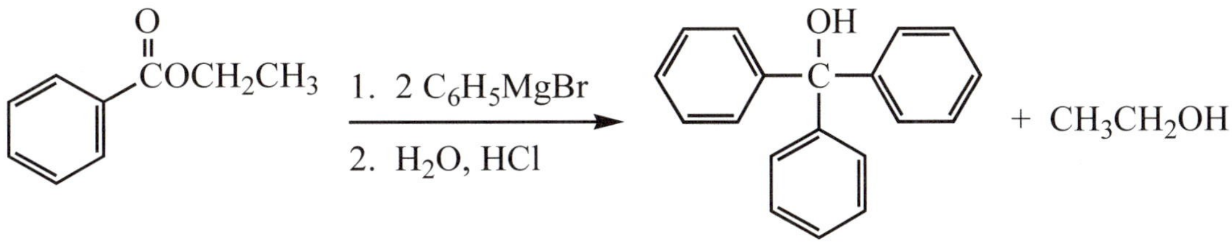

**15.26** *Show how to convert 2-hydroxybenzoic acid (salicylic acid) to these compounds.*

(a)

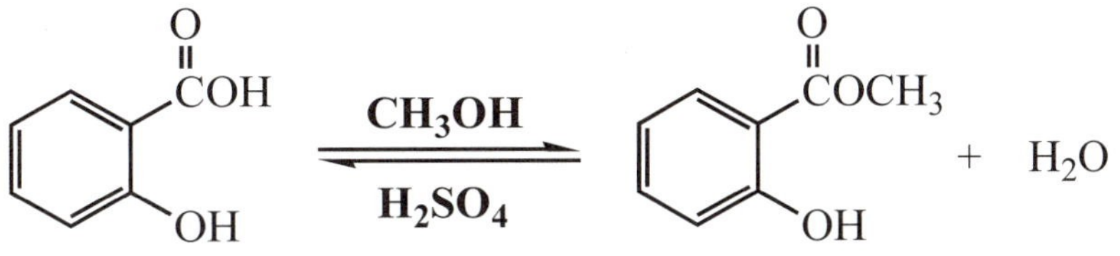

*Methyl salicylate*
*(Oil of wintergreen)*

(b)

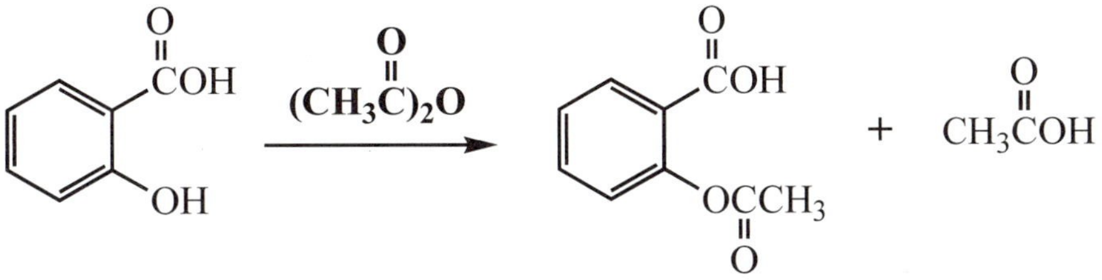

*Acetyl salicylic acid*
*(Aspirin)*

**15.27** *What product is formed when benzamide is treated with these reagents?*

(a)

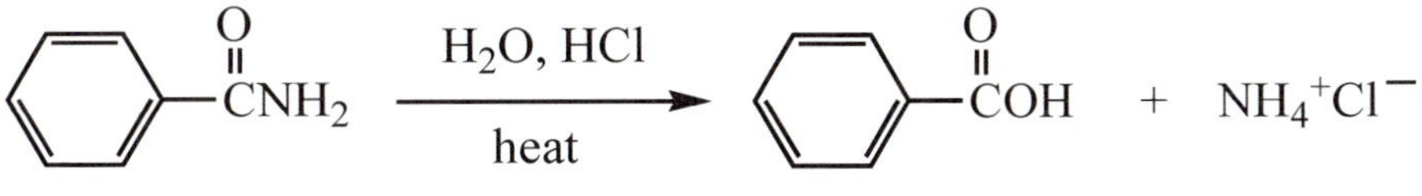

(b)

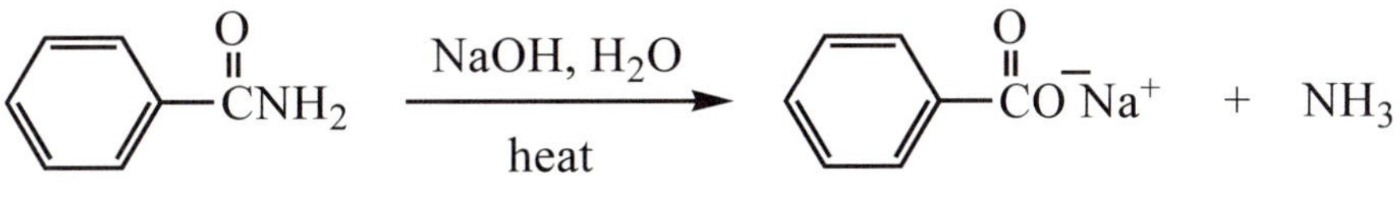

(c)  [structure: benzamide] $\xrightarrow[\text{2. H}_2\text{O}]{\text{1. LiAlH}_4}$ [structure: benzylamine CH$_2$NH$_2$]

**15.28**  *Treatment of γ-butyrolactone with two equivalents of methylmagnesium bromide followed by hydrolysis in aqueous acid gives a compound with the molecular formula C$_6$H$_{14}$O$_2$. Propose a structural formula for this compound.*

[structure: γ-butyrolactone] $\xrightarrow[\text{2. H}_2\text{O, HCl}]{\text{1. 2 CH}_3\text{MgBr}}$ [structure: HO—CH$_2$CH$_2$CH$_2$—C(CH$_3$)$_2$—OH]

$C_6H_{14}O_2$

**15.29**  *Show the product of treatment of γ-butyrolactone with each reagent.*

(a)  [structure: γ-butyrolactone] $\xrightarrow{\text{NH}_3}$ [structure: HO—CH$_2$CH$_2$CH$_2$—C(=O)—NH$_2$]

(b)  [structure: γ-butyrolactone] $\xrightarrow[\text{2. H}_2\text{O}]{\text{1. LiAlH}_4}$ [structure: HO—CH$_2$CH$_2$CH$_2$CH$_2$—OH]

(c)  [structure: γ-butyrolactone] $\xrightarrow{\text{NaOH, H}_2\text{O, heat}}$ [structure: HO—CH$_2$CH$_2$CH$_2$—C(=O)—O$^-$Na$^+$]

**15.30**  *Show the product of treatment of N-methyl-γ-butyrolactam with each reagent.*

(a)  [structure: N-methyl-γ-butyrolactam] $\xrightarrow[\text{heat}]{\text{H}_2\text{O, HCl}}$ [structure: H$_3$C—$\overset{+}{\text{N}}$H$_2$—CH$_2$CH$_2$CH$_2$—C(=O)—OH]  + Cl$^-$

(b)  [structure: N-methyl-γ-butyrolactam] $\xrightarrow[\text{heat}]{\text{NaOH, H}_2\text{O}}$ [structure: H$_3$C—NH—CH$_2$CH$_2$CH$_2$—C(=O)—O$^-$Na$^+$]

(c)  [structure: N-methyl-γ-butyrolactam] $\xrightarrow[\text{2. H}_2\text{O}]{\text{1. LiAlH}_4}$ [structure: pyrrolidine ring N—CH$_3$]

**15.31** *Complete these reactions:*

(a)

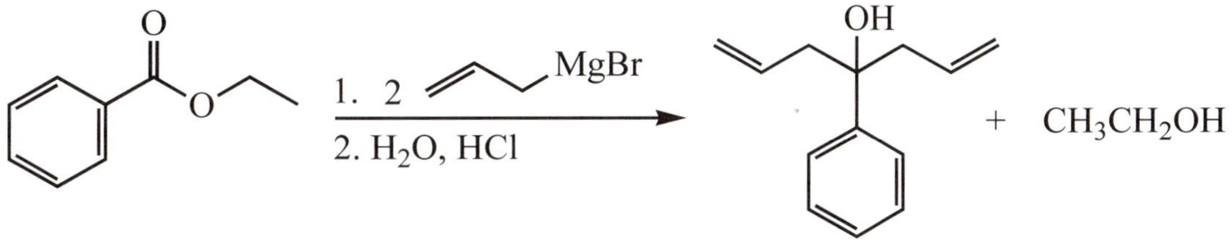

$$+ \quad CH_3CH_2OH$$

(b)

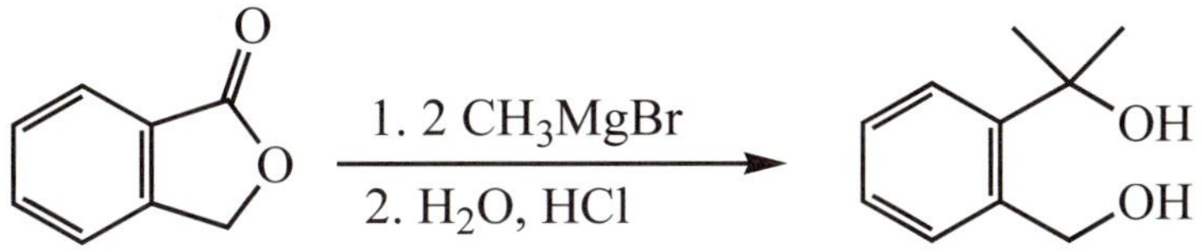

(c)

$$+ \quad CH_3CH_2OH$$

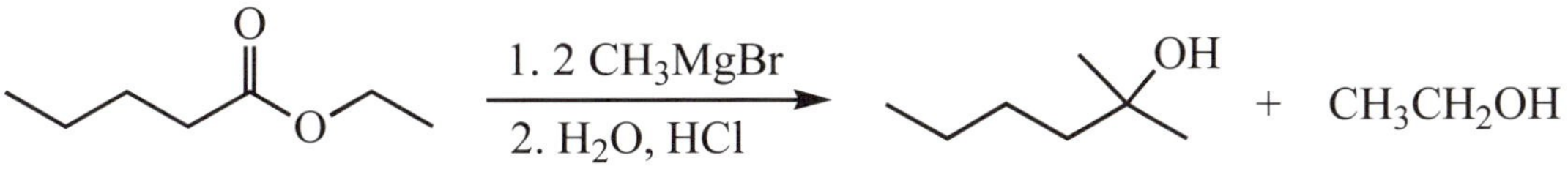

**15.32** *What combination of ester and Grignard reagent can be used to prepare each alcohol?*
*(a) 2-Methyl-2-butanol   (b) 3-Phenyl-3-pentanol     (c) 1,1-Diphenylethanol*

(a)

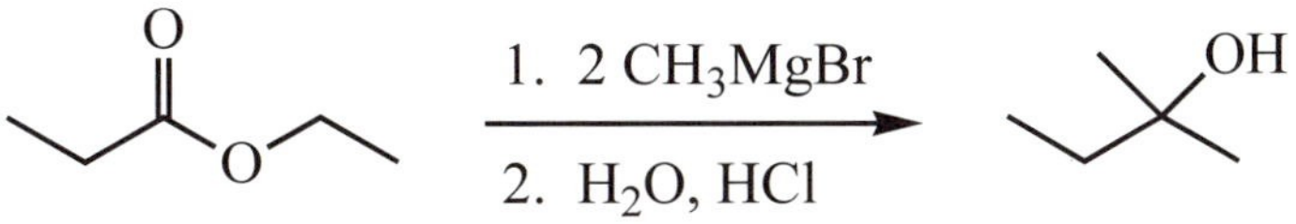

(b)

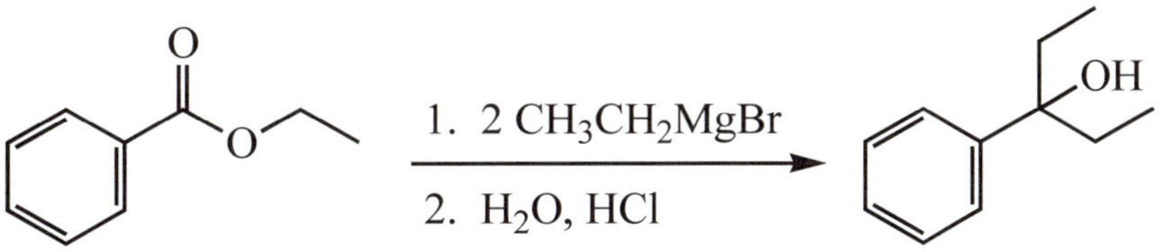

(c)

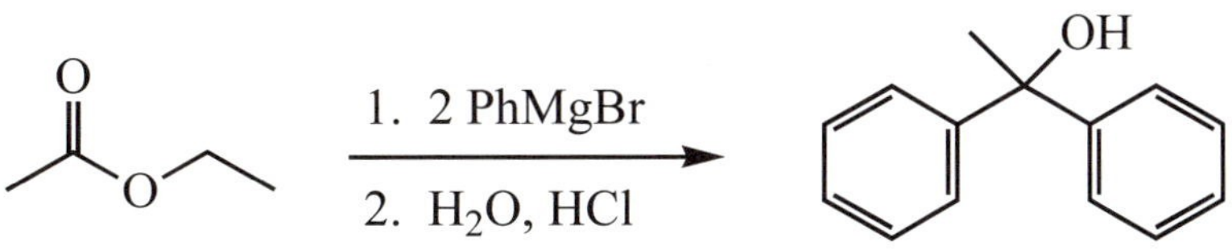

**15.33** *Reaction of a 1° or 2° amine with diethyl carbonate under controlled conditions gives a carbamic ester. Propose a mechanism for this reaction.*

Diethyl carbonate   +   1-Butanamine (Butylamine)   →   A carbamic ester   +   EtOH

**Step 1: Nucleophilic acyl attack by the amine forms an unstable tetrahedral intermediate.**

**Step 2: Collapse of the tetrahedral intermediate forms a protonated carbamic ester.**

**Step 3: Formation of carbamic ester is accomplished by the deprotonation of an ammonium hydrogen.**

**15.34** *Barbiturates are prepared by treatment of diethyl malonate or a derivative of diethyl malonate with urea in the presence of sodium ethoxide as a catalyst. Following is an equation for the preparation of barbital from diethyl 2,2-diethylmalonate and urea. Barbital, a long-duration hypnotic and sedative, is prescribed under a dozen or more trade names.*

Diethyl                 Urea                              5,5-Diethylbarbituric aicd
2,2-diethylmalonate                                              (Barbital)

*(a) Propose a mechanism for this reaction.*

**Step 1: Deprotonation of urea by ethoxide to a resonance-stabilized anion.**

**Step 2: Nucleophilic attack by the deprotonated urea on the ester forms a tetrahedral intermediate.**

**Step 3: The carbonyl forms after the tetrahedral intermediate collapses.**

**Step 4:  Deprotonation of an N-H by ethoxide yields a resonance-stabilized anion.  Only one resonance contributing structure is shown.  Try drawing the other one.**

**Step 5:  Cyclization of the anionic intermediate forms a tetrahedral intermediate.**

**Step 6:  Collapse of the tetrahedral intermediate yields barbital.**

*(b) The pK$_a$ of barbital is 7.4.  Which is the most acidic hydrogen in this molecule and how do you account for its acidity?*

**The most acidic hydrogen is the imide hydrogen.  The enhanced acidity relative to other N-H groups results from the inductive effects from the adjacent carbonyl groups and the resonance stabilization of the conjugate base anion.  The following are three resonance contributing structures for the barbiturate anion.**

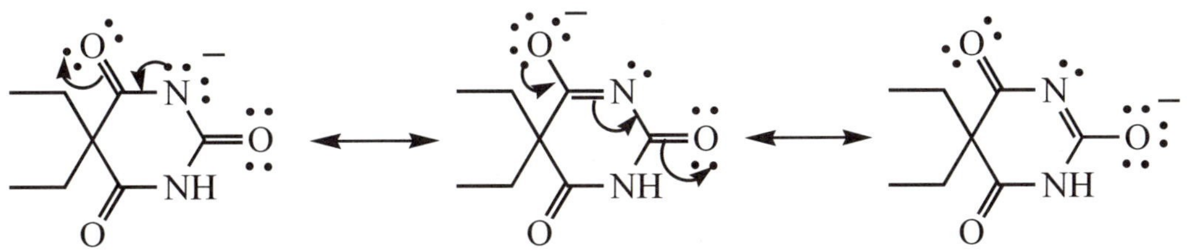

**15.35** *Name and draw structural formulas for the products of complete hydrolysis of meprobamate and phenobarbital in hot aqueous acid. Meprobamate is a tranquilizer prescribed under one or more of 58 different trade names. Phenobarbital is a long-acting sedative, hypnotic, and anticonvulsant. [Hint: Remember that when heated, β-dicarboxylic acids and β-ketoacids undergo decarboxylation (Section 14.9B).]*

(a)

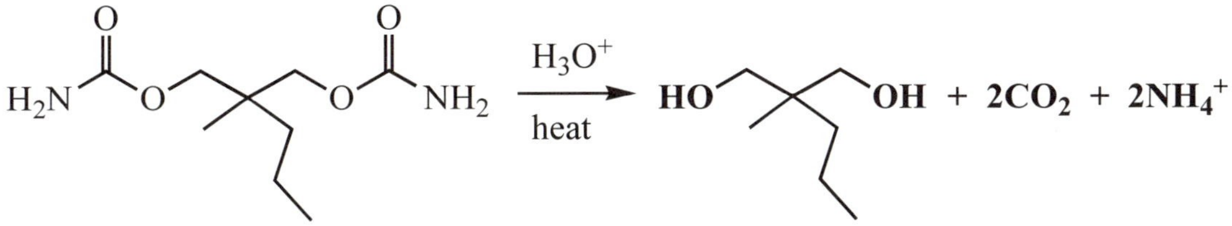

Meprobamate

2-Methyl-2-propyl-
1,3-propanediol

(b)

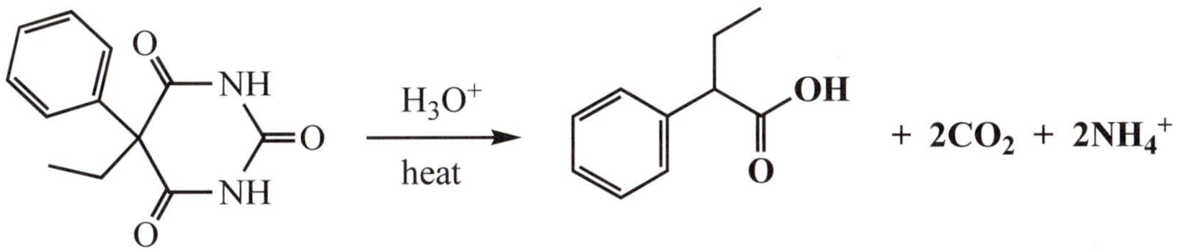

Phenobarbital

2-Phenylbutanoic acid

**Synthesis**

**15.36** *N,N-Diethyl-m-toluamide (DEET), the active ingredient in several common insect repellents, is synthesized from 3-methylbenzoic acid (m-toluic acid) and diethylamine. Show how this synthesis can be accomplished.*

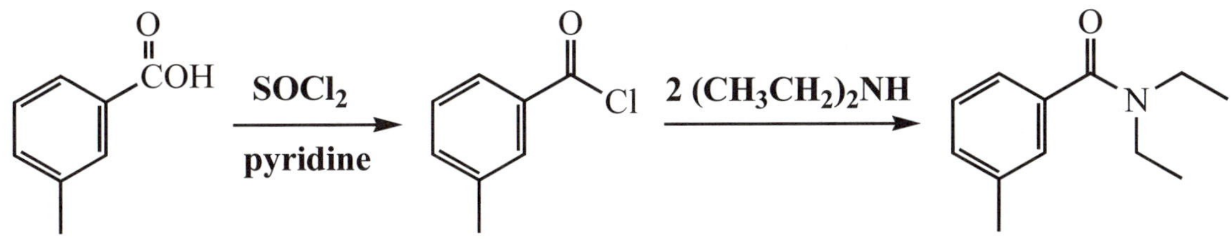

3-methylbenzoic acid
(m-toluic acid)

N,N-Diethyl-m-toluamide
(DEET)

**15.37** *Show how to convert ethyl 2-pentenoate to these compounds.*

(a)

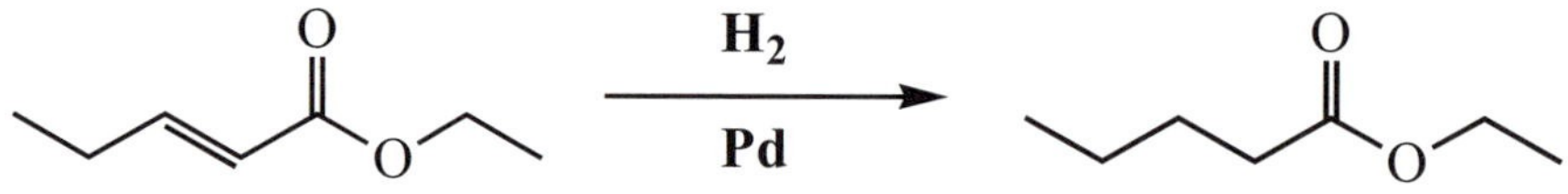

(b)

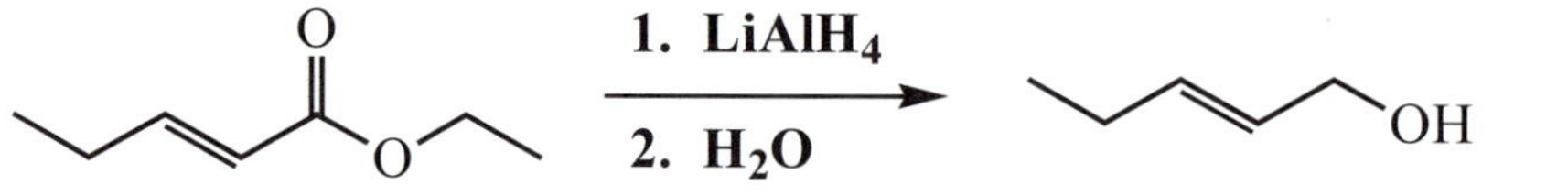

(c)

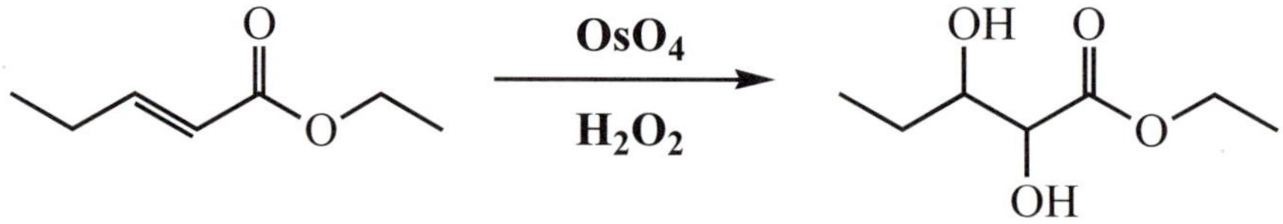

**15.38** *Procaine (its hydrochloride is marketed as Novocaine) was one of the first local anesthetics for infiltration and regional anesthesia. Show how to synthesize procaine using the three given reagents as the sources of carbon atoms.*

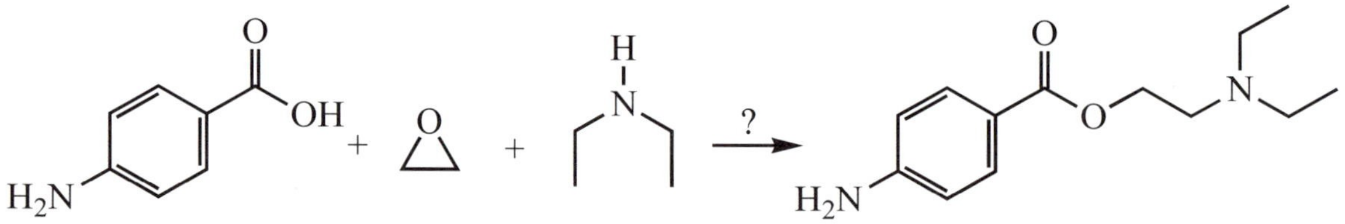

4-Aminobenzoic      Ethylene   Diethylamine                    Procaine
     acid            oxide

**The key to procaine synthesis is the formation of the ester. The alcohol is first synthesized by the reaction of diethylamine with ethylene oxide.**

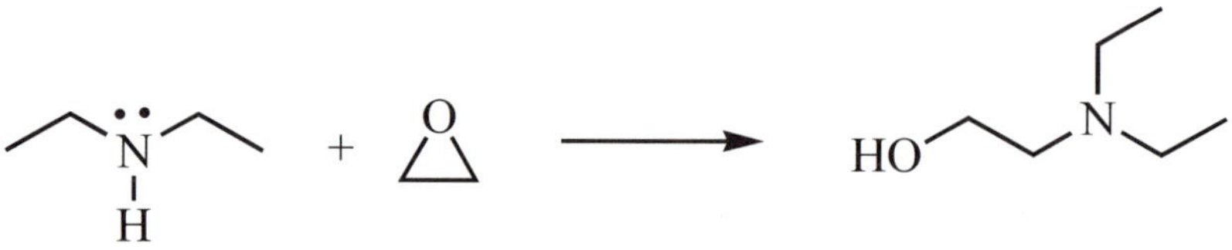

**The alcohol is then reacted with 4-aminobenzoic acid in a Fischer esterification to form the ester. Remember, because the esterification is carried out under acidic conditions, the amine groups will be protonated under the reaction conditions. Appropriate treatment of the product with a base such as NaOH will generate the free amines.**

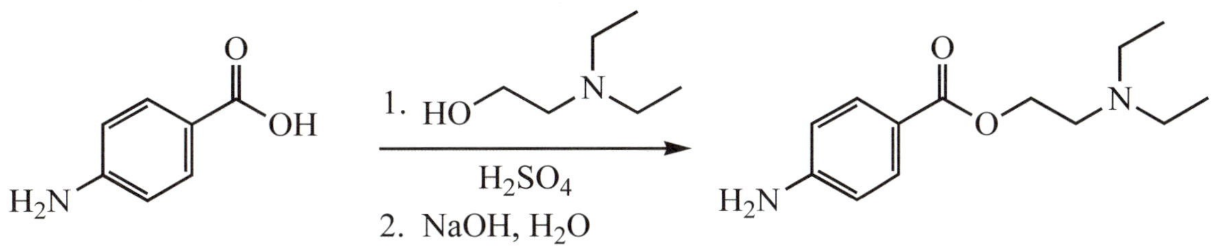

**15.39** *There are two nitrogen atoms in Procaine. Which of the two is the stronger base? Draw the structural formula for the salt formed when Procaine is treated with one mole of aqueous HCl.*

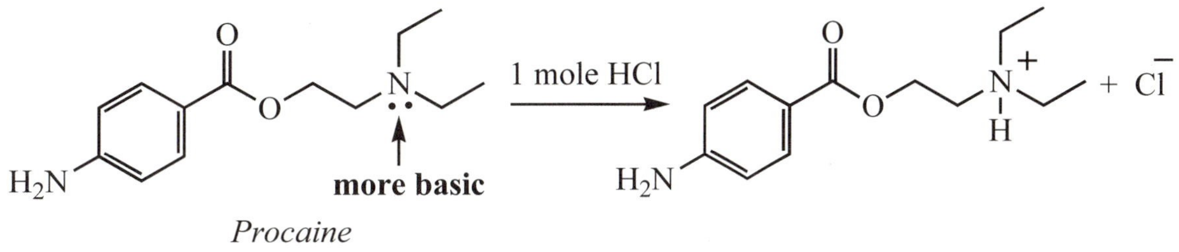

**more basic**

*Procaine*

**Alkylamines are more basic than arylamines. In arylamines, the nitrogen lone pair of electrons is delocalized into the benzene ring through resonance, leaving it less available for bonding with protons and electrophiles.**

**15.40** *Starting materials for the synthesis of the herbicide propranil, a weed killer used in rice paddies, are benzene and propanoic acid. Show reagents to bring about this synthesis.*

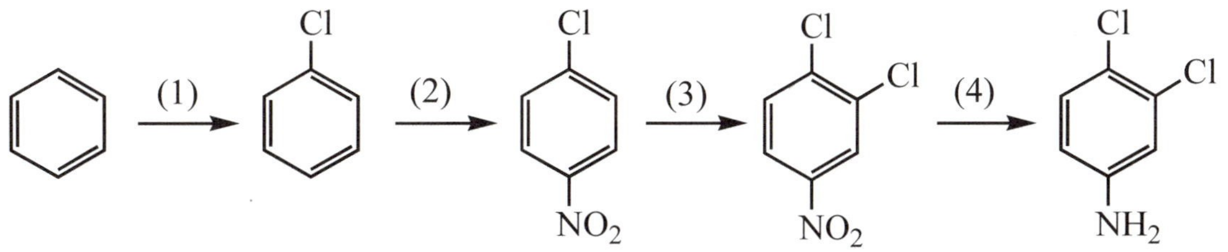

**(1) Cl₂, FeCl₃  (2) HNO₃, H₂SO₄  (3) Cl₂, FeCl₃  (4) H₂, Ni**

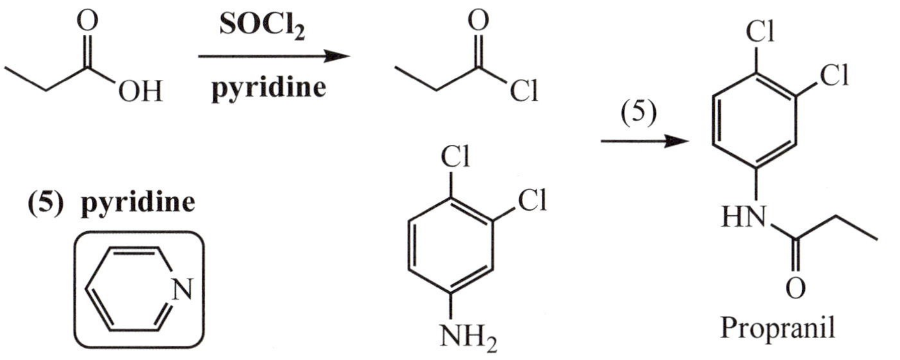

**(5) pyridine**

*Propranil*

**In Step 5, pyridine is used both as solvent and a non-nucleophilic base for the removal of the HCl produced from the reaction. If pyridine is not used, then two moles of 3,4-dichloroaniline must be used, with one mole for the substitution reaction and one mole for the removal of HCl.**

**15.41** *Following are structural formulas for three local anesthetics. Lidocaine was introduced in 1948 and is now the most widely used local anesthetic for infiltration and regional anesthesia. Its hydrochloride is marketed under the name Xylocaine. Etidocaine (its hydrochloride is marketed as Duranest) is comparable to lidocaine in onset, but its analgesic action lasts two to three times longer. Mepivacaine (its hydrochloride is marketed as Carbocaine) is faster and somewhat longer in duration than lidocaine.*

Lidocaine
(Xylocaine)

Etidocaine
(Duranest)

Mepivacaine
(Carbocaine)

*(a) Propose a synthesis of lidocaine from 2,6-dimethylaniline, chloroacetyl chloride (ClCH₂COCl), and diethylamine.*

*(b) Propose a synthesis of etidocaine from 2,6-dimethylaniline, 2-chlorobutanoyl chloride, and ethylpropylamine.*

*(c) What amine and acid chloride can be reacted to give mepivacaine?*

**15.42** *Following is the outline of a five-step synthesis for the anthelmintic (against worms) diethylcarbamazine. Diethylcarbamazine is used chiefly against nematodes, small cylindrical or slender threadlike worms such as the common roundworm, which are parasitic in animals and plants.*

Ethylene oxide

Diethylcarbamazine

*(a) Propose a reagent for Step 1. Which mechanism is more likely for this step; $S_N1$ or $S_N2$? Explain.*

**The reagent used for Step 1 is $NH(CH_3)_2$. It reacts with ethylene oxide in an $S_N2$ mechanism. Ethylene oxide is highly reactive toward moderate to good nucleophiles and methylamine is a moderate nucleophile. Many weaker nucleophiles require acidic conditions to react with ethylene oxide.**

*(b) Propose a reagent for Step 2.*

**Thionyl chloride ($SOCl_2$) will be effective for converting alcohol hydroxyls to chlorides (remember, HCl is formed in this reaction, thus protonating the amine), followed by treating the resulting hydrochloride salt with dilute NaOH solution to obtain the free amine.**

*(c) Propose a reagent for Step 3.*

**Step 3 requires NH₃ to complete the reaction.**

*(d) Ethyl chloroformate, the reagent for Step 4, is both an acid chloride and an ester. Account for the fact that Cl is displaced from this reagent rather than OCH₂CH₃.*

**Chloride is selectively displaced because it is a much better leaving group than ethoxide. The higher electronegativity of the chlorine atom relative to oxygen atom provides more stabilization of the negative charge on the chloride anion than oxygen stabilizes the negative charge in an ethoxide anion.**

**15.43**  *Following is an outline of a five-step synthesis for methylparaben, a compound widely used as a preservative in foods. Propose reagents for each step.*

Toluene

Methyl 4-hydroxybenzoate
(Methylparaben)

**(1) HNO₃, H₂SO₄   (2) Na₂Cr₂O₇, H₂SO₄   (3) CH₃OH, H₂SO₄**

**(4) H₂, Ni   (5) NaNO₂, H₂SO₄ T < 5 ûC then H₂O**

**Looking Ahead**

**15.44**  *Identify the most acidic proton in each of the following esters.*

*(a)*

**Most acidic**

**Resonance-stabilized conjugate base**

**Resonance-stabilized conjugate base**

The loss of the most acidic protons results in the most stabilized conjugate base, which is also the weakest conjugate base.

**15.45** *Does a nucleophilic acyl substitution occur between the ester and the nucleophile shown? Propose an experiment that would verify your answer?*

Label the oxygen in sodium methoxide with an isotope of oxygen such as oxygen 18 (O-18). Using excess O-18 labeled sodium methoxide, if the ester replaces the O-16 methoxy group with an O-18 methoxy group, then acyl substitution occurred.

**15.46** *Explain why a nucleophile, Nu, attacks not only the carbonyl carbon, but also the other β-carbon as indicated in the following α,β-unsaturated ester.*

Close inspection of the resonance structures of the α,β-unsaturated ester reveals electrophilic sites at the carbonyl carbon and the β-carbon. These electrophilic carbons are prime targets for nucleophiles.

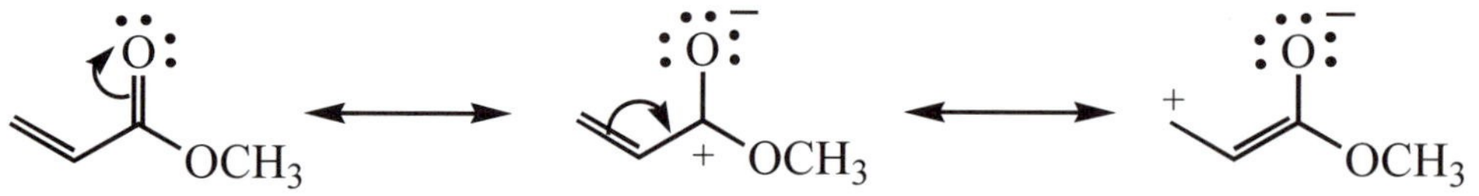

**Nucleophilic attack at the carbonyl carbon generates an anionic tetrahedral intermediate, with the negative charge stabilized by an electronegative oxygen. The tetrahedral intermediate can collapse by ejecting a methoxide anion.**

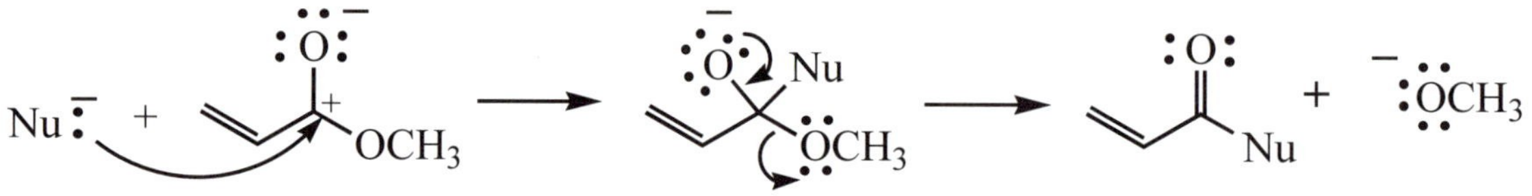

**Nucleophilic attack at the β-carbon generates a resonance-stabilized anion referred to as an enolate anion, with the negative charge stabilized through resonance.**

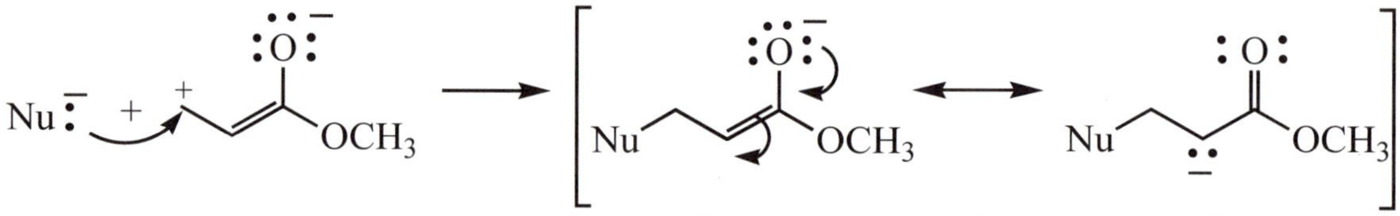

**Resonance-stabilized intermediate**

**15.47** *Explain why a Grignard reagent will not undergo nucleophilic acyl substitution with the following amide.*

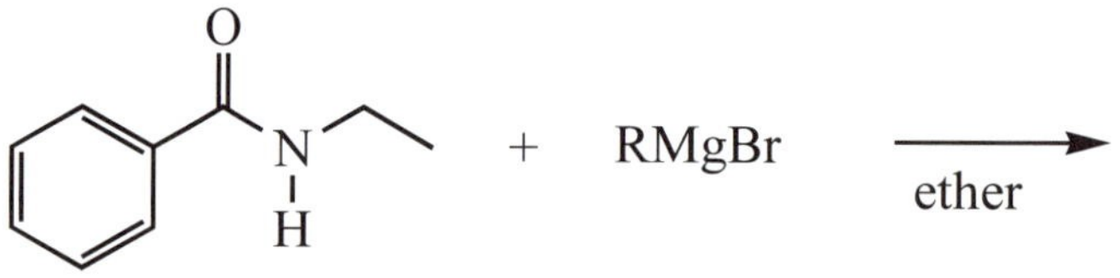

**The amide proton is relatively acidic to a Grignard reagent, therefore, the Grignard reagent will deprotonate the amide instead of undergoing acyl substitution.**

**15.48** *At low temperatures, the amide shown below exhibits cis-trans isomerism, while at higher temperatures it does not. Explain how this is possible.*

**The amide is stabilized by resonance and the hybrid structure has significant double bond character between the nitrogen and carbonyl carbon. This significant double bond character hinders bond rotation about the C-N bond. At low temperatures, there isn't enough energy to exceed the energy barrier of rotation, thus two different resonance structures can exist that are cis-trans isomers. At higher temperatures,**

**there is enough energy to exceed the barrier of rotation and amide C-N bond can freely rotate.**

## CHAPTER 16
### Solutions to Problems

**16.1**    *Identify the acidic hydrogens in each compound.*

(a) *Cyclohexanone*

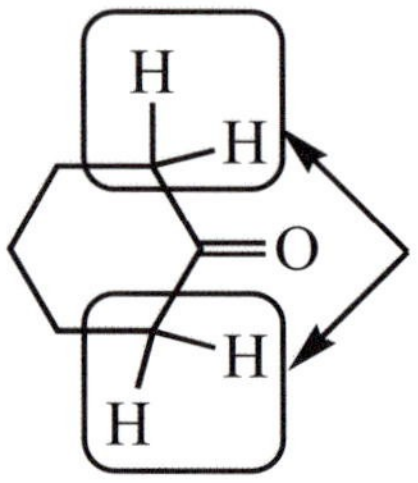

most acidic

(b) *Acetophenone*

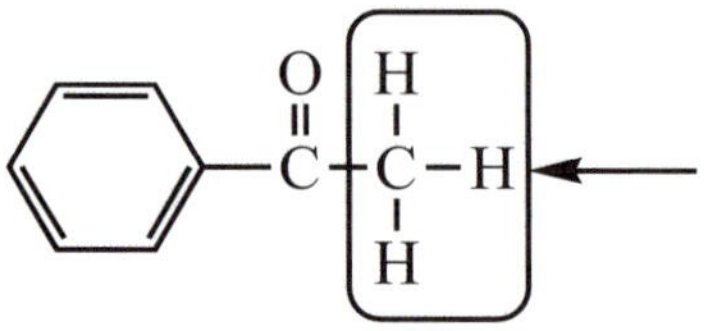

most acidic

**16.2**    *Draw the product of the base-catalyzed aldol reaction of each compound.*

(a)

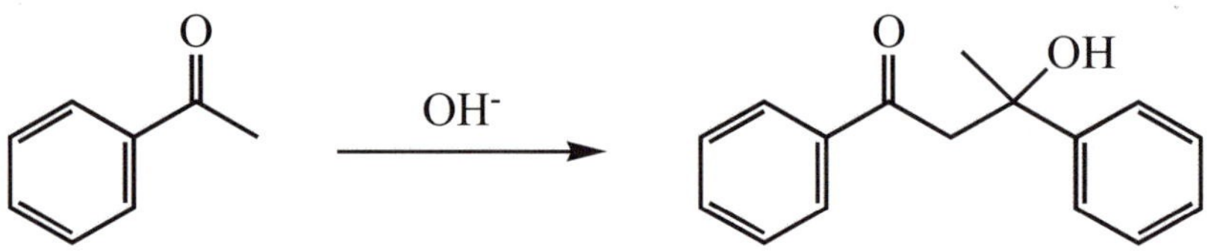

OH⁻

(b)

OH⁻

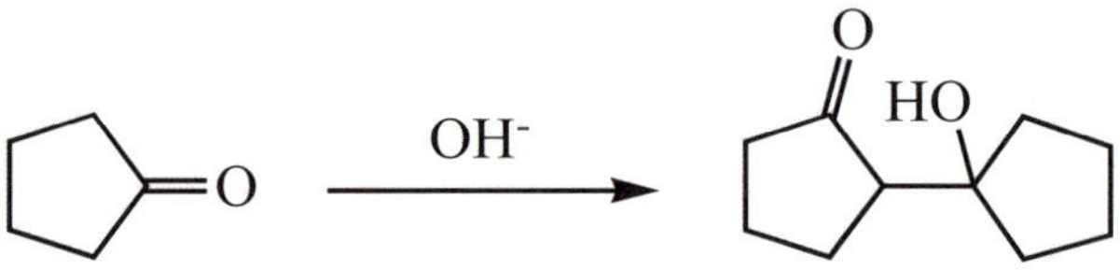

**16.3**    *Draw the product of base-catalyzed dehydration of each aldol product from Problem 16.2.*

(a)

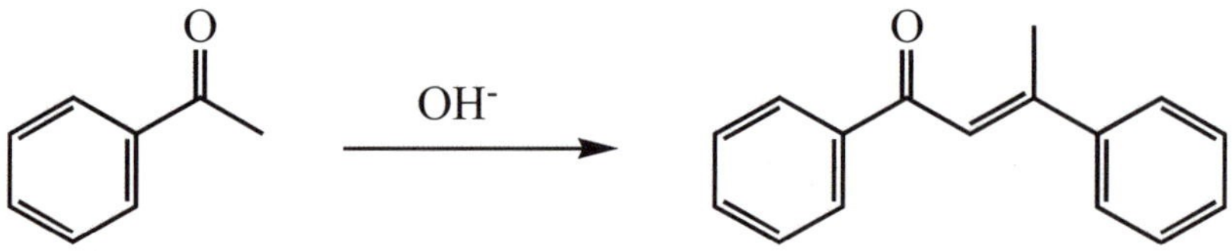

OH⁻

(b)

OH⁻

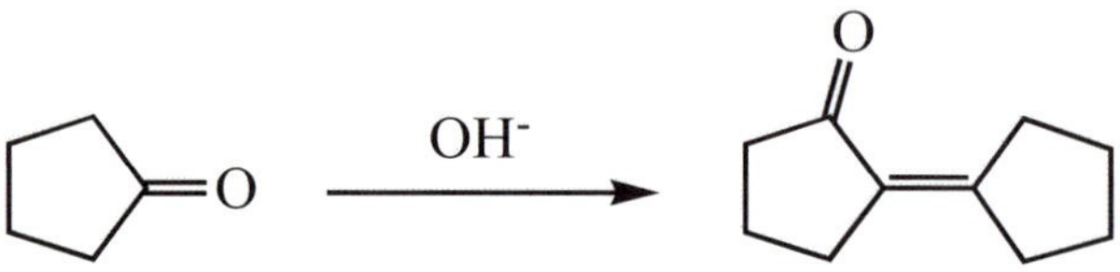

**16.4**   *Draw the product of the crossed aldol reaction between benzaldehyde and 3-pentanone and the product formed by its base-catalyzed dehydration.*

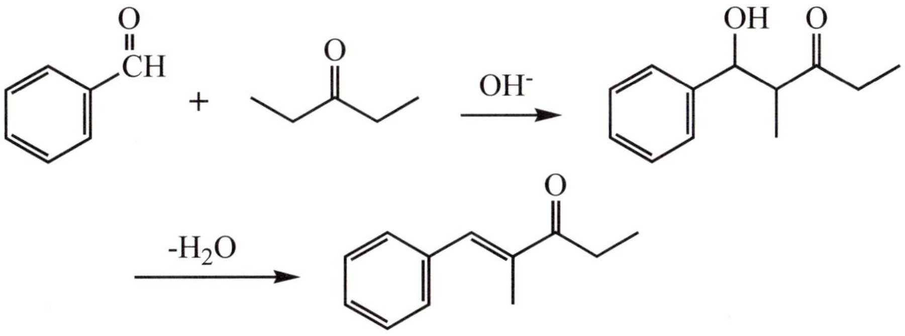

**16.5**   *Show the product of Claisen condensation of ethyl 3-methylbutanoate in the presence of sodium ethoxide.*

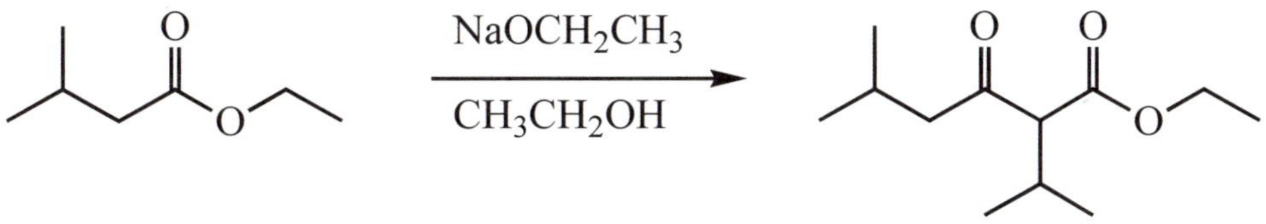

**16.6**   *Complete the equation for this crossed Claisen condensation:*

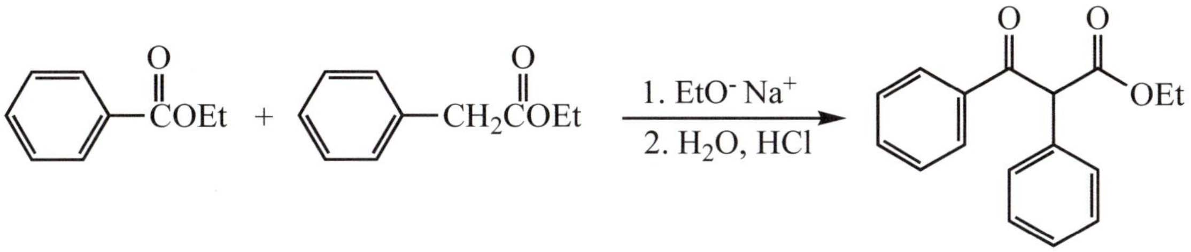

**16.7**   *Show how to convert benzoic acid to 3-methyl-1-phenyl-1-butanone using a Claisen condensation at some stage in the synthesis.*

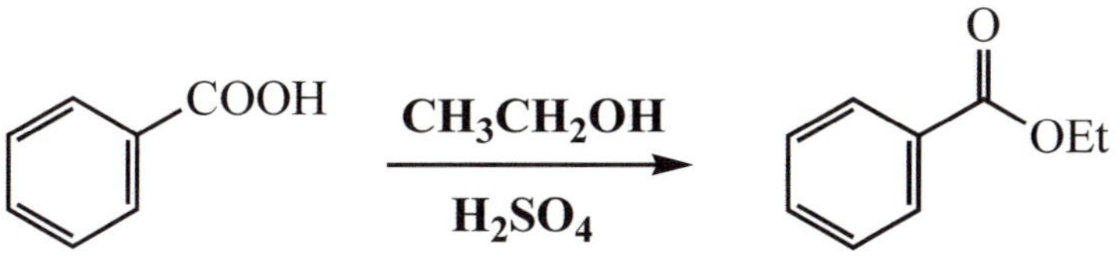

**Benzoic acid**

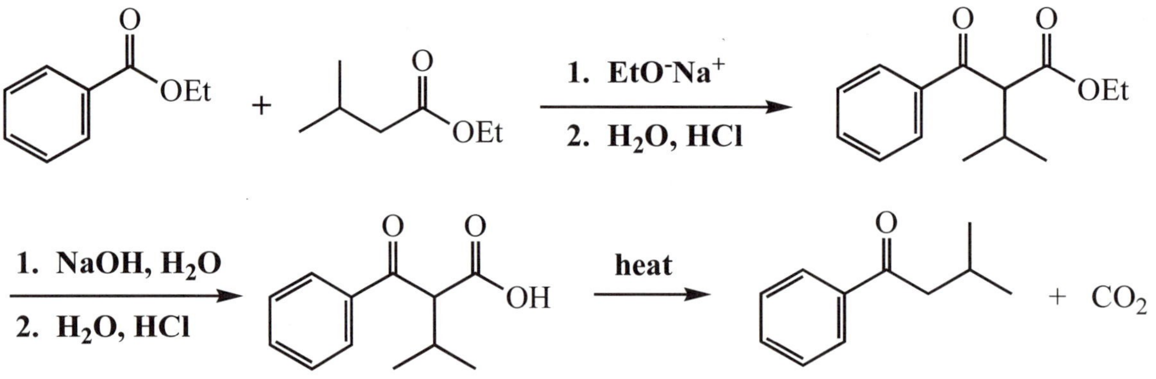

**3-Methyl-1-phenyl-1-butanone**

**16.8** *Show the product formed from each Michael product in the solution to Example 16.9 after (1) hydrolysis in aqueous NaOH, (2) acidification, and (3) thermal decarboxylation of each β-ketoacid or β-dicarboxylic acid. These reactions illustrate the usefulness of the Michael reaction for the synthesis of 1,5-dicarbonyl compounds.*

*(a)*

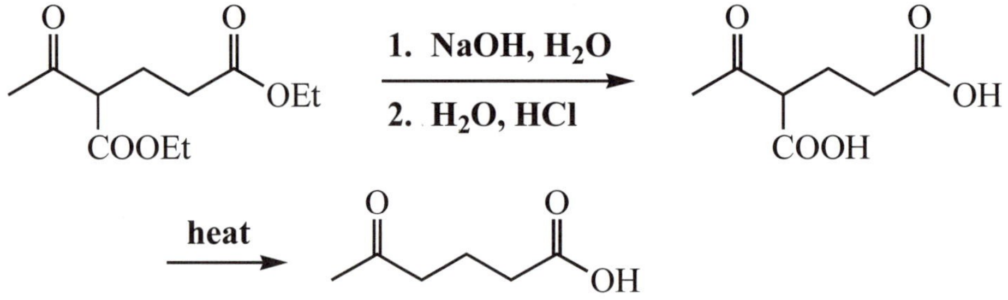

*(b)*

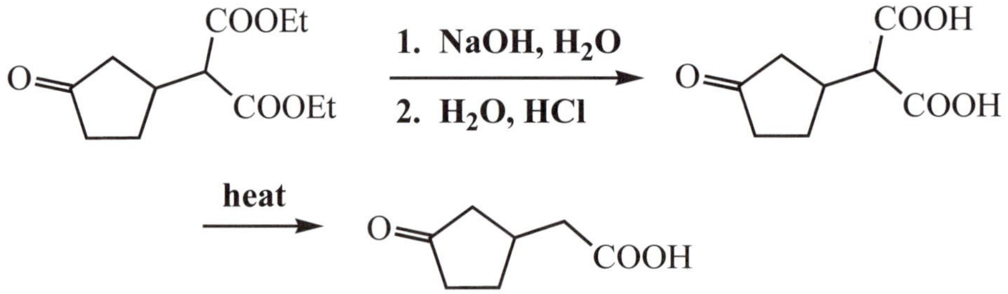

**16.9** *Show how the sequence of Michael reaction, hydrolysis, acidification, and thermal decarboxylation can be used to prepare pentanedioic acid (glutaric acid).*

**16.10** *The product of the double Michael reaction in Example 16.11 is a diester and, when treated with sodium ethoxide in ethanol, undergoes a Dieckmann condensation. Draw the structural formula for the product of this Dieckmann condensation followed by acidification with aqueous HCl.*

**The Aldol Reaction**

**16.11** *Estimate the pK$_a$ of each compound and then arrange them in order of increasing acidity.*

(a)   $CH_3\overset{O}{\overset{\|}{C}}CH_3$       (b)   $CH_3\overset{OH}{\overset{|}{C}H}CH_3$       (c) $CH_3CH_2\overset{O}{\overset{\|}{C}}OH$

      pK$_a$ ~ 20                pK$_a$ ~ 17               pK$_a$ ~ 5

**The order of acidity ranked from least to most acidic is:**

$$CH_3\overset{O}{\overset{\|}{C}}CH_3 \quad < \quad CH_3\overset{OH}{\overset{|}{C}H}CH_3 \quad < \quad CH_3CH_2\overset{O}{\overset{\|}{C}}OH$$

**16.12** *Identify the most acidic hydrogen(s) in each compound.*

**The most acidic proton is circled. In general, the order of acidity goes as follows:**

$$R-\overset{O}{\overset{\|}{C}}OH \; > \; Ph-OH \; > \; R-OH \; > \; R-\overset{O}{\overset{\|}{C}}-\overset{H}{\overset{|}{C}}-$$

(a) 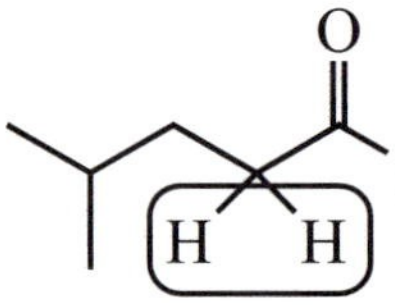   (b) 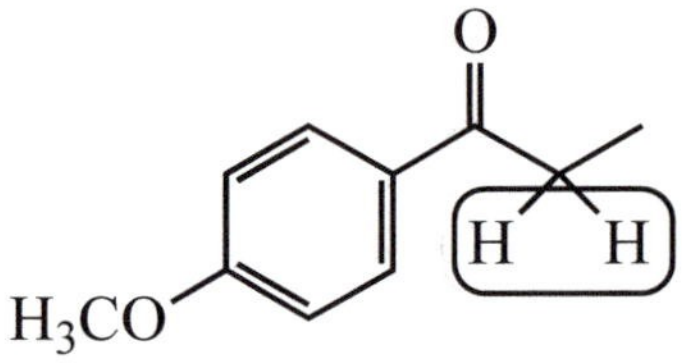   (c)

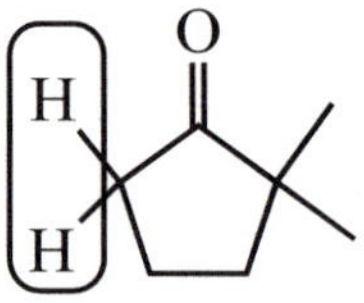

(d) 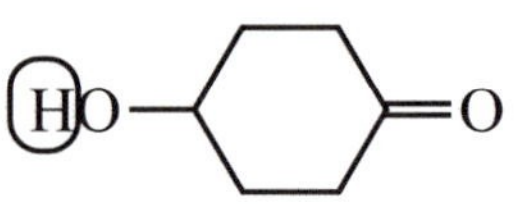   (e) 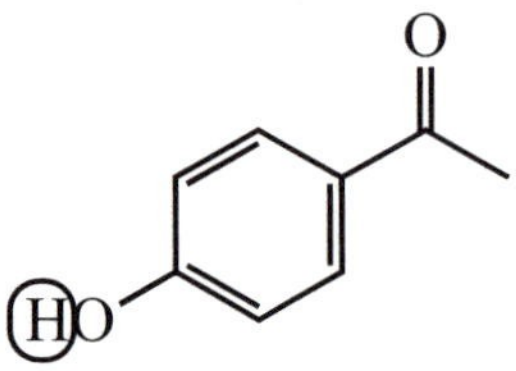   (f)

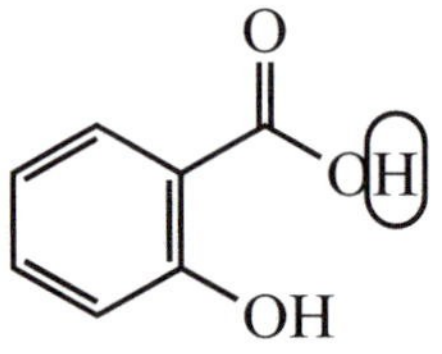

**16.13**   *Write a second contributing structure of each anion and use curved arrows to show the redistribution of electrons that gives your second structure.*

(a) 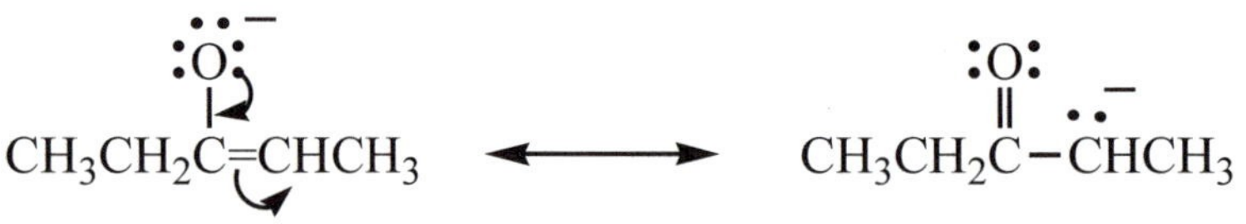   $CH_3CH_2C=CHCH_3$   $\longleftrightarrow$   $CH_3CH_2C-CHCH_3$

(b) 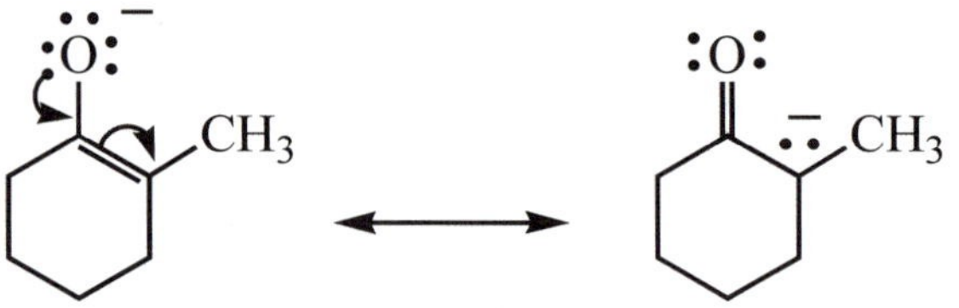   $CH_3$   $\longleftrightarrow$   $CH_3$

(c) 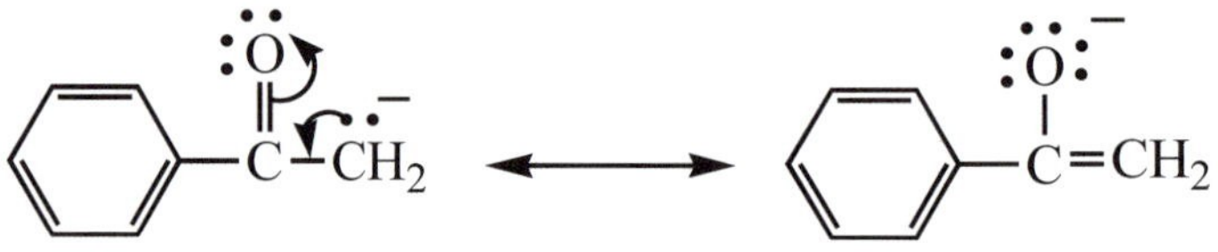   $C-CH_2$   $\longleftrightarrow$   $C=CH_2$

**16.14**   *Treatment of 2-methylcyclohexanone with base gives two different enolate anions. Draw the contributing structure for each that places the negative charge on carbon.*

**In 2-methylcyclohexanone, there are two different α-hydrogens. Deprotonation of either one leads to two isomeric enolates. Each enolate is stabilized by resonance and the resulting resonance hybrid is composed of two contributing structures where the negative charge is found on the oxygen and carbon atoms.**

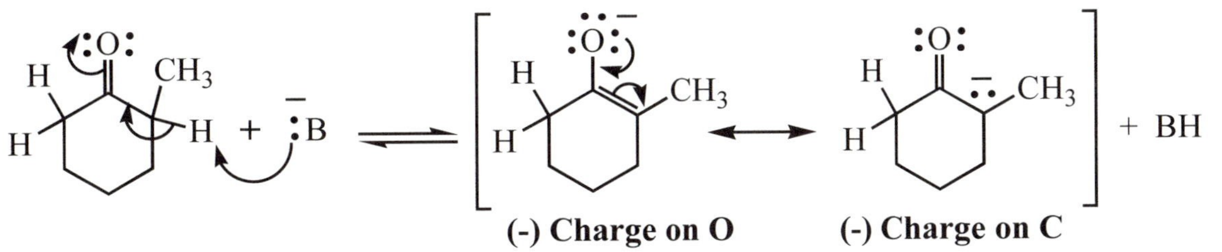

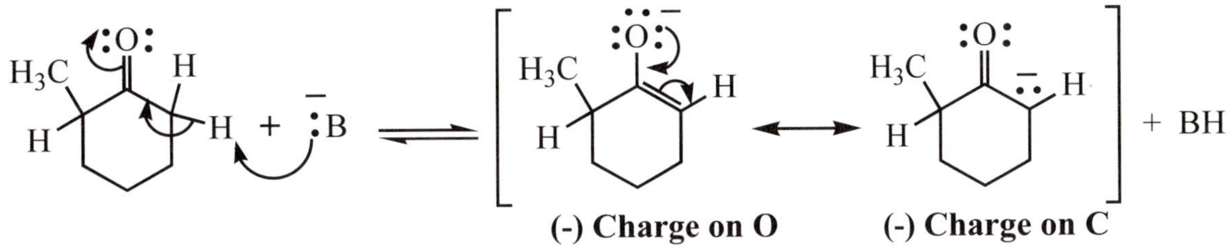

**(-) Charge on O**          **(-) Charge on C**

\+ BH

**(-) Charge on O**          **(-) Charge on C**

\+ BH

**16.15** *Draw a structural formula for the product of the aldol reaction of each compound and for the α,β-unsaturated aldehyde or ketone formed by dehydration of each aldol product.*

(a) $\xrightarrow{OH^-}$

$\xrightarrow{-H_2O}$          +

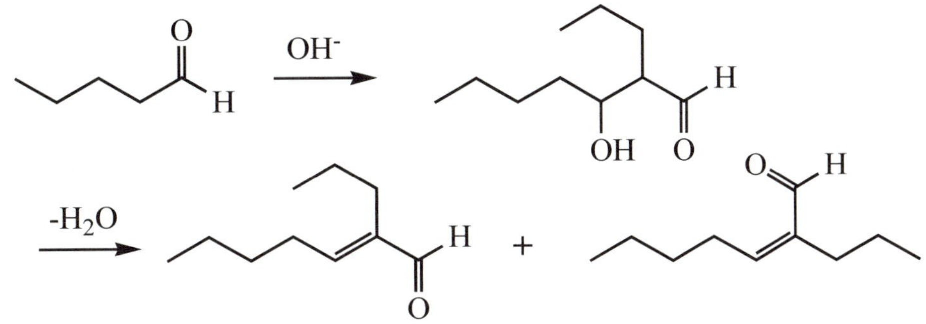

(b) $\xrightarrow{OH^-}$

$\xrightarrow{-H_2O}$

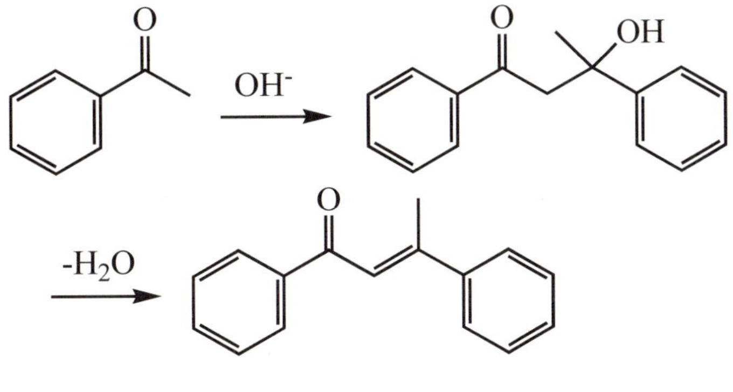

(c) $\xrightarrow{OH^-}$          $\xrightarrow{-H_2O}$

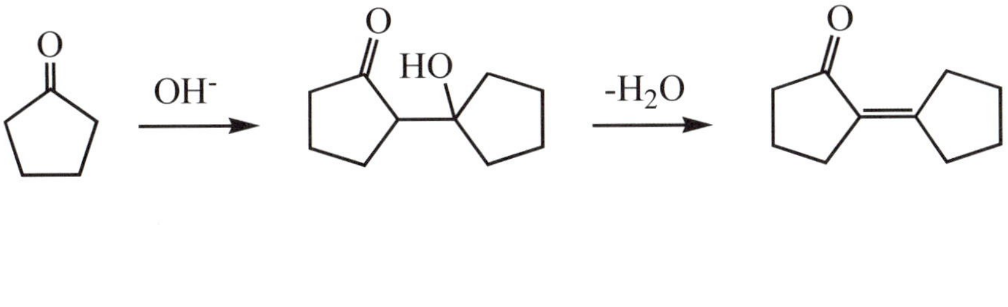

(d)

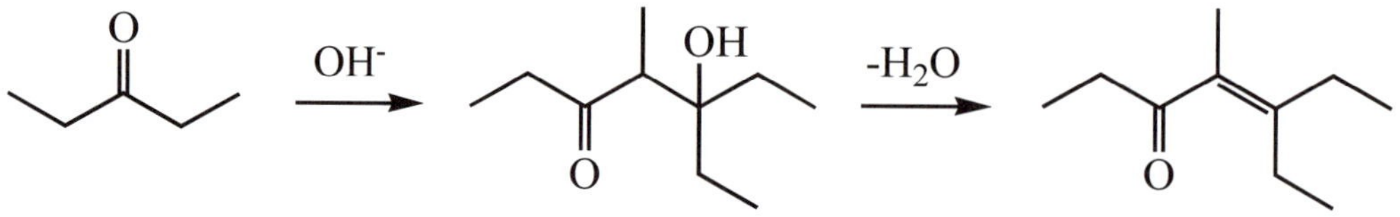

**16.16** *Draw a structural formula for the product of each crossed aldol reaction and for the compound formed by dehydration of each aldol product.*

(a) $(CH_3)_3CCH$ + $CH_3CCH_3$

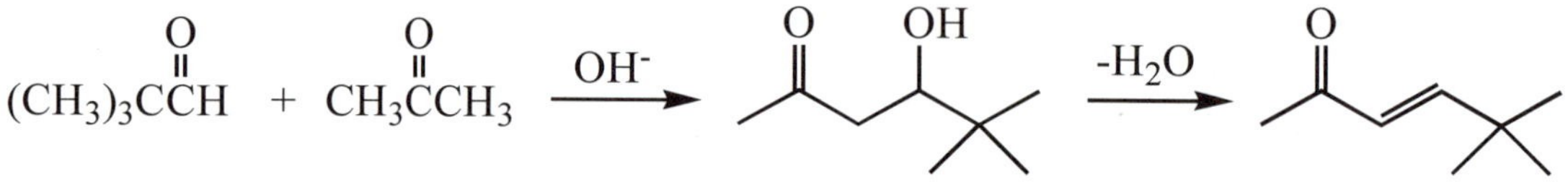

(b)

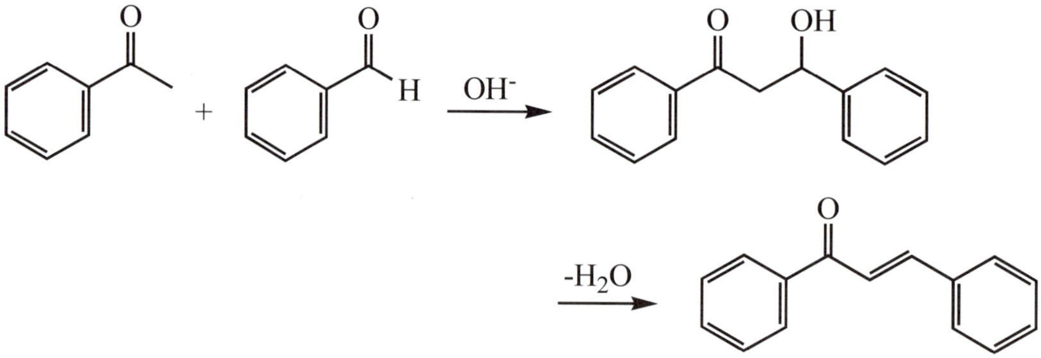

(c)

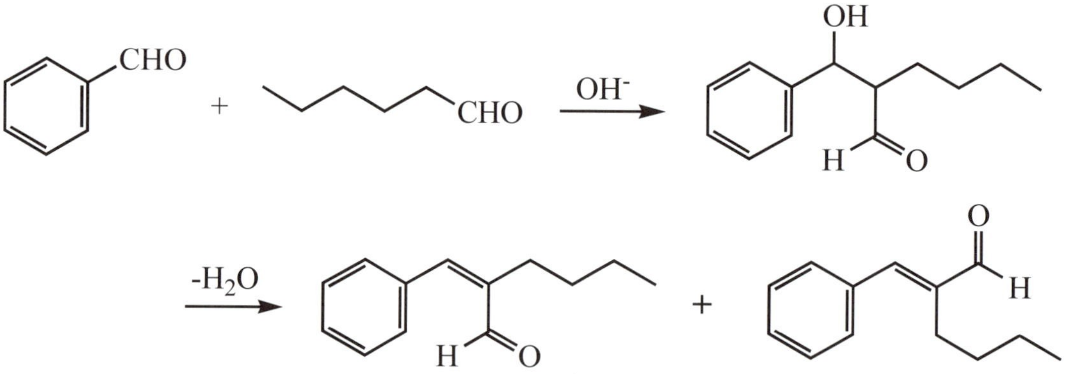

(d)

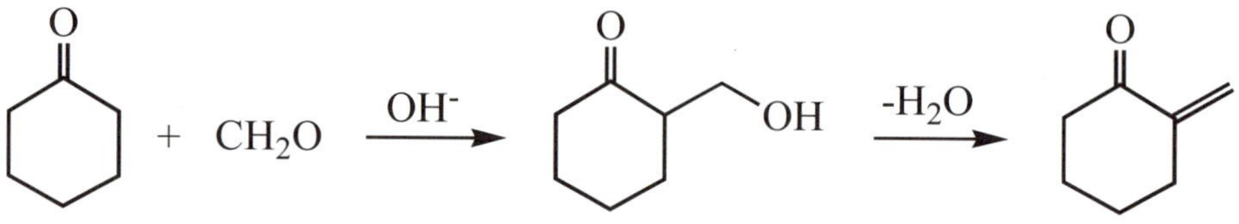

**16.17** *When a 1:1 mixture of acetone and 2-butanone is treated with base, five aldol products are possible. Draw a structural formula for each.*

**There are three different enolate anions that can be formed from two different ketones. Each of these enolates can react with each ketone to form six different products as shown.**

**16.18** *Show how to prepare each α,β-unsaturated ketone by an aldol reaction followed by dehydration of the aldol product.*

(a)

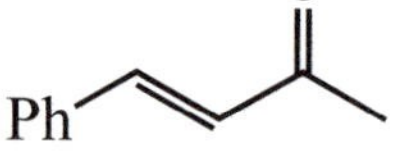

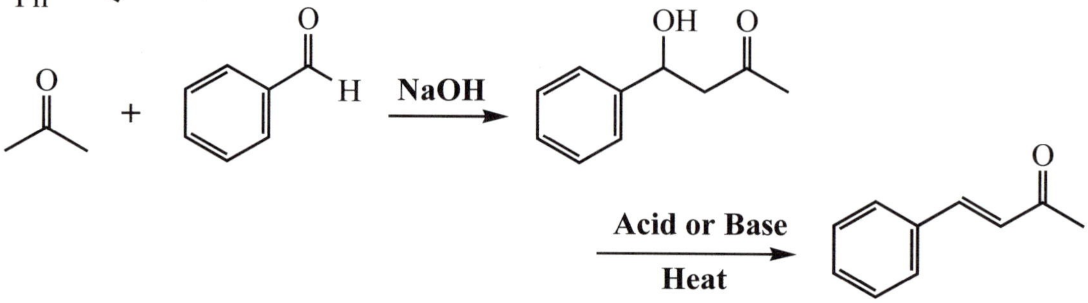

(b)

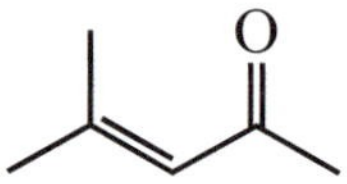

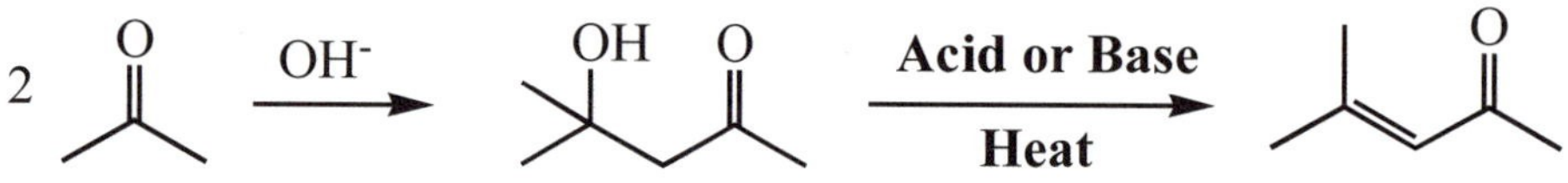

**16.19** *Show how to prepare each α,β-unsaturated aldehyde by an aldol reaction followed by dehydration of the aldol product.*

(a)

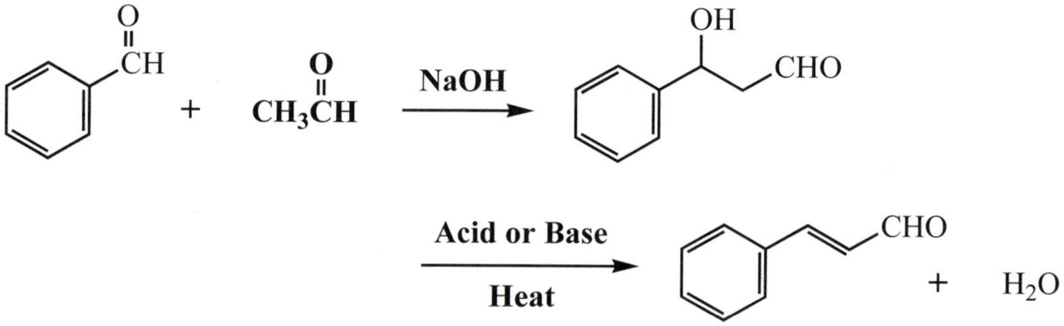

(b)

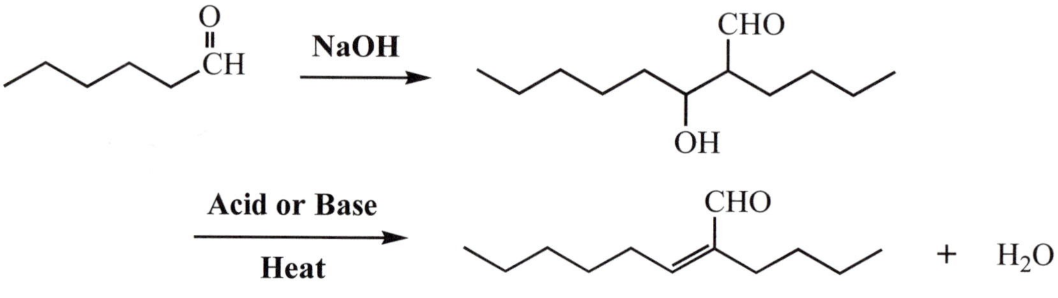

**16.20** *When treated with base, the following compound undergoes an intramolecular aldol reaction followed by dehydration to give a product containing a ring (yield 78%). Propose a structural formula for this product.*

**Look for intramolecular aldol condensations that will form five- or six-membered rings, as their formation is favored.**

$$\text{base} \longrightarrow C_{10}H_{14}O + H_2O$$

**There are three α-carbons that can form an enolate anion and then undergo an intramolecular aldol condensation. Two of these possible intramolecular aldol condensations result in three-membered rings, which are not favored because of high ring strain. The third possibility results in a stable five-membered ring aldol product.**

**Enolate anion forms at this carbon**

base

+ H₂O

**Nucleophilc enolate carbon attacks at the carbonyl carbon**

**16.21** *Propose a structural formula for the compound of molecular formula $C_6H_{10}O_2$ that undergoes an aldol reaction followed by dehydration to give this α,β-unsaturated aldehyde.*

$$C_6H_{10}O_2 \xrightarrow{\text{base}}$$

CHO　　+　　H₂O

1-Cyclopentenecarbaldehyde

**Hexanedial will undergo an intramolecular aldol condensation to yield 1-cyclopentenecarbaldeyde.**

1. NaOH; H₂O

2. NaOH; heat

+　H₂O

**Hexanedial**

**16.22** *Show how to bring about this conversion.*

**This reaction is an interesting intramolecular aldol condensation. An enolate can form on any one of the α-carbons and then undergo a nucleophilic attack on the other carbonyl. The result is the same aldol product no matter which α-carbon forms the enolate. One possibility involves an enolate formation on C2, which then attacks C6, yielding the bicyclic aldol product. The aldol product dehydrates upon heating with either acid or base to form the α,β-unsaturated ketone.**

**16.23** *Oxanamide, a mild sedative, is synthesized from butanal in these five steps.*

Butanal

2-Ethyl-2-hexenal          2-Ethyl-2-hexenoic acid

2-Ethyl-2-hexenoyl          2-Ethyl-2-hexenamide                Oxanamide
chloride

*(a) Show reagents and experimental conditions to bring about each step in this synthesis.*

**(1) NaOH, H$_2$O   (2) H$^+$, heat   (3) SOCl$_2$   (4) 2 mol NH$_3$   (5) CH$_3$CO$_3$H**

*(b) How many stereocenters are in oxanamide? How many stereoisomers are possible for this compound?*

**There are two stereocenters in oxanamide setting up the possibility of four stereoisomers.**

**16.24** *Propose a structural formula for each lettered compound.*

A                                   B

**The Claisen and Dieckmann Condensations**

**16.25** *Show the product of Claisen condensation of each ester.*

*(a) Ethyl phenylacetate in the presence of sodium ethoxide.*

*(b) Methyl hexanoate in the presence of sodium methoxide.*

**16.26** *Draw a structural formula for the product of saponification, acidification, and decarboxylation of each β-ketoester formed in Problem 16.25.*

*(a)*

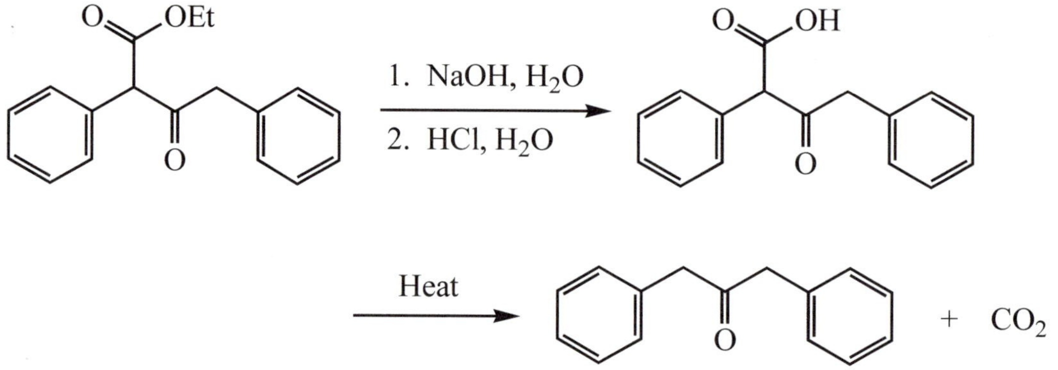
OEt
O
OH
1. NaOH, $H_2O$
2. HCl, $H_2O$
Heat
+ $CO_2$

*(b)*

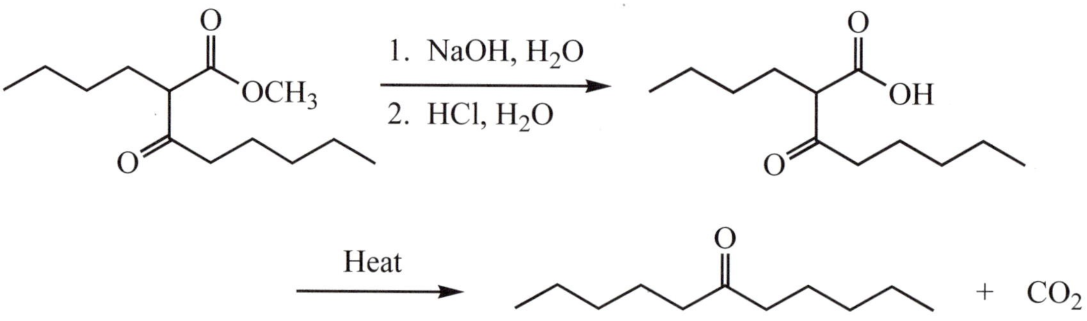
O
$OCH_3$
OH
1. NaOH, $H_2O$
2. HCl, $H_2O$
Heat
+ $CO_2$

**16.27** *When a 1:1 mixture of ethyl propanoate and ethyl butanoate is treated with sodium ethoxide, four Claisen condensation products are possible. Draw a structural formula for each product.*

**The solution to this problem is similar to the solution for problem 16.17. There are two different ester enolate anions that can be formed from two different esters. Each of these enolates can react with each ester to form four different products as shown.**

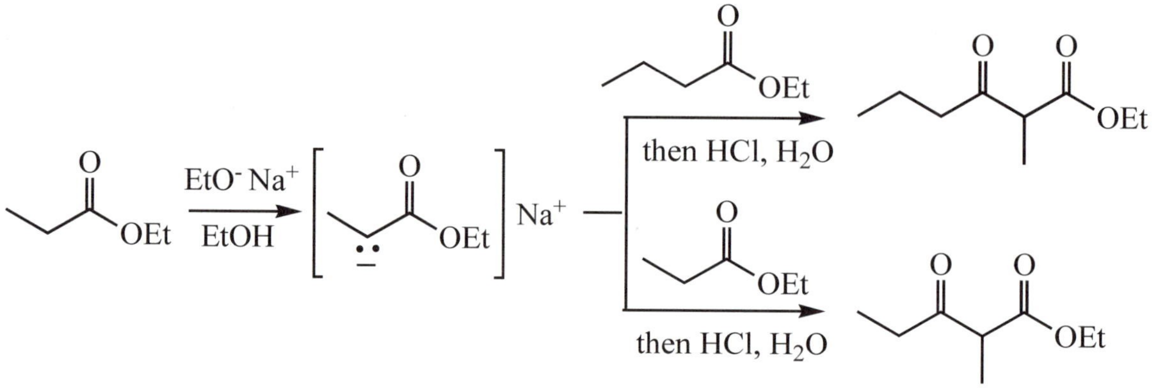
O
OEt
$EtO^- Na^+$
EtOH
$Na^+$
then HCl, $H_2O$
then HCl, $H_2O$

**16.28** *Draw a structural formula for the β-ketoester formed in the crossed Claisen condensation of ethyl propanoate with each ester:*

**16.29** *Complete the equation for this crossed Claisen condensation:*

**16.30**  *The Claisen condensation can be used as one step in the synthesis of ketones, as illustrated by this reaction sequence. Propose structural formulas for compounds A, B, and the ketone formed in this sequence.*

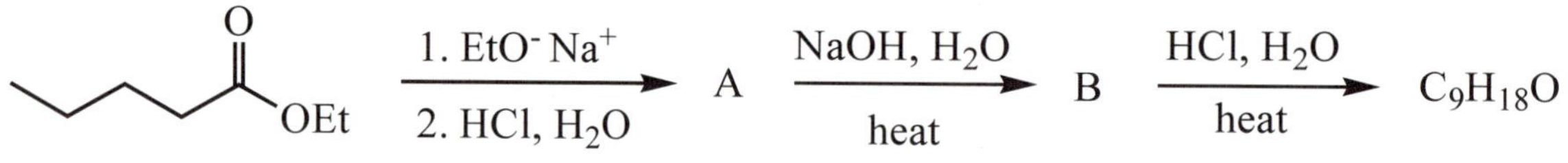

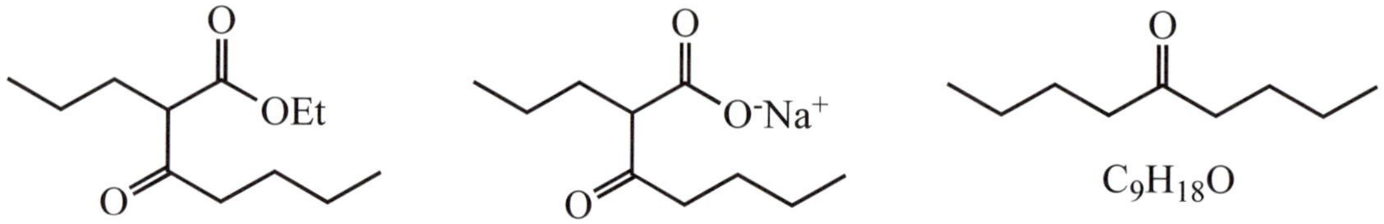

**16.31**  *Draw a structural formula for the product of treating each diester with sodium ethoxide followed by acidification with HCl. (Hint: These are Dieckmann condensations.)*

*(a)*

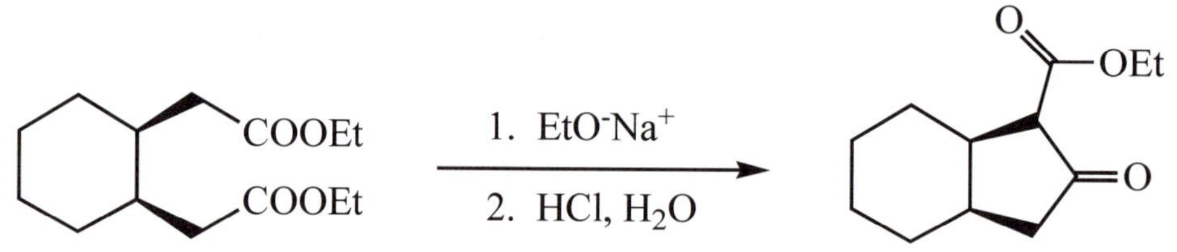

*(b)*

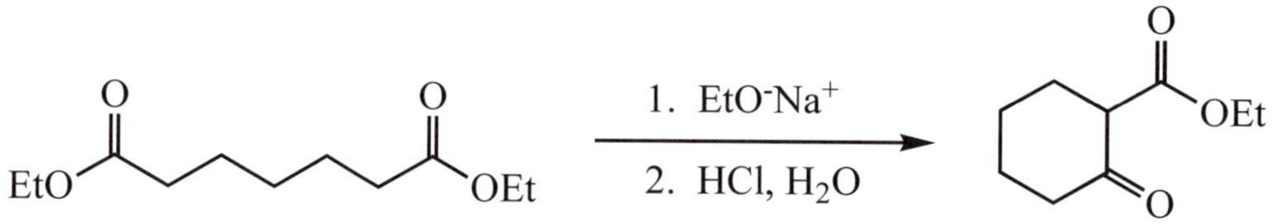

**16.32**  *Claisen condensation between diethyl phthalate and ethyl acetate followed by saponification, acidification, and decarboxylation forms a diketone, $C_9H_6O_2$. Propose structural formulas for compounds A, B, and the diketone.*

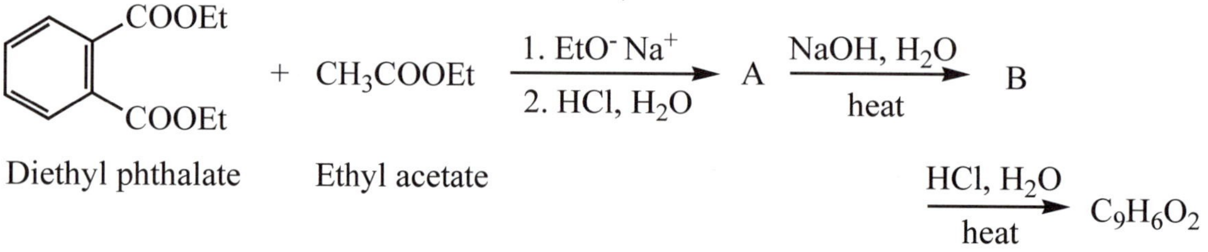

Diethyl phthalate          Ethyl acetate

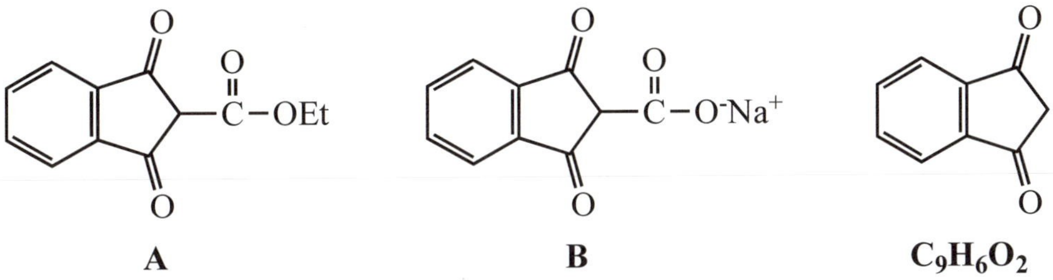

A                              B                              $C_9H_6O_2$

**16.33** *The rodenticide and insecticide pindone is synthesized by the following sequence of reactions. Propose a structural formula for pindone.*

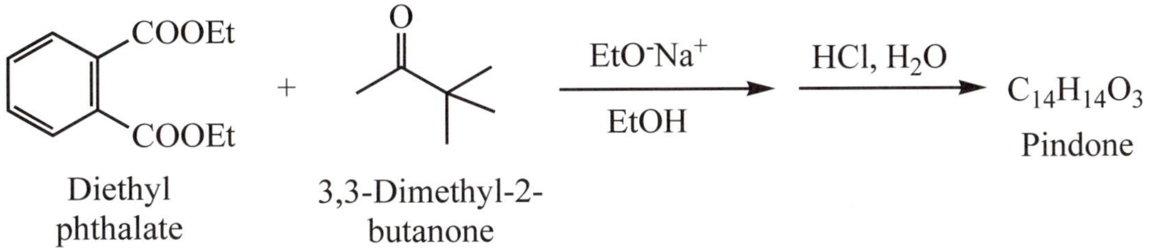

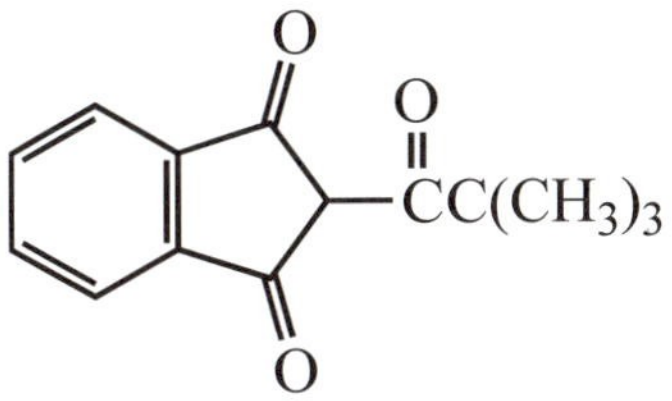

Diethyl                3,3-Dimethyl-2-
phthalate              butanone

$C_{14}H_{14}O_3$
Pindone

Pindone

**16.34** *Fentanyl is a non-opoid (non-morphine like) analgesic used for the relief of severe pain. It is approximately 50 times more potent in humans than morphine itself. One synthesis for fentanyl begins with 2-phenylethanamine.*

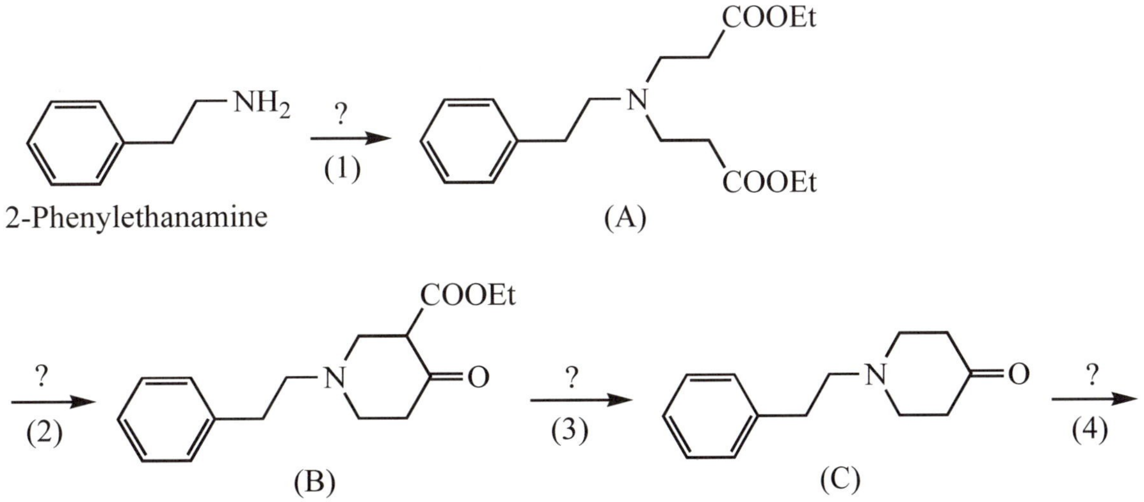

2-Phenylethanamine                              (A)

(B)                                    (C)

(E)                                                                      (F)

Fentanyl

*(a) Propose a reagent for Step 1. Name the type of reaction that occurs in this step.*

**The reaction for Step 1 is a Michael addition performed twice and involves the ester:**

*(b) Propose a reagent to bring about Step 2. Name the type of reaction that takes place in this step.*

**Step 2 is a Dieckmann condensation and requires the enolate formation using sodium ethoxide in ethanol followed by protonation with an aqueous mineral acid such as HCl in water.**

*(c) Propose a series of reagents that will bring about Step 3.*

**Step 3 involves a saponification followed by an acidification, and is finished by a decarboxylation. The reagents used to carry out these reactions include hydrolyzing the ester by heating in aqueous sodium hydroxide to generate the carboxylate salt. The carboxylate salt is converted to the β-ketocarboxylic acid using aqueous mineral acid. The β-ketocarboxylic acid undergoes decarboxylation by gentle heating to yield compound C.**

*(d) Propose a reagent for Step 4. Identify the imine (Schiff base) part of Compound E.*

**Step 4 requires the use of aniline (Ph-NH$_2$) followed by a dehydration to form the imine.**

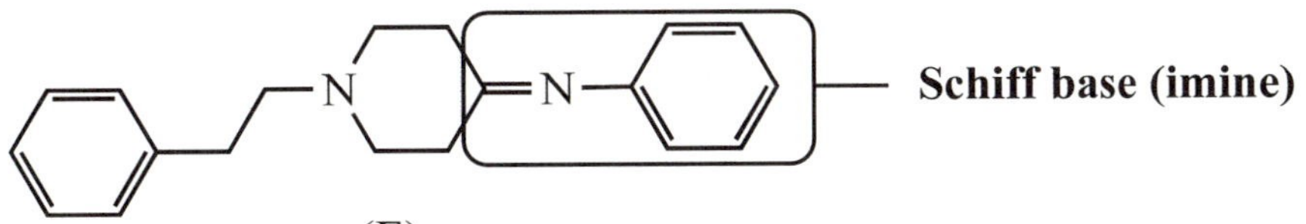

(E)

Schiff base (imine)

*(e) Propose a reagent to bring about Step 5.*

**Step 5 is a hydrogenation of an imine using H$_2$ and a nickel catalyst.**

*(f) Propose two different reagents, either of which will bring about Step 6.*

**The amide formation from compound F to fentanyl can be accomplished several ways. Two separate routes include:**

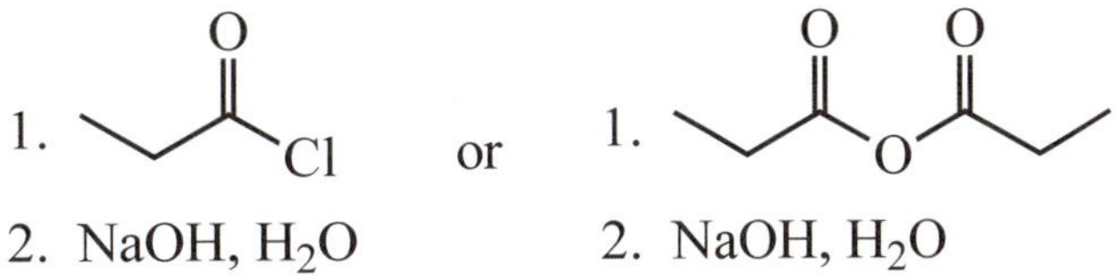

**Both routes generate acids in the first step that protonate the stronger 3° alkyl amine group to form the ammonium salt in the product. The sodium hydroxide used in the second part of Step 6 regenerates the free amine form of fentanyl.**

*(g) Is fentanyl chiral? Explain.*

**Fentanyl is not chiral because it has no stereocenters and possesses a mirror plane of symmetry.**

**16.35** *Meclizine is an antiemetic (it helps prevent or at least lessen the throwing up associated with motion sickness, including seasickness). Among the names of its over-the-counter preparations are Bonine, Sea-Legs, Antivert, and Navicalm.*

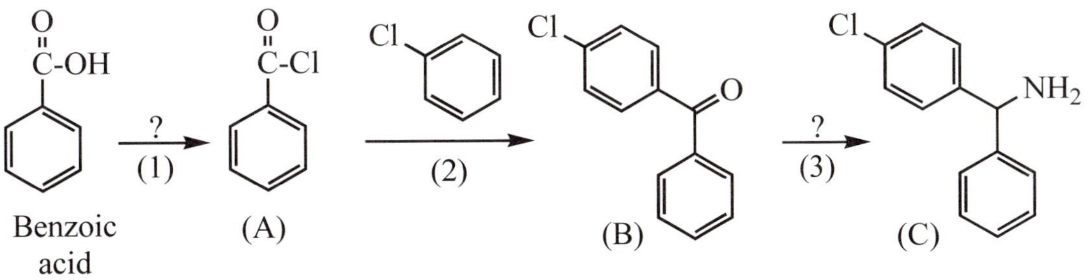

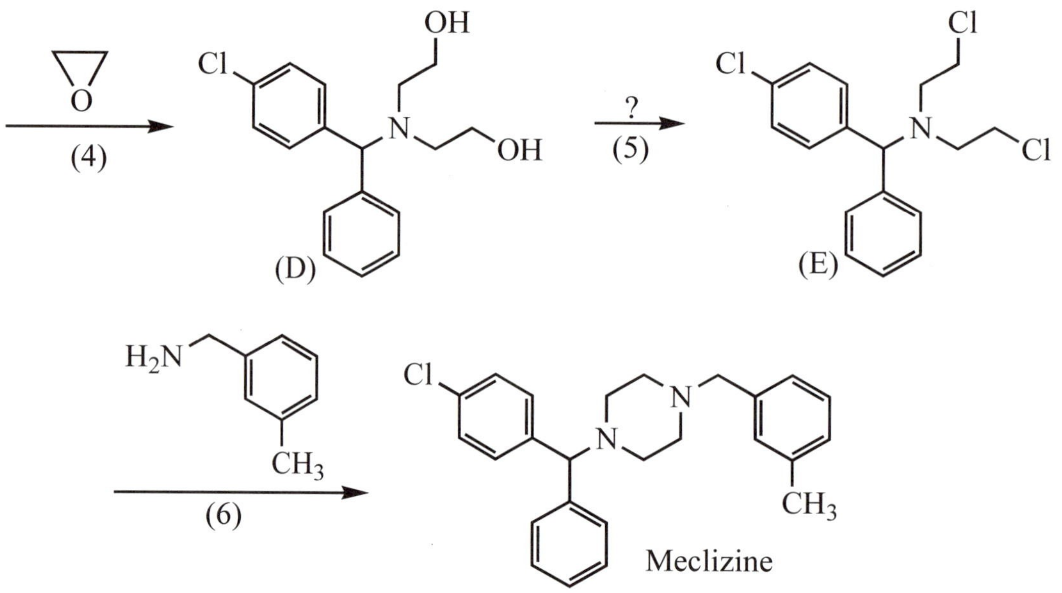

Meclizine

(a) *Name the functional group in (A). What reagent is most commonly used to convert a carboxyl group to this functional group?*

**The functional group in compound A is an acid chloride. Treatment of carboxylic acids with thionyl chloride (SOCl₂) is most commonly used to synthesize acid chlorides.**

(b) *The catalyst for Step 2 is aluminum chloride, AlCl₃. Name the type of reaction that occurs in this step. The product shown here has the orientation of the new group para to the chlorine atom of chlorobenzene. Supposing you were not told the orientation of the new group. Would you have predicted it to be ortho, meta, or para to the chlorine atom? Explain.*

**Step 2 is a Friedel-Crafts reaction. The chloro substituent is an ortho-para director, therefore, ortho and para products would be favored in this step.**

(c) *What set of reagents can be used in Step 3 to convert the C=O group to an -NH₂ group?*

**Ammonia (NH₃) and hydrogenation of the imine intermediate with H₂/Ni will accomplish the conversion of the ketone to a primary amine in Step 3.**

(d) *The reagent used in Step 4 is the cyclic ether ethylene oxide. Most ethers are quite unreactive to nucleophiles such as the 1° amine in this step. Ethylene oxide, however, is an exception to this generalization. What is it about ethylene oxide that makes it so reactive toward ring-opening reactions with nucleophiles?*

**The driving force behind the ring-opening reactions of epoxides with nucleophiles is the release of the high ring strain in the three-membered ring.**

*(e) What reagent can be used in Step 5 to convert each 1° alcohol to a 1° halide?*

**In Step 5, SOCl$_2$ can be used to convert each 1° alcohol to a 1° chloride.**

*(f) Step 6 is a double nucleophilic displacement.  Which mechanism is more likely for this reaction, S$_N$1 or S$_N$2?  Explain.*

**The nucleophilic displacement in Step 6 is an S$_N$2 reaction.  Primary carbocations are far too unstable to form for S$_N$1 reactions and the amine is a strong enough nucleophile to displace the 1° chloride.  Primary halides and amine nucleophiles favor S$_N$2.**

**16.36**  *2-Ethyl-1-hexanol is used for the synthesis of the sunscreen octyl p-methoxycinnamate (see Chemical Connections 15A).  This primary alcohol can be synthesized from butanal by the following series of steps.*

*(a) Propose a reagent to bring about Step 1. What name is given to this type of reaction?*

**Step 1 is an aldol condensation that is accomplished using aqueous sodium hydroxide.**

*(b) Propose a reagent for Step 2.*

**Step 2 is a dehydration reaction that requires heating the aldol product in a mineral acid such as sulfuric acid or base such as sodium hydroxide.**

*(c) Propose a reagent for Step 3.*

**In Step 3, hydrogenation with H$_2$ and Ni will reduce both the aldehyde and alkene.**

*(d) The following is a structural formula for the commercial sunblocking ingredient octyl-p-methoxycinnamate.  What carboxylic acid would you use to form this ester? How would you bring about the esterification reaction?*

**Fischer esterification of *p*-methoxycinnamic acid and 2-ethyl-1-hexanol catalyzed by sulfuric acid generates octyl *p*-methoxycinnamate.**

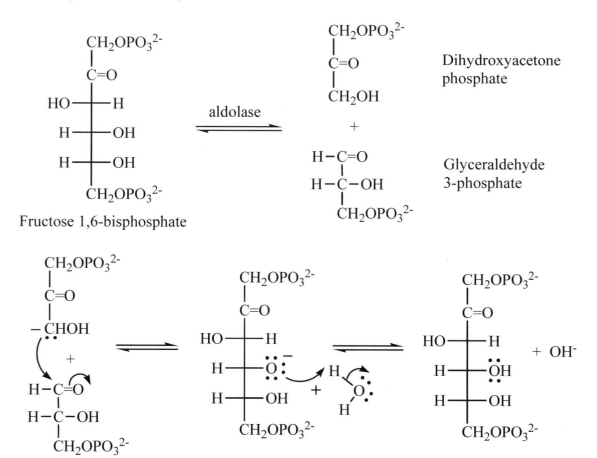

Octyl *p*-methoxycinnamate

## Looking Ahead

**16.37** *This reaction is one of the ten steps in glycolysis (Section 22.4), a series of enzyme-catalyzed reactions by which glucose is oxidized to two molecules of pyruvate. Show that this step is the reverse of an aldol reaction.*

**16.38** *This reaction is the fourth in the set of four enzyme-catalyzed steps by which the hydrocarbon chain of a fatty acid (Section 22.6) is oxidized, two carbons at a time, to acetyl-coenzyme A. Show that this reaction is the reverse of a Claisen condensation.*

β-Ketoacyl-CoA　　　　Coenzyme A　　　　An acyl-CoA　　　Acetyl-CoA

**The enzyme-catalyzed process is functionally the reverse of a crossed-Claisen condensation that takes place between the enolate anion of acetyl-CoA and an acyl-CoA.**

**16.39** *Steroids are a major type of lipid (Section 21.5) with a characteristic tetracyclic ring system. Show how the A ring of the steroid testosterone can be constructed from the precursors shown using a Michael reaction followed by an aldol reaction (with dehydration).*

Testosterone

**16.40** *The third step of the citric acid cycle (Section 22.7) involves the protonation of one of the carboxylate groups of oxalosuccinate, a β-ketoacid, followed by decarboxylation to form α-ketoglutarate.  Provide the structure of α-ketoglutarate.*

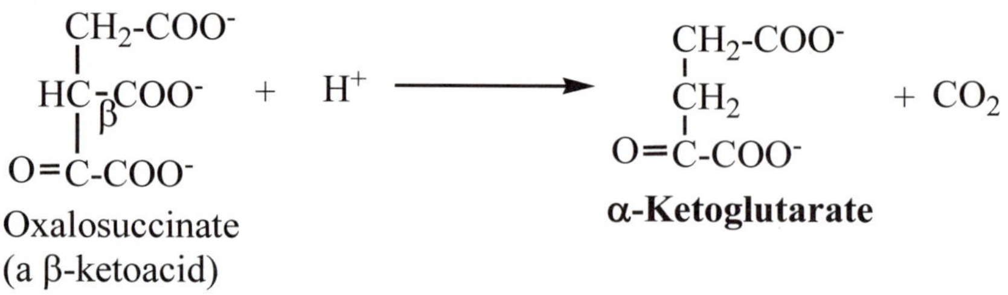

CH₂-COO⁻
|
HC–COO⁻     +   H⁺   ⟶
|  β
O=C-COO⁻

Oxalosuccinate
(a β-ketoacid)

CH₂-COO⁻
|
CH₂              + CO₂
|
O=C-COO⁻

**α-Ketoglutarate**

# CHAPTER 17
## *Solutions to Problems*

**17.1**  *Given the following structure, determine the polymer's repeat unit, redraw the structure using the simplified parenthetical notation, and name the polymer.*

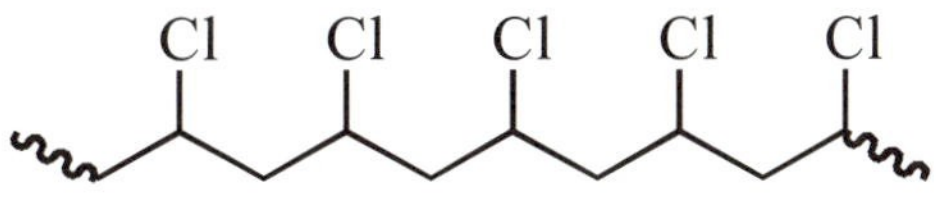

**This polymer is derived from vinyl chloride (chloroethene) and is therefore called poly(vinyl chloride), which is also known as PVC.**

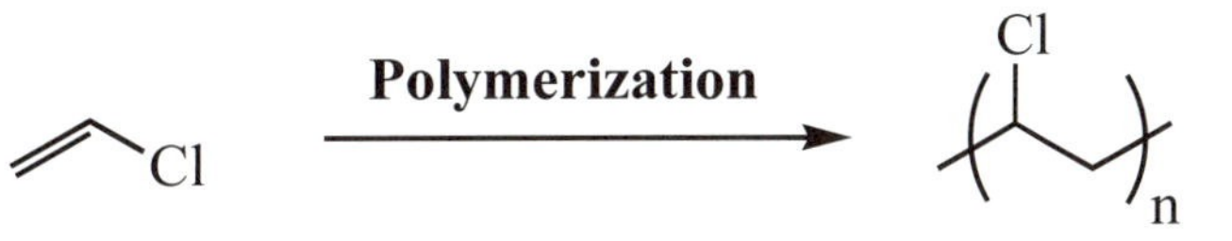

**Can also be represented as:**

**17.2**  *Write the repeating unit of the epoxy resin formed from the following reaction.*

**Polymerization occurs by a nucleophilic attack of the amine nitrogen on the least substituted carbon of the epoxide to give the following repeating structure:**

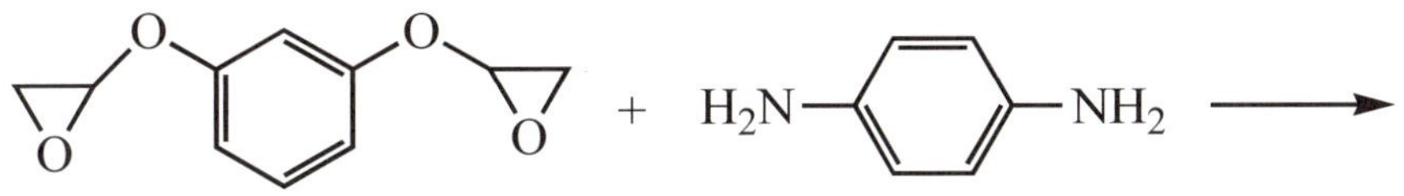

A diepoxide                          A diamine

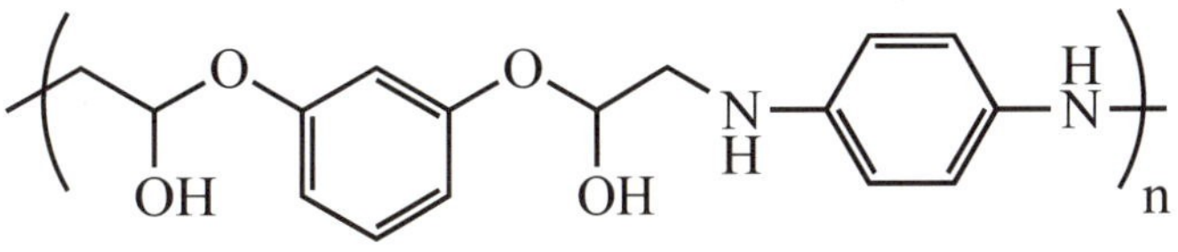

## Step-Growth Polymers

**17.3**    *Identify the monomers required for the synthesis of each step-growth polymer.*

(a)

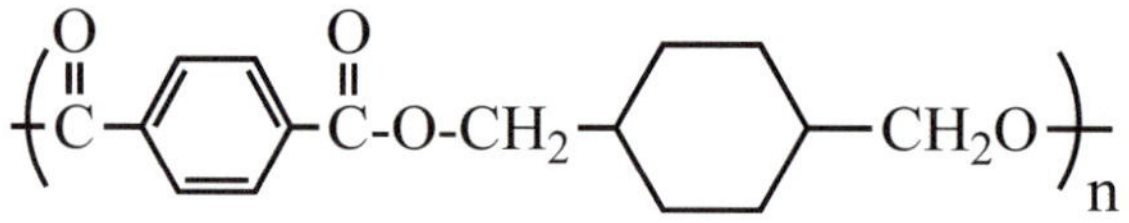

Kodel
(a polyester)

**Monomers:**    HOC—⬡—COH    and    HOCH$_2$—⬡—CH$_2$OH

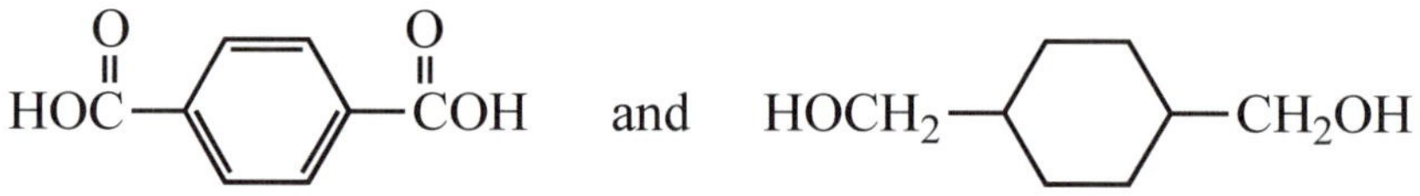

(b)

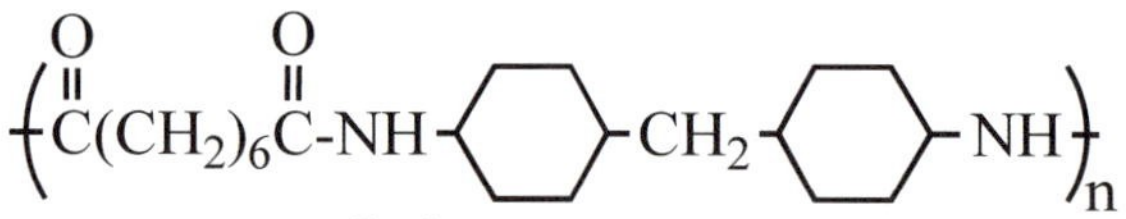

Quiana
(a polyamide)

**Monomers:**    HOC(CH$_2$)$_6$COH    and    H$_2$N—⬡—CH$_2$—⬡—NH$_2$

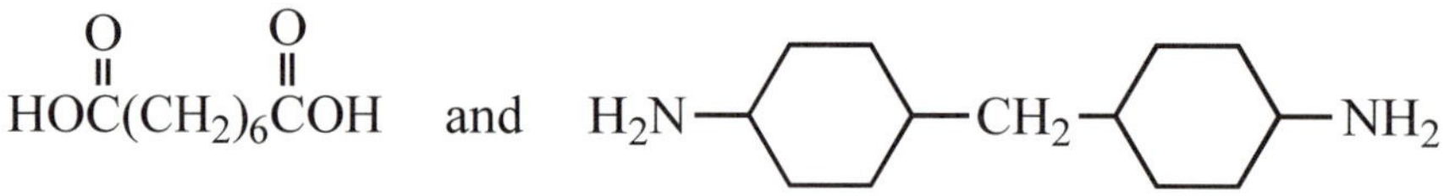

(c)

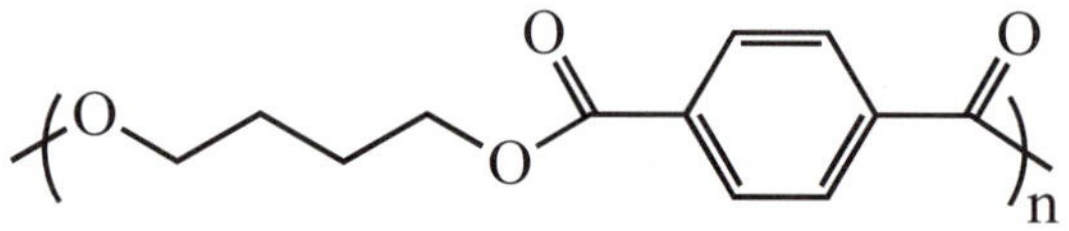

(a polyester)

**Monomers:**    HO⌒⌒OH    and    HOC—⬡—COH

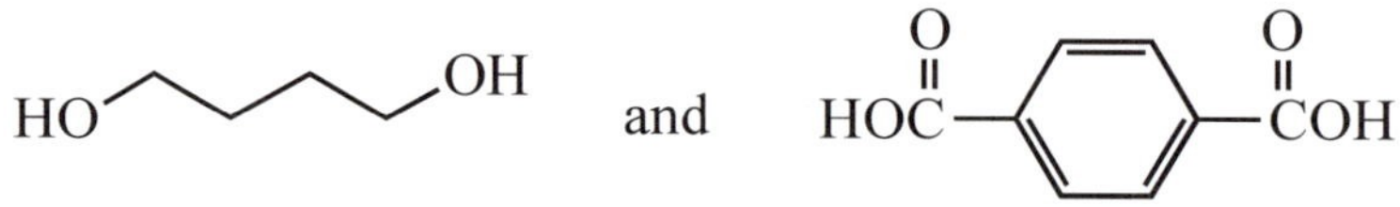

(d)

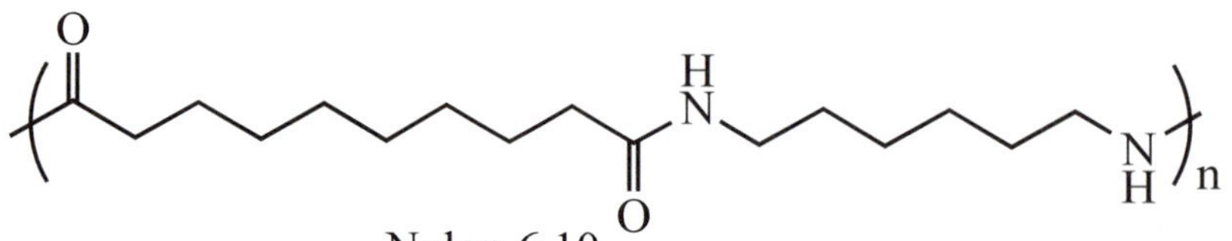

Nylon 6,10
(a polyamide)

**Monomers:**

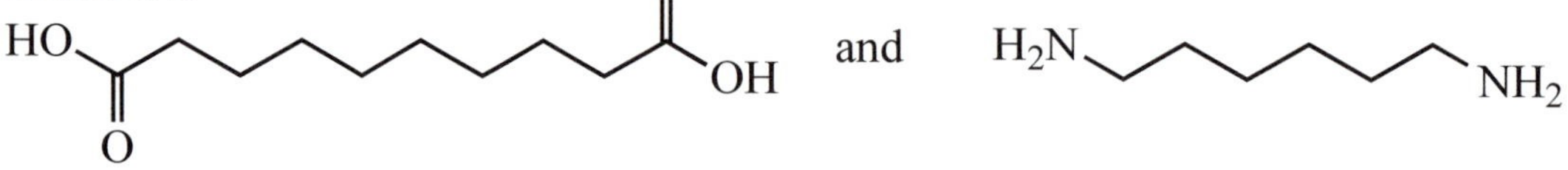

HO⌒⌒⌒OH    and    H$_2$N⌒⌒⌒NH$_2$

**17.4**  *Poly(ethylene terephthalate) (PET) can be prepared by this reaction.  Propose a mechanism for the step-growth reaction in this polymerization.*

Dimethyl terephthalate          Ethylene·glycol

Poly(ethylene terephthalate)          Methanol

**The polymer chain propagates by a series of transesterification reactions.  The high temperature of the reaction makes the use of a catalyst unnecessary and also allows the methanol to be removed from the equilibrium, favoring the polymer products. The following mechanism describes the first step of the reaction.**

**Step 1:  Nucleophilic attack of the alcohol hydroxyl on the ester carbonyl forms a protonated tetrahedral intermediate.**

**Step 2:  Collapse of the tetrahedral intermediate generates methoxide as a leaving group and the protonated glycol ester.**

**Step 3: Methoxide deprotonates the protonated ethylene glycol ester, forming methanol and ethylene glycol mono ester.**

**17.5**  *Currently about 30% of PET soft drink bottles are being recycled. In one recycling process, scrap PET is heated with methanol in the presence of an acid catalyst. The methanol reacts with the polymer, liberating ethylene glycol and dimethyl terephthalate. These monomers are then used as feedstock for the production of new PET products. Write an equation for the reaction of PET with methanol to give ethylene glycol and dimethyl terephthalate.*

**The following equation is the reverse reaction of the polymer's formation outlined in problem 17.4. This reaction can be classified as a transesterification.**

Poly(ethylene terephthalate)        Methanol

Dimethyl terephthalate        Ethylene glycol

**17.6**  *Nomex is an aromatic polyamide (aramid) prepared from polymerization of 1,3-benzenediamine and the acid chloride of 1,3-benzenedicarboxylic acid. The physical properties of the polymer make it suitable for high-strength, high-temperature applications such as parachute cords and jet aircraft tires. Draw a structural formula for the repeating unit of Nomex.*

**Nomex is a polyamide that results from the reaction between a diamine and a diacid chloride.**

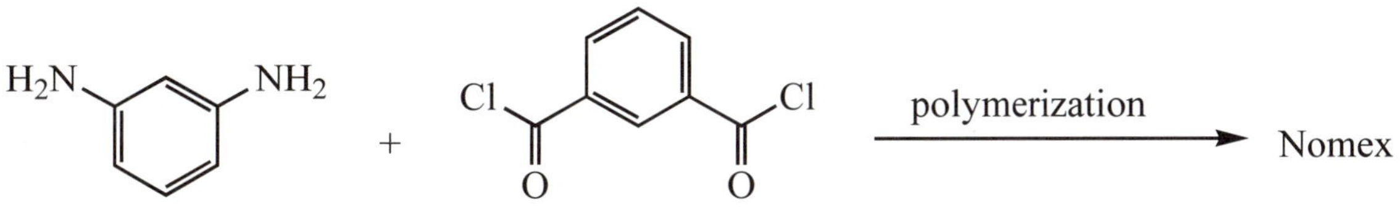

1,3-Benzenediamine              1,3-Benzene-
                                dicarbonyl chloride

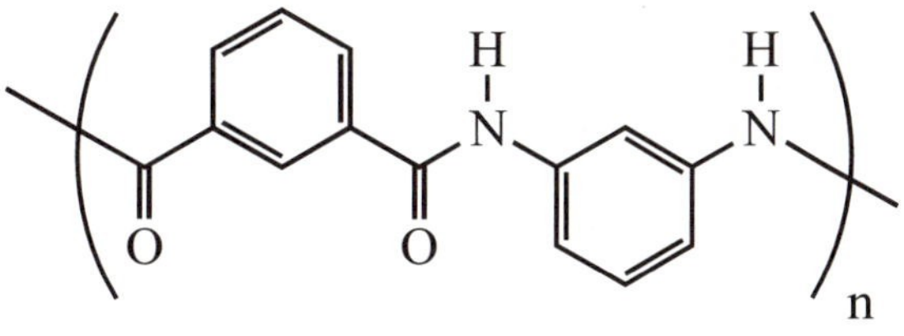

**Nomex**

**17.7**  *Nylon 6,10 [Problem 17.3(d)] can be prepared by reaction of a diamine and a diacid chloride.  Draw the structural formula of each reactant.*

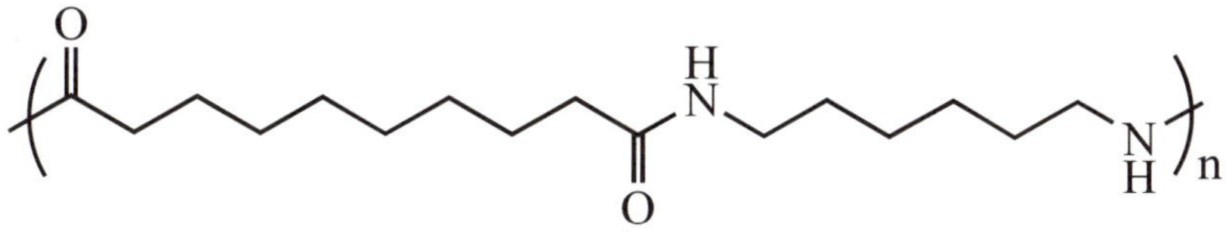

**Monomers:**

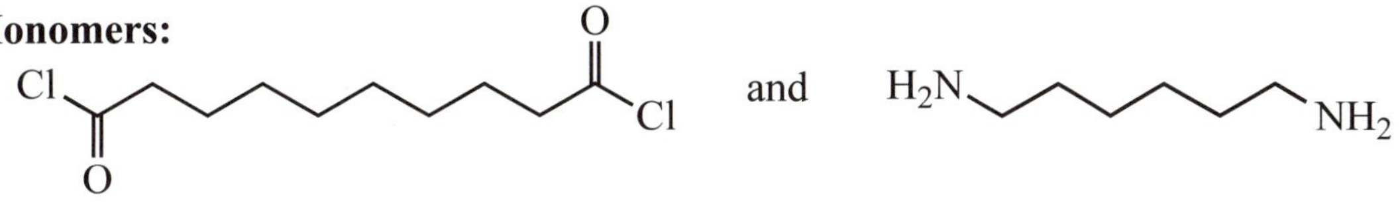

and

## Chain-Growth Polymerization

**17.8**  *Following is the structural formula of a section of polypropylene derived from three units of propylene monomer.*

$$\underset{\text{Polypropylene}}{-CH_2\overset{\overset{\displaystyle CH_3}{|}}{C}H-CH_2\overset{\overset{\displaystyle CH_3}{|}}{C}H-CH_2\overset{\overset{\displaystyle CH_3}{|}}{C}H-}$$

*Draw a structural formula for a comparable section of:*

*(a) Poly(vinyl chloride)*

$$\sim CH_2\overset{\overset{\displaystyle Cl}{|}}{C}H-CH_2\overset{\overset{\displaystyle Cl}{|}}{C}H-CH_2\overset{\overset{\displaystyle Cl}{|}}{C}H\sim \quad \text{or} \quad$$

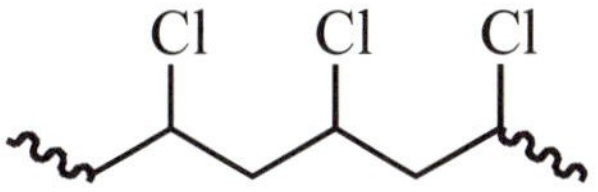

*(b) Polytetrafluoroethylene (PTFE)*

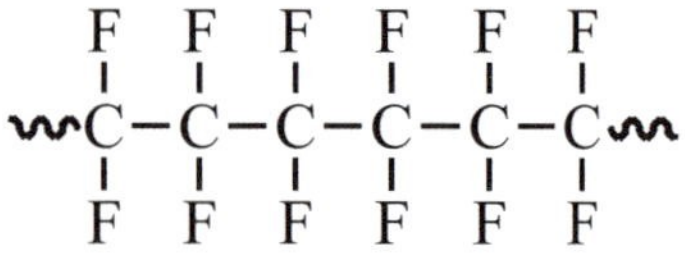

or

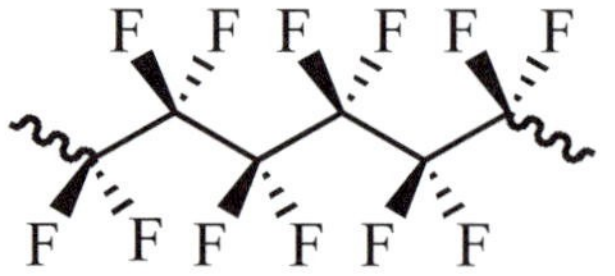

*(c) Poly(methyl methacrylate)*

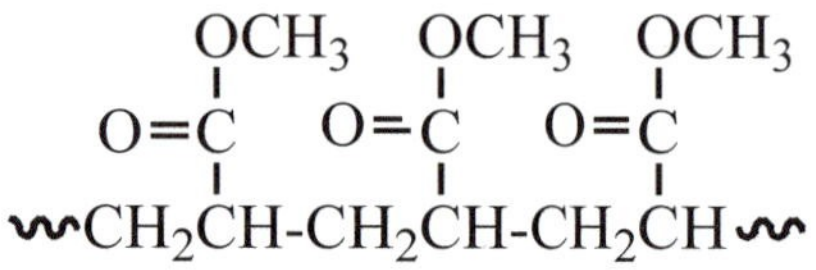

or

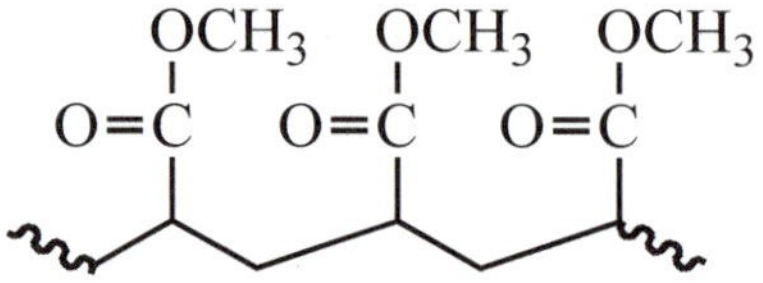

**17.9**  *Following are structural formulas for sections of three polymers.  From what alkene monomer is each derived?*

*(a)*

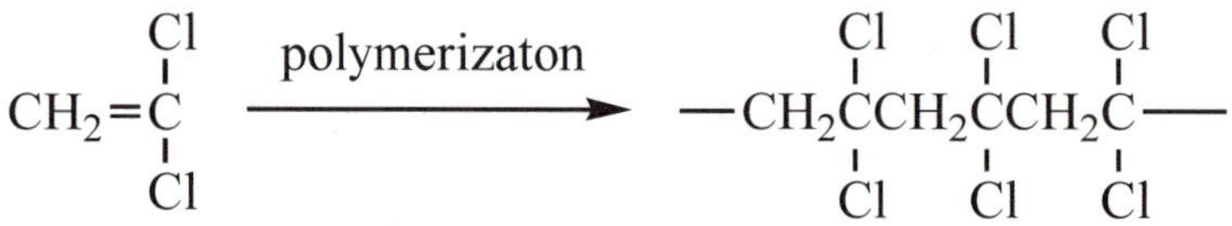

*(b)*

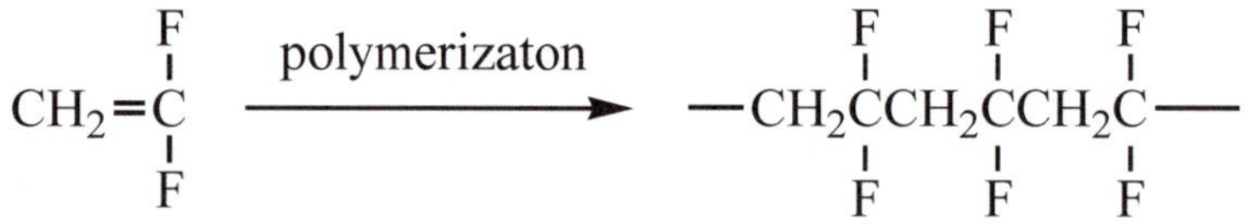

**17.10**  *Draw the structure of the alkene monomer used to make each chain-growth polymer.*

*(a)*          *(b)*

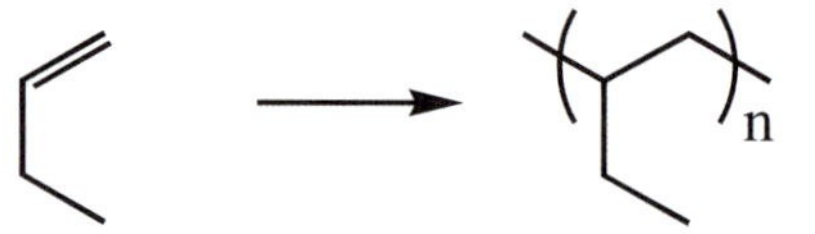

*(c)* 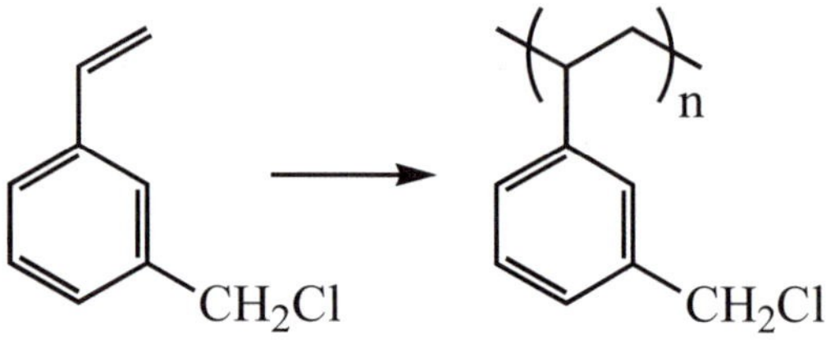          *(d)*

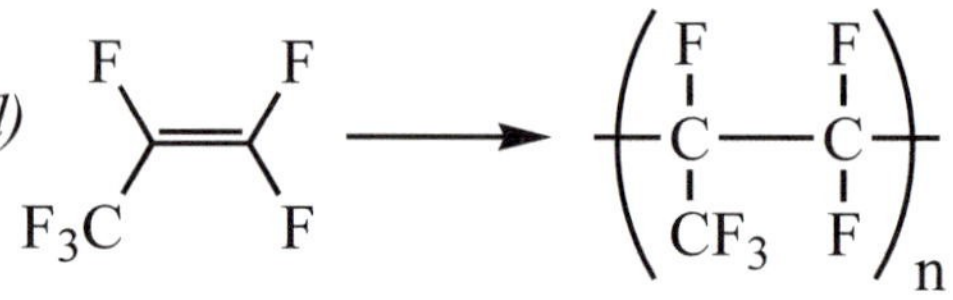

**17.11** *Low-density polyethylene (LDPE) has a higher degree of chain branching than high-density polyethylene (HDPE). Explain the relationship between chain branching and density.*

**The greater the degree of branching of the main monomer chains, the less efficient the chains pack together in the solid state and, therefore, the polymer density decreases.**

**17.12** *Compare the densities of low-density polyethylene (LDPE) and high-density polyethylene (HDPE) with the densities of the liquid alkanes listed in Table 3.4. How might you account for the differences between them?*

**As stated in this chapter, the density of LDPE is between 0.91 and 0.94 $g/cm^{-1}$, while the density of HDPE is 0.96 $g/cm^{-1}$. These values are considerably higher than the values of 0.626 to 0.730 $g/cm^{-1}$ for pentane through decane as listed in Table 3.4. The ratio of hydrogen atoms to carbon atoms is an important influence in the density of alkanes. Hydrogen atoms have such a low atomic weight compared to carbon atoms that the number of hydrogen atoms in a hydrocarbon affects the density of a hydrocarbon. As the hydrocarbon becomes larger, the hydrogen atom to carbon atom ratio (H:C) decreases. Thus polymers, which have the lowest H:C ratio, have a significantly larger density than smaller hydrocarbons.**

**17.13** *Polymerization of vinyl acetate gives poly(vinyl acetate). Hydrolysis of this polymer in aqueous sodium hydroxide gives poly(vinyl alcohol). Draw the repeat units of both poly(vinyl acetate) and poly(vinyl alcohol).*

$$\left[ \begin{array}{c} CH-CH_2 \\ | \\ OCCH_3 \\ \| \\ O \end{array} \right]_n \qquad\qquad \left[ \begin{array}{c} CH-CH_2 \\ | \\ OH \end{array} \right]_n$$

**Poly(vinyl acetate)**       **Poly(vinyl alcohol)**

**17.14** *As seen in the previous problem, poly(vinyl alcohol) is made by polymerization of vinyl acetate followed by hydrolysis in aqueous sodium hydroxide. Why is poly(vinyl alcohol) not made instead by polymerization of vinyl alcohol, $CH_2=CHOH$?*

**The proposed monomer for poly(vinyl alcohol) "vinyl alcohol" is unstable and exists as an enol in the enol-keto tautomerization equilibrium that favors the acetaldehyde tautomer.**

**Vinyl alcohol**
**(Enol tautomer)**

**Acetaldehyde**
**(Keto tautomer)**

**17.15** *As we know, the shape of a polymer chain affects its properties. Consider the three polymers shown. Which do you expect to be the most rigid? Which do you expect to be the most transparent (assume the same mol. weights)?*

Polymers A and B would be expected to be both rigid and opaque polymers. Polymer C would be expected to be a more flexible, transparent material. Both of these physical characteristics depend on the degree of crystallinity of the polymer. These three polymer chains all have repeating stereocenters. Both A and B are termed stereoregular; that is, the configurations of the stereocenters repeat in a consistent pattern over the length of the chain. In A, all stereocenters have the same configuration, whereas in B they alternative first *R*, then *S*, and so forth. Because of this stereoregular pattern, molecules of both polymers A and B pack well in the solid state with strong intermolecular interactions between molecules. Because of this form of packing, polymers A and B have a high degree of crystallinity and are rigid polymers. Polymer C, on the other hand, has a random orientation of stereocenters and, therefore, molecules of its chains do not pack as well in the solid state; it has a low degree of crystallinity. The lower the degree of crystallinity, the more transparent the polymer.

**Looking Ahead**

**17.16** *Cellulose, the principle component of cotton, is a polymer of D-glucose in which the monomeric unit repeats at the atoms indicated. Draw a 3-unit section of cellulose.*

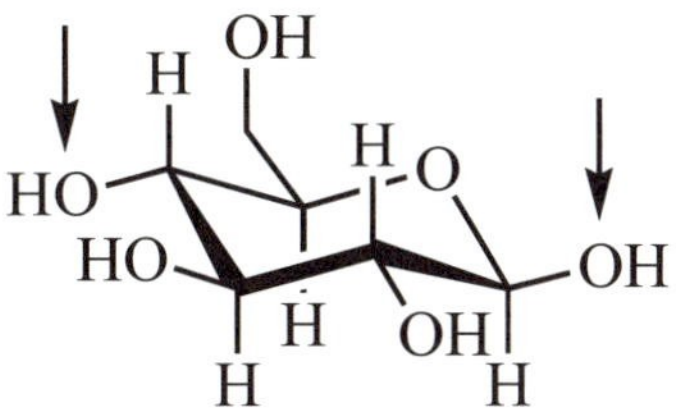

D-Glucose

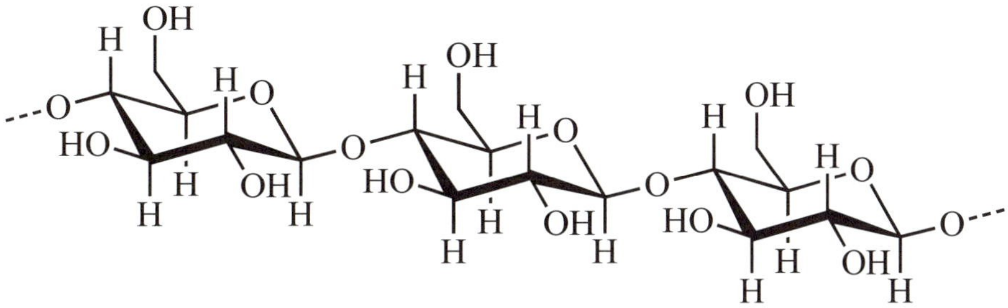

**17.17** *Is a repeating unit a requirement for a compound to be called a polymer?*

**Yes, the term "polymer" has historically referred to a large molecule with repeating units.**

**17.18** *Proteins are polymers of naturally occurring monomers called amino acids. Amino acids differ in the types of R groups available in nature. Explain how the following properties of a protein might be affected upon changing the R groups from -CH₂CH (CH₃)₂ to -CH₂OH.*

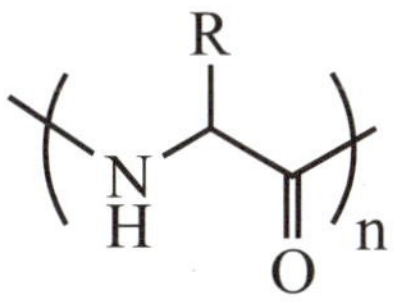

a protein

**The addition of a hydroxymethyl group in place of an isobutyl group increases the opportunity for intermolecular hydrogen bonding.**

*(a) solubility in water*

   The protein with a –CH$_2$OH side chain will be more soluble in water.  As the degree of intermolecular hydrogen bonding with water increases, the solubility in water also increases.

*(b) T$_m$*

   The protein with a –CH$_2$OH side chain will melt at a higher temperature.  As the degree of intermolecular hydrogen bonding increases, the melting point also increases.

*(c) crystallinity*

   The protein with a –CH$_2$OH side chain will be more crystalline.  As the degree of intermolecular hydrogen bonding increases, the lattice energy and order in the solid also increases.

*(d) elasticity*

   The protein with a –CH$_2$OH side chain will be less elastic.  As the degree of intermolecular hydrogen bonding increases between protein chains, the attractive interactions between protein chains also increase.  As the attractive forces between polymer chains increase, their motion relative to each other will decrease.

## CHAPTER 18
### Solutions to Problems

**18.1**   (a) Draw Fischer projections for all 2-ketopentoses.
            (b) Which are D-ketopentoses, which are L-ketopentoses, and which are enantiomers?

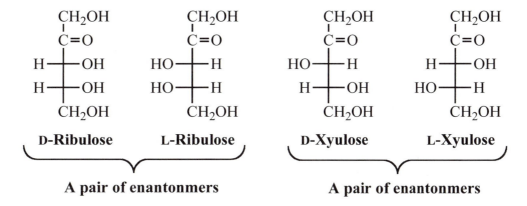

**18.2**   Mannose exists in aqueous solution as a mixture of α-D-mannopyranose and β-D-mannopyranose. Draw Haworth projections for these molecules.

**The only difference between α-D-mannopyranose and β-D-mannopyranose is the configuration at C1, the anomeric carbon.**

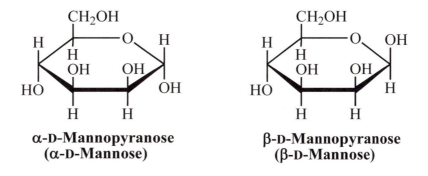

**18.3** *Draw chair conformations for α-D-mannopyranose and β-D-mannopyranose. Label the anomeric carbon atom in each.*

α-D-Mannopyranose β-D-Mannopyranose

**18.4** *Draw a structural formula for the chair conformation of methyl α-D-mannopyranoside (methyl α-D-mannoside). Label the anomeric carbon and the glycosidic bond.*

**Methyl α-D-mannopyranoside**

**18.5** *Draw a structural formula for the β-N-glycoside formed between β-D-ribofuranose and adenine.*

**The following are structural formulas for adenine, the monosaccharide hemiacetal, and the N-glycoside**

**18.6**  *NaBH₄ reduces D-erythrose to erythritol.  Do you expect the alditol formed under these conditions to be optically active or optically inactive?  Explain.*

**Erythritol is achiral because of a mirror plane in the molecule; therefore, the product will be optically inactive.**

**18.7**  *Draw Haworth and chair formulas for the α form of a disaccharide in which two units of D-glucopyranose are joined by a β-1,3-glycosidic bond.*

β-1,3-glycosidic bond

## Monosaccharides

**18.8**  *What is the difference in structure between an aldose and a ketose? Between an aldopentose and a ketopentose?*

**An aldose is a monosaccharide that has an aldehyde group and a ketose is a monosaccharide with a ketone group.  An aldopentose is a five-carbon monosaccharide containing an aldehyde group and a ketopentose is a five-carbon monosaccharide containing a ketone group.**

**18.9**  *Which hexose is also known as "dextrose"?*

**D-Glucose is also known as dextrose.**

**18.10**  *What does it mean to say that D- and L-glyceraldehyde are enantiomers?*

**D- and L-glyceraldehyde are enantiomers because they are nonsuperposable mirror images.**

**L-Glyceraldehyde          D-Glyceraldehyde**

**18.11**  *Explain the meaning of the designations D and L as used to specify the configuration of carbohydrates.*

**The designations D and L refer to the configuration at the stereocenter farthest from the aldehyde or ketone group of a monosaccharide. When a monosaccharide is drawn as a Fischer projection, a D-monosaccharide has the -OH of this carbon on the right; an L-monosaccharide has the -OH group on the left.**

**18.12**  *How many stereocenters are present in D-glucose? In D-ribose? How many stereoisomers are possible for each monosaccharide?*

**The stereocenters on D-glucose and D-ribose are indicated with an asterisk. D-Glucose has four stereocenters for a total of 16 stereoisomers. D-Ribose has three stereocenters for a total of eight stereoisomers.**

**D-Glucose          D-Ribose**

**18.13** *Which compounds are D-monosaccharides and which are L-monosaccharides?*

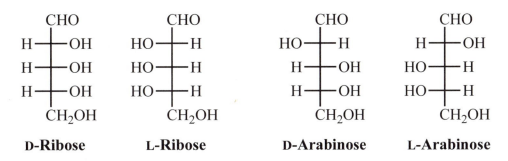

**Structures (a) and (c) are D-monosaccharides and structure (b) is an L-monosaccharide.**

**18.14** *Draw Fischer projections for L-ribose and L-arabinose.*

| | | | |
|---|---|---|---|
| CHO | CHO | CHO | CHO |
| H—OH | HO—H | HO—H | H—OH |
| H—OH | HO—H | H—OH | HO—H |
| H—OH | HO—H | H—OH | HO—H |
| CH₂OH | CH₂OH | CH₂OH | CH₂OH |
| **D-Ribose** | **L-Ribose** | **D-Arabinose** | **L-Arabinose** |

**L-Ribose and L-arabinose are mirror images of D-ribose and D-arabinose, respectively. The most common error in answering this question is to start with the Fischer projection for the D-sugar and then invert the configuration of C4 only. While the monosaccharide thus drawn is an L-sugar, it is not the correct one.**

**18.15** *Explain why all mono- and disaccharides are soluble in water.*

**Each carbon of a monosaccharide has a hydroxyl that is able to participate both in hydrogen bond accepting and hydrogen bond donating with water molecules.**

**18.16** *What is an amino sugar? Name the three amino sugars most commonly found in nature.*

**Amino sugars contain an –NH₂ group in place of an –OH group. Only three amino sugars are common in nature: D-glucosamine, D-mannosamine, and D-galactosamine, whose structures are shown below.**

CHO
H——NH₂
HO——H
H——OH
H——OH
CH₂OH

**D-Glucosamine**

CHO
H₂N——H
HO——H
H——OH
H——OH
CH₂OH

**D-Mannosamine**

CHO
H——NH₂
HO——H
HO——H
H——OH
CH₂OH

**D-Galactosamine**

**18.17** *2,6-Dideoxy-D-altrose, known alternatively as D-digitoxose, is a monosaccharide obtained on hydrolysis of digitoxin, a natural product extracted from purple foxglove (Digitalis purpurea). Digitoxin has found wide use in cardiology because it reduces pulse rate, regularizes heart rhythm, and strengthens heartbeat. Draw the structural formula of 2,6-dideoxy-D-altrose.*

CHO
H——H
H——OH
H——OH
H——OH
CH₃      **2,6-Dideoxy-D-altrose**

## The Cyclic Structure of Monosaccharides
**18.18** *Define the term anomeric carbon.*

**The anomeric carbon is the new carbon stereocenter created upon the formation of a cyclic structure. In terms of carbohydrate chemistry, the anomeric carbon is the hemiacetal carbon of the cyclic form of the monosaccharide.**

**18.19** *Explain the conventions for using α and β to designate the configuration of cyclic forms of monosaccharides.*

**The designation β means that the -OH group on the anomeric carbon of a cyclic hemiacetal is on the same side of the ring as the terminal -CH₂OH group. The designation α means that it is on the opposite side from the terminal -CH₂OH group.**

**18.20** *Are α-D-glucose and β-D-glucose anomers? Explain. Are they enantiomers? Explain.*

**α-D-Glucose and β-D-glucose are anomers. They only differ in the configuration at the hemiacetal carbon (the anomeric carbon). All of the other carbon stereocenters**

**have the same configuration.  Therefore, the anomers are not enantiomers; they are diastereomers.**

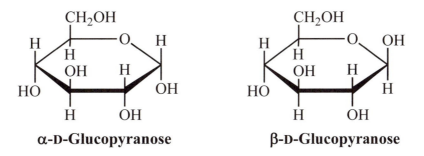

**α-D-Glucopyranose**          **β-D-Glucopyranose**

**18.21**  *Are α-D-Gulose and α-L-gulose anomers?  Explain.*

**No, they are not anomers; they are enantiomers.  Anomers differ only in the configuration at the anomeric carbon.  In going from α-D-gulose to α-L-gulose, all of the stereocenters have inverted their configuration.**

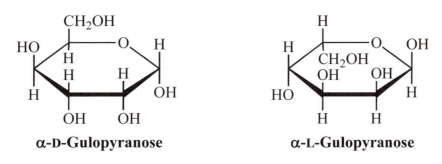

**α-D-Gulopyranose**          **α-L-Gulopyranose**

**18.22**  *In what way are chair conformations a more accurate representation of molecular shape of hexopyranoses than Haworth projections?*

**Hexopyranoses are six-membered rings in the chair conformation.  Haworth projections represent the six-membered rings as being planar, which they are not.**

**18.23**  *Draw α-D-glucopyranose (α-D-glucose) as a Haworth projection.  Now, using only the information given here, draw Haworth projections for these monosaccharides.*
   *(a)  α-D-mannopyranose (α-D-mannose).  The configuration of D-mannose differs from that of D-glucose only at carbon 2.*

(b) *α-D-gulopyranose (α-D-gulose). The configuration of D-gulose differs from that of D-glucose at carbons 3 and 4.*

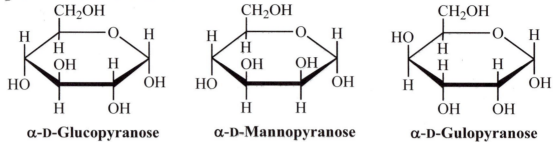

α-D-Glucopyranose          α-D-Mannopyranose          α-D-Gulopyranose

**18.24** *Convert each Haworth projection to an open-chain form and then to a Fischer projection. Name the monosaccharide you have drawn.*

(a) **In converting a Haworth projection to a Fischer projection, (1) the cyclic hemiacetal was broken into the component acyclic aldehyde, (2) the C4-C5 bond is rotated to position the terminal –CH₂OH group in a horizontal orientation, and (3) the structure is rotated 90° counterclockwise and stretched out vertically to a Fischer projection, placing the aldehyde group on top and the –CH₂OH on the bottom. Another way of looking at it is that the structure differs from D-glucose only at C3, therefore the structure represents D-allose.**

D-Allose

(b) **The conversion of the Haworth projection to a Fischer projection for the structure below proceeds by the same method used for D-allose. The structure differs from D-glucose at C2, C3, and C4, therefore the structure represents D-idose.**

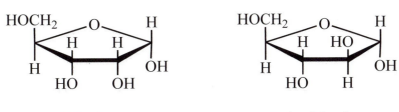

**D-Idose**

**18.25**  *Convert each chair conformation to an open-chain form and then to a Fischer projection. Name the monosaccharide you have drawn.*

(a)

CHO
H——OH
HO——H
HO——H
H——OH
CH₂OH

**D-Galactose**

(b)

CHO
H——OH
H——OH
H——OH
H——OH
CH₂OH

**D-Allose**

**18.26**  *The configuration of D-arabinose differs from the configuration of D-ribose only at carbon 2.  Using this information, draw a Haworth projection for α-D-arabinofuranose (α-D-arabinose).*

**α-D-Ribofuranose**                    **α-D-Arabinofuranose**

**18.27**  *Explain the phenomenon of mutarotation with reference to carbohydrates. By what means is it detected?*

**During mutarotation, the α and β anomers of a carbohydrate convert to an equilibrium mixture of the two cyclic forms and the open chain form. Mutarotation can be detected by observing the change on optical activity over time as the two forms equilibrate.**

**18.28**  *The specific rotation of α-D-glucose is +112.2°. What is the specific rotation of α-L-glucose?*

**α-L-Glucose is the enantiomer of α-D-glucose, therefore, the specific rotation of α-L-glucose is -112.2°.**

**18.29**  *When α-D-glucose is dissolved in water, the specific rotation of the solution changes from +112.2° to +52.7°. Does the specific rotation of α-L-glucose also change when it is dissolved in water? If so, to what value does it change?*

**Yes. The specific rotation of α-L-glucose changes to -52.7°. Remember that α-L-glucose and α-D-glucose are enantiomers and that the only difference between enantiomers in an achiral environment is the sign of their specific rotation of plane-polarized light.**

## Reactions of Monosaccharides

**18.30**  *Draw Fischer projections for the product(s) formed by reaction of D-galactose with the following. In addition state whether each product is optically active or optically inactive.*

(a)

D-Galactose          $\xrightarrow[\text{H}_2\text{O}]{\text{NaBH}_4}$          Galactitol
(meso; inactive)

```
        CHO                              COOH
   H ───── OH                      H ───── OH
  HO ───── H                      HO ───── H
(b)HO ───── H      AgNO₃         HO ───── H
   H ───── OH     ─────────>      H ───── OH
        CH₂OH      NH₃; H₂O            CH₂OH
```

$$\text{D-Galactose} \qquad \text{AgNO}_3,\ \text{NH}_3;\ \text{H}_2\text{O} \qquad \text{D-Galactonic acid}$$

**D-Galactose**

**D-Galactonic acid
(Chiral; optically active)**

**18.31**  *Repeat Problem 18.30 using D-ribose.*

```
        CHO                              CH₂OH
   H ───── OH                      H ───── OH
(a)H ───── OH      NaBH₄          H ───── OH
   H ───── OH      ─────────>     H ───── OH
        CH₂OH         H₂O              CH₂OH
```

**D-Ribose**

**Ribitol
(meso; inactive)**

```
        CHO                              COOH
   H ───── OH                      H ───── OH
(b)H ───── OH      AgNO₃          H ───── OH
   H ───── OH     ─────────>      H ───── OH
        CH₂OH      NH₃; H₂O            CH₂OH
```

**D-Ribose**

**D-Ribonic acid
(Chiral; optically active)**

**18.32**  *Reduction of D-fructose by NaBH$_4$ gives two alditols, one of which is D-sorbitol.  Name and draw a structural formula for the other alditol.*

**18.33**  *There are four D-aldopentoses (Table 18.1).  If each is reduced with NaBH$_4$, which yield optically active alditols?  Which yield optically inactive alditols?*

**D-Arabinose and D-lyxose yield optically active alditols. D-Ribose and D-xylose yield optically inactive (because they are meso compounds) alditols.**

**18.34** *Account for the observation that reduction of D-glucose with NaBH₄ gives an optically active alditol, whereas reduction of D-galactose with NaBH₄ gives an optically inactive alditol.*

**18.35** *Which two D-aldohexoses give optically inactive (meso) alditols on reduction with NaBH₄?*

**18.36** *Name the two alditols formed by NaBH₄ reduction of D-fructose.*

D-Fructose    NaBH₄    →    **D-Glucitol (D-Sorbitol)**  +  **D-Mannitol**

**18.37** *L-Fucose, one of several monosaccharides commonly found in the surface polysaccharides of animal cells (Chemical Connections 18B), is synthesized biochemically from D-mannose in the following eight steps:*

D-Mannose   (1) →   (2) →   (3) →   (4) →   (5) →   (6) →   (7) →   (8) →   L-Fucose

*(a) Describe the type of reaction (oxidation, reduction, hydration, dehydration, and the like) involved in each step.*

**(1) Formation of a cyclic hemiacetal from a carbonyl group and a 2° alcohol.**
**(2) Oxidation of a 2° alcohol to a ketone.**
**(3) Dehydration of a 1° alcohol to a carbon-carbon double bond.**
**(4) Reduction of a carbon-carbon double bond to a carbon-carbon single bond.**
**(5) Keto-enol tautomerism.**
**(6) Keto-enol tautomerism.**
**(7) Reduction of a ketone to a 2° alcohol.**
**(8) Opening of a cyclic hemiacetal to an aldehyde and an alcohol.**

*(b) Explain why this monosaccharide, which is derived from D-mannose, now belongs to the L series.*

**It is the configuration at C5 of the aldoses that determines whether it is of the D-series or the L-series.  The stereochemistry at C5 was lost in Step 3, then inverted from the original D-series to the L-series in Step 4.**

**Ascorbic Acid**

**18.38**  *Is ascorbic acid a biological oxidizing agent or a biological reducing agent?  Explain.*

**Ascorbic acid is a biological reducing agent.  It is very easily oxidized to L-dehydroascorbic acid by giving up the enediol hydroxyl hydrogen atoms when abstracted by free radicals.**

**L-Ascorbic acid**          **L-Dehydroascorbic acid**

**18.39**  *Ascorbic acid is a diprotic acid with the following acid ionization constants: $pK_{a1}$ 4.10 and $pK_{a2}$ 11.79.  The two acidic hydrogens are those connected with the enediol part of the molecule.  Which hydrogen has which ionization constant? (Hint:  Draw separately the anion derived by loss of one of these hydrogens and that formed by loss of the other hydrogen.  Which anion has the greater degree of resonance stabilization?)*

$$pK_a \; 4.10 \longrightarrow HO \qquad OH \longleftarrow pK_a \; 11.79$$

**Loss of the more acidic proton results in a conjugate base that is stabilized by three resonance structures delocalizing the negative charge over two oxygen atoms and a carbon atom.**

**Loss of the proton with the lesser of the two acidities results in the formation of a conjugate base with only two resonance structures delocalizing the charge of an oxygen atom and a carbon atom.**

**Disaccharides and Oligosaccharides**
**18.40**  *Define the term glycosidic bond.*

**A glycosidic bond is the bond from the anomeric carbon of a glycoside to an –OR group.**

**18.41**  *What is the difference in meaning between the terms glycosidic bond and glucosidic bond?*

**A glycoside bond is a bond from the anomeric carbon of a glycoside to an -OR group.  A glucoside is a glycoside bond from a glucoside to an -OR group.**

**18.42**  *Do glycosides undergo mutarotation?*

**Glycosides do not undergo mutarotation.  Glycosides, like their acyclic acetals, are stable in aqueous and alkaline conditions, therefore glycosides are not in equilibrium with their open chain forms.**

**18.43**  *In making candy or sugar syrups, sucrose is boiled in water with a little acid, such as lemon juice.  Why does the product mixture taste sweeter than the starting sucrose solution?*

**Sucrose is a disaccharide composed of the monosaccharides D-glucose and D-fructose linked by a glycosidic bond.  The acid catalyzes the hydrolysis of the glycosidic bond to give D-glucose and D-fructose.  D-Fructose has a relative sweetness of 174 compared with 100 for sucrose.  Thus, converting sucrose into fructose and glucose increases the sweetness of the mixture.**

**18.44**  *Which disaccharides are reduced by NaBH₄?*
    *(a) Sucrose          (b) Lactose                (c) Maltose*

**In order for these disaccharides to react with NaBH₄, they must contain at least one carbonyl group that is in equilibrium with the open chain form (a cyclic hemiacetal). Maltose and lactose contain a monosaccharide ring that is a hemiacetal, but sucrose does not.**

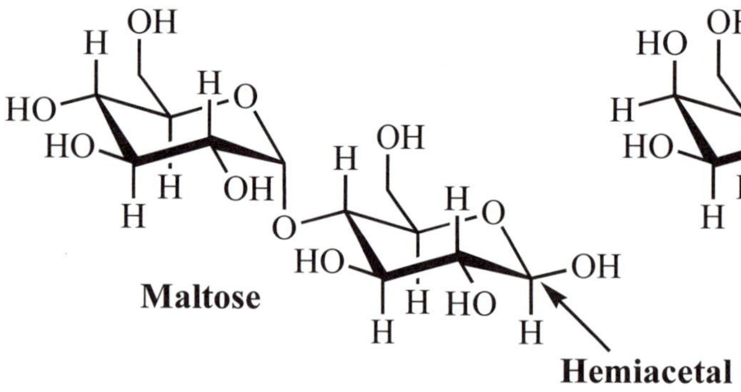

**Maltose**

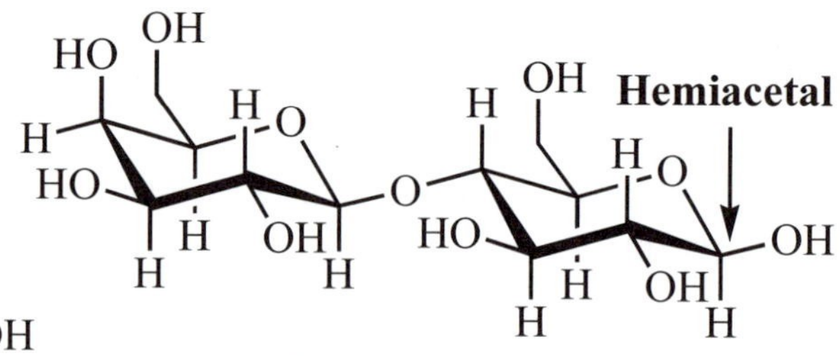

**Hemiacetal**

**Lactose**

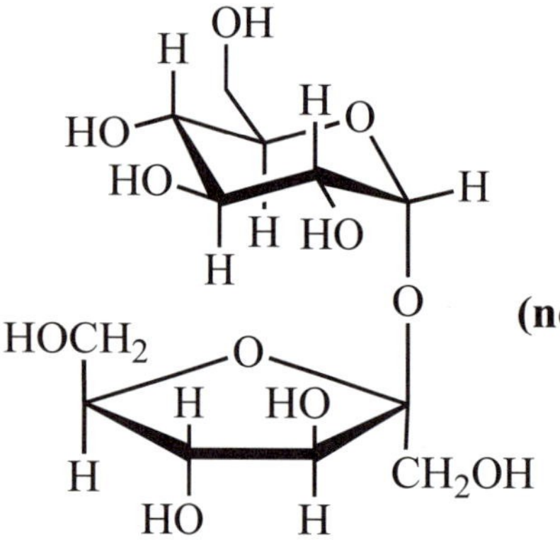

**Hemiacetal**

**Sucrose**
**(no cyclic hemiacetal)**

**18.45** *Trehalose is found in young mushrooms and is the chief carbohydrate in the blood of certain insects. Trehalose is a disaccharide consisting of two D-monosaccharide units, each joined to the other by an α-1,1-glycosidic bond.*

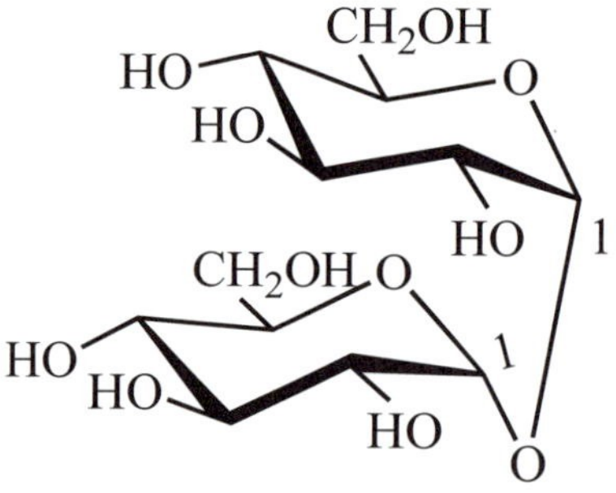

Trehalose

*(a) Is trehalose a reducing sugar?*

**Trehalose lacks a monosaccharide ring that is a cyclic hemiacetal. Instead, the anomeric carbon of each monosaccharide unit is involved in the formation of a glycosidic bond. The equilibrium between the cyclic form and the open chain carbonyl is therefore impossible, thus trehalose cannot be a reducing sugar.**

*(b) Does trehalose undergo mutarotation?*

**Non-reducing sugars do not undergo mutarotation.**

*(c) Name the two monosaccharide units of which trehalose is composed.*

**Trehalose is composed of two units of D-glucose.**

**18.46** *Hot water extracts of ground willow bark are an effective pain reliever. Unfortunately, the liquid is so bitter that most persons refuse it. The pain reliever in these infusions is salicin. Name the monosaccharide unit in salicin.*

β-D-Glucose unit

Salicin

**Polysaccharides**
**18.47** *What is the difference in structure between oligosaccharides and polysaccharides?*

**An oligosaccharide contains approximately 6-10 monosaccharide units. A polysaccharide contains more, generally many more than 10 monosaccharide units.**

**18.48** *Name three polysaccharides that are composed of units of D-glucose. In which are the glucose units joined by α-glycosidic bonds? In which are they joined by β-glycosidic bonds?*

**Three polysaccharides that are composed of D-glucose include cellulose, starch (composed of amylose and amylopectin), and glycogen. Cellulose links about 2200 glucose units by β-1,4-glycosidic bonds. The amylose in starch is an unbranched polymer with about 4000 glucose units linked by α-1,4-glycosidic bonds. Amylopectin also contains up to 10,000 glucose units linked by α-1,4-glycosidic bonds, but also contains branching that uses α-1,6-glycosidic bonds. Glycogen, like amylopectin, is a highly branched polysaccharide linking $10^6$ glucose units with α-1,4- and α-1,6 glycosidic bonds.**

**Cellulose:**

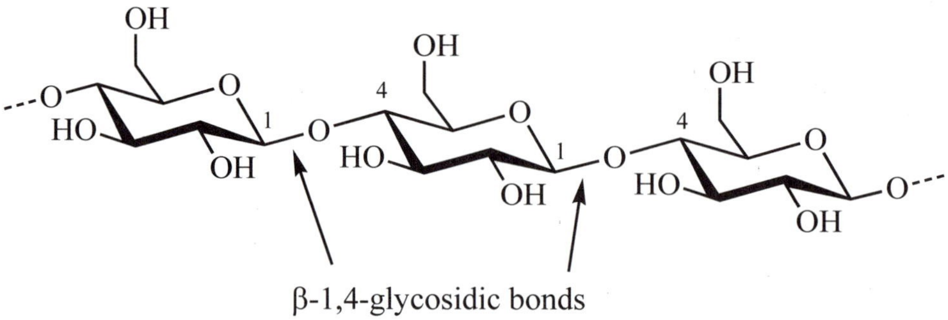

β-1,4-glycosidic bonds

**Starch
(amylose and amylopectin):**

α-1,4-glycosidic bonds

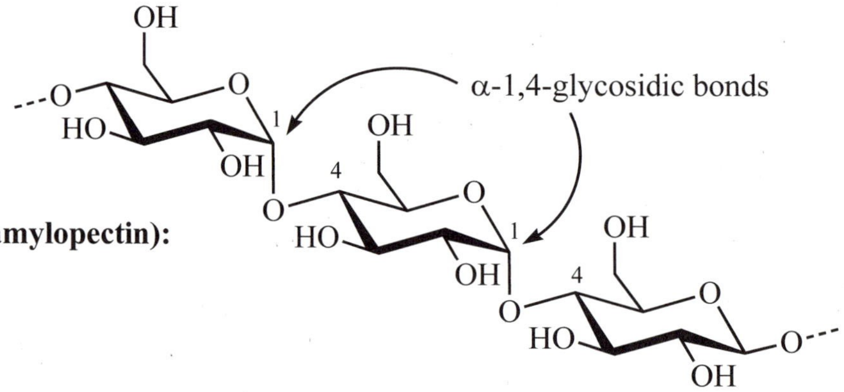

**18.49** *Starch can be separated into two principal polysaccharides, amylose and amylopectin. What is the major difference in structure between the two?*

**The difference is in the degree of chain branching. Amylose is composed of unbranched chains, whereas amylopectin is a branched network with branches starting at β-1,6 glycosidic bonds.**

**18.50** *A Fischer projection of N-acetyl-D-glucosamine is given in Section 18.2E.*

*(a) Draw Haworth and chair structures for the α- and β-pyranose forms of this monosaccharide.*

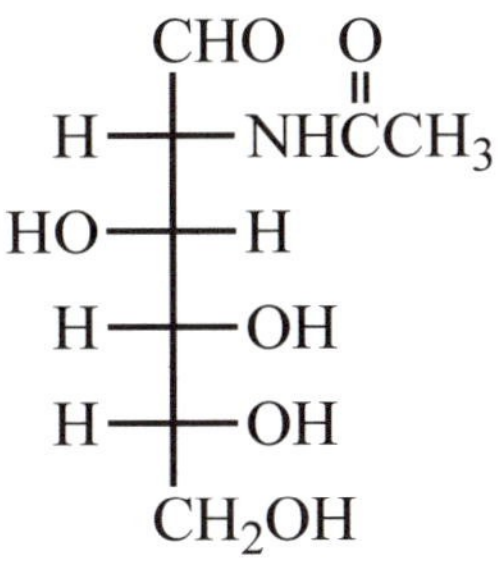

**α-anomer**

CHO

H———NHCCH$_3$ (O)

HO———H

H———OH

H———OH

CH$_2$OH

**N-Acetyl-D-glucosamine**

**β-anomer**

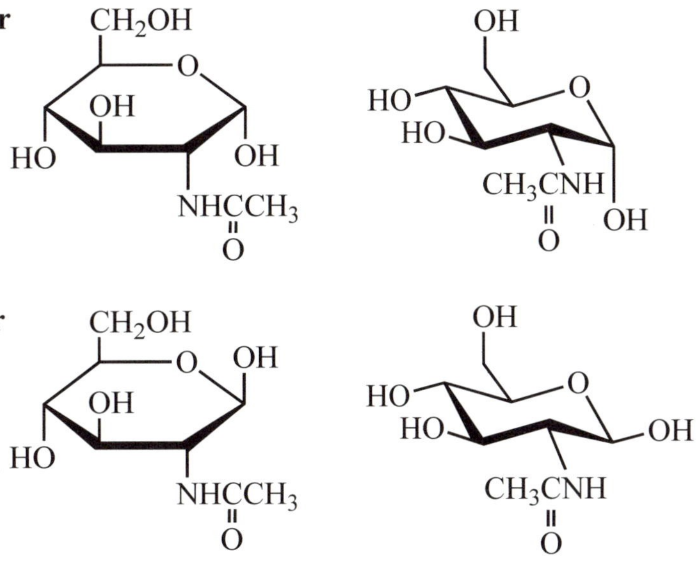

*(b) Draw Haworth and chair structures for the disaccharide formed by joining two units of the pyranose form of N-acetyl-D-glucosamine by a β-1,4-glucosidic bond. If you drew this correctly, you have the structural formula for the repeating dimer of chitin, the structural polysaccharide component of the shell of lobster and other crustaceans.*

**The following structures are the Haworth and chair formulas for the β-anomer of this disaccharide.**

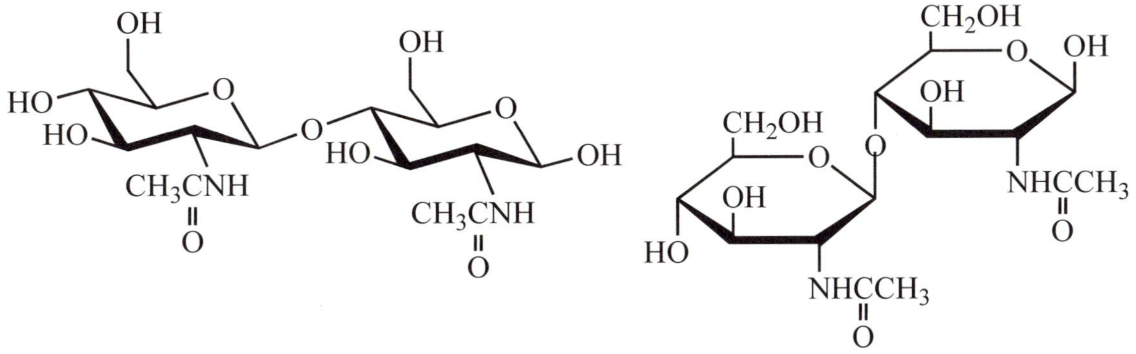

**18.51** *Propose structural formulas for the repeating disaccharide unit in these polysaccharides:*

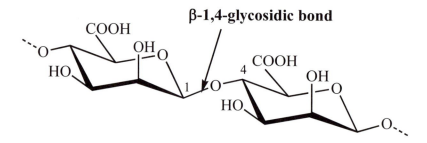

CHO
HO—H
HO—H
H—OH
H—OH
COOH

D-Mannuronic acid

CHO
H—OH
HO—H
HO—H
H—OH
COOH

D-Galacturonic acid

*(a) Alginic acid, isolated from seaweed, is used as a thickening agent in ice cream and other foods. Alginic acid is a polymer of D-mannuronic acid in the pyranose form joined by β-1,4-glycosidic bonds.*

**β-1,4-glycosidic bond**

*(b) Pectic acid is the main component of pectin, which is responsible for the formation of jellies from fruits and berries. Pectic acid is a polymer of D-galacturonic acid in the pyranose form joined by α-1,4-glycosidic bonds.*

**α-1,4-glycosidic bond**

**18.52** *Following is a Haworth projection and a chair conformation for the repeating disaccharide unit in chondroitin 6-sulfate. This biopolymer acts as a flexible connecting matrix between the tough protein filaments in cartilage. It is available as a dietary supplement, often combined with D-glucosamine sulfate. Some believe this combination can strengthen and improve joint flexibility.*

(a) *From what two monosaccharide units is the repeating disaccharide unit of chondroitin 6-sulfate derived?*

**The repeating units in chondroitin 6-sulfate are derivatives of the monosaccharides D-galactose and D-glucose.**

(b) *Describe the glycosidic bond between the two units.*

**There is a β-1,3- glycosidic bond between the D-glucose and D-galactose derivative units.**

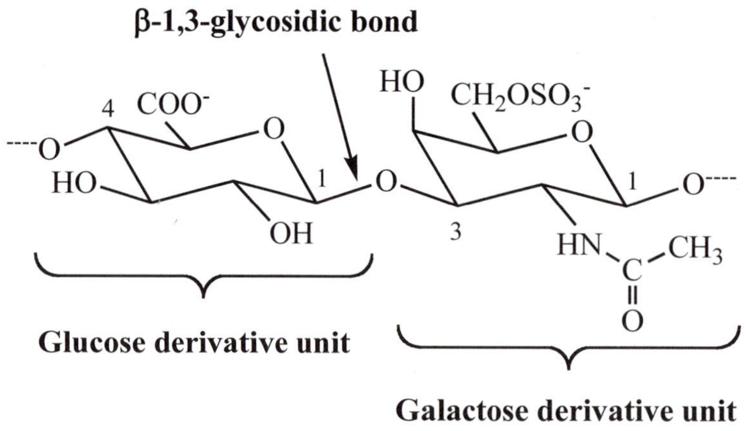

**β-1,3-glycosidic bond**

Glucose derivative unit

Galactose derivative unit

## Looking Ahead

**18.53**  *One step in glycolysis, the pathway that converts glucose to pyruvate (Section 22.4), involves an enzyme-catalyzed conversion of dihydroxyacetone phosphate to D-glyceraldehyde 3-phosphate.   Show that this transformation can be regarded as two enzyme-catalyzed keto-enol tautomerizations (Section 13.9).*

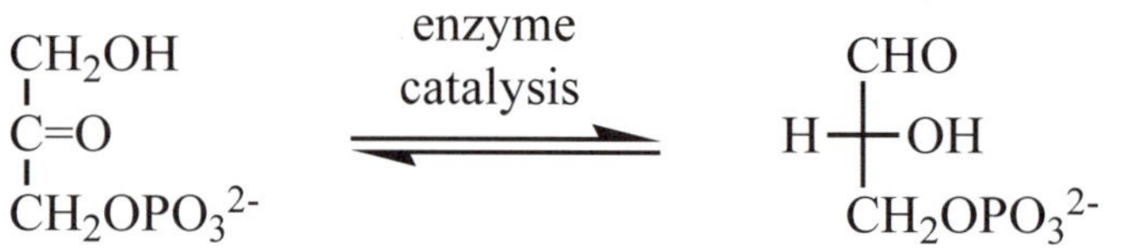

Dihydroxyacetone
phosphate

D-Glyceraldehyde
3-phosphate

**The first step involves the tautomerization of the hydroxy-ketone to the enediol. The enediol then undergoes a second tautomerization to the hydroxy-aldehydes (keto form).**

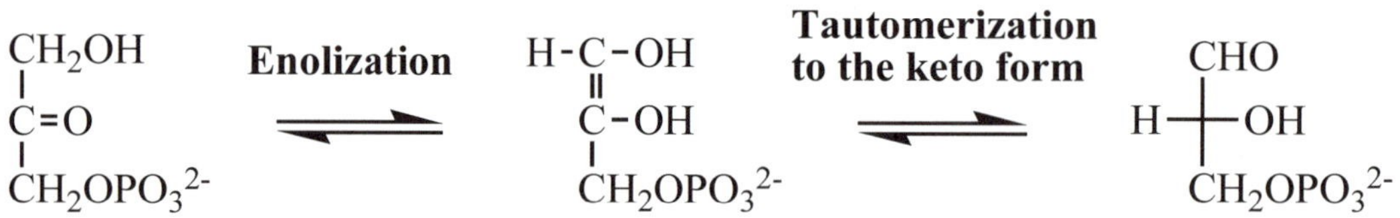

**Enediol intermediate**

**18.54**  *One pathway for the metabolism of glucose 6-phosphate is its enzyme-catalyzed conversion to fructose 6-phosphate.   Show that this transformation can be regarded as two enzyme-catalyzed keto-enol tautomerizations.*

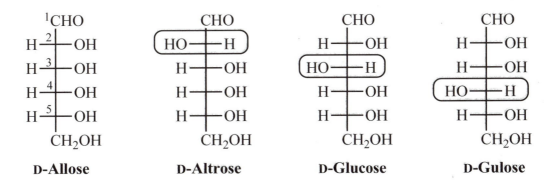

CHO
H——OH
HO——H
H——OH
H——OH
$CH_2OPO_3{}^{2-}$

**D-Glucose 6-phosphate**

enzyme catalysis

$CH_2OH$
C=O
HO——H
H——OH
H——OH
$CH_2OPO_3{}^{2-}$

**D-Fructose-6-phosphate**

**The first step involves the tautomerization of the hydroxy-ketone to the enediol. The enediol then undergoes a second tautomerization to the hydroxy-aldehydes (keto form).**

CHO
H——OH
HO——H
H——OH
H——OH
$CH_2OPO_3{}^{2-}$

**Enolization**

H–C–OH
‖
C–OH
HO——H
H——OH
H——OH
$CH_2OPO_3{}^{2-}$

**Tautomerization to the keto form**

$CH_2OH$
C=O
HO——H
H——OH
H——OH
$CH_2OPO_3{}^{2-}$

**18.55** *Epimers are carbohydrates that differ in configuration at only one stereocenter.*

*(a) Which of the aldohexoses are epimers of each other?*

> **There are a number of sets of epimers. One example set includes D-allose, D-altrose (differs at C2), D-glucose (differs at C3) and D-gulose (differs at C4).**

$^1$CHO
H$^2$—OH
H$^3$—OH
H$^4$—OH
H$^5$—OH
$CH_2OH$

**D-Allose**

CHO
(HO——H)
H——OH
H——OH
H——OH
$CH_2OH$

**D-Altrose**

CHO
H——OH
(HO——H)
H——OH
H——OH
$CH_2OH$

**D-Glucose**

CHO
H——OH
H——OH
(HO——H)
H——OH
$CH_2OH$

**D-Gulose**

*(b) Are all anomer pairs also epimers of each other? Explain. Are all epimers also anomers? Explain.*

> **Anomers always come in pairs; anomers of D-aldoses differ in configuration at C-1 in the cyclic hemiacetal form. Anomers of 2-ketoses differ in configuration**

**only at C-2 in the cyclic hemiacetal form. Epimers differ in configuration only at a carbon other than the anomeric carbon.**

**18.56** *Oligosaccharides are very valuable therapeutically, but are especially difficult to synthesize even though the starting materials are readily available. Shown is the structure of globotriose, the receptor for a series of toxins synthesized by some strains of E. coli. From left to right, globotriose consists of an α-1,4-linkage of galactose to galactose which is part of a β-1,4-linkage to glucose. The squiggly line indicates that the configuration at that carbon can be α or β. Suggest why it would be difficult to synthesize this trisaccharide, for example by first forming the galactose-galactose glycosidic bond and then forming the glycosidic bond to glucose.*

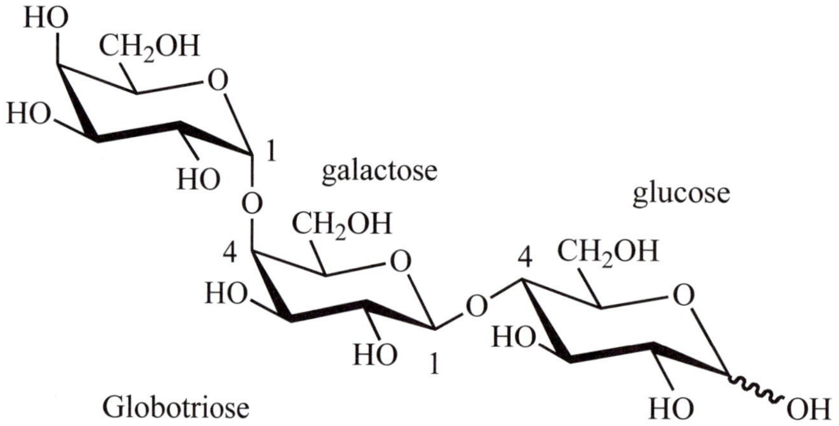

galactose

galactose

glucose

Globotriose

**Starting from galactose-galactose glycosidic bond, it will be difficult to control the stereochemistry of the 1,4-glycosidic bond between the galactose-glucose on either end of the trisaccharide.**

## CHAPTER 19
### Solutions to Problems

**19.1**     *Of the 20 protein-derived amino acids shown in Table 19.1, which contain:*

*(a) no stereocenter?*

**Only glycine does not contain a stereocenter.**

$$H-\underset{\underset{NH_3^+}{|}}{\overset{\overset{H}{|}}{C}}-COO^-$$

**Glycine**

*(b) two stereocenters?*

**Isoleucine and threonine each contain two stereocenters.**

**Isoleucine**                    **Threonine**

**19.2**     *The isoelectric point of histidine is 7.64. Toward which electrode does histidine migrate on paper electrophoresis at pH 7.0?*

**An amino acid will have at least a partial positive charge at any pH that is below its isoelectric point. Because the isoelectric point of histidine is 7.64, it will have a partial positive charge at pH 7.0 and migrate toward the negative electrode.**

**19.3**     *Describe the behavior of a mixture of glutamic acid, arginine, and valine on paper electrophoresis at pH 6.0.*

**At pH 6.0, glutamic acid (pI 3.08) has a net negative charge and will move toward the positive electrode. At this pH, arginine (pI 10.76) will have a net positive charge and move toward the negative electrode; valine (pI 6.00) will be neutral (have no net charge) and, therefore, remains at the origin.**

**19.4**     *Draw a structural formula for Lys-Phe-Ala. Label the N-terminal amino acid and the C-terminal amino acid.  What is the net charge on this tripeptide at pH 6.0?*

Due to the presence of the basic lysine residue, this tripeptide will have a net positive charge at pH 6.0.

**19.5**     *Which of these tripeptides are hydrolyzed by trypsin? By chymotrypsin?*
*(a) Tyr-Gln-Val     (b) Thr-Phe-Ser     (c) Thr-Ser-Phe*

Trypsin catalyzes the hydrolysis of peptide bonds formed by the carboxyl groups of lysine and arginine. Trypsin will not cleave any of these peptides because there are no arginine or lysine residues present.  Chymotrypsin catalyzes the hydrolysis of peptide bonds formed by the carboxyl groups of phenylalanine, tyrosine, and tryptophan. Chymotrypsin will cleave between Tyr and Gln residues in (a) and between Phe and Ser residues of (b).

**19.6**  *Deduce the amino acid sequence of an undecapeptide (11 amino acids) from the experimental results shown in the table.*

| Experimental Procedure | Amino Acids Determined from Procedure |
|---|---|
| **Amino Acid Analysis of Undecapeptide** | Ala, Arg, Glu, $Lys_2$, Met, Phe, Ser, Thr, Trp, Val |
| **Edman degradation** | Ala |
| **Trypsin-Catalyzed Hydrolysis** | |
| Fragment E | Ala, Glu, Arg |
| Fragment F | Thr, Phe, Lys |
| Fragment G | Lys |
| Fragment H | Met, Ser, Trp, Val |
| **Chymotrypsin-Catalyzed Hydrolysis** | |
| Fragment I | Ala, Arg, Glu, Phe, Thr |
| Fragment J | $Lys_2$, Met, Ser, Trp, Val |
| **Treatment with Cyanogen Bromide** | |
| Fragment K | Ala, Arg, Glu, $Lys_2$, Met, Phe, Thr, Val |
| Fragment L | Trp, Ser |

**Based on the Edman degradation result, alanine is the *N*-terminal residue of the peptide. Fragment E must have Arg on the *C*-terminal end because it is a peptide produced by trypsin cleavage. Because we know Ala is the *N*-terminal residue, Fragment E must have the sequence Ala-Glu-Arg. There must be two lysine residues or an arginine and lysine residue adjacent to each other based on the appearance of a single lysine residue as Fragment G. Since Fragment J has two lysines and no arginine residues, the two lysine residues must be adjacent to each other. From the chymotrypsin cleavage, we know that the C-terminal residues of Fragments I and J must be Phe and Trp, respectively. The sequence of Fragment L must therefore be Ser-Trp. Generation of Fragment L by CNBr cleavage indicates the last three amino acids in the peptide are Met-Ser-Trp. Thus, the sequence of Fragment H must be Val-Met-Ser-Trp. This information, combined with the knowledge that there are two lysine residues adjacent to each other, indicates that Fragment J has the sequence Lys-Lys-Val-Met-Ser-Trp. Because we know that the C-terminal residue of Fragment I must be Phe and that Fragment I must start with Ala-Glu-Arg, the entire sequence of fragment I must be Ala-Glu-Arg-Thr-Phe. Putting Fragments I and J together gives the following sequence for the entire peptide:**

**Ala-Glu-Arg-Thr-Phe-Lys-Lys-Val-Met-Ser-Trp**

**19.7**    *At pH 7.4, with what amino acid side chains can the side chain of lysine form salt linkages?*

**At pH 7.4, the only negatively charged side chains are the carboxylates of glutamic acid and aspartic acid. Therefore, these are the amino acid side chains that will form salt bridges with lysine side chains.**

**Amino Acids**

**19.8**    *What amino acid does each abbreviation stand for?*

   (a) Phe  **Phenylalanine**       (b) Ser  **Serine**         (c) Asp  **Aspartic acid**
   (d) Gln  **Glutamine**           (e) His  **Histidine**      (f) Gly  **Glycine**
   (g) Tyr  **Tyrosine**

**19.9**    *Configuration of the stereocenter in α-amino acids is most commonly specified using the D,L convention.  It can also be identified using the R,S convention (Section 6.4).  Does the stereocenter in L-serine have the R or the S configuration?*

**L-Glyceraldehyde**      **L-Serine**        **Line-angle**
                         **Fischer**          **drawing**
                       **projection**

**If you remember from carbohydrates, the D,L convention is based on the configuration of D and L glyceraldehyde.  Draw the Fischer projection of L-serine based on the configuration of L-glyceraldehyde.  The Fisher projection can be rotated sideways to give the line-angle drawing.  The configuration of the stereocenter can be determined using the Fischer projection or line-angle drawing. The stereocenter in L-serine has the *S* configuration.**

**19.10**  *Assign an R or S configuration to the stereocenter in each amino acid.*
         (a) L-Phenylalanine        (b) L-Glutamic acid       (c) L-Methionine

**The configuration of the stereocenter in each of the above amino acids is *S*.**

**19.11**  *The amino acid threonine has two stereocenters. The stereoisomer found in proteins has the configuration 2S,3R about the two stereocenters. Draw a Fischer projection of this stereoisomer and also a three-dimensional representation.*

**Threonine**

**19.12**  *Define the term zwitterion.*

**A zwitterion is a molecule that has a full positive charge and a full negative charge. The two charges cancel each other, thus a zwitterion has no net charge.**

**19.13**  *Draw zwitterion forms of these amino acids.*

*(a)*  **Valine**

*(b)*  **Phenylalanine**

*(c)*  **Glutamine**

**19.14**  *Why are Glu and Asp often referred to as acidic amino acids?*

**Glutaric acid (Glu) and aspartic acid (Asp) are referred to as acidic amino acids because their side chains contain carboxylic acid groups. Note that both Glu and Asp are negatively charged at neutral pH.**

**19.15**  *Why is Arg often referred to as a basic amino acid? Which two other amino acids are also basic amino acids?*

**Arg is referred to as a basic amino acid because its side chain contains a basic group (an -NH$_2$ group). His and Lys also have basic side chains and are considered basic amino acids as well.**

**19.16**  *What is the meaning of the alpha as it is used in α-amino acid?*

**The alpha in α-amino acid indicates that the amino group is on the carbon that is α to the carboxyl group.**

**19.17**  *Several β-amino acids exist.  There is a unit of β-alanine, for example, contained within the structure of coenzyme A (Section 22.2D).  Write the structural formula of β-alanine.*

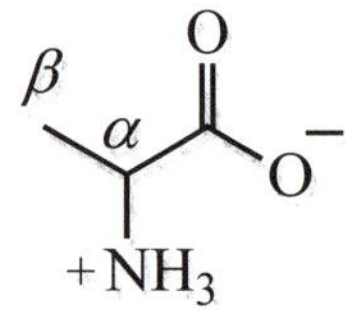

*α*-Alanine

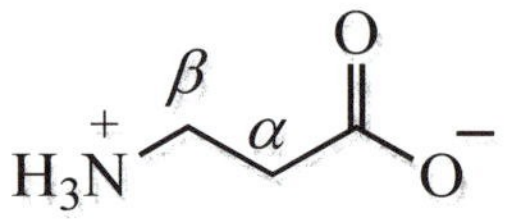

*β*-Alanine

**19.18**  *Although only L-amino acids occur in proteins, D-amino acids are often a part of the metabolism of lower organisms.  The antibiotic actinomycin D, for example, contains a unit of D-valine, and the antibiotic bacitracin A contains units of D-asparagine and D-glutamic acid.  Draw Fischer projections and three-dimensional representations for these three D-amino acids.*

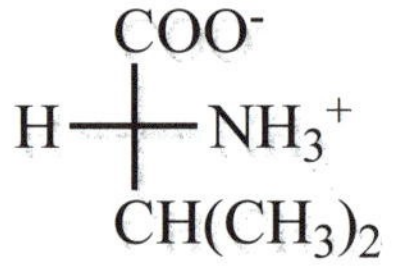

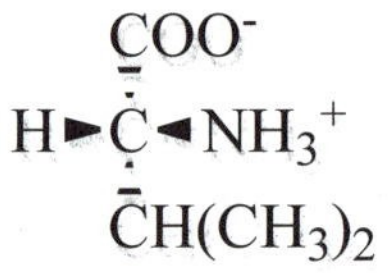

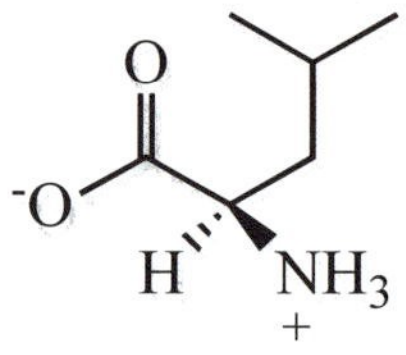

D-Valine

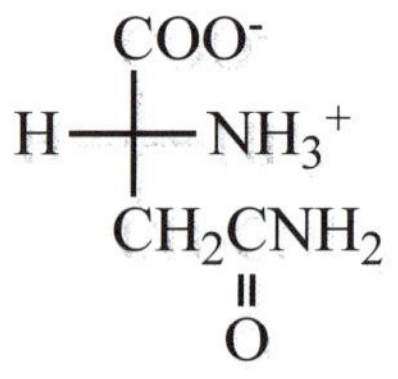

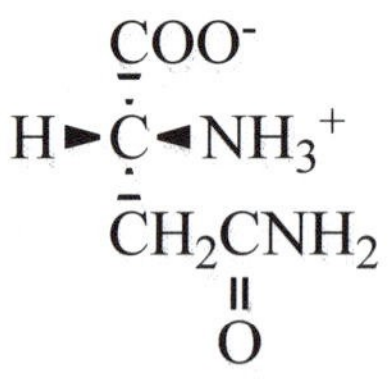

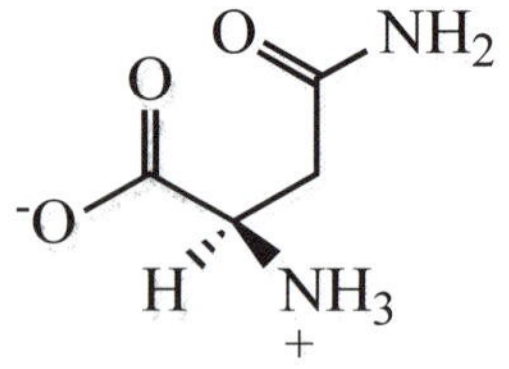

D-Asparagine

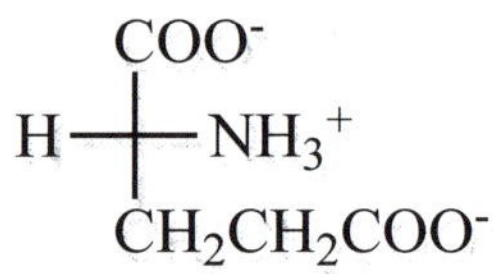

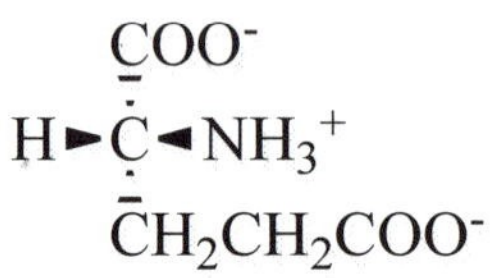

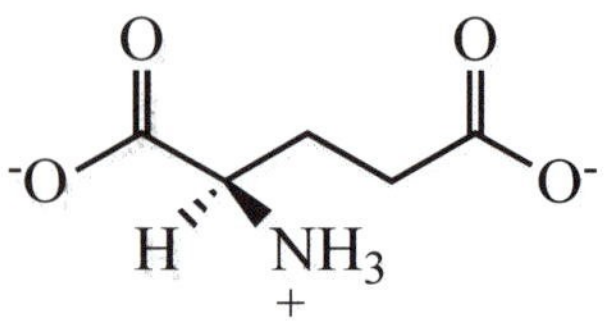

D-Glutamic acid

**19.19** *Histamine is synthesized from one of the 20 protein-derived amino acids. Suggest which amino acid is its biochemical precursor and the type of organic reaction(s) involved in its biosynthesis (e.g., oxidation, reduction, decarboxylation, nucleophilic substitution).*

Histidine is the biological precursor to histamine and a biosynthetic decarboxylation is involved in the conversion of His to histamine. Both the histamine and histidine are drawn in the form present at pH 7.4, the pH of blood.

**19.20** *Both norepinephrine and epinephrine are synthesized from the same protein-derived amino acid. From which amino acid are they synthesized and what types of reactions are involved in their biosynthesis?*

Norepinephrine and epinephrine are derived from the amino acid tyrosine. The bolded bonds outline the structural similarities to tyrosine. In both cases, biosynthesis of these molecules involves decarboxylation, aromatic hydroxylation ortho to the original phenolic –OH group (oxidation), and hydroxylation of the benzylic methylene group (oxidation). Epinephrine is also methylated on the α-amino group (nucleophilic substitution). All of the molecules in this problem are drawn in the form present at pH 7.4.

**19.21** *From which amino acid are serotonin and melatonin synthesized and what types of reactions are involved in their biosynthesis?*

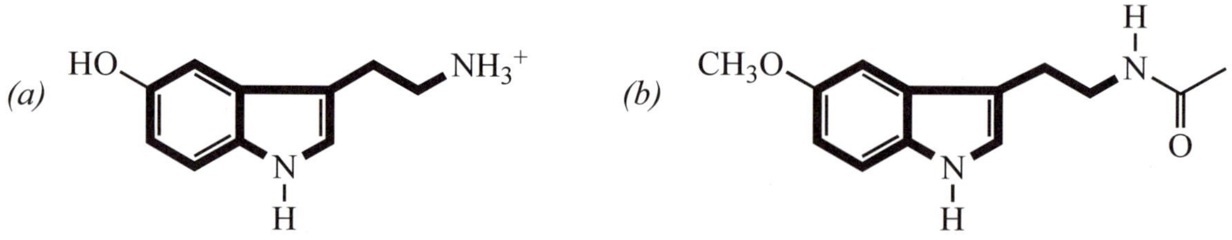

(a) Serotonin

(b) Melatonin

**Serotonin and melatonin are derived from the amino acid tryptophan. The bolded bonds outline the structural similarities to tryptophan. In both cases, biosynthesis of these molecules involves decarboxylation. In the case of serotonin, there is also an aromatic hydroxylation (oxidation). For melatonin, there is an aromatic methoxy group added (oxidation followed by a substitution reaction) and the amine group is acetylated (acyl substitution). All of the molecules in this problem are drawn in the form present at pH 7.4.**

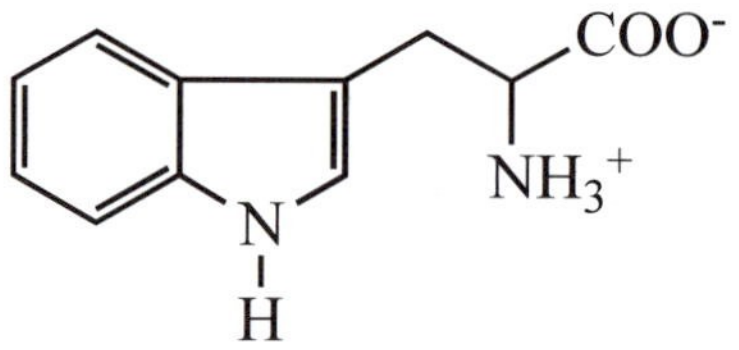

## Acid-Base Behavior of Amino Acids

**19.22** *Draw a structural formula for the form of each amino acid most prevalent at pH 1.0.*

(a) Threonine

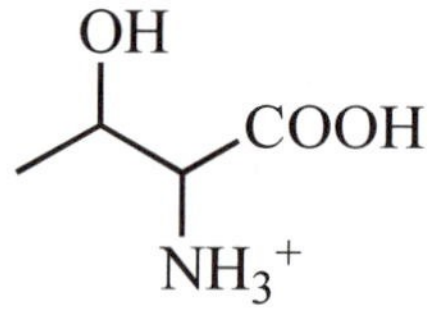

(b) Arginine

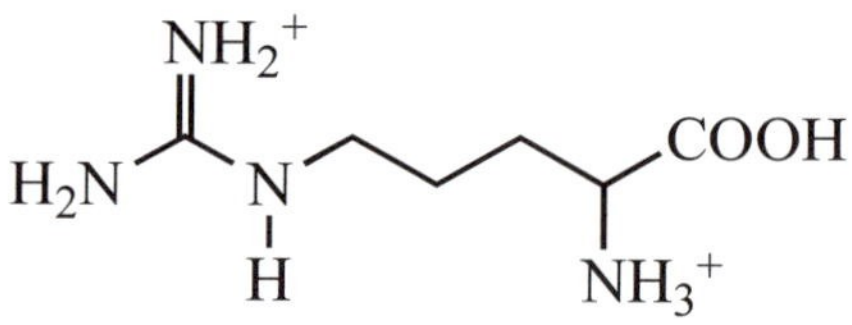

(c) Methionine

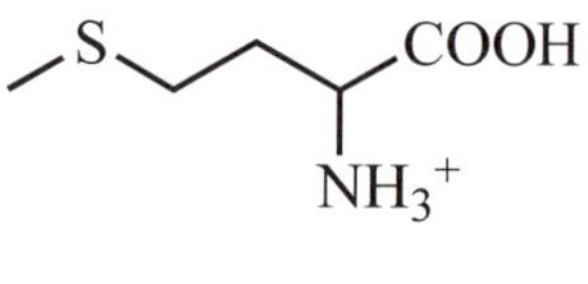

(d) Tyrosine

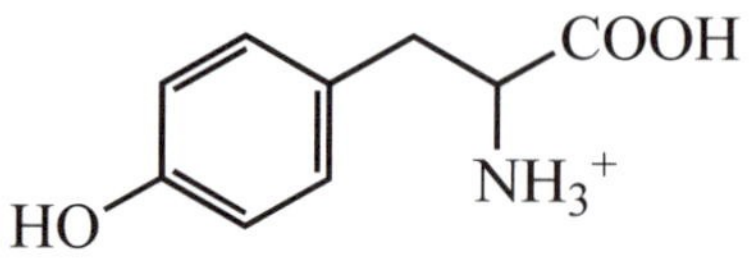

**19.23** *Draw a structural formula for the form of each amino most prevalent at pH 10.0.*

(a) *Leucine*

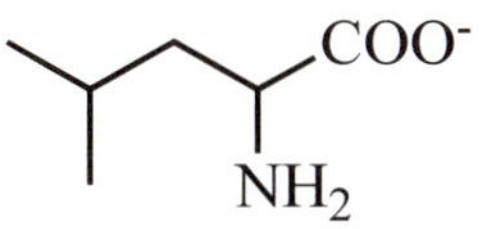

(b) *Valine*

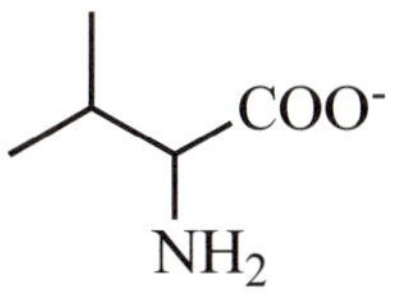

(c) *Proline*

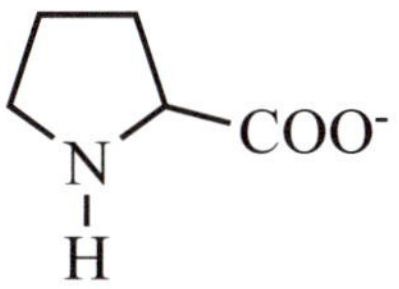

(d) *Aspartic acid*

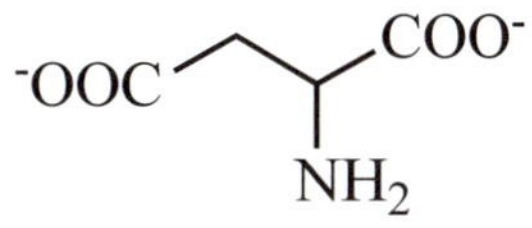

**19.24** *Write the zwitterion form of alanine and show its reaction with*

(a) $\xrightarrow{\text{1 mole NaOH}}$    $+$ Na$^+$ $+$ H$_2$O

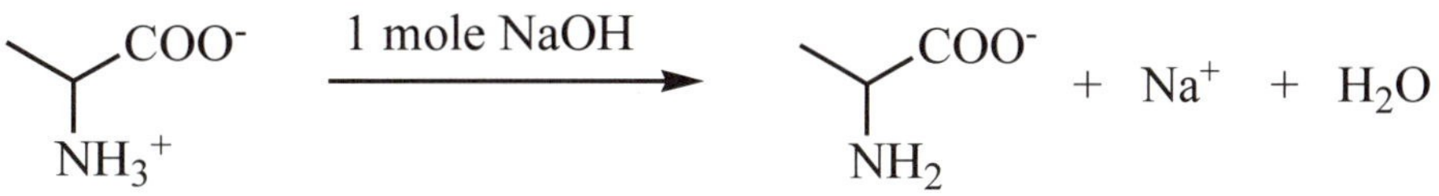

(b) $\xrightarrow{\text{1 mole HCl}}$    $+$ Cl$^-$

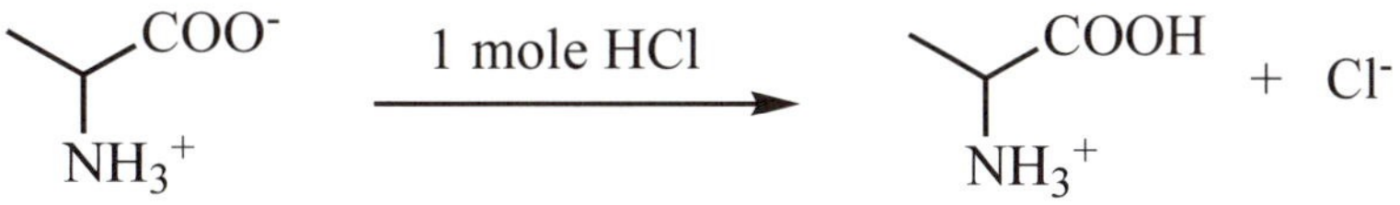

**19.25** *Write the form of lysine most prevalent at pH 1.0 and then show its reaction with the following.  Consult Table 19.2 for pK$_a$ values of the ionizable groups in lysine.*

**At pH 1.0, for the most prevalent form of lysine, both amino groups as well as the carboxylic acid group are protonated with a total charge of +2.**

(a)  pH 1.0   $\xrightarrow{\text{1 mole NaOH}}$

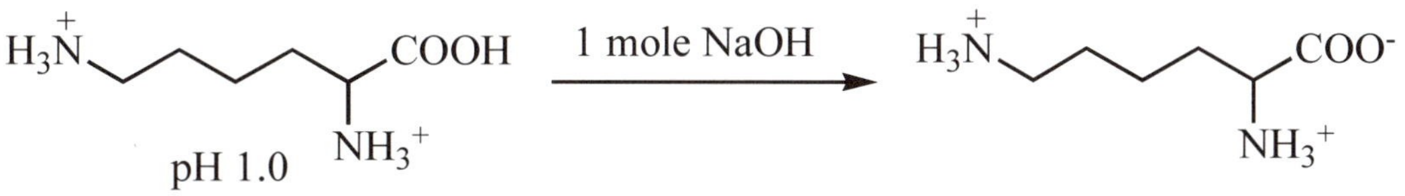

(b)  pH 1.0   $\xrightarrow{\text{2 mole NaOH}}$

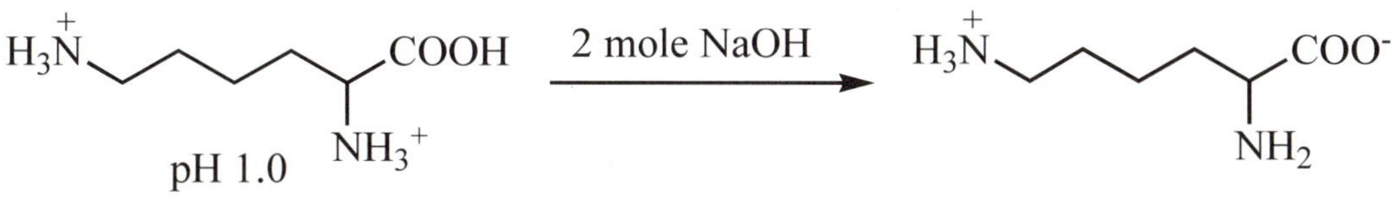

(c)

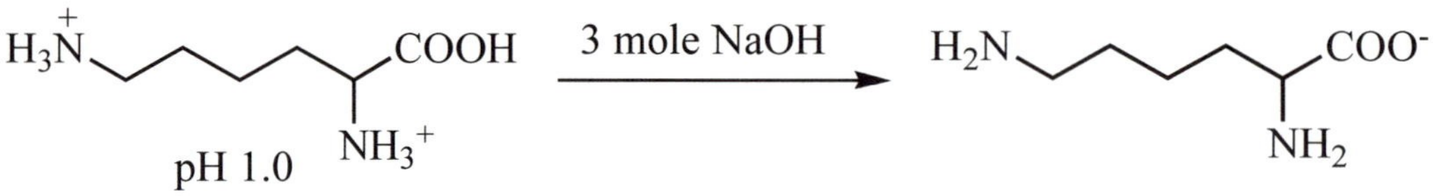

**19.26** *Write the form of aspartic acid most prevalent at pH 1.0 and then show its reaction with the following. Consult Table 19.2 for pK_a values of the ionizable groups in aspartic acid.*

**At pH 1.0, the most prevalent form of aspartic acid has both carboxylic acid groups as well as the amino group protonated and a total charge of +1.**

(a)

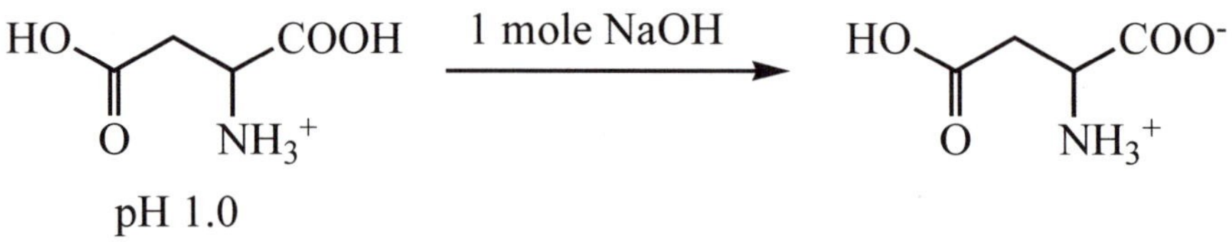

(b)

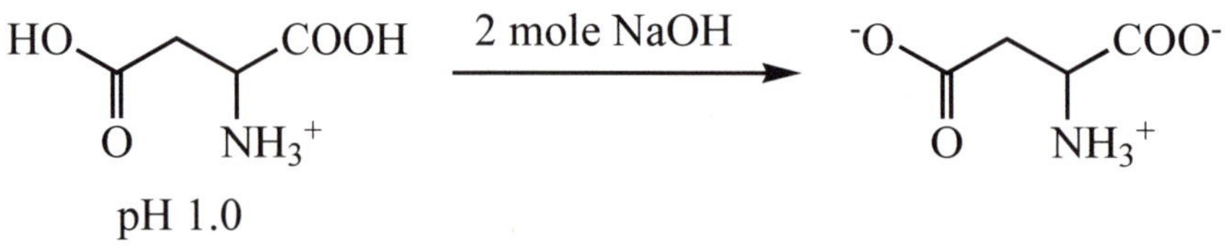

(c)

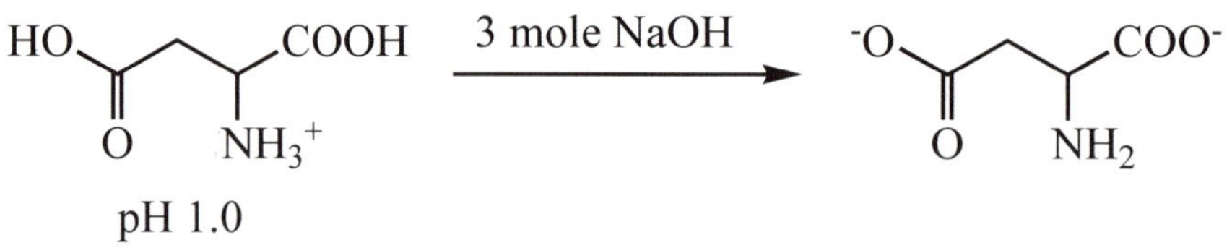

**19.27**  *Given pK$_a$ values for ionizable groups from Table 19.2, sketch curves for the titration of*

*(a) Glutamic acid with NaOH*

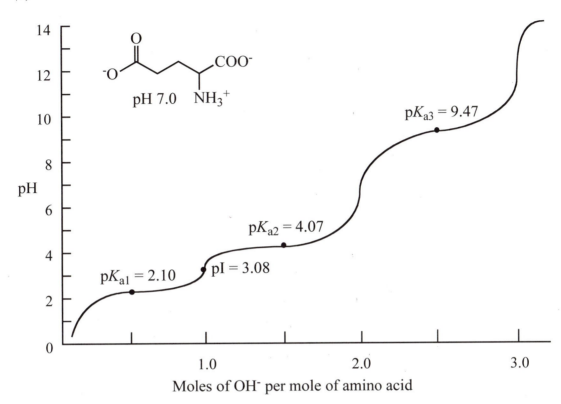

*(b) Histidine with NaOH.*

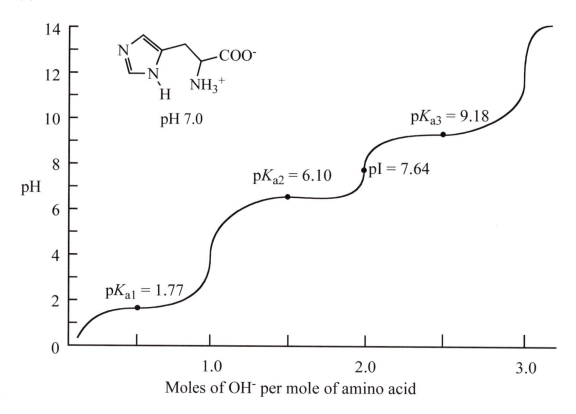

**19.28** *Draw a structural formula for the product formed when alanine is treated with the following reagents:*

**19.29** *Account for the fact that the isoelectric point of glutamine (pI 5.65) is higher than the isoelectric point of glutamic acid (pI 3.08).*

pH 7.0                                   pH 7.0

**Glutamic acid**                     **Glutamine**

**Amino acids have no net charge at their pI. Glutamic acid possesses an acidic side chain, so in order for Glu to be neutral, the net charge on the carbonyls must be −1. Thus the pI for glutamic acid will fall between the $pK_a$ values of each carboxyl group:**

$$pI = 1/2(pK_a \text{ } \alpha\text{-COOH} + pK_a \text{ -COOH})$$
$$pI = 1/2(2.10 + 4.07) = 3.08$$

**The amide side chain on glutamine is already uncharged near neutral pH, so the pI of the amino acid is determined by the $pK_a$ values for the only ionizable groups, namely the $\alpha$-carboxyl and the $\alpha$-amino groups. Glutamine's pI can be determined using the following equation:**

$$pI = 1/2(pK_a \text{ } \alpha\text{-COOH} + pK_a \text{ } \alpha\text{-NH}_3{}^+)$$
$$pI = 1/2(2.17 - 9.03) = 5.65$$

**19.30** *Enzyme-catalyzed decarboxylation of glutamic acid gives 4-aminobutanoic acid (Section 19.2D). Estimate the pI of 4-aminobutanoic acid.*

**There is little, if any inductive effect operating between the amino and carboxyl groups of the 4-aminobutanoic acid because there are three methylene groups between them and inductive effects decrease with increasing distance between the two groups. Thus, $pK_a$ of the amino group of 4-aminobutanoic acid is like that of a simple amino group, near 10. Similarly, the $pK_a$ of the carboxyl group is like that of a simple carboxyl group, near 4.5. Given these estimates for the $pK_a$ values, the pI will be:**

$$pI = 1/2(pK_a \text{ } \alpha\text{-COOH} + pK_a \text{ } \alpha\text{-NH}_3{}^+)$$
$$pI = 1/2(4.5 - 10.0) = 7.25$$

**19.31**  *Guanidine and the guanidino group  present in arginine are two of the strongest organic bases known.  Account for their basicity.*

**The basicity of guanidine and the guanidino group is due to the large resonance stabilization of the protonated guanidinium ion.**

**19.32**  *At pH 7.4, the pH of blood plasma, do the majority of protein-derived amino acids bear a net negative charge or a net positive charge?*

**The majority of the amino acids have pI values between 5 and 6, therefore they will bear a net negative charge at pH 7.4.  The only exceptions are arginine, histidine, and lysine (the three basic amino acids).**

**19.33**  *Do the following compounds migrate to the cathode or to the anode on electrophoresis at the specified pH?*

*(a) Histidine at pH 6.8*

**pI = 7.64:  At  pH 6.8, histidine has a net positive charge and migrates toward the negative electrode (cathode).**

*(b) Lysine at pH 6.8*

**pI = 9.74:  At  pH 6.8, lysine has a net positive charge and migrates toward the negative electrode (cathode).**

*(c) Glutamic acid at pH 4.0*

**pI = 3.08:  At  pH 4.0, glutamic acid has a net negative charge and migrates toward the positive electrode (anode).**

*(d) Glutamine at pH 4.0*

**pI = 5.65:  At  pH 4.0, glutamine has a net positive charge and migrates toward the negative electrode (cathode).**

*(e) Glu-Ile-Val at pH 6.0*

**At pH 4.0, the carboxyl group on the side chain of glutamic acid will largely be deprotonated, as will the α-carboxyl group on valine. The α-amino group (on glutamic acid) will mostly be protonated. Thus, Glu-Ile-Val will have a net negative charge and will migrate toward the positive electrode (anode).**

*(f) Lys-Gln-Tyr at pH 6.0*

**At pH 6.0, the α-amino acid and the side chain amino group of lysine will both be protonated, while the carboxyl group of tyrosine will be deprotonated. Thus the molecule will have a net positive charge and will migrate toward the negative electrode (cathode).**

**19.34** *At what pH would you carry out an electrophoresis to separate the amino acids in each mixture?*

**Remember that amino acids at pH values below their isoelectric points will have some degree of positive charge. Amino acids at pH values above their isoelectric points will have some degree of negative charge. Amino acids at pH values equal to their isoelectric points will have no net charge.**

*(a) Ala, His, Lys*

**Electrophoresis can be performed at pH 7.64, the isoelectric point of histidine (His). At this pH, the histidine is neutral and will not move. Lysine (Lys) will be positively charged and move toward the negative electrode while alanine (Ala) will have a slight negative charge and move toward the positive electrode.**

*(b) Glu, Gln, Asp*

**Electrophoresis can be performed at pH 3.03, the isoelectric point of glutamic acid (Glu). At this pH, glutamic acid is neutral and will not move. Glutamine (Gln) will be positively charged and move toward the negative electrode while aspartic acid (Asp) will have a slight negative charge and move toward the positive electrode.**

*(c) Lys, Leu, Tyr*

**Electrophoresis can be performed at pH 6.04, the isoelectric point of leucine (Leu). At this pH, leucine is neutral and will not move. Lysine (Lys) will be positively charged and move toward the negative electrode while tyrosine (Tyr) will have a slight negative charge and move toward the positive electrode.**

**19.35**  *Examine the amino acid sequence of human insulin (Figure 19.13) and list each Asp, Glu, His, Lys, and Arg in this molecule.  Do you expect human insulin to have an isoelectric point nearer that of the acidic amino acids (pI 2.0 - 3.0), the neutral amino acids (pI 5.5 - 6.5), or the basic amino acids (pI 9.5 - 11.0)?*

|  |  | No. |
|---|---|---|
| **Amino acid residues with carboxylic acid side chains** | Aspartic acid (Asp) | 0 |
|  | Glutamic Acid (Glu) | 4 |
| **Amino acid residues with amino side chains** | Histidine (His) | 2 |
|  | Lysine (Lys) | 1 |
|  | Arginine (Arg) | 1 |

**Insulin has an equal number of acidic (4) and basic (4) side chains.  Its isoelectric point must be at a pH in which all the acidic groups are deprotonated  (pH > 4) and all the basic groups are protonated (pH<6).  The pI of insulin should fall between 5.5 and 6.5, which is similar to that of a neutral amino acid.  The experimentally determined pI value for insulin is 5.30-5.35.**

## Primary Structure of Polypeptides and Proteins

**19.36**  *If a protein contains four different SH groups, how many different disulfide bonds are possible if only a single disulfide bond is formed?  How many different disulfides are possible if two disulfide bonds are formed?*

**If only one disulfide bond were formed from the four different cysteine residues, then a total of six different disulfide bonds are possible.  Three different disulfides are possible when two disulfide bonds are formed.**

**19.37**  *How many different tetrapeptides can be made if*

*(a) The tetrapeptide contains one unit each of Asp, Glu, Pro, and Phe?*

**Who ever said organic chemistry didn't involve math?  Any of the four residues can be in the first position, any of the remaining three amino acids in the second position, and so on, thus, with four amino acid residues, there are:**

*4 x 3 x 2 x 1 = 24 possible tetrapeptides.*

*(b) All 20 amino acids can be used, but each only once?*

**Using the same logic, with 20 amino acids used, there are:**

*20 x 19 x 18 x 17 = 116,280 tetrapeptides*

**19.38**  *A decapeptide has the following amino acid composition:*

$$Ala_2, Arg, Cys, Glu, Gly, Leu, Lys, Phe, Val$$

*Partial hydrolysis yields the following tripeptides:*

$$Cys\text{-}Glu\text{-}Leu \ + \ Gly\text{-}Arg\text{-}Cys \ + \ Leu\text{-}Ala\text{-}Ala \ + \ Lys\text{-}Val\text{-}Phe \ + \ Val\text{-}Phe\text{-}Gly$$

*One round of Edman degradation yields a lysine phenylthiohydantoin. From this information, deduce the primary structure of this decapeptide.*

**The Edman degradation result indicates that the Lys residue must be at the *N*-terminus.  Given this information, the rest of the peptide sequence can be deduced from the overlap among the tripeptide sequences as shown below:**

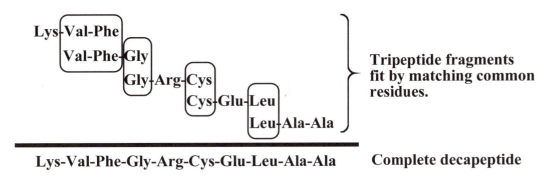

**Lys-Val-Phe-Gly-Arg-Cys-Glu-Leu-Ala-Ala          Complete decapeptide**

**19.39**  *Following is the primary structure of glucagon, a polypeptide hormone of 29 amino acids.  Glucagon is produced in the α-cells of the pancreas and helps maintain blood glucose levels in a normal concentration range.*

```
1             5                   10                  15
His-Ser-Glu-Gly-Thr-Phe-Thr-Ser-Asp-Tyr-Ser-Lys-Tyr-Leu-Asp-Ser-Arg-Arg-
                        20                  25                  29
                  Ala-Gln-Asp-Phe-Val-Gln-Trp-Leu-Met-Asn-Thr
```

**Glucagon**

*Which peptide bonds are hydrolyzed when this polypeptide is treated with each reagent?*

*(a) Phenyl isothiocyanate*

**Phenyl isothiocyanate only cleaves the *N*-terminal amino acid, so the His-Ser bond would be hydrolyzed.**

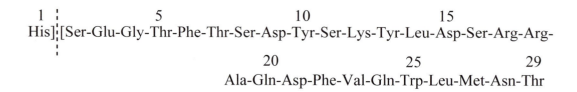

```
   1  :              5                    10                    15
 His]:[Ser-Glu-Gly-Thr-Phe-Thr-Ser-Asp-Tyr-Ser-Lys-Tyr-Leu-Asp-Ser-Arg-Arg-
    :
    :
                                20               25               29
                      Ala-Gln-Asp-Phe-Val-Gln-Trp-Leu-Met-Asn-Thr
```

*(b) Chymotrypsin*

**Chymotrypsin only cleaves peptide bonds on the carboxyl side of phenylalanine, tyrosine, and tryptophan.**

```
   1              5             :         10     :          :    15
 His-Ser-Glu-Gly-Thr-Phe]:[Thr-Ser-Asp-Tyr]:[Ser-Lys-Tyr]:[Leu-Asp-Ser-Arg-Arg-
                          :                :            :
                            20               :    25     :          29
                      Ala-Gln-Asp-Phe]:[Val-Gln-Trp]:[Leu-Met-Asn-Thr
                                     :            :
```

*(c) Trypsin*

**Trypsin only cleaves peptide bonds on the carboxyl side of arginine and lysine.**

```
   1              5                    10           :          15          :    :
 His-Ser-Glu-Gly-Thr-Phe-Thr-Ser-Asp-Tyr-Ser-Lys]:[Tyr-Leu-Asp-Ser-Arg]:[Arg:
                            20               '    25               '29
                      Ala-Gln-Asp-Phe-Val-Gln-Trp-Leu-Met-Asn-Thr
```

*(d) Br-CN*

**Cyanogen bromide cleaves on the *C*-terminal side of methionine residues.**

```
   1              5                    10                    15
 His-Ser-Glu-Gly-Thr-Phe-Thr-Ser-Asp-Tyr-Ser-Lys-Tyr-Leu-Asp-Ser-Arg-Arg-
                                20               25       :          29
                      Ala-Gln-Asp-Phe-Val-Gln-Trp-Leu-Met]:[Asn-Thr
                                                          :
```

**19.40**  *A tetradecapeptide (14 amino acid residues) gives the following peptide fragments on partial hydrolysis. From this information, deduce the primary structure of this polypeptide. Fragments are grouped according to size.*

<u>Pentapeptide Fragments</u>
*Phe-Val-Asn-Gln-His*
*His-Leu-Cys-Gly-Ser*
*Gly-Ser-His-Leu-Val*

<u>Tetrapeptide Fragments</u>
*Gln-His-Leu-Cys*
*His-Leu-Val-Glu*
*Leu-Val-Glu-Ala*

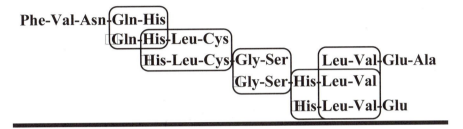

**Complete peptide:**

**Phe-Val-Asn-Gln-His-Leu-Cys-Gly-Ser-His-Leu-Val-Glu-Ala**

**19.41**  *Draw a structural formula of these tripeptides. Mark each peptide bond, the N-terminal amino acid, and the C-terminal amino acid.*

*(a) Phe-Val-Asn*

**Phe-Val-Asn**

*(b) Leu-Val-Gly*

**N-terminal AA**          **C-terminal AA**

peptide bonds

**Leu-Val-Gly**

**19.42**  *Estimate the pI of each tripeptide on Problem 19.41.*

**The pI values can be estimated by using the $pK_a$ of the amino group for the N-terminal amino acid, and the $pK_a$ of the carboxylic acid group for the C-terminal amino acid. The carboxylic acid will have a $pK_a$ 4.8, a value similar to an unsubstituted carboxylic acid. The values for the appropriate amino groups are listed in Table 19.2**

*(a) Phe-Val-Asn*

**pI = 1/2(9.4 + 4.8) = 7.2**

*(b) Leu-Val-Gln*

**pI = 1/2(9.76 + 4.8) = 7.3**

**19.43**  *Glutathione (G-SH), one of the most common tripeptides in animals, plants, and bacteria, is a scavenger of oxidizing agents. In reacting with oxidizing agents, glutathione is converted to G-S-S-G.*

Glutathione     $\overset{+}{H_3}NCHCH_2CH_2\overset{\displaystyle O}{\overset{\|}{C}}NHCH\overset{\displaystyle O}{\overset{\|}{C}}NHCH_2COO^-$
                      $\underset{\displaystyle COO^-}{|} \qquad\qquad \underset{\displaystyle CH_2SH}{|}$

*(a) Name the amino acids in this tripeptide.*

**The amino acids in glutathione are glutamic acid (Glu), cysteine (Cys), and glycine (Gly),**

*(b) What is unusual about the peptide bond formed by the N-terminal amino acid?*

**The *N*-terminal glutamic acid is linked to the next residue by an amide bond with the carboxyl group of the glutamic acid side chain, not the α-carboxyl group.**

*(c) Is glutathione a biological oxidizing agent or a biological reducing agent?*

$$2\text{G-SH} \quad \xrightarrow{\text{- 2H}} \quad \text{G-S-S-G}$$

**The glutathione loses 2 hydrogens in the process, so it is oxidized. Therefore glutathione is a biological reducing agent**

*(d) Write a balanced equation for reaction of glutathione with molecular oxygen, $O_2$, to form G-S-S-G and $H_2O$. Is molecular oxygen oxidized or reduced in this process?*

$$2\text{G-SH} \quad + \quad 1/2 O_2 \quad \longrightarrow \quad \text{G-S-S-G} \quad + \quad H_2O$$

**The glutathione is oxidized and the oxygen reduced in this process.**

**19.44** *Following is a structural formula for the artificial sweetener aspartame. Each amino acid has the L configuration.*

Aspartame

*(a) Name the two amino acids in this molecule.*

**Aspartame is composed of aspartic acid (Asp) attached via a peptide bond to the methyl ester of phenylalanine (Phe).**

*(b) Estimate the isoelectric point of aspartame?*

$$pI = 1/2(9.82 + 3.86) = 6.84$$

*(c) Draw structural formulas for the products of hydrolysis of aspartame in 1 M HCl.*

**Esters are far more easily hydrolyzed than amides, therefore, the ester is selectively hydrolyzed while leaving the amide intact.  If the reaction were to be run under more forcing conditions, the amide bond would also hydrolyze, yielding the amino acids Phe and Asp.**

## Three-Dimensional Shapes of Polypeptides and Proteins

**19.45**  *Examine the α-helix conformation.  Are amino acid side chains arranged all inside the helix, all outside the helix, or randomly?*

**All amino acid side chains extend outside of the helix.**

**19.46**  *Distinguish between intermolecular and intramolecular hydrogen bonding between the backbone groups on polypeptide chains.  In what type of secondary structure do you find intermolecular hydrogen bonds?  In what type do you find intramolecular hydrogen bonding?*

**Intramolecular hydrogen bonding is responsible for the formation and stability of the α-helix secondary structures.  Intermolecular hydrogen bonding is possible in β-sheet secondary structures.**

**19.47**  *Many plasma proteins found in aqueous environment are globular in shape.  Which amino acid side chains would you expect to find on the surface of a globular protein and in contact with the aqueous environment?  Which would you expect to find inside, shielded from the aqueous environment?  Explain.*
*(a) Leu          (b) Arg          (c) Ser          (d) Lys          (e) Phe*

**Polar, acidic, and basic side chains will prefer to be on the outside of the protein surface, in contact with the aqueous environment to maximize hydrophilic interactions.  Of the choices, (b) Arg, (c) Ser, and (d) Lys best fit this criterion.**

>Nonpolar side chains will prefer to avoid contact with the aqueous environment, turning inward to the protein to maximize hydrophobic interactions.  Of the choices, these are (a) Leu and (e) Phe.

**Looking Ahead**

**19.48**  *Heating can disrupt the 3° structure of a protein.  Explain the chemical processes that occur upon heating a protein.*

>Heating a protein destroys the weak interactions responsible for tertiary structure, primarily hydrogen bonding.  Although heating does not usually break the amino acid sequences responsible for primary structure, heating can provide enough energy to break the hydrogen bonds that maintain structures such as the α-helix and β-sheets.

**19.49**  *Some amino acids cannot be incorporated into proteins because they are self-destructive. Homoserine, for example, can use its side chain OH group in an intramolecular nucleophilic acyl substitution to cleave the peptide bond and form a cyclic structure on one end of the chain.  Draw this structure and explain why serine does not suffer the same fate.*

continuing
chain

Homoserine
residue

? + $H_3N$⌇⌇⌇COO⁻

Serine residue

no reaction

>**The homoserine residue undergoes a cyclization to form a lactone and an amino acid.**

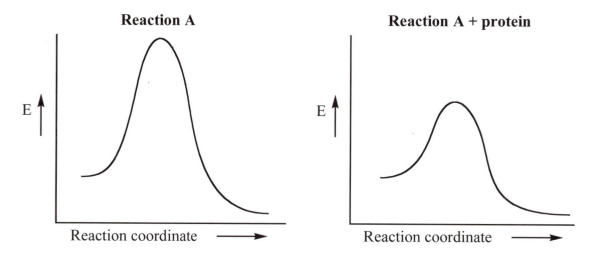

**The proposed cyclization for the serine residue would involve the formation of a four-membered lactone ring, which is highly unstable due to ring strain and therefore, unlikely to occur.**

**19.50** *Would you expect a decapeptide of only isoleucine residues to form an α-helix? Explain.*

Bulky
side chain

NH₃⁺

**Isolucine**

**Isoleucine has a large bulky side chain, which would destabilize and prevent the α-helix formation of a decapeptide composed entirely of this amino acid.**

**19.51** *Which type of protein would you expect to achieve the following?*

| Reaction A | Reaction A + protein |
|---|---|

E ↑

Reaction coordinate ⟶

E ↑

Reaction coordinate ⟶

**The process shown in the second energy diagram is a lowering of activation energy. This is associated with catalysis and a function of proteins known as enzymes.**

**CHAPTER 20**
*Solutions to Problems*

**20.1**   *Draw a structural formula for 2'-deoxythymidine 3'-monophosphate*

**20.2**   *Draw a structural formula for the section of DNA that contains the base sequence CTG and is phosphorylated at the 3' end only.*

**20.3**   *Write the complementary DNA base sequence for 5'-CCGTACGA-3'.*

**The complimentary sequence is 5'-TCGTACGG-3'**

**20.4**   *Here is a portion of the nucleotide sequence in phenylalanine tRNA.*

*3'-ACCACCUGCUCAGGCCUU-5'*

*Write the nucleotide sequence of its DNA complement.*

**Remember that the base uracil (U) in RNA is the complement to adenine (A) in DNA. The complement DNA sequence is 5'-TGGTGGACGAGTCCGGAA-3'**

**20.5**   *The following section of DNA codes for oxytocin, a polypeptide hormone:*

*3'-ACG-ATA-TAA-GTT-TTA-ACG-GGA-GAA-CCA-ACT-5'*

*(a) Write the base sequence of the mRNA synthesized from this section of DNA.*

**5'-UGC-UAU-AUU-CAA-AAU-UGC-CCU-CUU-GGU-UGA-3'**

*(b) Given the sequence of bases in part (a), write the primary structure of oxytocin.*

**The amino acid sequence of oxytocin is:**

**Amino terminus-Cys-Tyr-Ile-Gln-Asn-Cys-Pro-Leu-Gly-Carboxyl terminus**

**Note how the last codon, UGA, does not code for an amino acid, but rather is a stop signal.**

**20.6**   *The following is another section of the bovine rhodopsin gene. Which of the endonucleases given in Example 20.6 will catalyze cleavage of this section?*

*5'-ACGTCGGGTCGTCGTCCTCTCGCGGTGGTGAGTCTTCCGGCTCTTCT-3'*

**The Sac1, F*nu*DII, and HpaII cleavage sites are:**

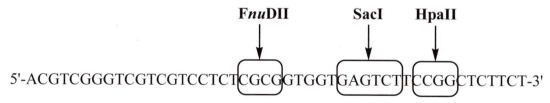

## Nucleosides and Nucleotides

**20.7**   *Two drugs used in the treatment of acute leukemia are 6-mercaptopurine and 6-thioguanine. In each drug, the oxygen at carbon 6 of the parent molecule is replaced by divalent sulfur. Draw structural formulas for the enethiol (the sulfur equivalent of an enol) forms of 6-mercaptopurine and 6-thioguanine.*

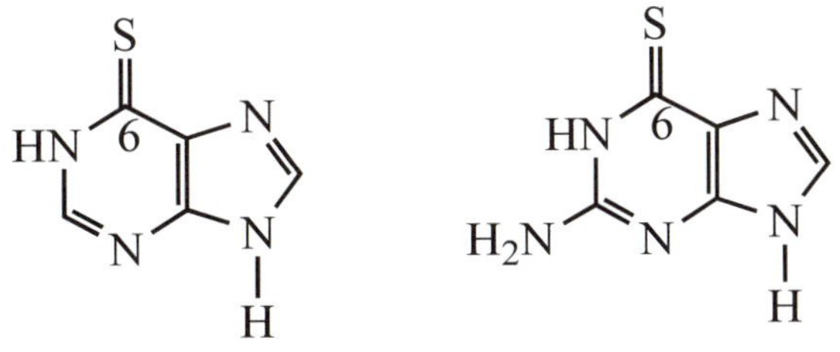

6-Mercaptopurine         6-Thioguanine

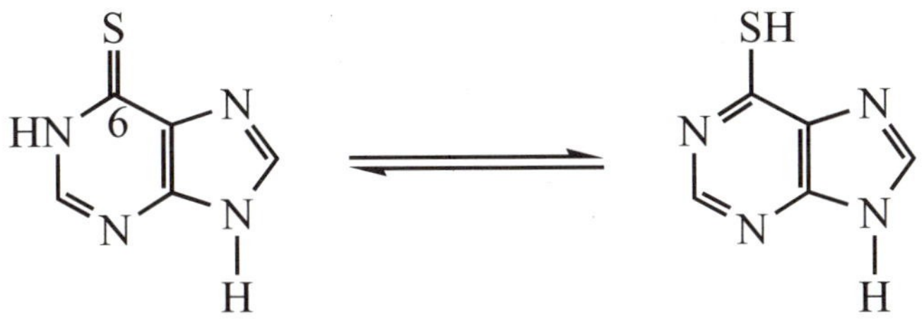

**Enethiol**

**Enethiol**

**20.8**   *Following are structural formulas for cytosine and thymine. Draw two additional tautomeric forms for cytosine and three additional tautomeric forms for thymine.*

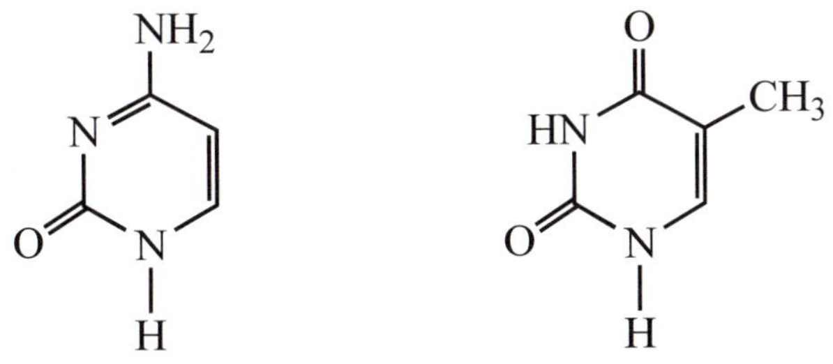

Cytosine (C)         Thymine (T)

**There are three additional tautomeric forms of cytosine:**

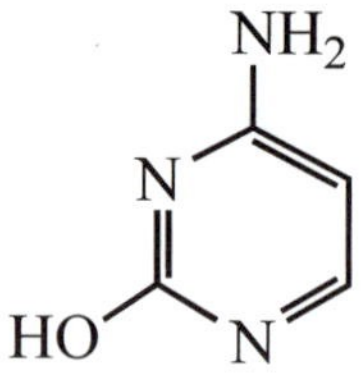

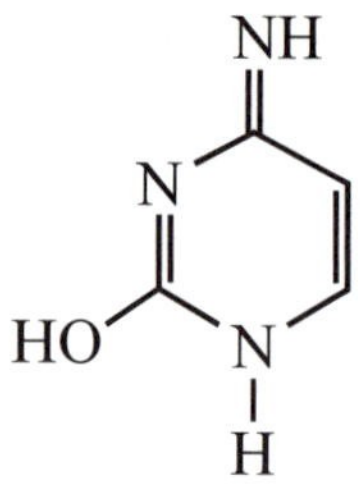

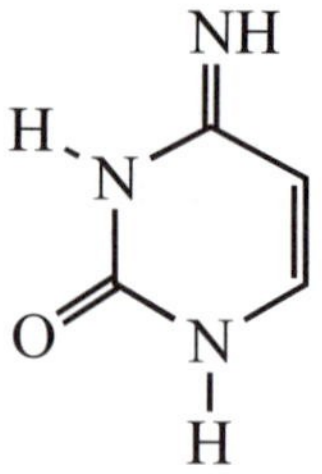

**There are three additional tautomeric forms of thymine:**

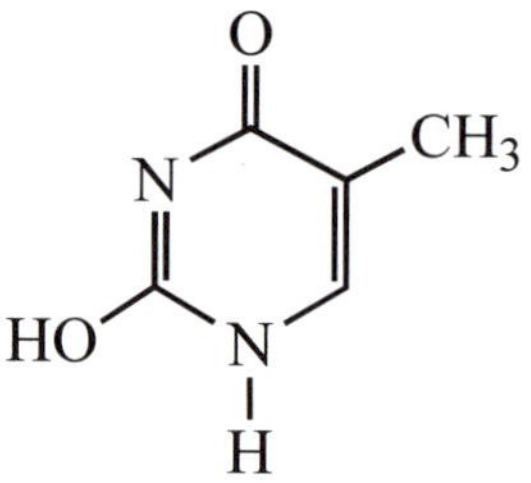

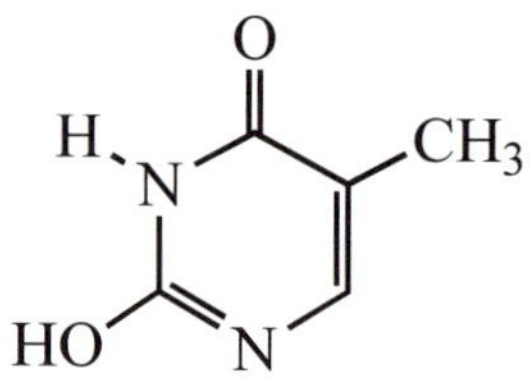

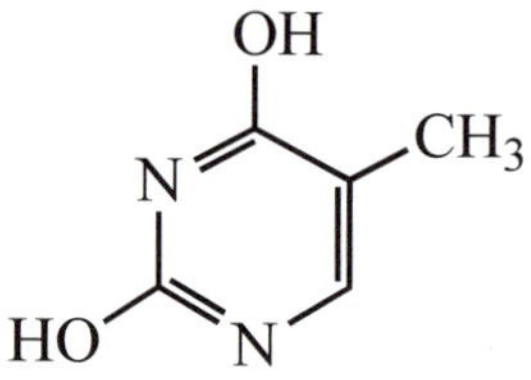

**20.9**    *Draw a structural formula for a nucleoside composed of*

    *(a) β-D-Ribose and adenine*        *(b) β-2-Deoxy-D-ribose and cytosine*

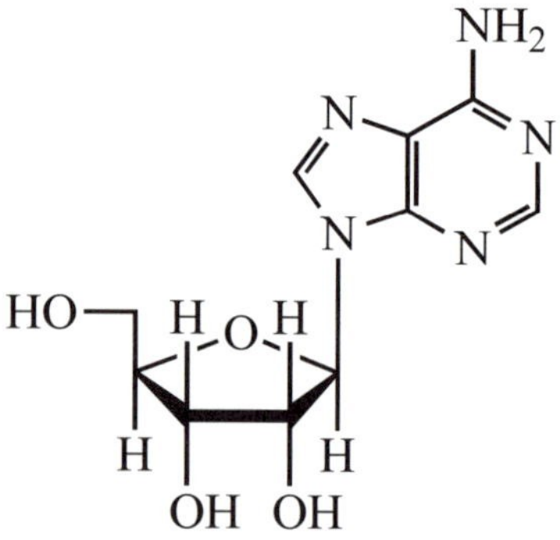

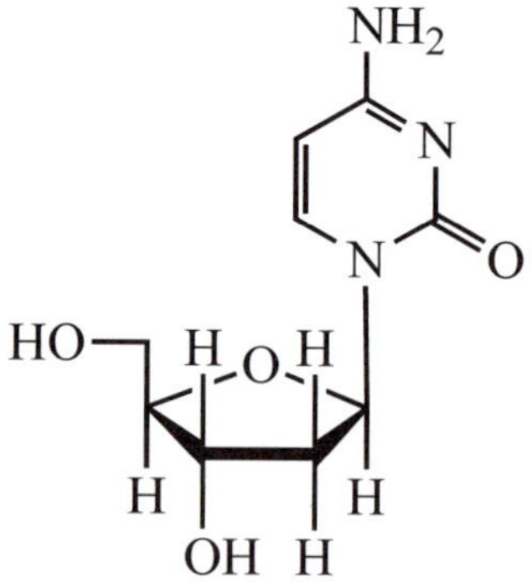

**20.10** *Nucleosides are stable in water and in dilute base. In dilute acid, however, the glycosidic bond of a nucleoside undergoes hydrolysis to give a pentose and a heterocyclic aromatic amine base. Propose a mechanism for this acid-catalyzed hydrolysis.*

**Step 1:**

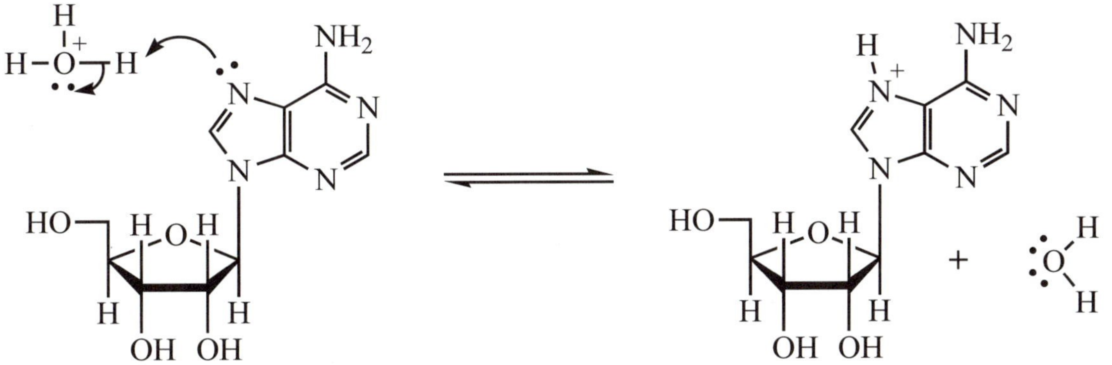

**Step 2:**

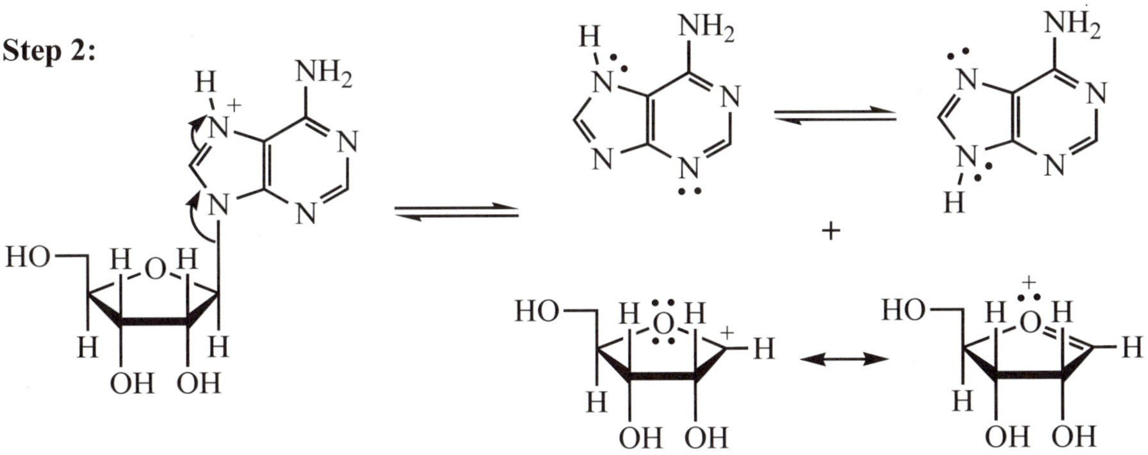

**Step 3:**

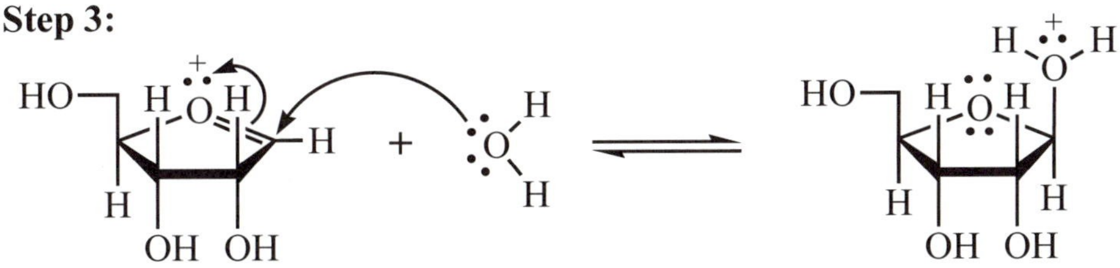

**Step 4:**

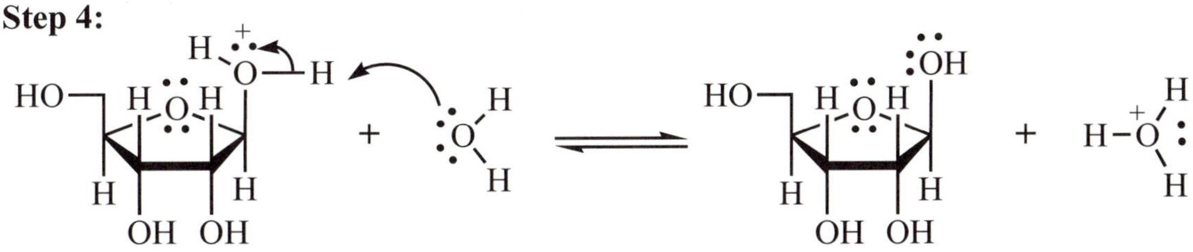

**20.11** *Explain the difference in structure between a nucleoside and a nucleotide.*

**Nucleotides differ from nucleosides in that nucleotides contain phosphate groups at the 5' or 3' position.**

**20.12** *Draw a structural formula for each nucleotide and estimate its net charge at pH 7.4, the pH of blood plasma.*

*(a) 2'-Deoxyadenosine 5'-triphosphate (dATP)*

**The values for the first three $pK_a$'s of dATP are all below 5.0, so these are fully deprotonated at pH 7.4.  The fourth $pK_a$ of dATP is 7.0, so that at pH 7.4, the overall charge will be about midway between –3 and –4.**

*(b) Guanosine 3'-monophosphate (GMP)*

**The two $pK_a$ values for GMP are well below 7.4, so both are fully deprotonated giving an overall charge of -2.**

*(c) 2'-Deoxyguanosine 5'-diphosphate (dGDP)*

**The values for the first two p$K_a$'s of dGDP are both below 5.0, so these are fully deprotonated at 7.4. The third p$K_a$ of dGDP is 6.7, so it will be mostly deprotonated at pH 7.4, giving the molecule the net charge close to -3.**

**20.13** *Cyclic-AMP, first isolated in 1959, is involved in many diverse biological processes as a regulator of metabolic and physiological activity. In it, a single phosphate group is esterified with both the 3' and 5' hydroxyls of adenosine. Draw a structural formula of cyclic-AMP.*

**Cyclic AMP**

## The Structure of DNA

**20.14** *Why are deoxyribonucleic acids called "acids"? What are the acidic groups in their structure?*

**Deoxyribonucleic acids are called acids because the phosphodiester groups of the backbone are acidic. At neutral pH, they are fully deprotonated, leading to the anionic nature of DNA.**

**20.15**  *Human DNA contains approximately 30.4% A. Estimate the percentages of G, C, and T and compare them with the values presented in Table 20.1.*

**Because A pairs with T, there must also be 30.4% T in human DNA. Thus, A and T comprise 60.8% of human DNA. The remainder (39.2%) must be G and C, split equally to give 19.6% G and 19.6% C. These values agree very well with the experimental values found in Table 20.1.**

**20.16**  *Draw a structural formula for the DNA tetranucleotide 5'-A-G-C-T-3'. Estimate the net charge on this tetranucleotide at pH 7.0. What is the complementary tetranucleotide to this sequence?*

**As shown in the structure, there is a net charge of -5 on this tetranucleotide at pH 7.0. The complimentary tetranucleotide has the sequence 3'-T-C-G-A-5'**

**20.17**  *List the postulates of the Watson-Crick model of DNA secondary structure.*

1) **DNA consists of two antiparallel polynucleotide strands coiled in a right-handed manner about the same axis to form a double helix.**

2) **Purine and pyrimidine bases project inward toward the axis of the helix and are always paired in a very specific manner, A with T and G with C.**

3) **Base pairs are stacked one on top of the other with a distance of 3.4 Å between base pairs and with ten base pairs in one complete turn of the helix.**

4) **There is one complete turn of the helix every 34 Å.**

**20.18**  *The Watson-Crick model is based on certain experimental observations of base composition and molecular dimensions.  Describe these observations and show how the Watson-Crick model accounts for each.*

**Chargaff found that in different organisms, the amount of A is very close to the amount of T and the amount of G very close to the amount of C, even though different organisms have different ratios of A to G.  The base-pairings postulates of the Watson-Crick model fully explain the observed ratios of bases.  The geometry of the Watson-Crick model is consistent with the periodicity and thickness observed in the X-ray diffraction data of Franklin and Wilkins.**

**20.19**  *Compare the α-helix of proteins and the double helix of DNA in these ways:*

*(a)  The units that repeat in the backbone of the polymer chain.*

**The repeating units in the backbone of the α-helices of proteins are α-amino acids linked by amide bonds, while those in DNA are repeating units of β-2'-deoxy-D-ribose linked via 3',5'-phosphodiester bonds.**

*(b)  The projection in space of substituents along the backbone (the R groups in the case of amino acids; purine and pyrimidine bases in the case of double-stranded DNA) relative to the axis of the helix.*

**The R groups of amino acids in α-helices point outward from the helix, whereas the purine and pyrimidine bases of a DNA double helix point inward and away from the aqueous cellular environment.**

**20.20**  *Discuss the role of the hydrophobic interactions in stabilizing double-stranded DNA.*

**The relatively hydrophobic bases of the DNA double helix are stacked on the inside, surrounded by the relatively hydrophilic sugar-phosphate backbone, that is directed outward toward its aqueous cellular environment.  The stacking of the hydrophobic bases minimizes their contact with water.**

**20.21**  *Name the type of covalent bond(s) joining monomers in these biopolymers.*

*(a) Polysaccharides    (b) Polypeptides         (c) Nucleic acids*

*(a)* **Glycosidic bonds join monomer units in polysaccharides.**

*(b)* **Peptide (amide) bonds join monomer units in polypeptides.**

*(c)* **Phosphate ester bonds join monomer units in nucleic acids.**

**20.22**  *In terms of hydrogen bonding, which is more stable, an A-T base pair or a G-C base pair?*

**A G-C base pair is held together by three hydrogen bonds, while an A-T base pair is held together by only two hydrogen bonds.  Thus, a G-C base pair is more stable than an A-T base pair.**

**20.23**  *At elevated temperatures, nucleic acids become denatured; that is, they unwind into single-stranded DNA.  Account for the observation that the higher the G-C content of a nucleic acid, the higher the temperature required for its thermal denaturation.*

**G-C base pairs have 3 hydrogen bonds, whereas A-T base pairs have only two hydrogen bonds (see Problem 20.22).  The more hydrogen bonds, the more energy required to break them.  Thus, nucleic acids with more G-C content will require more energy (in the form of heat) to denature.**

**20.24**  *Write the DNA complement for 5'-ACCGTTAAT-3'.  Be certain to label which is the 5' end and which is the 3' end of the complement strand.*

**The complimentary sequence is:  3'-TGGCAATTA-5'**

**20.25**  *Write the DNA complement for 5'-TCAACGAT-3'.*

**The complimentary sequence is: 3'-AGTTGCTA-5'**

**Ribonucleic Acids**

**20.26** *Compare the degree of hydrogen bonding in the base pair A-T found in DNA with that in the base pair A-U found in RNA.*

**The only difference between T and U is the presence of a methyl group on T. The methyl group's effect on hydrogen bonding is minimal, therefore the extent of hydrogen bonding between A and T compared to that of U and A is essentially the same.**

**20.27** *Compare DNA and RNA is these ways.*
   *(a) Monosaccharide units*          *(b) Principal purine and pyrimidine bases*
   *(c) Primary structure*            *(d) Location in the cell*
   *(e) Function in the cell*

   *(a)* **In DNA, the monosaccharide repeating units are β-2'-deoxy-D-ribose, whereas in RNA they are β-D-ribose.**

   *(b)* **Both DNA and RNA use the same purines, G and C. They differ in that RNA uses U, while DNA uses T. Both use the pyrimidine A.**

   *(c)* **The primary structure of DNA consists of pairs of A, T, G, and C, whereas that of RNA consists of single strands of A, U, G, and C.**

   *(d)* **RNA is found in the cytoplasm (the space between the cell wall and the nucleus) of cells while DNA is found in the nucleus of cells.**

   *(e)* **DNA's function is information storage, while RNA's function is translation and transcription.**

**20.28** *What type of RNA has the shortest lifetime in cells?*

**Messenger RNA has the shortest lifetime in cells, usually on the order of a few minutes or less. This short lifetime is thought to allow for a very tight control over how much protein is synthesized in the cell at any one time.**

**20.29**  *Write the mRNA complement for 5'-ACCGTTAAT-3'.  Be certain to label which is the 5' end and which is the 3' end of the mRNA strand.*

**The mRNA compliment is 3'-UGGCAAUUA-5'**


**20.30**  *Write the mRNA complement for 5'-TCAACGAT-3'.*

**The mRNA compliment is 3'-AGUUGCUA-5'**


**The Genetic Code**

**20.31**  *What does it mean to say that the genetic code is degenerate?*

**Degenerate refers to the fact that several amino acids are coded for by more than one triplet (also referred to as a codon).  There are 64 different triplets, 61 triplets code for the 20 amino acids and three of the triplets represent stop signals.**


**20.32**  *Aspartic acid and glutamic acid have carboxyl groups on their side chains and are called acidic amino acids. Compare the codons for these two amino acids.*

**All of the codons for these two acidic amino acids begin with GA.  The codons for aspartic acid are GAU and GAC, while the codons for glutamic acid are GAA and GAG.**


**20.33**  *Compare the structural formulas of the aromatic amino acids phenylalanine and tyrosine.  Compare also the codons for these two amino acids.*

**Phe and Ala differ in their side chains.  Although both have aromatic rings in their side chains, Phe contains a benzene ring, whereas Tyr contains a phenol.  The codons for phenylalanine are UUU and UUC, while the codons for tyrosine are UAU and UAC.  Their codons differ only in the second position.**


**20.34**  *Glycine, alanine, and valine are classified as nonpolar amino acids. Compare their codons.  What similarities do you find? What differences do you find?*

**Glycine:**     **GGU, GGC, GGA, GGG**
**Alanine:**     **GCU, GCC, GCA, GCG**
**Valine:**      **GUU, GUC, GUA, GUG**

**All of these amino acids have four mRNA codons, all codons start with G, and in each case, the first two bases of the codon are identical for a given amino acid.  For**

coding purposes, this makes the last base in the sequence for each amino acid irrelevant.

**20.35**  *Codons in the set CUU, CUC, CUA, and CUG all code for the amino acid leucine. In this set, the first and second bases are identical, and the identity of the third base is irrelevant. For what other sets of codons is the third base also irrelevant, and for what amino acid(s) does each set code?*

As mentioned in Problem 20.34, the last base in the codon for glycine (GGX), alanine (GCX), and valine (GUX) is irrelevant (X can be any of the bases for the third position). Other codons where the third base is irrelevant include: arginine (CGX), serine (UCX), proline (CCX), and threonine (ACX).

**20.36**  *Compare the codons with a pyrimidine, either U or C, as the second base. Do the majority of the amino acids specified by these codons have hydrophobic or hydrophilic side chains?*

The majority of amino acids with a pyrimidine in the second position of their codons are hydrophobic. This set contains phenylalanine, leucine, isoleucine, methionine, valine, proline, and alanine. Only serine and threonine have a pyrimidine in the second position and also have hydrophilic side chains.

**20.37**  *Compare the codons with a purine, either A or G, as the second base. Do the majority of the amino acids specified by these codons have hydrophilic or hydrophobic side chains?*

With the exception of tryptophan and glycine, all the codons with an A or G as the second base code for amino acids with polar, hydrophilic side chains.

**20.38**  *What polypeptide is coded for by this mRNA sequence?*
            *5'-GCU-GAA-GUC-GAG-GUG-UGG-3'*

The mRNA sequence codes for the polypeptide:
            Amino *N*-terminus- Ala-Glu-Val-Glu-Val-Trp -Carboxy *C*-terminus

**20.39** *The alpha chain of human hemoglobin has 141 amino acids in a single polypeptide chain. Calculate the minimum number of bases on DNA necessary to code for the alpha chain. Include in your calculation the bases necessary for specifying termination of polypeptide synthesis.*

**A polypeptide of 141 amino acids requires one triplet codon for each amino acid plus one stop codon. Therefore, the minimum number of DNA bases required to code for the alpha chain is (141+1) x 3 = 426 bases.**

**20.40** *In HbS, the human hemoglobin found in individuals with sickle-cell anemia, glutamic acid at position 6 in the beta chain is replaced by valine.*

*(a) List the two codons for glutamic acid and the four codons for valine.*

**Glutamic acid codons:  GAA and GAG**
**Valine codons:  GUU, GUC, GUA, and GUG**

*(b) Show that one of the glutamic acid codons can be converted to a valine codon by a single substitution mutation; that is, by changing one letter in one codon.*

**Both glutamic acid codons can be converted to two valine codons by replacing the central A with a U residue.**

**Looking Ahead**

**20.41** *The loss of 3 consecutive Ts from the gene that codes for CFTR, a trans-membrane conductance regulator protein, results in the disease known as cystic fibrosis. Which amino acid is missing from CFTR to cause this disease?*

**The loss of three consecutive units of T on a gene would cause the loss on the complementary mRNA strand of AAA.  This codon triplet codes for lysine, which is the amino acid missing from CFTR in cystic fibrosis.**

**20.42** *The following compounds have been researched as potential anti-viral agents.  Suggest how each might block the synthesis of RNA or DNA.*
*(a) Cordycepin (3'-deoxyadenosine)*

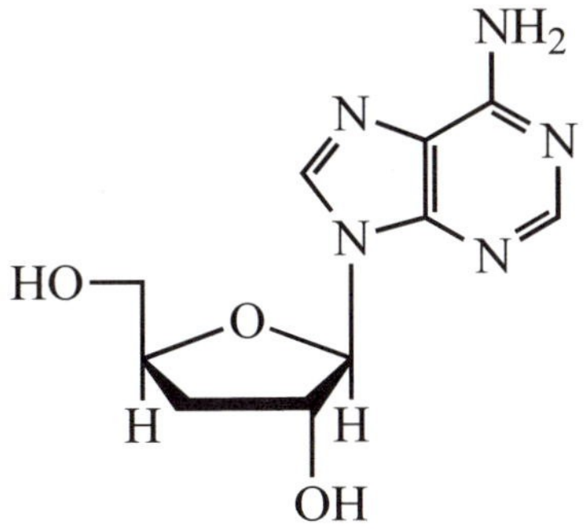

**The pentose unit (β-3-deoxy-□-ribose) in cordycepin mimics the β-□-ribose unit used to build RNA.  When incorporated in the RNA strand, the absence of the 3' hydroxyl in cordycepin terminates the RNA synthesis of the strand and destroys the virus.**

*(b)*

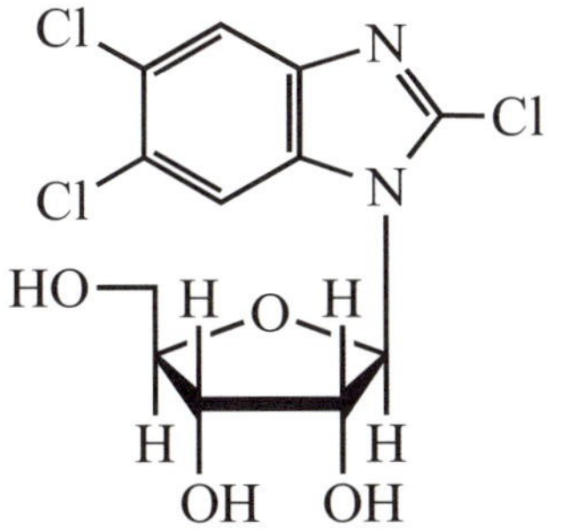

2,5,6-Trichloro-1-(β-D-ribofuranosyl)benzimidazole

**The nucleotide purine base analog 1,5,6-trichlorobenzimidazole on the β-□-ribose mimics a RNA nucleotide and will interfere with RNA polymerase, therefore block the synthesis of RNA.**

*(c)*

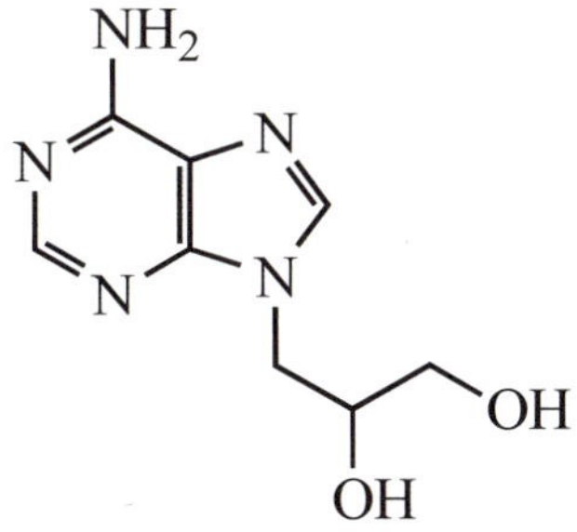

9-(2,3-Dihydroxypropyl)adenine

**9-(2,3-Dihydroxypropyl)adenine is similar to the adenosine nucleotide, except lacking a 5'-hydroxyl.  When incorporated in the RNA strand, the absence of the**

**5' hydroxyl in 9-(2,3-dihydroxypropyl)adenine terminates the RNA synthesis of the strand and destroys the virus.**

**20.43** *The ends of chromosomes are called telomeres. Telomeres can form unique and nonstandard structures. One example is the presence of base pairs between units of guanosine. Show how the guanine base can pair with another via hydrogen bonding.*

Guanine

**The dimer forms via two hydrogen bonds to form the guanine base pair.**

**20.44** *One synthesis of zidovudine (AZT) involves the following reaction (DMF is the solvent N,N-dimethylformamide). What type of reaction is this?*

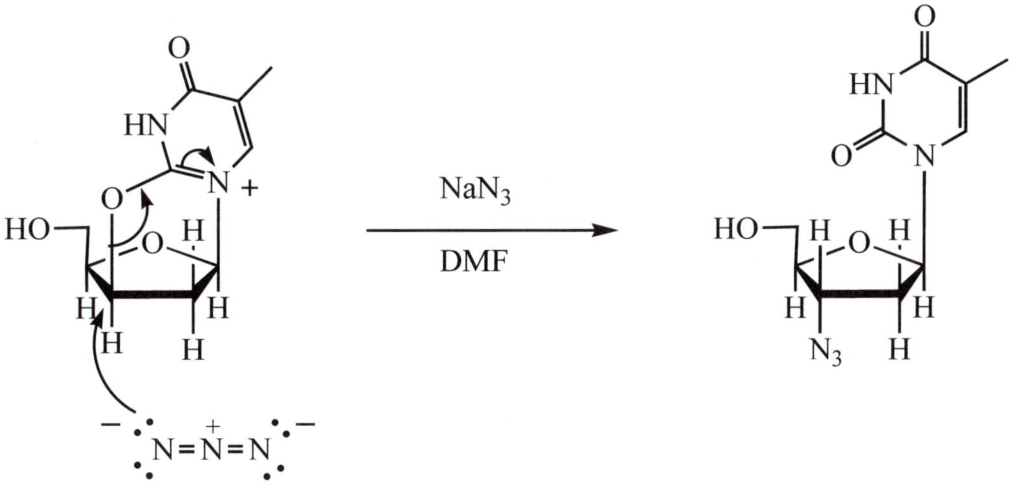

**This reaction will proceed by an S$_N$2 reaction mechanism. Azide anion is weakly basic, but a good nucleophile. The sugar substrate has the leaving group on a 2° carbon. DMF is a polar aprotic solvent. These conditions support an S$_N$2 mechanism. The inversion of stereochemistry at the reaction site is evidence that also supports an S$_N$2 reaction.**

## CHAPTER 21
### *Solutions to Problems*

**21.1**   (a) *How many constitutional isomers are possible for a triglyceride containing one molecule each of palmitic acid, oleic acid, and stearic acid?*

**There are three constitutional isomers possible; with the difference being which fatty acid is placed in the middle carbon of the glycerin substrate.**

$$
\begin{array}{ccc}
& O & \\
& \| & \\
H_2C-O-C-\text{palmitic} \\
| \quad O & \\
\quad \| & \\
HC^*-O-C-\text{steric} \\
| \quad O & \\
\quad \| & \\
H_2C-O-C-\text{oleic}
\end{array}
\qquad
\begin{array}{ccc}
& O & \\
& \| & \\
H_2C-O-C-\text{palmitic} \\
| \quad O & \\
\quad \| & \\
HC^*-O-C-\text{oleic} \\
| \quad O & \\
\quad \| & \\
H_2C-O-C-\text{steric}
\end{array}
\qquad
\begin{array}{ccc}
& O & \\
& \| & \\
H_2C-O-C-\text{steric} \\
| \quad O & \\
\quad \| & \\
HC^*-O-C-\text{palmitic} \\
| \quad O & \\
\quad \| & \\
H_2C-O-C-\text{oleic}
\end{array}
$$

(b) *Which of these constitutional isomers are chiral?*

**All of the above triglycerides are chiral. The stereocenters are indicated with an asterisk. Each constitutional isomer has one enantiomer, for three pairs of enantiomers.**

## Fatty Acids and Triglycerides
**21.2**   *Define the term hydrophobic.*

**Hydrophobic literally means, "water fearing." Hydrophobic species do not dissolve in water.**

**21.3**   *Identify the hydrophobic and hydrophilic region(s) of a triglyceride.*

**Hydrophobic region** ———

$$
\begin{array}{l}
\quad\quad\quad O \\
\quad\quad\quad \| \\
H_2C{-}O{-}C{-}(CH_2)_{14}CH_3 \\
| \quad\quad O \\
\quad\quad\quad \| \\
HC{-}O{-}C{-}(CH_2)_{14}CH_3 \\
| \quad\quad O \\
\quad\quad\quad \| \\
H_2C{-}O{-}C{-}(CH_2)_{14}CH_3
\end{array}
$$

——— **Hydrophobic region**

**Hydrophilic region**

**As an example, the above triglyceride of palmitic acid illustrates the hydrophilic and hydrophobic regions of a triglyceride. Notice that most of the molecule is hydrophobic, which explains why triglycerides are insoluble in water and so hydrophobic overall.**

**21.4**    *Explain why the melting points of unsaturated fatty acids are lower than those of saturated fatty acids.*

**Intermolecular dispersion forces involving the long hydrocarbon chains hold fatty acid molecules together in the solid state. Stronger intermolecular forces involve dipole-dipole attractions and hydrogen bonding between the carboxylic acid groups. Saturated fatty acids pack close together, maximizing the attractive forces. A cis carbon-carbon double bond in the long hydrocarbon chain introduces a kink in the carbon chain of unsaturated fatty acids, forcing non-ideal molecular packing thus diminishing the effects of intermolecular attractive forces. Therefore, there are weaker attractive forces between unsaturated fatty acid molecules than in their fully saturated counterparts, and as a result, unsaturated fatty acids have lower melting points.**

**21.5**    *Which would you expect to have the higher melting point, glyceryl trioleate or glyceryl trilinoleate?*

**Each oleic acid has one cis double bond, while each linoleic acid unit has two (see Table 21.1). Glyceryl trioleate has the higher melting point because its fatty acid components contain fewer carbon-carbon double bonds. The fewer carbon-carbon double bonds a fatty acid component contains, the more compact its structure becomes and the greater the ability of its molecules to pack together.**

**21.6**    *Draw a structural formula for methyl linoleate. Be certain to show the correct configuration of groups about each carbon-carbon double bond.*

**Methyl linoleate**

**21.7**    *Explain why coconut oil is a liquid triglyceride, even though most of its fatty acid*
          *components are saturated.*

          **Coconut oil is one of the few saturated liquid triglycerides because it contains mostly
          lower-molecular-weight fatty acids.  According to Table 21.2, 45% of the
          triglycerides in coconut oil contain a lauric acid component, which is a C12 fatty
          acid.  Triglycerides rich in lower-molecular-weight fatty acids have weaker
          dispersion forces between their molecules and therefore, have lower melting points.**


**21.8**    *It is common now to see "contains no tropical oils" on cooking oil labels, meaning that
          the oil contains no palm or coconut oil.  What is the difference between the composition
          of tropical oils and that of vegetable oils, such as corn oil, soybean oil, and peanut oil?*

          **Tropical oils contain triglycerides that consist of mostly lower-molecular-weight
          saturated fatty acids, while vegetable oils contain triglycerides that consist of mostly
          unsaturated acids.**


**21.9**    *What is meant by the term hardening as applied to vegetable oils?*

          **Hardening of vegetable oils is the catalytic hydrogenation of the C=C bonds in
          polyunsaturated vegetable oils.  By removing the cis C=C bonds, the saturated
          triglyceride molecules pack better, and thus melt at a higher temperature.  One
          consequence of hydrogenating vegetable oils is the isomerization of the cis C=C
          bonds to trans C=C.  Triglycerides with trans fatty acids are believed to contribute
          to an increased risk of heart disease, and soon food companies will be required to
          declare the trans-fatty acid content of their foods.**


**21.10**   *How many moles of $H_2$ are used in the catalytic hydrogenation of one mole of a
          triglyceride derived from glycerol, stearic acid, linoleic acid, and arachidonic acid?*

The triglyceride has six C=C bonds, and would therefore require six moles of hydrogen per mole of unsaturated triglyceride to complete the reduction to a saturated triglyceride.

**21.11** *Characterize the structural features necessary to make a good synthetic detergent.*

A good synthetic detergent should have a long hydrocarbon chain with a very polar head group at one end of the molecule. This combination will allow for the formation of micelle structures in aqueous solution that will dissolve hydrophobic substances such as dirt, grease, and oil. The polar head group should not form insoluble salts with the ions normally found in hard water, such as $Ca^{2+}$, $Mg^{2+}$, and $Fe^{2+}$.

**21.12** *Following are structural formulas for a cationic detergent and a neutral detergent. Account for the detergent properties of each.*

$$CH_3(CH_2)_6CH_2\overset{\overset{\displaystyle CH_3}{\displaystyle +|}}{\underset{\displaystyle CH_2C_6H_5}{\displaystyle N}}CH_3 \; Cl^-$$

$$\underset{HOH_2C}{\overset{HOH_2C}{HOCH_2\overset{|}{\underset{|}{C}}CH_2O\overset{O}{\overset{\|}{C}}(CH_2)_{14}CH_3}}$$

Benzyldimethyloctylammonium chloride
(a cationic detergent)

Pentaerythrityl palmitate
(a neutral detergent)

Both detergents have a hydrophobic long hydrocarbon tail attached to a very polar hydrophilic head. This combination allows for the formation of micelle structures in aqueous solution that will dissolve hydrophobic substances such as dirt, grease, and oil. In the case of benzyldimethyloctylammonium chloride, the polar group is a positively-charged ammonium group.

**21.13** *Identify some of the detergents used in shampoos and dish-washing liquids. Are they primarily anionic, neutral, or cationic detergents?*

The detergents used as shampoos and dishwashing liquids are alkylbenzenesulfonates and have the general formula:

$$R-\langle\text{benzene ring}\rangle-SO_3^-Na^+$$

where R represents a long hydrocarbon chain. They are primarily anionic detergents.

**21.14** *Show how to convert palmitic acid (hexadecanoic acid) into the following:*

*(a) Ethyl palmitate*

$$CH_3(CH_2)_{14}\overset{\overset{\displaystyle O}{\|}}{C}OH \quad + \quad CH_3CH_2OH \quad \underset{}{\overset{H^+}{\rightleftharpoons}} \quad CH_3(CH_2)_{14}\overset{\overset{\displaystyle O}{\|}}{C}OCH_2CH_3$$

*(b) Palmitoyl chloride*

$$CH_3(CH_2)_{14}\overset{\overset{\displaystyle O}{\|}}{C}OH \quad + \quad SOCl_2 \quad \longrightarrow \quad CH_3(CH_2)_{14}\overset{\overset{\displaystyle O}{\|}}{C}Cl$$

*(c) 1-Hexadecanol (cetyl alcohol)*

$$CH_3(CH_2)_{14}\overset{\overset{\displaystyle O}{\|}}{C}OH \quad \xrightarrow[\textbf{2. H}_2\textbf{O}]{\textbf{1. LiAlH}_4\textbf{, ether}} \quad CH_3(CH_2)_{14}CH_2OH$$

*(d) 1-Hexadecanamine*

$$CH_3(CH_2)_{14}\overset{\overset{\displaystyle O}{\|}}{C}OH \quad \xrightarrow{\textbf{SOCl}_2} \quad CH_3(CH_2)_{14}\overset{\overset{\displaystyle O}{\|}}{C}Cl \quad \xrightarrow{\textbf{2NH}_3}$$

$$CH_3(CH_2)_{14}\overset{\overset{\displaystyle O}{\|}}{C}NH_2 \quad \xrightarrow[\textbf{2. H}_2\textbf{O}]{\textbf{1. LiAlH}_4\textbf{, ether}} \quad CH_3(CH_2)_{14}CH_2NH_2$$

*(e) N,N-Dimethylhexadecanamide*

$$CH_3(CH_2)_{14}\overset{\overset{\displaystyle O}{\|}}{C}OH \quad \xrightarrow{\textbf{SOCl}_2} \quad CH_3(CH_2)_{14}\overset{\overset{\displaystyle O}{\|}}{C}Cl$$

$$\xrightarrow{\textbf{2HN(CH}_3\textbf{)}_2} \quad CH_3(CH_2)_{14}\overset{\overset{\displaystyle O}{\|}}{C}N(CH_3)_2$$

**21.15** *Palmitic acid (hexadecanoic acid, 16:0) is the source of the hexadecyl (cetyl) group in the following compounds. Each is a mild surface-acting germicide and fungicide and is used as a topical antiseptic and disinfectant.*

Cetylpyridinium chloride

Benzylcetyldimethylammonium chloride

*(a) Cetylpyridinium chloride is prepared by treating pyridine with 1-chlorohexadecane (cetyl chloride). Show how to convert palmitic acid to cetyl chloride.*

*(b) Benzylcetyldimethylammonium chloride is prepared by treating benzyl chloride with N,N-dimethyl-1-hexadecanamine. Show how this tertiary amine can be prepared from palmitic acid.*

**Phospholipids**

**21.16** *Draw the structural formula of a lecithin containing one molecule each of palmitic acid and linoleic acid.*

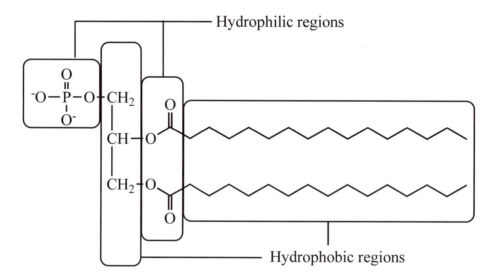

**21.17** *Identify the hydrophobic and hydrophilic region(s) of a phospholipid.*

**21.18** *The hydrophobic effect is one of the most important noncovalent forces directing the self-assembly of biomolecules in aqueous solution. The hydrophobic effect arises from tendencies of biomolecules (1) to arrange polar groups so that they interact with the aqueous environment by hydrogen bonding and (2) to arrange nonpolar groups so that they are shielded from the aqueous environment. Show how the hydrophobic effect is involved in directing*

**Molecules such as phospholipids and soaps/detergents have a hydrophilic head and a hydrophobic tail. They are involved in the formation of micelles and lipid bilayers in aqueous environments by assembling so that the hydrophobic tails associate together and are directed inward away from the aqueous environment and the hydrophilic heads turn outward towards water.**

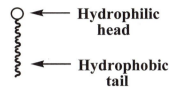

*(a)  Formation of micelles by soaps and detergents.*

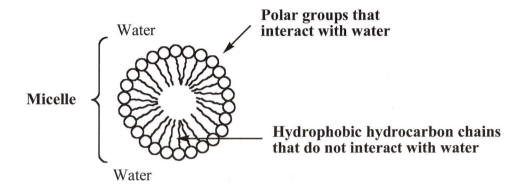

**Micelles are formed when the hydrophobic tails of the molecules associate with each other in the interior of a spherical structure, while the polar heads of the molecules are located on the outside surface where they can interact with water.**

*(b)   Formation of lipid bilayers by phospholipids.*

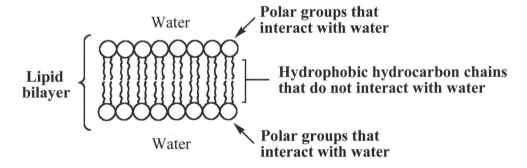

**Lipid bilayers are similar to micelles in how they assemble with hydrophobic tails directed inward and hydrophobic heads directed outward.  The difference is that lipid bilayers have water on both sides of the structure, while micelles have a hydrophobic environment on the inside with hydrophilic groups positioned on the outside interacting with the aqueous environment.**

*(c)  Formation of the DNA helix.*

**The relatively hydrophobic bases of the DNA double helix are stacked on the inside, held together by hydrogen bonding between the bases.   The bases are surrounded by the relatively hydrophilic sugar-phosphate backbone that is**

directed outward toward its aqueous cellular environment. The stacking of the hydrophobic bases minimizes their contact with water.

**21.19**  *How does the presence of unsaturated fatty acids contribute to the fluidity of biological membranes?*

**The presence of unsaturated fatty acids lowers melting points relative to saturated fatty acids because their cis C=C bonds create kinks in the chains that make them harder to pack closer together. In biological membranes, the cis double bonds in unsaturated fatty acids will likewise prevent dense packing within the hydrophobic inner layer, resulting in greater membrane fluidity.**

**21.20**  *Lecithins can act as emulsifying agents. The lecithin of egg yolk, for example, is used to make mayonnaise. Identify the hydrophobic part(s) and the hydrophilic part(s) of a lecithin. Which parts interact with the oils used in making mayonnaise? Which parts interact with the water?*

Hydrophilic region interacts with water

Hydrophobic region interacts with oil

**Steroids**

**21.21**  *Draw the structural formula for the product formed by treatment of cholesterol with $H_2/Pd$; with $Br_2$.*

**Each reaction is stereoselective. Hydrogen adds from the least hindered side of the carbon-carbon double bond, which in this case, approaches from the side opposite the OH on ring A and the CH$_3$ group at the junction of rings A and B. Bromination occurs via a cyclic bromonium cation resulting in anti addition of Br$_2$. Anti additions to cyclohexene rings correspond to trans diaxial additions, therefore, added bromine atoms are both axial.**

$$\xrightarrow[\text{Pd}]{H_2}$$

$$\xrightarrow{Br_2}$$

**21.22**   *List several ways in which cholesterol is necessary for human life. Why do many people find it necessary to restrict their dietary intake of cholesterol?*

       **Cholesterol is an important component of biological membranes where it serves to modulate membrane fluidity. In addition, cholesterol is an important precursor to a variety of steroid hormones. Cholesterol esters are an important component of atherosclerotic plaque, so restricting intake of dietary cholesterol is helpful for limiting atherosclerosis.**

**21.23**   *Both low-density lipoproteins (LDL) and high-density lipoproteins (HDL) consist of a core of triacylglycerols and cholesterol esters surrounded by a single phospholipid layer. Draw the structural formula of cholesteryl linoleate, one of the cholesterol esters found in this core.*

**21.24** *Examine the structural formulas of testosterone (a male sex hormone) and progesterone (a female sex hormone). What are the similarities in structure between the two? What are the differences?*

**Testosterone**             **Progesterone**

**Testosterone and progesterone are almost identical in structure. Both have the steroid skeleton and share the same substituents and stereochemistry, except for the substituent at C17. Testosterone has a hydroxyl at C17 as opposed to progesterone, which has an acetyl group at C17.**

**21.25** *Examine the structural formula of cholic acid and account for the ability of this and other bile salts to emulsify fats and oils and thus aid in their digestion.*

**Cholic acid as the carboxylate anion**
**Structural formula**

**Cholic acid as the carboxylate anion**
**Conformational formula**

**Cholic acid possesses the steroid A-D ring system with a cis ring fusion between rings A and B. It is able to emulsify fats oils because its steroid skeleton acts as a hydrophobic region and its carboxylate and hydroxyl groups act as a hydrophilic region much like the analogous regions of fatty acids. As can be seen on the conformational formula of cholic acid, all three of the hydroxyls point in the same direction, as do the methyl groups, except in the opposite direction of the hydroxyls. So, not only does cholic acid have a hydrophilic head with the carboxyl group and a hydrophobic tail with the steroid skeleton, but it also has a hydrophilic face with the hydroxyl groups and a hydrophobic face with the methyl groups.**

**21.26** *Following is a structural formula for cortisol (hydrocortisone). Draw a stereorepresentation of this molecule showing the conformations of the five- and six-membered rings.*

**Cortisol (Hydrocortisone)**
**Conformational formula**

**21.27** *Because some types of tumors need an estrogen to survive, compounds that compete with the estrogen receptor on tumor cells are useful anticancer drugs. The compound tamoxifen is one such drug. To what part of the estrone molecule is the shape of tamoxifen similar?*

Tamoxifen                                          Estrone

**The bolded bonds indicate similar structural elements between the two molecules.**

Tamoxifen                                          Estrone

## Prostaglandins

**21.28** *Examine the structure of PGF$_{2\alpha}$ and*

**cis-trans isomerization possible**

**PGF$_{2\alpha}$**

*(a) identify all stereocenters.*

**There are five carbon stereocenters, each marked with an asterisk.**

*(b) identify all double bonds about which cis,trans isomerism is possible.*

**There are two carbon-carbon double bond stereocenters.**

*(c) state the number of stereoisomers possible for a molecule of this structure.*

**There are $2^5 \times 2^2 = 128$ stereoisomers possible for the constitution of PGF$_{2\alpha}$.**

**21.29** *Following is the structure of unoprostone, a compound patterned after the natural prostaglandins (Section 21.6). Rescula, the isopropyl ester of unoprostone, is an antiglaucoma drug used to treat ocular hypertension. Compare the structural formula of this synthetic prostaglandin with that of PGF$_{2\alpha}$.*

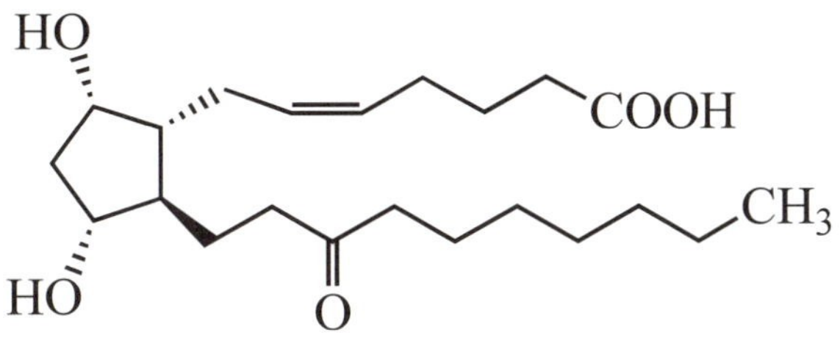

Unoprostone
(antiglaucoma)

**The synthetic prostaglandins unoprostone and Rescula have similar carbon skeletons to PGF$_{2\alpha}$. They differ in that the synthetic analogs do not have a C=C bond at C13, the analogs have two extra carbons (C21, C22), and instead of an alcohol at C15, the synthetics possess a carbonyl group.**

**Fat-Soluble Vitamins**

**21.30** *Examine the structural formula of vitamin A, and state the number of cis,trans isomers possible for this molecule.*

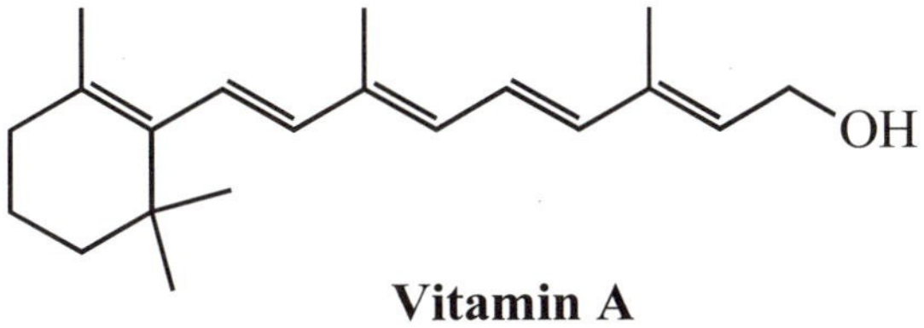

Vitamin A

**As shown in the above structure, vitamin A has four double bonds that can either be cis or trans. Therefore, there are $2^4$ or 16 possible cis,trans isomers. Note that the double bond in the ring cannot exist as a trans double bond.**

**21.31** *The form of vitamin A present in many food supplements is vitamin A palmitate. Draw the structural formula of this molecule.*

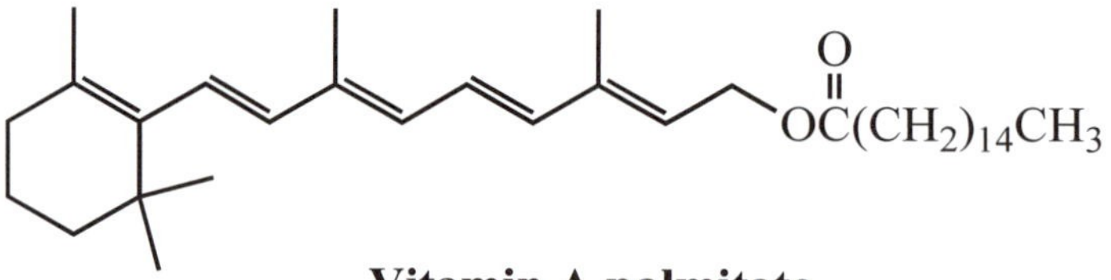

**Vitamin A palmitate**

**21.32** *Examine the structural formulas of vitamin A, 1,25-dihydroxy-D₃, vitamin E, and vitamin K₁. Do you expect them to be more soluble in water or in dichloromethane? Do you expect them to be soluble in blood plasma?*

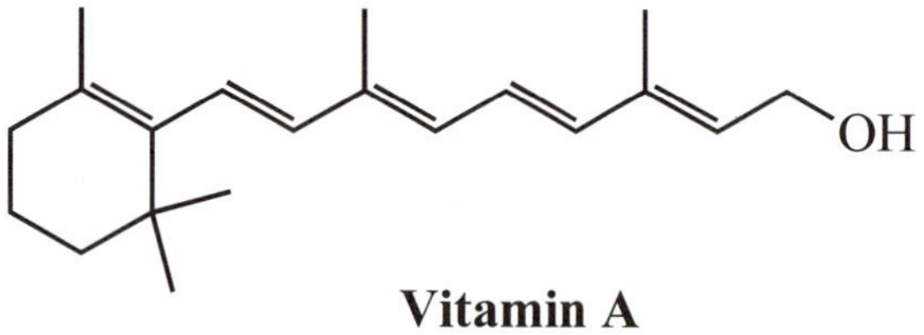

**Vitamin A**

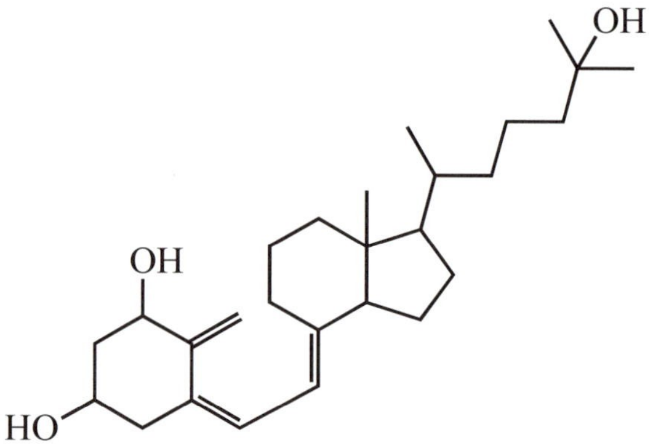

**1,25-Dihydroxyvitamin D₃**

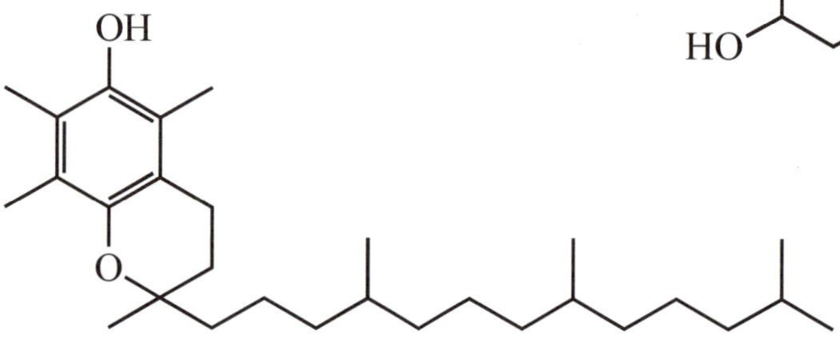

**Vitamin E**
**(α-Tocopherol)**

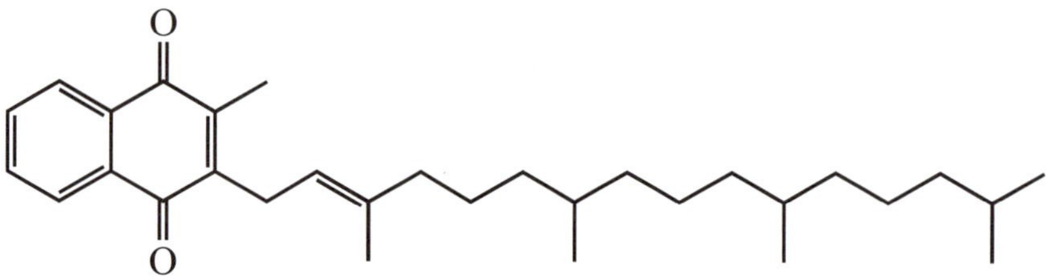

**Vitamin K**

**All of the above structures are extremely hydrophobic, so they will be more soluble in organic solvents, such as dichloromethane, than in polar solvents such as water.**

**Blood plasma is an aqueous solution; therefore, these vitamins will only be sparingly soluble in blood plasma.**

## Looking Ahead

**21.33** *Shown is a glycolipid, a class of lipid that contains a sugar residue. Glycolipids are found in cell membranes. Which part of the molecule would you expect to reside on the extracellular side of the membrane? Which monosaccharide is attached to the lipid in the illustration?*

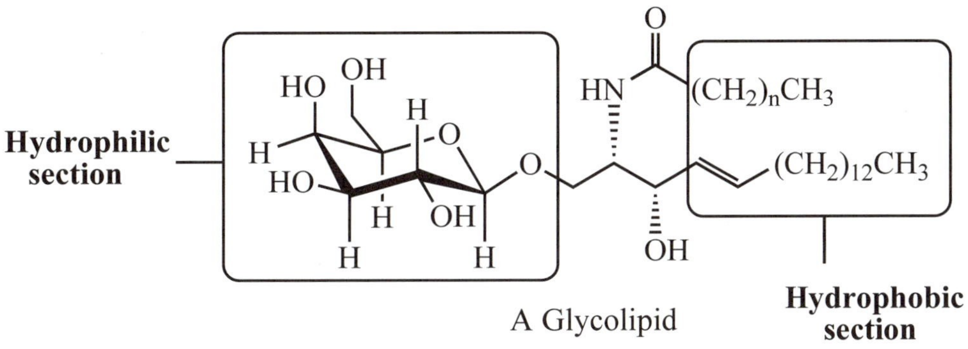

A Glycolipid

*(a)* **Because sugars are water-soluble molecules, expect the sugar residue of the glycolipid to lie on the extracellular side of the membrane.**

*(b)* **D-Galactose**

**21.34** *How would you expect temperature to effect fluidity in a cell membrane?*

**As temperature increases, so does molecular motion. As the molecular motion increases, the fluidity of the cell membrane increases.**

**21.35** *Which type of lipid movement is most favorable in cell membranes? Explain.*

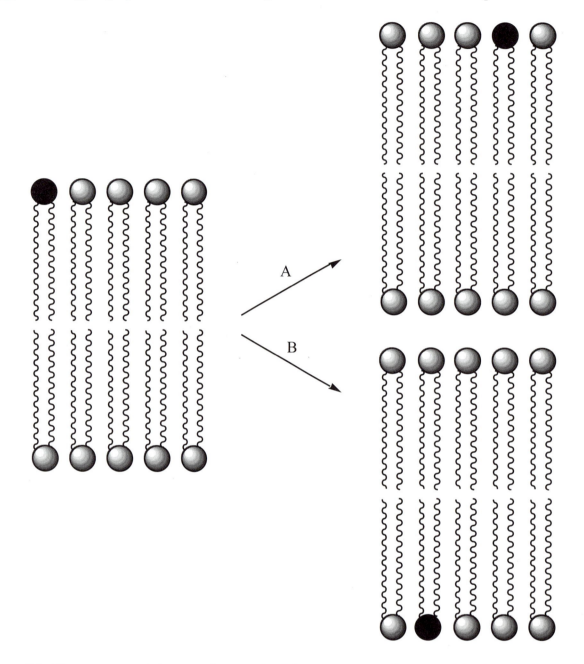

**Lipid movement A is most favorable. The movement illustrated in B would require the polar head group of the lipid to pass through the nonpolar part of the membrane.**

**21.36** *Aspirin works by transferring an acetyl group to the side chain of the 530$^{th}$ amino acid in the protein prostaglandin H$_2$ synthase-1. Draw the product of this reaction.*

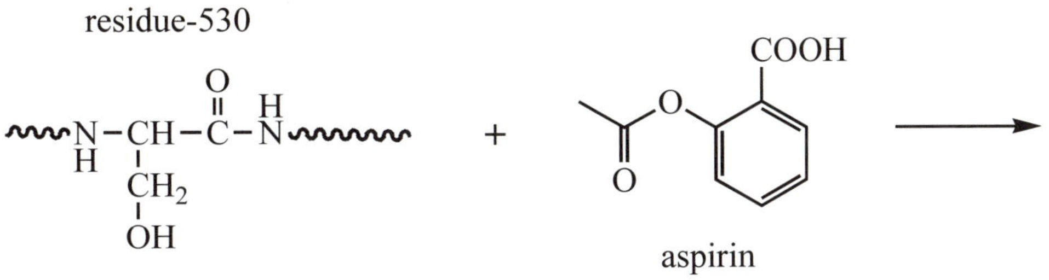

residue-530

aspirin

**Product:**

Acetyl group transfered

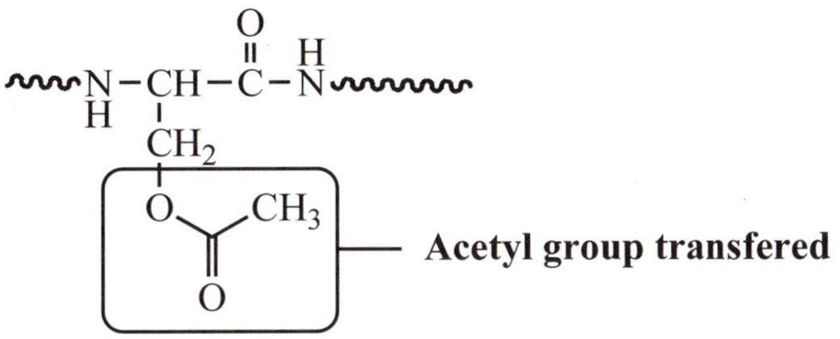

**The acetyl group is transferred to the hydroxyl instead of the amide-linkage nitrogens because the nitrogen atoms are less nucleophilic due to the resonance and inductive effects of the adjacent carbonyl group.**

**CHAPTER 22**
*Solutions to Problems*

**22.1**   *Under anaerobic (without oxygen) conditions, glucose is converted to lactate by a metabolic pathway called anaerobic glycolysis or, alternatively, lactate fermentation. Is anaerobic glycolysis a net oxidation, a net reduction, or neither?*

$$C_6H_{12}O_6 \xrightarrow{\text{anaerobic glycolysis}} 2CH_3CHCOO^- + 2H^+$$

Glucose                           Lactate

**Because there is no change in the number of hydrogen or oxygen atoms in the reaction, the conversion of glucose to two molecules of lactate is neither an oxidation nor reduction.**

**Glycolysis**

**22.2**   *Does lactate fermentation result in an increase or decrease in blood pH?*

**Lactate fermentation leads to an increase in the $H^+$ concentration in the bloodstream, therefore the bloodstream pH decreases.**

**22.3**   *Name one coenzyme required for glycolysis. From what vitamin is it derived?*

**The coenzyme nicotinamide adenine dinucleotide ($NAD^+$), which is derived from the vitamin niacin, is required for the oxidation steps of glycolysis.**

**22.4**   *Number the carbons of glucose 1 through 6. Which carbons of glucose become the carboxyl groups of the two pyruvates?*

**D-Glucose**                                              **Pyruvate**

By numbering the carbon atoms of glucose and following the different atoms through the pathway, it can be seen that the carboxyl group carbon atoms are derived from carbons C3 and C4 of glucose.

**22.5**   *How many moles of lactate are produced from three moles of glucose?*

$$C_6H_{12}O_6 \xrightarrow[\text{glycolysis}]{\text{anaerobic}} 2CH_3\overset{\overset{\displaystyle OH}{|}}{C}HCOO^- + 2H^+$$

Glucose                           Lactate

**During anaerobic glycolysis, two moles of lactate are formed for every one mole of glucose consumed, therefore, three moles of glucose give six moles of lactate.**

**22.6**   *Although glucose is the principal source of carbohydrates for glycolysis, fructose and galactose are also metabolized for energy.*

*(a) What is the main dietary source of fructose?  Of galactose?*

**The main dietary source of D-fructose is the disaccharide sucrose (table sugar), which is composed of D-glucose and D-fructose monosaccharide units.  The main dietary source of D-galactose is the disaccharide lactose (present in milk), which is composed of D-glucose and D-galactose.**

*(b) Propose a series of reactions by which fructose might enter glycolysis.*

**D-Fructose**                                    **D-Fructose 6-phosphate**

**Fructose is converted to fructose 6-phosphate and enters glycolysis at reaction 3, where it will be converted to fructose 1,6-bisphosphate.**

*(c) Propose a series of reactions by which galactose might enter glycolysis.*

**D-Galactose can be converted to D-glucose by enzyme-catalyzed inversion of configuration at C4, and then enters glycolysis at step 1.**

**D-Galactose**                                    **D-Glucose**

Enzyme-catalyzed inversion of configuration

**22.7**    *How many moles of ethanol are produced per mole of sucrose through the reactions of glycolysis and alcoholic fermentation? How many moles of $CO_2$ are produced?*

**By the reactions of glycolysis and alcoholic fermentation, one mole of sucrose gives four moles of ethanol and four moles of carbon dioxide. This can be determined by remembering that one mole of the disaccharide sucrose is first hydrolyzed to one mole of glucose and one mole of fructose. Each of these six-carbon monosaccharides enters glycolysis to give two moles of pyruvate, so a total of four moles of pyruvate are produced for each mole of sucrose used. Each mole of pyruvate is converted to one mole of ethanol and one mole of carbon dioxide, so a total of four moles of ethanol and four moles of carbon dioxide are produced for each mole of sucrose.**

**22.8**    *Glycerol derived from hydrolysis of triglycerides and phospholipids is also metabolized for energy. Propose a series of reactions by which the carbon skeleton of glycerol might enter glycolysis and be oxidized to pyruvate.*

**Glycerol enters glycolysis through the following enzyme catalyzed steps that lead to glyceraldehydes 3-phophate, which is converted to pyruvate via the normal glycolysis pathway.**

**Glycerol**            **Glycerol 1-phosphate**            **Dihydroxyacetone phosphate**

keto-enol tautomerism

**Glyceraldehyde 3-phosphate**

**22.9**   *Write a mechanism to show the role of NADH in the reduction of acetaldehyde to ethanol.*

**In the following mechanism, the steps are numbered 1-5.  The key feature of this mechanism is a hydride (H:⁻) transfer in Step 3 from NADH to the carbonyl group of acetaldehyde.**

**Arrows 1-2: Electrons within the ring flow from nitrogen.**

**Arrow 3:**  **Transfer of a hydride ion from the -CH₂- of the six-membered ring to the carbonyl carbon creates the new C-H bond to the carbonyl carbon of acetaldehyde.**

**Arrow 4:**  **The C=O pi bond breaks as the new C-H bond forms.**

**Arrow 5:**  **An acidic group, -BH, on the surface of the enzyme transfers a proton to the newly-formed alkoxide ion to complete formation of the hydroxyl group of ethanol.**

**22.10**  *Ethanol is oxidized in the liver to acetate ion by NAD⁺.*

*(a) Write a balanced equation for this oxidation.*

$$CH_3CH_2OH \ + \ 2\,NAD^+ \ + \ H_2O \ \longrightarrow \ CH_3\overset{\displaystyle O}{\overset{\|}{C}}O^- \ + \ 2\,NADH \ + \ 3H^+$$

*(b) Do you expect the pH of blood plasma to increase, decrease, or remain the same as a result of metabolism of a significant amount of ethanol?*

☐☐☐☐☐☐☐☐☐☐☐☐☐☐☐☐☐ **will drop due to the protons produced by the metabolism of significant amounts of ethanol.**

**22.11** *When pyruvate is reduced to lactate by NADH, two hydrogens are added to pyruvate: one to the carbonyl carbon, the other to the carbonyl oxygen.  Which of these hydrogens is derived from NADH?*

**The hydrogen added to the carbonyl carbon comes from NADH in the form of a hydride anion (H:⁻).  This species is highly nucleophilic and reacts with the electrophilic carbon of a carbonyl.**

**22.12** *Why is glycolysis called an anaerobic pathway?*

**Glycolysis is called an anaerobic pathway because no oxygen is involved.  Glycolysis was probably used by organisms that appeared before there was oxygen in the environment during the first billion years or so of biological evolution on earth.**

**22.13** *Which carbons of glucose end up in $CO_2$ as a result of alcoholic fermentation?*

**As shown in the answer to Problem 22.4, carbons C3 and C4 of D-glucose end up as carboxylic acid carbons of pyruvate.  These same two carbon atoms end up as $CO_2$ through decarboxylation as the result of alcoholic fermentation.**

**22.14** *Which steps in glycolysis require ATP?  Which steps produce ATP?*

**Reactions 1 and 3 of glycolysis (Figure 22.1) require ATP, while reactions 7 and 10 produce ATP.**

## β-Oxidation

**22.15**  *Write structural formulas for palmitic, oleic, and stearic acids, the three most abundant fatty acids.*

**Palmitic acid (C16)**

**Stearic acid (C18)**

**Oleic acid (C18)**

**22.16**  *A fatty acid must be activated before it can be metabolized in cells. Write a balanced equation for the activation of palmitic acid.*

**Activation of a fatty acid involves the formation of a thioester with coenzyme A. The proton is derived from the thiol group of CoA-SH.**

$$CH_3(CH_2)_{14}\overset{O}{\overset{\|}{C}}O^- \; + \; CoA\text{-}SH \; + \; ATP \longrightarrow CH_3(CH_2)_{14}\overset{O}{\overset{\|}{C}}SCoA \; + \; AMP$$

$$+ \; P_2O_7{}^{4-} \; + \; H^+$$

**22.17**  *Name three coenzymes necessary for β-oxidation of fatty acids. From what vitamin is each derived?*

**The three coenzymes and the vitamins from which they are derived are FAD (riboflavin), NAD$^+$ (niacin), and coenzyme A (pantothenic acid). All three coenzymes contain the heterocyclic aromatic amine base adenosine.**

**22.18**  *We have examined β-oxidation of saturated fatty acids, such as palmitic acid and stearic acid. Oleic acid, an unsaturated fatty acid, is also a common component of dietary fats and oils. This unsaturated fatty acid is degraded by β-oxidation but, at one stage in its degradation, requires an additional enzyme named enoyl-CoA isomerase. Why is this enzyme necessary, and what isomerization does it catalyze? (Hint: Consider both the* *configuration of the carbon-carbon double bond in oleic acid and its position in the carbon chain.)*

**After three rounds of β-oxidation, oleic acid (C18) is degraded to an unsaturated *cis*-C12 fatty acid and three moles of acetyl-CoA.  An enzyme-catalyzed rotation of the cis double bond would give a trans double bond, but it would be in the wrong position for β-oxidation to continue.  Thus the enoyl-CoA-isomerase must also change the position of the double bond so that it is in conjugation with the carbonyl group, as shown in the following scheme.**

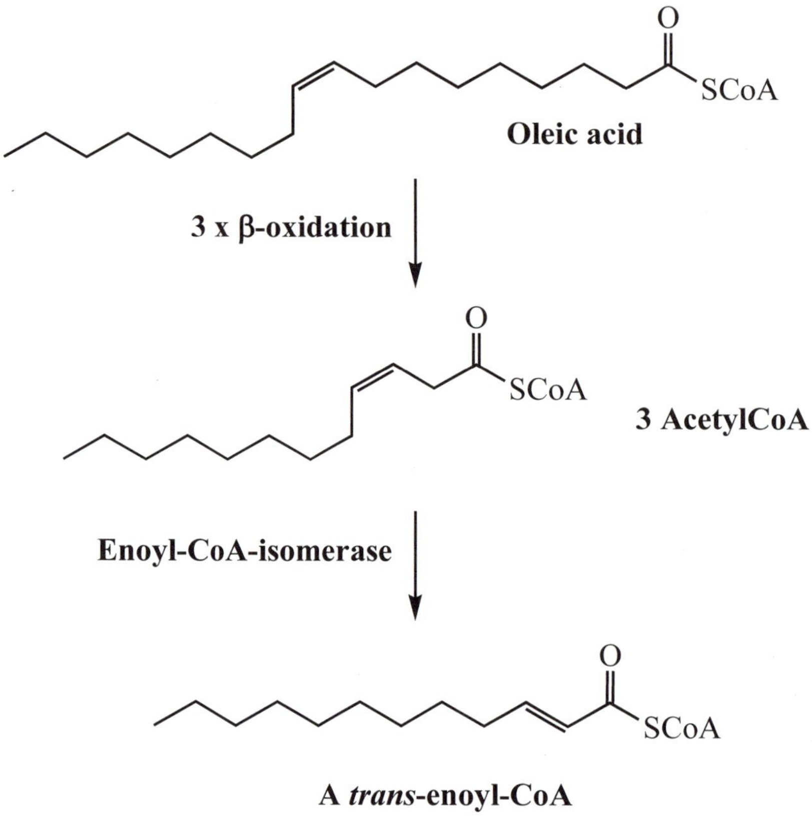

**Oleic acid**

**3 x β-oxidation**

**3 AcetylCoA**

**Enoyl-CoA-isomerase**

**A *trans*-enoyl-CoA**

## Citric Acid Cycle

**22.19**  *What is the main function of the citric acid cycle?*

**The main function of the citric acid cycle is to produce reduced coenzymes (NADH and FADH₂).  Their re-oxidation during respiration is coupled with the production of energy in the form of ATP.**

**22.20**  *Which steps in the citric acid cycle involve:*

*(a) formation of new carbon-carbon bonds*    **Steps 1 and 6**

**Step 1 is an aldol condensation.**
**Step 6 involves the formation of a new pi-bond through a dehydrogenation.**

(b) *breaking of carbon-carbon bonds*              **Steps 3, 4, and 7**

**Steps 3 and 4 are decarboxylations.**
**Step 7 involves a hydration of a carbon-carbon double bond.**

(c) *oxidation by NAD$^+$*                         **Steps 3, 4, and 8**

(d) *oxidation by FAD*                             **Step 6**

(e) *decarboxylation*                              **Steps 3 and 4**

(f) *creation of new stereocenters*               **Steps 1, 2, and 6**

**Step 1 is an aldol condensation that creates a new stereocenter.**
**Step 2 is an isomerization via a tautomerization.**
**Step 6 is a dehydrogenation that forms a double bond stereocenter.**

**22.21**  *What does it mean to say that the citric acid cycle is catalytic; that is, that it does not produce any new compounds?*

**The citric acid cycle accepts acetate units in the form of acetyl coenzyme A and generates two molecules of carbon dioxide per entering acetyl group.  It also accepts NAD$^+$ and FADH and generates NADH and FADH$_2$.  Other than these chemical transformations, there is no other net change in any of the intermediates in the cycle.**

**Additional Problems**

**22.22**  *Review the oxidation reactions of glycolysis, β-oxidation, and the citric acid cycle and compare the types of functional groups oxidized by NAD$^+$ with those oxidized by FAD.*

**NAD$^+$ oxidizes secondary alcohols to ketones, aldehydes to carboxylic acids, and carboxylic acids to carbon dioxide.  FAD oxidizes carbon-carbon single bonds to carbon-carbon double bonds**

**22.23**  *The "respiratory quotient" (RQ) is used in studies of energy metabolism and exercise physiology.  It is defined as the ratio of the volume of carbon dioxide produced to the volume of oxygen used:*

$$RQ = \frac{\text{Volume } CO_2}{\text{Volume } O_2}$$

*(a) Show that RQ for glucose is 1.00. (Hint: Look at the balanced equation for complete oxidation of glucose to carbon dioxide and water.)*

$$C_6H_{12}O_6 \;+\; 6O_2 \longrightarrow 6CO_2 \;+\; 6H_2O \qquad RQ = \frac{6CO_2}{6O_2} = 1.00$$

*(b) Calculate RQ for triolein, a triglyceride of molecular formula $C_{57}H_{104}O_6$.*

**In the balanced equation for the oxidation of one mole of triolein, 80 moles of $O_2$ are used to produce 57 moles of $CO_2$.**

$$C_{57}H_{104}O_6 \;+\; 80O_2 \longrightarrow 57CO_2 \;+\; 52H_2O \qquad RQ = \frac{57CO_2}{80O_2} = 0.71$$

*(c) For an individual on a normal diet, RQ is approximately 0.85. Would this value increase or decrease if ethanol were to supply an appreciable portion of caloric needs?*

**In the balanced equation for the oxidation of ethanol, 3 moles of $O_2$ are used and 2 moles of $CO_2$ are produced. The RQ = 2/3 = 0.67. Thus an individual's RQ would decrease if ethanol were to supply an appreciable portion of caloric needs.**

$$C_2H_6O \;+\; 3O_2 \longrightarrow 2CO_2 \;+\; 3H_2O$$

**22.24** *Acetoacetate, β-hydroxybutyrate, and acetone are commonly referred to within the health sciences as "ketone bodies", in spite of the fact that one of them is not a ketone at all. They are products of human metabolism and are always present in blood plasma. Most tissues, with the notable exception of the brain, have the enzyme systems necessary to use them as energy sources. Synthesis of ketone bodies occurs by the following enzyme-catalyzed reactions. Enzyme names are: (1) thiolase, (2) β-hydroxy-β-methylglutaryl-CoA synthase, (3) β-hydroxy-β-methylglutaryl-CoA lyase, and (5) β-hydroxybutyrate dehydrogenase. Reaction (4) is spontaneous and uncatalyzed.*

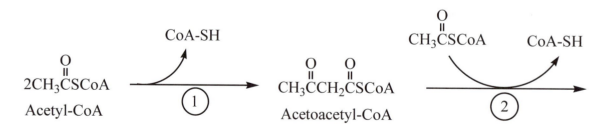

*Describe the type of reaction involved in each step.*

**Step 1: Claisen Condensation**          **Step 2: A carbonyl condensation reaction**

**Step 3: Reverse aldol condensation**    **Step 4: Decarboxylation**

**Step 5: Reduction of a ketone**

**22.25** *A connecting point between anaerobic glycolysis and β-oxidation is formation of acetyl-CoA. Which carbon atoms of glucose appear as methyl groups of acetyl-CoA? Which carbon atoms of palmitic acid appear as methyl groups of acetyl-CoA?*

**Carbons C1 and C6 of glucose appear as methyl groups of acetyl-CoA. Palmitic acid undergoes β-oxidation to produce acetyl-CoA, so the even carbon atoms 2, 4, 6, 8, 10, 12, 14, and 16 of palmitic acid appear as methyl groups of acetyl-CoA.**

**22.26** *Which of the steps in the following biochemical pathways use molecular oxygen as the oxidizing agent?*

(a) *Glycolysis*

**Glycolysis is considered an anaerobic pathway; therefore, no oxygen is used as an oxidizing agent. Instead, NAD$^+$ is the oxidizing agent.**

(b) *β-Oxidation*

**β-Oxidation does not use oxygen as an oxidizing agent. Instead, it uses NAD$^+$ and FAD as oxidizing agents.**

(c) *The citric acid cycle*

**T□□ □□□□□□ □cid cycle does not use oxygen as an oxidizing agent. Instead, it uses NAD$^+$ and FAD as oxidizing agents.**

**Looking Back**

**22.27** *Compare biological (enzyme-catalyzed) reactions to laboratory reactions in terms of:*

(a) *efficiency of yields*

**Enzyme-catalyzed reactions give 100% of the desired product; laboratory reactions rarely approach this efficiency.**

(b) *regiochemical outcome of products*

**Enzyme-catalyzed reactions are 100% regioselective. Although some laboratory reactions approach this efficiency, most do not.**

(c) *stereochemical outcome of products*

**Enzyme-catalyzed reactions are 100% stereoselective. Although some laboratory reactions approach this efficiency, most do not.**

**22.28** *Comment on the importance of stereochemistry in the synthesis of new drugs.*

**Biological processes involve the use of enzymes and receptors that are chiral. In many cases, one enantiomer may be biologically active for desirable results while the other enantiomer may not be active or may have negative results. The use of single-enantiomer drugs can insure maximum efficacy with the minimum amount of drug when compared to the use of racemic mixtures.**

**22.29** *Of the functional groups that we have studied, which are affected by the acidity of biological environments (biological pH)?*

**Of the functional groups we have studied in this chapter, carboxylic acids and amines are affected by the acidity of their biological environment; their degree of protonation or deprotonation, as the case may be, depends on the pH of their environment.**

**22.30** *Can you think of any aspect of your day to day life that does not involve or is not affected by organic chemistry? Explain.*

**All biological processes involve organic chemical reactions. All combustion reactions of fossil fuels that release energy for heat, electricity and transportation involve the oxidation of organic molecules. The synthesis of medicines, household goods, and clothing involves the use of organic compounds. The list of organic reactions that are necessary to make every day possible is disproportionably large compared to those that don't.**